Benchmark Papers in Geology

Series Editor: Rhodes W. Fairbridge
Columbia University

Published Volumes and Other Volumes in Preparation

ENVIRONMENTAL GEOMORPHOLOGY AND LANDSCAPE CONSERVATION, VOLUME I: Prior to 1900 / Donald R. Coates
RIVER MORPHOLOGY / Stanley A. Schumm
SPITS AND BARS / Maurice L. Schwartz
TEKTITES / Virgil E. and Mildred A. Barnes
GEOCHRONOLOGY / C. T. Harper
SLOPE MORPHOLOGY / Stanley A. Schumm and M. Paul Mosley
EVAPORITES / Douglas Kirkland and Robert Evans
ENVIRONMENTAL GEOMORPHOLOGY AND LANDSCAPE CONSERVATION, VOLUME III: Non-Urban / Donald R. Coates
GLACIAL ISOSTASY / John T. Andrews
WEATHERING AND SAPROLITE / Dan Yaalon
EARTHQUAKES: Geological Effects / Clarence R. Allen
ROCK CLEAVAGE / Dennis S. Wood
ENVIRONMENTAL GEOMORPHOLOGY AND LANDSCAPE CONSERVATION, VOLUME II: Urban / Donald R. Coates
BARRIER ISLANDS / Maurice L. Schwartz
VERTEBRATE PALEONTOLOGY / J. T. Gregory
PALEOMAGNETISM / Kenneth Creer
PALEOECOLOGY / John Chronic, Jr.
SOLID STATE GEOPHYSICS / Thomas J. Ahrens
OCEANOGRAPHY / Takashi Ichiye
PERMAFROST / J. D. Ives
PLAYAS AND CLAY-PANS / James L. Neal
GEOCHEMISTRY OF SULFUR / Isaac Kaplan
MODERN CARBONATES / Gerald Friedman
PHILOSOPHY OF GEOLOGY / Claude C. Albritton

Benchmark Papers in Geology

A *BENCHMARK*™ Books Series

TEKTITES

Edited by
VIRGIL E. and MILDRED A. BARNES
The University of Texas at Austin

Dowden, Hutchinson & Ross, Inc.
Stroudsburg, Pennsylvania

Library of Congress Catalog Card Number: 72-95942
ISBN: 0-87933-027-9

Library of Congress Cataloging in Publication Data

Barnes, Virgil Everett, 1903- comp.
Tektites.

(Benchmark papers in geology)
Bibliography: p.
1. Tektite--Addresses, essays, lectures. I. Barnes, Mildred, joint comp. II. Title.
QE399.B34 523.5'1 72-95942
ISBN 0-87933-027-9

Manufactured in the United States of America.

Exclusive distributor outside the United States and Canada: John Wiley & Sons, Inc.

Acknowledgments and Permissions

ACKNOWLEDGMENT

Gauthier-Villars—*Comptes rendus hebdomadaires des séances de l'Académie des Sciences de Paris*
"Les Tectites sans formes figurées de l'Indochine"

PERMISSIONS

The following papers have been reprinted with the permission of the authors and the copyright owners.

The Mineralogical Society—*Mineralogical Magazine*
"Silica-Glass from the Libyan Desert"

The University of Texas at Austin—*University of Texas Publication 3945*
"North American Tektites"

National Academy of Sciences—*Proceedings of the National Academy of Sciences*
"Origin of Indochinite Tektites"

American Association for the Advancement of Science—*Science*
"Detrital Mineral Grains in Tektites"
"Coesite Discovered in Tektites"
"Zircon and Chromite Crystals in a Muong Nong-Type Tektite"
"Rubidium–Strontium Age of the Bosumtwi Crater Area, Ghana, Compared with the Age of the Ivory Coast Tektites"
"Tektites That Were Partially Plastic after Completion of Surface Sculpturing"
"Glassy Objects (Microtektites?) from Deep-Sea Sediments near the Ivory Coast"
"Tektites Are Terrestrial"
"Tektite Glass *Not* in Apollo 12 Sample"
"Tektites from the Earth"

Microforms International Marketing Corporation—*Geochimica et Cosmochimica Acta*
"New Data on Georgia Tektites"
"New Data on the Nickel–Iron Spherules from Southeast Asian Tektites and Their Implications"
"Synthetic Metallic Spherules in a Philippine Tektite"
"Baddeleyite Inclusion in the Martha's Vineyard Tektite"
"Devitrification of Glass around Collapsed Bubbles in Tektites"
"Population Polygons of Tektite Specific Gravity for Various Localities in Australasia"
"Rhenium and Osmium Abundances in Tektites"
"The Distribution of Some Noble Metals in Meteorites and Natural Materials"
"The Radioactivity of the Ivory Coast Tektites and the Formation of the Bosumtwi Crater (Ghana)"
"Isotopic Composition of Lead from Tektites"
"Isotopic Comparison of Lead in Tektites with Lead in Earth Materials"
"On the Simultaneous Origin of Tektites and Other Natural Glasses"
"New Fission Track Ages of Tektites and Related Glasses"
"Tektite Sculpturing"

"Chemical Composition of Ivory Coast Microtektites"
"Aerodynamic Analysis of the Tektite Problem"
"Genetic Significance of the Chemical Composition of Tektites: A Review"

Macmillan Journals Ltd.—*Nature*
"Comparison of the Chemical Composition and Magnetic Properties of Tektites and Glasses Formed by Fusion of Terrestrial Rocks"
"A Possible Origin of Tektites by Soil Fusion at Impact Sites"

American Geophysical Union—*Journal of Geophysical Research*
"Libyan Desert Glass: Its Viscosity and Some Comments on Its Origin"
"Correlations between O^{18}/O^{16} Ratios and Chemical Compositions of Tektites"
"Sculpturing of Moldavites and the Problem of Micromoldavites"
"Physical and Chemical Properties of Australasian Microtektites"
"Australasian Tektite Geographic Pattern, Crater and Ray of Origin, and Theory of Tektite Events"

Pergamon Press Ltd.—*International Dictionary of Geophysics*
"Tektites"

North-Holland Publishing Co.—*Earth and Planetary Science Letters*
"A Second Tektite Fall in Australia"
"Comparison of the Chemical Variation in a Flanged Australite with the Chemical Variation among 'Normal' Australasian Microtektites"

The Geological Society of America, Inc.—*Geological Society of America Bulletin*
"Age of Asian Tektites"

Geological Society of Australia—*Journal of the Geological Society of Australia*
"Stratigraphical Evidence for the Terrestrial Age of Australites"

Yale University—*American Journal of Science*
"The Role of Aerodynamical Phenomena in Shaping and Sculpturing Australian Tektites"

Massachusetts Institute of Technology—*Proceedings of the Second Lunar Science Conference*
"Surface Micrography of Lunar Fines Compared with Tektites and Terrestrial Volcanic Analogs"

VEB Gustav Fischer Verlag—*Chemie der Erde*
"Description and Origin of Large Tektite from Thailand"

National Museum of Victoria (Australia)—*Memoirs of the National Museum of Victoria, Melbourne*
"Tektites"

Series Editor's Preface

The philosophy behind the "Benchmark Papers in Geology" is one of collection, sifting, and rediffusion. Scientific literature today is so vast, so dispersed, and, in the case of old papers, so inaccessible for readers not in the immediate neighborhood of major libraries that much valuable information has been ignored by default. It has become just so difficult, or so time consuming, to search out the key papers in any basic area of research that one can hardly blame a busy man for skimping on some of his "homework."

This series of volumes has been devised, therefore, to make a practical contribution to this critical problem. The geologist, perhaps even more than any other scientist, often suffers from twin difficulties—isolation from central library resources and immensely diffused sources of material. New colleges and industrial libraries simply cannot afford to purchase complete runs of all the world's earth science literature. Specialists simply cannot locate reprints or copies of all their principal reference materials. So it is that we are now making a concerted effort to gather into single volumes the critical material needed to reconstruct the background of any and every major topic of our discipline.

We are interpreting "Geology" in its broadest sense: the fundamental science of the Planet Earth, its materials, its history, and its dynamics. Because of training and experience in "earthy" materials, we also take in astrogeology, the corresponding aspect of the planetary sciences. Besides the classical core disciplines such as mineralogy, petrology, structure, geomorphology, paleontology, and stratigraphy, we embrace the newer fields of geophysics and geochemistry, applied also to oceanography, geochronology, and paleoecology. We recognize the work of the mining geologists, the petroleum geologists, the hydrologists, the engineering and environmental geologists. Each specialist needs his working library. We are endeavoring to make his task a little easier.

Each volume in the series contains an Introduction prepared by a specialist (the volume editor)—a "state of the art" opening or a summary of the objects and content of the volume. The articles, usually some thirty to fifty reproduced either in their entirety or in significant extracts, are selected in an attempt to cover the field, from the key papers of the last century to fairly recent work. Where the original works are in foreign languages, we have endeavored to locate or commission translations. Geologists, because of their global subject, are often acutely aware of the oneness of our world. The selections cannot, therefore, be restricted to any one country, and whenever possible an attempt is made to scan the world literature.

To each article, or group of kindred articles, some sort of "Highlight Commentary" is usually supplied by the volume editor. This should serve to bring that article into historical perspective and to emphasize its particular role in the growth of the field. References, or citations, wherever possible, will be reproduced in their entirety—for by this means the observant reader can assess the background material available to that particular author, or, if he wishes, he too can double check the earlier sources.

A "benchmark," in surveyor's terminology, is an established point on the ground, recorded on our maps. It is usually anything that is a vantage point, from a modest hill to a mountain peak. From the historical viewpoint, these benchmarks are the bricks of our scientific edifice.

Rhodes W. Fairbridge

Preface

More than a thousand papers have been published on tektites. Because of space limitations some difficult choices were made in selecting papers for this volume, and publications that otherwise would have been included in full are abstracted or mentioned in the appropriate place. Important early articles were chosen, as well as papers in the period of very intensive research on tektites, from about 1955 to the present time. Because of space limitations and the desire to bring to the reader's attention as many additional publications as possible, most of the papers reproduced in this book are not discussed in the introduction.

The contents have been roughly grouped, although some papers also cover several subjects. Twelve papers are primarily concerned with the geology, petrology, and mineralogy of tektites, eleven with their physical and chemical characteristics, and the next thirteen are almost equally divided on the subjects of age determination, sculpture, and microtektites. The next eight papers concentrate on the origin of tektites, and the final item reproduced is the bibliography from Baker's (1959) Memoir.†

We wish to take this opportunity to thank the National Science Foundation and the Bureau of Economic Geology, The University of Texas at Austin, for support of the senior author's tektite research, and to acknowledge the editorial help of Richard Adloff in preparing the translation of Lacroix's paper. In addition we wish to acknowledge with thanks the cooperation and assistance, in our world-wide tektite research, of various governmental organizations, universities, and individuals too numerous to list here. We also thank colleagues for exchange of tektite specimens, data, and reprints, and for their generosity in providing their papers for reproduction in this book.

Where the affiliations and addresses of authors are omitted in the commentaries, they will be found in the reproduced papers.

Virgil E. Barnes and Mildred A. Barnes
January, 1973

†In order to shorten the list of references, the reader is referred to Baker's (1959) bibliography for papers published prior to 1959; this bibliography is the last item reproduced in this volume. All references marked with an asterisk are reproduced in this volume.

Contents

IV. SCULPTURE OF TEKTITES

V. MICROTEKTITES

VI. ORIGIN OF TEKTITES

Contents by Author

Introduction

The origin of tektites has aroused the curiosity of man for more than a millennium. Primitive peoples, often believing that tektites fell from the sky, regarded them as objects of mystery and magic and collected them as talismans. Because these curiously shaped, distinctive, glassy objects appeared to be unrelated to the rocks and pebbles with which they were associated, many scientists considered them to be meteorites. Suess (1900)† derived the name of tektite for these exotic bodies from the Greek word *tektos* (melted).

Tektites are most often about the size of a walnut, but they range from considerably smaller to blocky specimens weighing as much as 28 pounds (12.8 kg). They appear to be black, but when held up to the light some are translucent. Those from Czechoslovakia are most often a beautiful clear light green, and those from the Libyan Desert are generally the color of straw, i.e., a light greenish yellow. Some of these may range to smoky gray or milky white. Even tektites which are opaque in whole specimens, such as bediasites, philippinites, indochinites, and australites, are translucent and yellowish brown in thin plates.

In form, when unbroken, they may be the shape of a teardrop, a dumbbell, a rod, a sphere, a disk, or a peculiarly flanged button. More often the original form is not discernable because of modification by spalling, breaking, or etching. A special type of tektite which is blocky and layered was first described by Lacroix (1935a)* from near Muong Nong, Laos, and was confirmed by Barnes and Pitakpaivan (1962)* and Barnes (1964) to be chemically identical to the type of tektite referred to as "splash-form."

---

†In order to shorten the list of references for this introduction, the reader is referred to Baker's (1959)* bibliography for papers published prior to 1959; this bibliography is the last item reproduced in this volume. All references marked with an asterisk are reproduced in this volume.

Rinehart (1958) described the manner in which a column of spurted or splashed melt would have collapsed to form spheres, teardrops, and dumbbells and it is these, and related forms and their fragments, that we refer to as "splash-form" tektites. This type of tektite is limited in size, controlled by the surface tension of the melt from which it formed. The largest splash-form tektite known is a philippinite the size of a grapefruit, weighing 1069.6 g, or 2.36 lb (Beyer, 1962, p. 130), whereas the largest known Muong Nong-type layered tektite weighs 12.8 kg or 28 lb (Barnes, 1971)*. This specimen is a segment from a much larger mass not yet recovered, but reported by Pitakpaivan (1972) to be a sheet covering several square meters.

Tektites are found in widespread groups in definite areas (strewn-fields) in various parts of the world (Fig. 1), each group having arrived at widely separated times with no individuals as yet found to have arrived between such times. The groups vary in age from perhaps as young as 15,000 years for those in Australia to about 40,000,000 years for North American tektites.

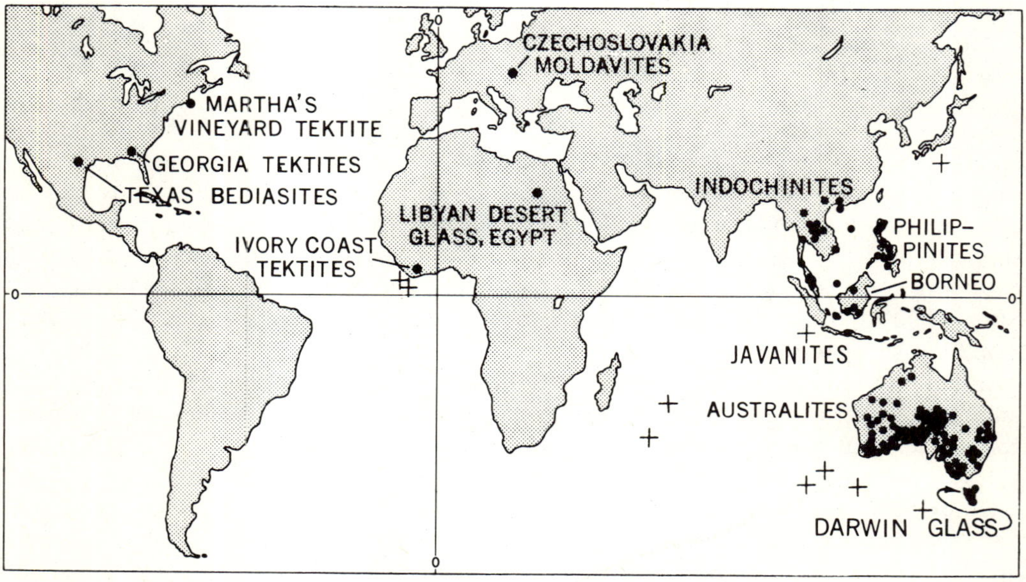

Fig. 1. Distribution of tektites, microtektites (+), and Darwin glass on earth. Adapted from Barnes (1963).

Most occurrences have been given distinctive names, such as "moldavites" (vltaviny), named for the Moldau (Vltava) River, for those found in Czechoslovakia, and "bediasites," named for the Bedias Indians, for those found in Texas. Others are named after the place or country in which they were found, e.g., australites, javanites, billitonites, indochinites, and philippinites. A few groups, such as the Ivory Coast tektites, Libyan Desert glass, Georgia tektites, and the lone specimen from Martha's Vineyard, Massachusetts, are not specifically named. Macusanite, formerly called americanite (Martin, 1934b), is obsidian and not tektite as originally thought (Barnes et al., 1970). Darwin Glass is considered to be an impactite in some way related to the australites (Fleischer et al., 1969).

The first known account of tektites is reported in a paper by Lee Da-ming (1963), in which he states that they were accurately described in China as early as in the T'ang dynasty (middle tenth century) by Liu Sun in his book *Ling Piao Lu Yi,* or "Notes on the wonders beyond the Nanling Mountains in Kwangtung." Liu Sun wrote "In Leichow, after sudden rainstorms, people collect black stones in the field which they call *lei-gong-mo* (inkstone of the Thunder-god). These stones make a ringing sound when struck and have a lovely brilliant luster" (Barnes, 1969).

The first mention of tektites in scientific literature described them as a special kind of volcanic glass (Mayer, 1788). Others believing in the igneous origin of tektites include Darwin (1844, 1851), who compared australite buttons to volcanic bombs, Streich (1893), Stelzner (1893a), and Dunn (1908b, 1912b). Dunn postulated that australite buttons were blebs of glass that formed at the base of bubbles blown in the throats of volcanoes.

Some people who speculated on tektites thought that they might be the product of prehistoric glass manufacture (Lindaker, 1792, p. 272; Makowsky, 1881a, b, c) or from burning coal seams (Lindaker, 1792). Others thought that they might be fulgurites produced by lightning (Rutley, 1885); outcasts from moon volcanoes (Verbeek, 1897a, b); a kind of meteorite (Suess, 1898a, b, 1900; Walcott, 1898); aerial fulgurites caused by lightning fusing dust during dust storms (Gregory, 1912; Chapman, 1929, 1933); or dessication of silicate gel masses (Easton, 1921; Van Lier, 1933).

Others suggested that tektites were produced from the tails of comets by oxidation in the atmosphere (Michel, 1922; Goldschmidt, 1924); that they are plastic sweepings from a meteorite passing through the atmosphere (Hardcastle, 1926; Hanus, 1928a, b), or debris from a celestial body or planet similar to earth blanketed by sedimentary rocks (Linck, 1926a; Barnes, 1951; Stair, 1954); or that they were formed by oxidation of light metal meteorites plunging into the atmosphere, resulting in the production of glass (Lacroix, 1932b; Beyer, 1934; Suess, 1935; Fenner, 1935a; Michel, 1939).

Now it is generally accepted that tektites are the result of some kind of impact on earth (meteorite impact—Spencer, 1933a, b, c, and Krinov, 1958; comet impact —Urey, 1957) or on the moon (Nininger, 1940, 1941, 1943, 1947; O'Keefe, 1963).

For an exhaustive treatment of postulated origins of tektites prior to 1959 the reader is referred to Baker's* monograph (1959, pp. 120–147). Since 1959, variations of previously suggested origins have been introduced. Cohen (1961, 1963) postulated that tektites resulted from an asteroid or comet impact on earth, and Chapman (1971)* postulated that Australasian tektites were splashed from the moon to earth along one particular ray of the impact crater Tycho.

O'Keefe and Adams (1965) postulated segregation of quartz in lunar welded tuffs as parent material for Darwin glass and Libyan Desert glass, although no such segregation has been observed to take place in welded tuffs on earth. Yagi (1966) points out that the difference in density between quartz and associated glass is insufficient to allow effective separation, and states that high concentration of quartz is possible only through chemical weathering followed by mechanical sorting.

O'Keefe (1970) reverted to moon volcanoes as the source of tektites when it was found that the quartz content of surficial lunar rocks is insignificant. However, volcanic explosions violent enough to throw tektites to earth would have spread ash of tektite composition over the entire moon, and surely some of this material would have been collected during the Apollo missions.

Lyttleton (1964) analysed the results of the passage of the earth through a large comet. He suggested that the gravitational action of the earth would have deflected inward those particles of the comet that otherwise would have passed by to the sides, and would have focused them into an axial line behind the earth. There, a high-density region would have been built up, with the particles melting from impaction with each other and aggregating to form tektites. The pull of the earth would have produced a downward jet of tektites moving radially inward.

Adams and Huffaker (1964)* postulated the formation of tektites from a hypothetical parent body which carried out a skipping flight in the earth's atmosphere.

Sun (1963) postulated that the grazing impact of a comet melted loess in the Yellow River valley of China, dragging the melt southward. This theory of tektite origin has relevance in that loess would make an excellent tektite parent material (Vogt, 1935).

Lin (1966, 1967) suggested that a large impact might "blow a bubble through the earth's atmosphere and hence propel small objects over intercontinental distances." Chapman and Gault (1967) disagreed.

Barnes (1961 and later) favored a comet or asteroid impact on earth to produce tektites; he further suggested (Barnes, 1971)* that the interaction of the coma of comets with the atmosphere might produce sufficient energy to melt surficial rocks and form puddles of melt which were whipped into the air by tornadic winds caused by the impact, to form splash-form tektites and microtektites. The residual melt would have formed the Muong Nong-type tektites.

Glassy spherules resembling microtektites, found from time to time notably in samples from drilled wells, usually can be explained as accidentally introduced contaminants related to man's activities. A glass spherule with a long tail found at Scott base, Antarctica, in residue from melted ice is typical of spherules found in glass wool; however, Shima (1966) found that the composition of this spherule is that of a tektite, a very unlikely composition for glass wool. He considered that this specimen might be a microtektite. Undoubted microtektites were first reported in cores from south of Australia and Sumatra (Fig. 1) by Glass (1967) and Glass and Heezen (1967).

Leake (1970) in examining microtektite compositional trends found that tektites and microtektites have a common history. O'Keefe (1969) suggested that: "Microtektites may represent the unwelded component of a lunar ash flow, of which the Muong Nong tektites are the welded component."

Glass (1972) cmmpared lunar glass spherules and microtektites. Although there are similarities, the two are distinguishably different.

Storzer and Gentner (1970) found by fission-track studies that the Bavarian molasse contains bentonitic glasses of the age of moldavites and of Nordlinger Ries glass. The bentonitic glasses are considered to be corroded relicts of micromoldavites.

Age determinations on tektites using the potassium-40–argon-40 method (Suess et al., 1951; Suess, 1951) are reviewed by Zahringer (1963a, b). In general, geologic ages and K^{40}–Ar^{40} and fission-track ages agree; however, for australites geologic ages appear to be much younger (Gill, 1970; Lovering et al., 1972*) than the 0.7 million years found by the other two methods.

The first fission-track dating of tektites was reported by Fleischer and Price (1964).

Smith (reported in Barnes, 1971*), using nannofossils, found that an indochinite had rested on the floor of the South China Sea for about one million years on the basis of a 3-million-year duration of the Pleistocene.

Durrani et al. (1970) have used tektite thermoluminescence for dating tektites, with limited success.

Boeckl (1972) investigated carbon-14 in tektites, and none was found outside of statistical fluctuations.

Schnetzler and Pinson (1963) cited summaries of earlier work (Barnes, 1940*; Baker, 1959), summarized chemical investigations, and compiled chemical data not previously published. They did not include Darwin glass, Libyan Desert glass, and various obsidians that at one time or another have been compared to tektites. In general they find that tektites are in a class of their own and that they are distinct from acid igneous rocks. Tektites are of high calcium, magnesium, and iron content and of low sodium and potassium content when compared with acid igneous rocks in the general silica range of 70–80 per cent. Also, in contrast to the case of acid igneous rocks, their potassium content is higher than their sodium content.

In an article on tektites in the International Dictionary of Geophysics, Barnes (1967)* includes another 38 analyses in his summary. More than 200 major element analyses have been published since 1963, and numerous minor and trace element analyses have been made. Zahringer (1963b) reviewed isotopes in tektites, and additional papers on isotopes have been published since. Publications which contain a large proportion of the post-1963 chemical data are listed with the discussion preceding the article on "Tektites" (Barnes, 1967)*.

Zahringer and Gentner (1963) found that the isotopic composition of argon in bubbles in gas-rich, splash-form tektites, Muong Nong-type tektites, and Darwin glass is the same as in air argon, thus demonstrating that gas in tektite bubbles is atmospheric in origin.

The possible role of vapor fractionation in altering the composition of tektites during their formation was reviewed by Philpotts and Pinson (1966). Walter (1967) discussed vapor fractionation in detail but was undecided whether or not it indicated a lunar versus a terrestrial origin. Walter and Giutronich (1967) found that the silica content in one silicate melt decreased from 82 to 45 per cent at temperatures of 2600–2800°C for 45 min; and that the volatility of the components of the melt is: $K_2O > SiO_2 > Na_2O > FeO > MgO \geqslant Al_2O_3 \geqslant CaO$. Walter and Clayton (1967) concluded that the compositional trend in bediasites is caused by vapor fractionation.

Taylor and Kaye (1969*) concluded: "Deductions based on vaporisation experiments must be treated with caution because of our lack of understanding of conditions at the impact site, where near-vacuum conditions are commonly assumed. . . . The chemistry of all classes of tektites is comparable to that of greywackes, subgreywackes or arkoses, and the internal variations in composition are consistent with this interpretation."

As a basis of interpreting observed evidence of tektite spallation and fragmentation, Centolanzi and Chapman (1965) analyzed thermally induced stresses residual in tektite glass. Centolanzi (1969) concluded that the natural stress in tektites, combined with stress caused by aerodynamic loading, would have limited the size of spherical tektites to the order of 1 kg, and that the largest cylinder likely to survive atmospheric entry is estimated to be of the order of 100 kg.

Gault and Wedekind (1969), because of micrometeoroid impact on tektites, estimate the survival time of tektites in circular heliocentric orbits would have been of the order of 10^3 and 10^4 years for objects 1 cm and 10 cm in diameter, respectively, thus accounting for the lack of tektites falling to earth between showers. Paddack (1969) estimates that solar radiation pressure can cause rotational bursting of tektites in about 60,000 years. Gault and Wedekind point out that the two different mechanisms would complement each other.

The question of where and how tektites originated is still a matter for controversy, but most workers agree that they resulted from impact. Some may still argue that the impacts occurred on the moon and that tektites are droplets of melted lunar rock that splashed onto the earth. However, it has been found that none of the materials brought back from the moon by the various Apollo missions could be the source of tektite glass, whereas many studies relate them, chemically and otherwise, to earth materials.

The Australian and Javanese tektites, which show unmistakable aerodynamic ablation, remain the most puzzling groups. If they originated from the earth, the event which formed them must have been so powerful as to blast out a section of the atmosphere, permitting these tektites to be ejected beyond the atmosphere and ablated upon their return to the earth.

References†

*Adams, E. W., and Huffaker, R. M. (1964). Aerodynamic analysis of the tektite problem. *Geochim. Cosmochim. Acta,* **28,** 881–892.

*Baker, G. (1959). Tektites. *Mem. Natl. Museum Victoria, Melbourne,* No. 23, 313 pp.

Barnes, V. E. (1961). Temperature of tektite formation (abstract 37). *CERN Conference on Fission and Spallation Phenomena and Their Application to Cosmic Rays,* Sept. 26–29, Geneva.

Barnes, V. E. (1963) Tektite strewn-fields. *In* "Tektites" (J. A. O'Keefe, ed.), Chapter 2, pp. 25–50. Univ. Chicago Press, Chicago, Illinois.

Barnes, V. E. (1964). Variation of petrographic and chemical characteristics of indochinite tektites within their strewn-field. *Geochim. Cosmochim. Acta,* **28,** 893–913.

*Barnes, V. E. (1967). Tektites. *In* "International Dictionary of Geophysics" (K. Runcorn, ed.), Vol. 2, pp. 1507–1518. Pergamon Press, Oxford.

Barnes, V. E. (1969). Progress of tektite studies in China. *E⊕S (Trans. Amer. Geophys. Union),* **50,** 704–709.

*Barnes, V. E. (1971). Description and origin of large tektite from Thailand. *Chem. Erde,* **30,** 13–19.

†In order to shorten the list of references for this introduction, the reader is referred to Baker's* bibliography for papers published prior to 1959; this bibliography is the last item reproduced in this volume. All references marked with an asterisk are reproduced in this book.

Barnes, V. E., Edwards, G., McLaughlin, W. A., Friedman, I., and Joensuu, O. (1970). Macusanite occurrence, age, and composition, Macusani, Peru. *Geol. Soc. Amer. Bull*, **81,** 1539–1546.

*Barnes, V. E., and Pitakpaivan, K. (1962). Origin of Indochinite tektites. *Proc. Natl. Acad. Sci. (U.S.)*, **48,** 947–955.

Beyer, H. O. (1962). "Philippine Tektites." Vol. 1, Part 2, 290 pp. Univ. Philippines Publications in Natural History and Space Science, Quezon City and Manila.

Boeckl, R. S. (1972). Search for carbon-14 in tektites. *J. Geophys. Res.*, **77,** 367–368.

Centolanzi, F. J. (1969). Maximum tektite size as limited by thermal stress and aerodynamic loads. *J. Geophys. Res.*, **74,** 6723–6736.

Centolanzi, F. J., and Chapman, D. R. (1965). Analysis of thermally induced stresses residual in tektite glass (abstract). *Trans. Amer. Geophys. Union,* **46,** p. 118.

*Chapman, D. R. (1971). Australasian tektite geographic pattern, crater and ray of origin, and theory of tektite events. *J. Geophys. Res.*, **76,** 6309–6338.

Chapman, D. R., and Gault, D. E. (1967). Critique of "Cometary impact and the origin of tektites." *J. Geophys. Res.*, **72,** 2695–2699.

Cohen, A. J. (1961). A semi-quantitative hypothesis of tektite origin by asteroid impact (abstract). *J. Geophys. Res.*, **66,** 2521.

Cohen, A. J. (1963). Asteroid- or comet-impact hypothesis of tektite origin: the moldavite strewn-fields. *In* "Tektites" (J. A. O'Keefe, ed.), pp. 189–211. Univ. Chicago Press, Chicago, Illinois.

Durrani, S. A., Christodoulides, C., and Ettinger, K. V. (1970). Thermoluminescence in tektites. *J. Geophys. Res.*, **75,** 983–995.

Fleischer, R. L., and Price, P. B. (1964). Glass dating by fission fragment tracks. *J. Geophys. Res.*, **69,** 331–339.

Fleischer, R. L., Price, P. B., Viertl, J. R. M., and Woods, R. T. (1969). Ages of Darwin glass, Macedon glass, and Far Eastern tektites. *Geochim. Cosmochim. Acta, 33,* 1071–1074.

Gault, D. E., and Wedekind, J. A. (1969). The destruction of tektites by micrometeoroid impact. *J. Geophys. Res.*, **74,** 6780–6794.

Gill, E. D. (1970). Age of australite fall. *J. Geophys. Res.*, **75,** 996–1002.

Glass, B. (1967). Microtektites in deep-sea sediments. *Nature,* **214,** 372–374.

Glass, B. (1972). Comparison between lunar glass spherules and microtektites (abstract). *E⊕S (Trans. Amer. Geophys. Union),* **53,** 428.

Glass, B., and Heezen, B. C. (1967). Tektites and geomagnetic reversals. *Nature,* **214,** 372.

Krinov, E. L. (1958). Some considerations on tektites. *Geochim. Cosmochim. Acta,* **14,** 259–266.

Leake, B. E. (1970). Some paradoxes in Australasian microtektite compositional trends. *J. Geophys. Res.*, **75,** 349–356.

Lee Da-ming (1963). A preliminary survey and study of the tektites, lei-gong-mo, from Leichow Peninsula and Hainan Island, China. *Scientia Geologica Sinica,* No. 1, pp. 42–49.

Lin, C. S. (1966). Cometary impact and the origin of tektites. *J. Geophys. Res.*, **71,** 2427–2437.

Lin, C. S. (1967). Reply [to Chapman and Gault's (1967) critique]. *J. Geophys. Res.*, **72,** 2700–2703.

*Lovering, J. F., Mason, B., Williams, G. E., and McColl, D. H. (1972). Stratigraphical evidence for the terrestrial age of australites. *J. Geol. Soc. Australia,* **18,** 409–418.

Lyttleton, R. A. (1964). A cometary mechanism for the formation of tektites. *Geochim. Cosmochim. Acta,* **28,** 807–820.

O'Keefe, J. A. (1963). The origin of tektites. *In* "Tektites" (J. A. O'Keefe, ed.), pp. 167–188. Univ. Chicago Press, Chicago, Illinois.

O'Keefe, J. A. (1969). The microtektite data: implications for the hypothesis of the lunar origin of tektites. *J. Geophys. Res.*, **74,** 6795–6804.

O'Keefe, J. A. (1970). Tektite glass in Apollo 12 sample. *Science,* **168,** 1209–1210.

O'Keefe, J. A., and Adams, E. W. (1965). Tektite structure and lunar ash flows. *J. Geophys. Res.*, **70,** 3819–3829.

Paddack, S. J. (1969). Rotational bursting of small celestial bodies: effects of radiation pressure. *J. Geophys. Res.*, **74,** 4379–4381.

Philpotts, J. A., and Pinson, W. H., Jr. (1966). New data on the chemical composition and origin of moldavites. *Geochim. Cosmochim. Acta,* **30,** 253–266.
Pitakpaivan, K. (1972). Letter of June 11.
Rinehart, J. S. (1958). Impact effects and tektites. *Geochim. Cosmochim. Acta,* **14,** 287–290.
Schnetzler, C. C., and Pinson, W. H., Jr. (1963). The chemical composition of tektites. *In* "Tektites" (J. A. O'Keefe, ed.), pp. 95–129. Univ. Chicago Press, Chicago, Illinois.
Shima, M. (1966). Glassy spherules (microtektites?) found in ice at Scott Base, Antarctica. *J. Geophys. Res.,* **71,** 3595–3596.
Storzer, D., and Gentner, W. (1970). Micromoldavites from the Bavarian molasse (abstract). *Meteoritics,* **5,** 225.
Sun, M. S. (1963). The origin of Asia-Australian tektites (abstract). *Trans. Amer. Geophys. Union,* **44,** 93–94.
*Taylor, S. R., and Kaye, M. (1969). Genetic significance of the chemical composition of tektites: a review. *Geochim. Cosmochim. Acta,* **33,** 1083–1100.
Walter, L. S. (1967). Tektite compositional trends and experimental vapor fractionation of silicates. *Geochim. Cosmochim. Acta,* **31,** 2043–2063.
Walter, L. S., and Clayton, R. N. (1967). Oxygen isotopes: experimental vapor fractionation and variation in tektites. *Science,* **156,** 1357–1358.
Walter, L. S., and Giutronich, J. E. (1967). Vapor fractionation of silicate melts at high temperatures and atmospheric pressures. *Solar Energy,* **11** (3 and 4), 163–169.
Yagi, K. (1966). Discussion of paper by J. A. O'Keefe and E. W. Adams "Tektite structure and lunar ash flows." *J. Geophys. Res.,* **71,** 5492–5493.
Zahringer, J. (1963a). K–Ar measurements of tektites. *In* "Radioactive Dating," pp. 289–305. International Atomic Energy Agency, Vienna, Austria.
Zahringer, J. (1963b) Isotopes in tektites. *In* "Tektites" (J. A. O'Keefe, ed.), pp. 137–149. Univ. Chicago Press, Chicago, Illinois.
Zahringer, J., and Gentner, W. (1963). Radiogenic and atmospheric argon content of tektites. *Nature,* **199,** 583.

Geology, Petrology, and Mineralogy

I

Editors' Comments on Paper 1

Clayton and Spencer*, in addition to reporting for the first time the occurrence of Libyan Desert glass, restated Spencer's (1933a) meteorite impact hypothesis for tektite origin. Spencer (1933b, 1937, 1939) later published additional information on Libyan Desert glass.

The discovery of the Libyan Desert silica glass strewn-field by P. A. Clayton in 1932 introduced further evidence concerning the origin of tektites that has since all too frequently been disregarded by scientists theorizing on the problem. Libyan Desert glass is a dominantly silica, layered tektite that occurs in an environment that is equally rich in silica, indicating that the glass was formed locally. The final proof of its local origin would be to find an outcrop of sandstone, the upper part of which is Libyan Desert glass.

The area of Libyan Desert glass had not been revisited since the explorations of Ralph Bagnold and of Clayton and Spencer during the 1930s until June 3, 1971, when James Underwood, Ali Sbeta, and V. E. Barnes spent two hours in the westernmost part of the strewn-field, where they recovered two dozen pieces of glass. In addition they found an Egyptian airplane that had made a forced landing April 11, 1968, with nine men aboard who soon perished from lack of water (Barnes and Barnes, 1972). From this incident one senses the remoteness and inaccessibility of the area and the reason geologists have not in general explored the area in search of clues to the origin of the glass.

Patrick Andrew Clayton (1896–1962), Inspector, Desert Surveys of Egypt (1920–1938), previously served in the Royal Field Artillery in Greece and Turkey. He wrote various monographs on his work in the Western Desert and, in conjunction with Dr. Spencer, published on the occurrence of silica glass. Following two years in Tanganyika, Colonel Clayton was recalled to duty under General Wavell in Libya, where he served in the Long Range Desert Group. He received the Founder's Medal of the Royal Geographical Society for his maps used in the 1940–1943 campaigns in Libya. Copies of these maps were used by Underwood, Barnes, and Sbeta in 1971 during their visit to the area of Libyan Desert glass occurrence.

Leonard James Spencer (1870–1959) held office in the Mineral Department of the British Museum for more than forty years until his retirement in 1935, but intimate connection with the department lasted for another twenty-four years, to the time of his death. Dr. Spencer joined Col. Clayton on an expedition to the area of Libyan Desert glass occurrence in December, 1934, just prior to his retirement at age sixty-five. It was after Dr. Spencer's retirement that we corresponded about tektites and meteorites, and that Libyan Desert glass was exchanged for bediasites. We are grateful to him for pointing out errors in the bibliography in Barnes' "North American tektites," some of which were carried over into Baker's (1959)* bibliography. These errors are noted in the commentary preceding Baker's (1959)* bibliography.

References

*Baker, G. (1959). Tektites. *Mem. Natl. Museum Victoria, Melbourne,* No. 23, 313 pp.

Barnes, V. E. (1940). North American Tektites. *Univ. Texas Pub. 3945,* pp. 477–582.

Barnes, V. E., and Barnes, M. A. (1972). World-wide investigation of tektites continued. *Lapidary J.,* **26,** 18–48.

Spencer, L. J. (1933a). Origin of tektites. *Nature,* **131,** 117–118, 876.

Spencer, L. J. (1933b). Two new gem stones. *The Gemnologist,* **3,** 110–113.

Spencer, L. J. (1937). The tektite problem. *Mineral. Mag.,* **24** (156), 503–506.

Spencer, L. J. (1939). Tektites and silica-glass. *Mineral. Mag.,* **25** (167), 425–440.

Reprinted from *Mineralogical Magazine*, **23**, 501–508 (1934)

1

Silica-glass from the Libyan Desert.[1]

By P. A. Clayton, F.G.S.
Inspector, Desert Surveys, Survey of Egypt,
and L. J. Spencer, M.A., Sc.D., F.R.S.
Keeper of Minerals, British Museum.

[Read November 9, 1933.]

Discovery and occurrence (P. A. Clayton).

An occurrence of silica-glass in lumps up to 10 lb. in weight, smaller fragments being scattered over an area at least 80 by 25 kilometres, has been discovered in the course of the work of the Egyptian Desert Surveys. A Survey expedition under Mr. P. A. Clayton was sent out in December 1932 to work in the unexplored area of the Sand Sea and Gilf Kebir, west of Major R. A. Bagnold's 1930 route, and found the glass on December 29, 1932. It occurs lying on the surface of the Nubian sandstone in the sand-free corridors between the high north-south dune ranges of the south-west of the Sand Sea (fig. 1). The largest lumps were found in approximately latitude 25° 25′ N., longitude 25° 30′ E. The district is not easily accessible, as direct travel from Abu Mungar or Dakhla by car is impracticable. Water is only available at Ain Dalla, reached by a winding route some 500 km. long through the Sand Sea, or at Kufara Oasis some 300 km. away in Italian Cyrenaica.

A hurried preliminary reconnaissance has so far not revealed anything in the geology of the district which throws light on the occurrence of the silica-glass, and no crater has been found in the district. Some 30 per cent. of the area is hidden by dune ranges nearly 100 m. high, but the dunes are so regular that it is probable that the surface on which they lie is featureless. The dunes run in parallel north-south ranges 2 to 5 km. apart, steep on the eastern face, with a long soft sand slope on the west, and the underlying Nubian sandstone plain, generally free of sand, has an almost imperceptible slope averaging 1 in 1000 from south to north.

[1] Portions of this paper were published in the Geographical Journal, October, 1933, vol. 82, pp. 375–377.

The surface of this plain in the southern part of the silica-glass area is hard quartzite, overlain in places with reddish sandstone, and also in parts by a reddish alluvial deposit probably derived from weathering of the sandstone. Some of this reddish material can be seen adhering to the specimens of glass which were lying on it. Such specimens were not found more than half embedded, and though holes were dug where the glass lay thickest, none was found below the surface.

Farther north the surface is a coarse quartz gravel, the pebbles of which are very rounded, while the glass shows no signs of having been rolled, though its surface is wind-eroded. As one goes north the gravel becomes finer, till at the northernmost point where glass was definitely found the surface is a thin crust of fine gravel over a white angular quartz sand. Small pieces of clear quartz are common in the gravel, and not easy to distinguish when searching from fragments of glass 3–5 mm. long.

No well-worked tools of silica-glass were found, though a few flakes were certainly made by man. Only two small fragments, in company with a definite artifact of quartzite, were found actually on the sand ridges, and these must have been carried up there and dropped. Near the largest lumps a number of spots were found where groups of fairly large (10 cm.) fragments occur within a square metre, as though a large lump had been broken up. Attempts to fit the pieces together were unsuccessful, possibly due to the way in which the glass flies to pieces on fracture.

The extent of the area over which it is distributed is only vaguely known. None was seen south of the old Kufara to Abu Mungar camel route, and the first pieces were seen about 4 km. north of this, For about 26 km. farther north up the central corridor in which it occurs pieces are visible every few yards, even from a moving car, while farther north one has to stop and search, sometimes for minutes before a single fragment is found. 84 km. north of the old camel road it was only after considerable search that a few tiny fragments were found. The distribution in the adjacent corridors on either side was similar, though the lumps were smaller and less plentiful than in the central corridor, while pieces were also picked up in the next corridors east and west.

About a hundredweight of the silica-glass was collected and brought to Cairo, though more could have been collected if transport were available.

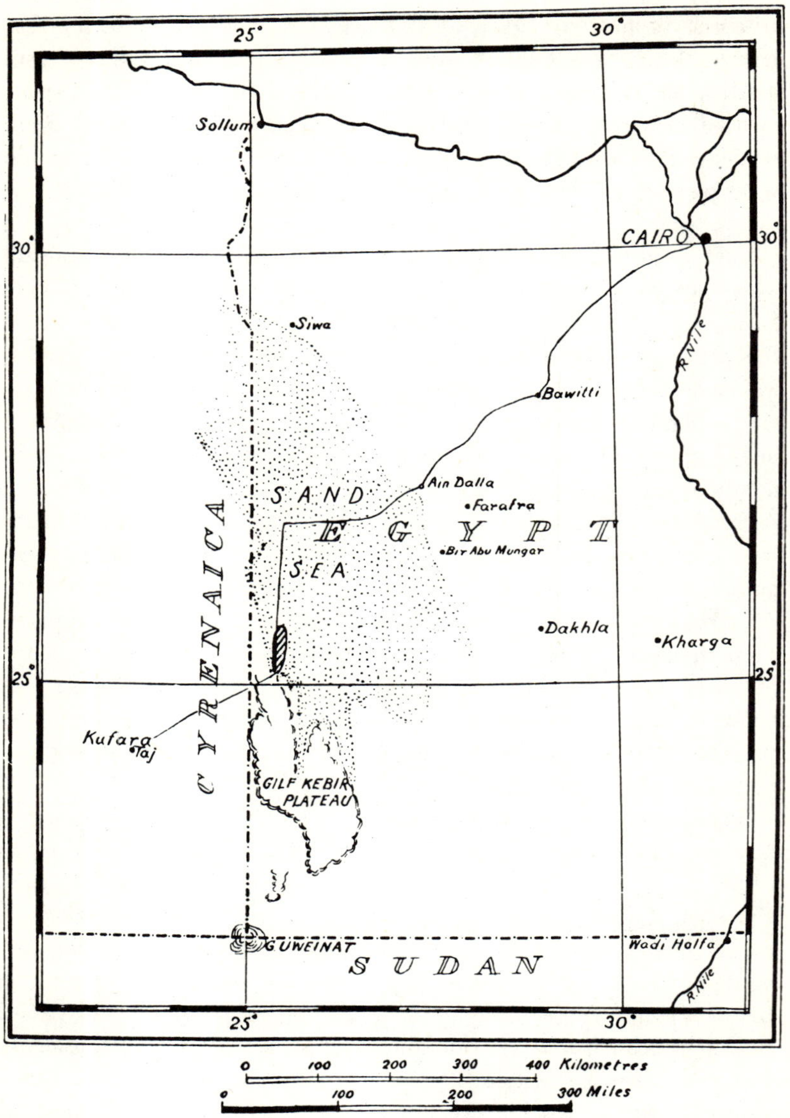

= Distribution of Silica-glass in the Libyan Desert

FIG. 1. Sketch-map showing the locality of the silica-glass.

Subsequently sorting over the collection of small pieces, it was surprising to find how many of them (at least 10 %) showed old fractures which might have been produced by an intentional blow (fig. 2). Mr. George Murray and the Rev. P. Bovier-Lapierre were prepared to accept most of these as having been chipped off; but Dr. John Ball and Mr. Brunton, regarding the curved bump as a conchoidal fracture and not a bulb of percussion, will admit only

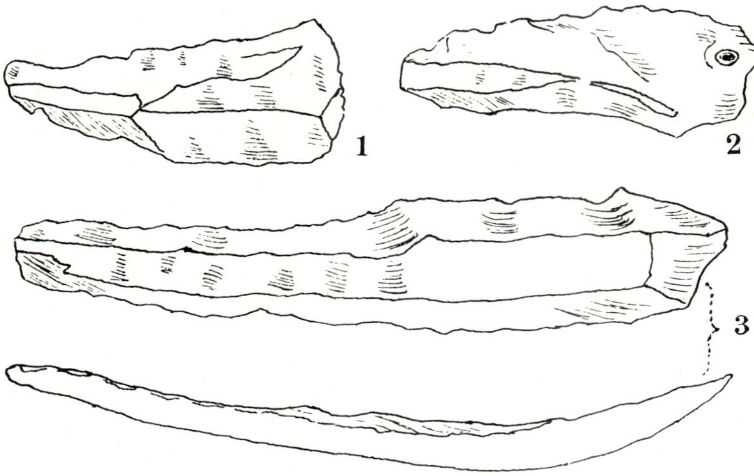

FIG. 2. Worked flakes of silica-glass. No. 2 is perforated by a hole in which a sand grain is wedged.

about 1 %. Three of these authorities agree in calling the flakes 'late neolithic or pre-dynastic'. A possible explanation of the groups of fragments found on the surface of the desert, as mentioned above, is that they were broken by human hands.

Characters and origin (L. J. Spencer).

Seven pieces of this remarkable material have been generously presented by the Survey of Egypt for the Mineral Collection of the British Museum (Natural History), and are now displayed in the Recent Addition Case in the Mineral Gallery at South Kensington, together with other material to which they appear to be most closely related, namely tektites from Bohemia, Moravia, French Indo-China, Australia, and Colombia. The masses are irregular in shape and are wind-worn on all sides (figs. 3 and 4). They range in weight from 90·4 to 2279 grams ($3\frac{1}{4}$ oz. to 5 lb.). The colour is pale greenish-yellow and the material is in part quite clear and transparent. Some parts are cloudy owing to the presence of minute bubbles, and when these are still more abundant the material has a milky white appearance. One piece shows smoke-grey streaks in the greenish-yellow glass, and indications of a flow structure are also shown in other pieces.

The clear material when cut and polished (in a step-cut form) makes quite an effective gemstone. One cut for the Museum weighs 10·57 grams (52·86 carats).[1] The specific gravity of this was determined by hydrostatic weighing to be 2·21. On the refractometer it gave the following values for the refractive index. Less reliable values were obtained by the prism method through a 60°-prism cut from the glass, owing to the scattering of the rays of light by the minute bubbles. The dispersion n_{Tl-Li} is very low.

Light.	Silica-glass, Libyan Desert. Refractometer.	Prism.	Pure Silica-glass.[2]
Red (Li)	1·4595	1·459	1·45605
Yellow (Na)	1·4624	1·462	1·45845
Green (Tl)	1·4645	1·465	1·46053
n_{Tl-Li}	0·0050	0·006	0·00448

The hardness is 6 on Mohs's scale; the material readily scratches window-glass and is itself readily scratched by quartz.

In thin sections under the microscope the glass is colourless and optically isotropic. The minute bubbles do not exceed 0·1 mm. across, and the larger ones are usually collected into groups. They are round or elongated, but the direction of elongation is variable, and no indications of flow structure are seen in the sections. A very fine brownish dust pervades the whole glass, and there are a few small brown patches of irregular outline, and very few minute (0·02 mm.) black spots.

Small white spherulites, apparently due to devitrification, are occasionally present. In one specimen these are more abundant and are arranged in approximately parallel lines, like strings of small ($\frac{3}{4}$ mm.) beads. They have a minutely hummocky surface, and in

[1] L. J. Spencer, Gemmologist, London, 1933, vol. 3, p. 111. [Min. Abstr., **5**-398.]

[2] Values (for Li and Tl by interpolation) given by R. B. Sosman, International Critical Tables, 1929, vol. 6, p. 341.

thin section are seen to consist of a radial aggregate of comparatively coarse acicular crystals of quartz. This specimen is also peculiar in showing a clean-cut cylindrical bore completely traversing the mass of glass; it is $3\frac{1}{2}$ cm. long and 6 mm. in diameter, and approximately

FIG. 3. Silica-glass from the Libyan Desert. Wind-worn mass of 5 lb. Half actual size.

parallel to the lines of spherulites (fig. 4). Two other such bore-holes with rounded ends penetrate for only a short distance in the same general direction. Occasionally small pits are seen on the surface of the specimens, in which a grain of sand is sometimes wedged (compare fig. 2).

The largest spherulite, $1\frac{1}{2}$ mm. in diameter, occurring alone and partly projecting from a small fragment of clear glass, afforded an opportunity of detaching a fragment for examination. This has specific gravity 2·35, and is very weakly birefringent with refractive index less than 1·49, suggesting cristobalite.

A chemical analysis of the glass made by Mr. M. H. Hey in the Museum laboratory gave:

SiO_2.	TiO_2.	Al_2O_3.	Fe_2O_3.	FeO.	CaO.	Na_2O.	H_2O.	Total.
97·58	0·21	1·54	0·11	0·23	0·38	0·34	0·10	100·49

Traces of NiO, MnO, and MgO are also present. K_2O is absent. Of the water 0·05 % was lost at 110° C., and the remainder on ignition. For the main analysis 1·0339 gram was used, and for FeO and

FIG. 4. Silica-glass from the Libyan Desert. Actual size. Spherulites are shown more distinctly through the flat polished surface. The rod marks a cylindrical bore which is penetrated sideways at one point. Another shorter bore with rounded end is seen towards the bottom.

alkalis separate portions of about 1 gram and ½ gram respectively. A special test was made for nickel on 2·97 grams, and the amount of NiO is probably less than 0·001 %, certainly less than 0·005. The specific gravity of the solid lump was found to be D_4^{20} 2·206 ± 0·002, and of the fine powder D_4^{20} 2·208 ± 0·002, corrected to vacuum, indicating that the bubbles present are not more than 0·6 % of the material by volume.

This silica-glass in its characters and mode of occurrence is most closely allied to tektites. But it is found in much larger masses and in larger quantity than any tektites previously known, and also it is different in colour. Tektites are usually black and in the mass opaque, but those from Bohemia and Moravia are a clear green, and small ones from Colombia are pale smoky-grey to almost colourless.

The origin of tektites is still an unsolved problem.[1] The generally accepted view at the present time is that they are of cosmic origin, but none has been actually observed to fall from the sky. The presence of silica-glass in large quantity around the meteorite craters recently discovered at Wabar in Arabia and Henbury in Central Australia, suggests that tektites may have been formed by the fusion of desert sand in the heat developed by the fall of gigantic meteorites. No evidence of this, however, has been detected in the Libyan Desert; and, moreover, the wide distribution (80 × 25 km.) of the material is rather against this view. Mr. Clayton remarks that from the surface geology of the district there are no indications that the material might have been transported by water, as is undoubtedly the case in the distribution of tektites.

The origin of this silica-glass on the Libyan Desert thus remains an unsolved problem. More observations are needed.

A group of craters in the Libyan Desert at about 22° 20′ N., 25° 30′ E., between Gebel 'Uweinat and Gilf Kebir, was photographed from the air by Sir Robert Clayton[2] in 1932 and was visited later in the same year by Major R. A. Bagnold's expedition.[3] They are described by Dr. K. S. Sandford[4] as explosion craters with walls of 'steeply tilted, hardened, and locally fused Nubian sandstone'. Within the craters are 'plugs of white trachytic rock and adjacent to them the fusion of the sandstone walls was very marked'. Mention is also made of an 'irony sinter'. Such a description is suggestive of meteorite craters rather than volcanic craters. Some of the silica-glass from Meteor Crater in Arizona would answer to the field description of a 'white trachytic rock', and the 'irony sinter' may perhaps be meteoritic iron-shale. 'A dark basalt-like rock of much the same specific gravity as the trachyte' is rather suggestive of the silica-glass from the meteorite craters at Henbury and Wabar. Although this locality is 200 miles to the south of the area where the lumps of clear silica-glass are found, it seems possible that there may be some connexion between the two occurrences.

[1] Nature, London, 1933, vol. 131, pp. 117, 876; vol. 132, pp. 571, 678. [Min. Abstr., **5**–303, 304, 407.]

[2] Geogr. Journ., 1933, vol. 81, p. 257, and plate facing p. 252.

[3] Geogr. Journ., 1933, vol. 82, pp. 106, 215.

[4] K. S. Sandford, Volcanic craters in the Libyan Desert. Nature, London, 1933, vol. 131, pp. 46–47.

Editors' Comments on Paper 2

Although stray pieces of the kind of glass found in quantity near Muong Nong, Laos, had been known previously, Lacroix was the first to describe such a deposit in detail and relate it to tektites in general. This kind of layered glass has come to be known as the Muong Nong type of tektite. The manner in which the Muong Nong-type indochinites and Libyan Desert glass formed must be taken into consideration in any theory of tektite origin. Evidence now suggests that this glass may have formed as puddles of soil melt that accumulated in low places, rather than being meteoritic in origin (Barnes, 1971)*.

Previously, Lacroix (1934) recorded another "first" in tektite research by reporting the occurrence of tektites in the Ivory Coast. David et al. (1927) had postulated that all tektite occurrences were meteoritic in origin and were located on one great circle of the earth. The finding of Ivory Coast tektites 40 degrees from the great circle weakened this postulate. La Paz (1938) later postulated two great circles. However, neither of these great circles passed near the locality of North American bediasites reported two years later (Barnes, 1940)*.

François Antoine Alfred Lacroix (1863–1948) was a mineralogist, petrologist, volcanologist, and geologist of world renown. He became Professor of Mineralogy at the Musée National d'Histoire Naturelle, Paris, in 1893 and was elected to the Académie des Sciences de l'Institut de France in 1904. Lacroix described most of the internal features of tektites, leaving only positive identification and interpretation of a few of these features to others. A complete list of his publications on tektites is included in Baker (1959)*.

References

*Baker, G. (1959). Tektites. *Mem. Natl. Museum Victoria, Melbourne,* No. 23. 313 pp.

*Barnes, V. E. (1940) North American tektites. *Univ. Texas Pub. 3945,* pp. 475–582.

*Barnes, V. E. (1971). Description and origin of large tektite from Thailand. *Chem. Erde,* **30,** 13–19.
David, T. W. E., Summers, H. S., and Ampt, G. A. (1927). The Tasmanian tektite–Darwin glass. *Proc. Roy. Soc. Victoria,* **39** (2), 167–190.
Lacroix, A. (1934). Sur la découverte de tectites à la Côte d'Ivoire. *C. R. Acad. Sci. Paris,* **199**, 1539–1542.
La Paz, L. (1938). The great-circle distribution of the tektites. *Popular Astron.,* **46,** 224–230.

2

The Unshaped Tektites of Indochina†

ALFRED LACROIX

One of the characteristics of the tektites of Indochina and adjacent areas, as well as those of all the other regions where these fragments of black glass are found, is that they are small.

> They consist, in more or less abundance, of shaped pieces (teardrops, very elongated pears, spheres, discs, etc.) modeled in the course of the vertical fall of rather fluid material, and of the debris of the latter, mixed with fragments of irregular form: plates, little twisted pieces, and so on. These shaped pieces often contain gaseous bubbles, spherical or elongated, up to some centimeters in diameter. Very characteristic are the intersecting convex fracture surfaces of crestlike appearance or more complicated form, but always convex and with curved ridges; they are those of a tempered glass, as is indicated by the phenomena of secondary birefringence; these surfaces make it possible to perceive, in thick layers, the contortions of the interior movements which the glass underwent before its consolidation.

The object of the present note is to describe another type of Indochinese tektite, of which, until now, I have observed only a few specimens [Smach; Attopeu; Samrong near Kratié (Cambodia)] scattered among the shaped forms, and of which a remarkable deposit has just been discovered in Moi country in lower Laos. There, instead of being exceptional, this type is the rule. The deposit is situated some kilometers to the south of Muong Nong, on the left bank of Sé La Nong, in the province of Savannakhet, quite near its border with Saravane province.

---

†Translated from Les Tectites sans formes figurées de l'Indochine by Alfred Lacroix, *Comtes rendus hebdomadaires des séances de l'Académie des Sciences de Paris*, **200**, 2129–2132 (1935). Translated by Mildred A. Barnes.

In 1932 the chief of the district of Muong Nong brought to M. Détrie, an official of the province who was making a circuit, a block of black glass weighing about 2 kg; this block had been found the preceding year, a year of food shortage, by a native woman who was looking for edible roots in the forest. The nature of this specimen not having been recognized, it was broken into pieces and I received a piece of it. Later, M. Détrie agreed to send more of the specimens to me and provided information on the locality of their deposit which enabled M. J.-H. Hoffet, of the Geological Service of Indochina, to find it and to make the following observations; they are of intense interest because, for the first time, it has been possible to study a primary deposit of tektites independent of alluvium and intentionally uncovered.

The following section was obtained by trenching: under a superficial bed of recent alluvium 1.05 meters thick, an ancient lateritic clay was found, resting on the upper part of Liassic sandstone. The tektites are localized in the first 10 cm of this clay, disseminated in exceptional density. A surface area of 1 m^2 was excavated and searched, and furnished 362 fragments weighing a total of 9.7 kg; among them should be noted three large pieces weighing, respectively, 560, 530, and 220 g.

In a neighboring area of 9 m^2, more than 2000 fragments were collected, many of which resulted from the breakage of larger pieces in the course of the excavation work: M. Hoffet estimated at around 1500 the number of pieces originally whole. Among them, one of 3.2 kg should be cited; it is the largest tektite that has ever been found. Together they weighed 20.6 kg. These figures are given only as proof of the richness of this very localized deposit; no tektites were found elsewhere in the surrounding area. It measures about 100 m^2 and has yielded, in all, 67.5 kg of specimens.

None of these pieces has a shaped form; large or small, all are angular, some are simply chips but have the same appearance; their surface, sometimes shiny, is covered with erosion pits, hemispherical, small, crowded together, and separated from their neighbors by sharp ridges. Thanks to this peculiarity, they make one think of moldavites, with the difference that the corrosion pits are not deep, worm-like grooves of large size, as in the latter.

This rough surface, harsh to the touch, distinguishes the tektites of Muong Nong from those of other indochinite deposits, where the etching is more spread out and gives a smoother feel to the surface.

The glass is not contorted, but somewhat ribbony. This texture does not cause the production of elongate corrosion sculpture common to tektites from other deposits, but the ribboning is more accentuated on certain particularly corroded and rough surfaces.

The edges of the pieces are often jagged, bristling with sharp points which suggest that certain small fragments of the same form collected in the laterite were detached from larger pieces by corrosion.

The fracture is not in the form of a crest; it is flat or conchoidal. The glass, therefore, is not tempered since, because of its mass, it was cooled less quickly than the shaped pieces. Optical examination confirms this opinion. Finally, let us note the absence of large gas bubbles and the abundance of very small bubbles, most often perfectly spherical. They are regularly distributed.

Their chemical composition and their index of refraction cannot be distinguished from those of other Indochinese tektites. The heterogeneity of the glass, manifested by the variations of color in thin plates, is greater.

Finally, it should be pointed out that these tektites are often encrusted with lateritic limonite, embedded in their cavities; as this lateritic limonite encloses some grains of quartz, it is certain that the laterite was secondarily deposited, and was not produced from the tektite.

To sum up: here one sees not large drops of glass modeled by fusion in the course of their fall, but the debris of a rather considerable mass originating, like classic stony meteorites, from the breaking up of still larger blocks, probably fractured by their impact against the ground; their debris is distributed around the point of fall and all of the material was attacked by the same chemical corrosion as in the case of normal tektites. The concentration of all these fragments on one small surface, the disorderly association of pieces of such various dimensions, and their angular forms with sharp-edged ridges exclude any notion of transport and justify the idea formulated above of a deposit in place.

In conclusion, I shall point out the discovery in the south of Cambodia, in the course of excavations made by M. Dalet in the archeological site called Tûol Prah Théat (the mound of the Sainte Relique), situated near Kompong Speu, of a block of tektite analogous to those described here, but less corroded; it certainly came from a deposit comparable to that described above. It was given to the Museum by M. Coedès, director of l'Ecole française d'archéologie d'Extrême-Orient, who provided with it the following information:

> This tektite was found at a depth of 1 m, along with various fragments of sculptures and ceramics coming from the sanctuary, which must have fallen into the pit at the time that the sacred repository was violated by the pillagers of ruins whose sordid activities have spared very few of the monuments of Cambodia.
>
> Repositories of relics in the foundations of sanctuaries usually contain leaves of gold, often associated with fragments of quartz or other hard stones."

This discovery is thus analogous to that made at Prasar Trapéang Thual (northern Cambodia), the knowledge of which I owe to M. Parmentier. It seems to indicate not only that the ancient Khmers knew these fragments of black glass, but that they valued them and ascribed some religious significance to them, either because of the origin which they attributed to them, or because of the rarity and exceptional dimensions of the object which they collected and preserved.

These findings are moreover interesting from another point of view. An inscription engraved on a shoulder-blade of a statue of a Buddha, found in the same excavation, indicates that the recently explored monument dates from the 7th century of our era. Therefore, the origin of this tektite necessarily precedes that epoch.

Editors' Comments on Paper 3

Only a few papers mentioning tektites had been published in the United States when "North American tektites" appeared in print. This paper reported for the first time the presence of tektites in North America, included a review of world tektite literature, and served as a basic reference during the upswing in tektite research created by the space program.

Because of more recent information on tektites covered in Barnes (1967)* and in other papers in this volume, only the Introduction and excerpts containing information on the occurrence of lechatelierite are reproduced. The discovery that the glassy particles in tektites are lechatelierite provided another unifying property distinguishing tektites from all other natural glasses except fulgurites and impactites. The presence of lechatelierite also shows that quartz was present in the source material for tektites, and suggests that fusion was rapid, that heating was very intense, and that cooling was rapid. The character of the lechatelierite, i.e., the degree of irregularity of the surface, amount of deformation, and the degree of bubbliness, has been used since to judge variation of temperature within a strewn-field (Barnes and Pitakpaivan, 1962; Barnes 1967)*.

Other important findings reported in this paper are that glass identical in petrographic character to tektite glass was formed by a broken power line arcing through sandy soil; and that high-temperature fusions of sedimentary material contained lechatelierite. To learn more about how lechatelierite forms, experiments were performed at The University of Texas in which quartz as well as other minerals were incorporated in synthetic tektite glass and melted in a solar furnace. In every experiment the quartz changed to lechatelierite and most other minerals diffused within seconds into the synthetic tektite glass melt (Barnes, 1971)*.

All references cited in "North American tektites," except Bowen (1915), are in Baker's bibliography.

References

*Baker, G. (1959). Tektites. *Mem. Natl. Museum, Victoria, Melbourne,* No. 23, 313 pp.

*Barnes, V. E. (1967). Tektites. *In* "International Dictionary of Geophysics" (K. Runcorn, ed.), Vol. 2, pp. 1507–1518. Pergamon Press, Oxford.

*Barnes, V. E. (1971). Description and origin of large tektite from Thailand. *Chem. Erde,* **30,** 13–19.

*Barnes, V. E., and Pitakpaivan, K. (1962). Origin of indochinite tektites. *Proc. Natl. Acad. Sci. (U.S.),* **48,** 947–955.

Bowen, N. L. (1915). The later stages of the evolution of the igneous rocks. *J. Geol.,* **23,** 10.

Reprinted from *University of Texas Publication 3945*, 477–582 (1940)

3

CONTRIBUTIONS TO GEOLOGY, 1939

Part 2

NORTH AMERICAN TEKTITES

Virgil E. Barnes

INTRODUCTION

During April, 1936, several curiously shaped black glass objects from Grimes County, Texas, were brought to the attention of the personnel of the Bureau of Economic Geology. Dr. H. B. Stenzel recognized them as being tektites. This is the first time, so far as can be found by scanning the literature, that tektites have been found in North America.[1]

Sporadic glasses from other places have been named after the locality in which they were found. The word "tektite," originated by Suess (194)[2] in 1900, is a group name used to cover the three known occurrences at that time. He states: "Als gemeinschaftlichen Namen für die ganze Gruppe habe ich nach der Eigenschaft der Körper, welche im Gegensatze zu den übrigen Meteoriten ganzlich durchgeschmolzene Massen sind, die Bezeichnung 'Tektite' gewählt. (τήχειν, Schmelzen von Metallen und anderen harten Massen; τήχτος, geschmolzen.)" It is desirable that the Grimes County tektites receive a name suggestive of their locale; therefore, the word "bediasite" is suggested. Bedias is a place name common in Grimes County, Texas, traceable to the Bedias Indians who formerly lived there.

* * * * * * *

In order to study the interior structure of the bediasites with greater facility, plates about a millimeter thick were cut out of bediasites Nos. 2, 10, 21, 23, 34, 50, and 53. These plates, which are of a brownish color, show flow structure very well. The flow structure viewed by transmitted light appears to the unaided eye much like arrested heat waves. In all seven glass plates the flow structure is contorted, and it does not conform to the exterior surface.

The glass plates contain bubbles and some lechatelierite (fused quartz) particles. Particle is used advisedly in this case for want

[1]La Paz (116a)[2] states: "Tektites are exhibited occasionally in placer mining camps in the United States. However, it is the author's experience that persistent questioning disclosed always that such specimens came originally from the tektite-sprinkled gold fields of Australia."

[2]Numbers in the text in parentheses, as (116a, 194), refer to entries under corresponding numbers in the bibliography.

Issued June, 1940.

of a more precise word. The bubbles are spherical, indicating that they formed or assumed their shape after the flow structure had been developed. A statistical study was made of both the bubbles and the lechatelierite particles in the plates from bediasites Nos. 34 and 50. The data are presented in Table II.

Table II. Estimates of amount of bubbles and inclusions contained in bediasites.

	No. 50		No. 34	
	Bubbles	Lechatelierite Particles	Bubbles	Lecha-telierite Particles
Size in mm. Max.	.065	.48 x .06	.336	.35 x .19
Size in mm. Min.	.002–	.015– x .005–	.002–	.015 x .005–
Size in mm. Ave.	.015	.136 x .032	.058	.125 x .057
Number per cc.	1305	690	14,260	5980
Volume cc.	.0000023	.000048	.0015	.0012
Per cent by volume	.00023	.0048	.15	.12

The general variation in the number, size, and amount of bubbles and lechatelierite particles to be expected is well shown in this table. It is readily seen that No. 34 has about 650 times as much volume occupied by bubbles and about 25 times as much volume occupied by lechatelierite particles, as does No. 50. The average size of the lechatelierite particles does not vary widely between the two, whereas the average bubble size is much larger in No. 34. Bediasite No. 55, Plate 23, contains a very large bubble or, more strictly speaking, a compound bubble composed of three parts. The present shape of the hole is such that three bubbles must have been in contact except for very thin diaphragms of glass. These diaphragms have been broken, and all that is left are two raised rings at the point where the bubbles met. Four other bediasites have been found containing large bubbles, one of which is a centimeter in diameter. In each case the bubbles have been exposed by removal of material by weathering.

The lechatelierite particles are microscopic and vary widely in shape (Pl. 25). Most of them are linear, and most of them are aligned with the flow structure. Many have embayed surfaces, suggestive of fusion. Several are hooked at one end. Their color is a little deeper than that of the glass, and they have a definitely pinkish cast. The particles, however, are colorless when placed in an immersion liquid of nearly the same index, thus suggesting that the color seen is due to reflections from the enclosing glass. The index of refraction is near 1.464 and is definitely below 1.472 and above 1.453 as determined by the immersion method. To obtain lechatelierite particles for index determination a portion of one tektite was crushed. Very few particles were found, and only a very few of these were in such a position at the edge of the glass grain so that the Becke line between the liquid and the particle could be seen.

Between crossed nicols the lechatelierite particles are isotropic and rarely show extremely weak anisotropism. A few of the lechatelierite particles have an abnormal concentration of bubbles in and about them. Lechatelierite has an index of refraction of 1.458, which is slightly lower than the index of refraction determined for this mineral. Quartz is seldom pure, and any impurity contained would raise the index of refraction of the melt. Most quartz contains gaseous inclusions. These gases are retained in the melts from such quartz and are very difficult to remove. The abnormal concentration of gases, therefore, in and about some of these lechatelierite particles is a further indication helping to establish their identity. Anderson (3) in a study of fulgurites has noted groups of bubbles within partially fused grains of feldspar and quartz which he attributes to "expansion of gas or fluid cavities, or air spaces and cracks, preëxisting in quartz and decomposed feldspars." Lechatelierite is found mostly in fulgurites, and it is also known about volcanoes. It has also been found in the Canyon Diablo meteorite crater of Arizona.

Each of the lechatelierite grains shows strain effects. Each grain has both dilatational and compressional strains, situated in alternate quadrants. This strain, in combination with the general lenticularity of the grains, causes them to show interference figures between crossed nicols and without convergent light, Plate 25, figures 6 and 7.

The presence of these lechatelierite particles indicates limited liquid miscibility. Bowen (20) states: "There is only one kind of phenomenon which could be considered as definite proof of the occurrence of limited liquid miscibility and it would, without doubt, be very commonly observed if limited miscibility among rock-forming silicates were a fact. The observed phenomenon would be the occurrence in glassy, or partly glassy, extrusive lavas of distinct globules of material still partly glass, large or small, according to their opportunity for aggregation, and of composition different from that of the main mass. Here would lie indisputable proof of the formation of immiscible liquid globules." The above statement should be enlarged somewhat to cover the deformation of globules after they have formed. In the case of tektites, strongly developed flow structure is present (Pl. 26, figs. 1–6), and any globules that might have been present would have been deformed and drawn out in the direction of flow. The case for limited miscibility in these tektites is far from complete.

There is a more plausible explanation for the presence of these lechatelierite particles. If the original material contained quartz grains before it was melted and if fusion was rapid, followed by rapid cooling, it is possible that the quartz would be fused into lechatelierite, but because of its viscosity it would not be able to mix with the rest of the glass in the time available. This would be a case of limited miscibility of a type but not in the same sense as stated by Bowen, in which the enclosed globules are formed by aggregation while the material is molten.

The glass chip taken from the DeWitt County tektite contains two lechatelierite particles. On this basis it is probably safe to assume that the DeWitt County tektite and the Grimes County bediasites are of the same origin.

All glass plates cut from the bediasites show bands of anisotropism between crossed nicols. These bands correspond to the flow structure and are probably anisotropic because of strain. By the use of the sensitive tint plate the strain is better seen, and in areas where anisotropism is not recognized by crossed nicols strain is also observed. Strain is distributed irregularly throughout the plates with alternation of tensional and compression zones. Wright (240) has postulated the distribution of strain to be expected in rounded glass objects depending on the heat conditions through which they have passed. According to him three general conditions might be expected, namely: (1) the object might be without strain or only slightly strained, having been derived from the interior of a mass that cooled slowly and consequently had enough time to be annealed; (2) the object was extremely hot and cooled rapidly, such as would be true in the case of the formation of a volcanic bomb; or (3) the body was very cold and the exterior was heated very rapidly, such as would be the case with a glass meteorite.

* * * * * * *

DESCRIPTION AND EXAMINATION OF OTHER GLASSES

Some perlites from Globe, Arizona, contained in the Bureau of Economic Geology collection, faintly resemble tektites in appearance. Their index of refraction varies between about 1.485 and 1.500, and their density varies between about 2.34 and 2.37. Under the microscope, however, they are easily distinguished from tektites by the large number of small lath-shaped crystallites. Before the blowpipe this perlite is very refractory, becoming vesicular only with prolonged heating. One phenomenon which is worthy of comment is that the original surfaces became frothy much more readily than did freshly broken surfaces. This suggests that changes have taken place at the surface of these perlites making them much less refractory. Another perlite from Arizona, received from Mr. A. R. Allen of Trinidad, Colorado, was examined. Mr. Allen remarked on its similarity to some of the tektites. This perlite has a specific gravity of 2.342, an index of refraction of 1.495, and contains lath-shaped crystallites.

A sample of glass from Freestone County, Texas, was submitted to the Bureau. This glass was formed by a broken power line arcing through the soil upon which it fell. The connotation of this glass was not immediately noted, and even the name of the donor is not remembered. This glass is in part similar to fulgurite tubes and is composed of an inner glassy surface surrounded by a highly vesicular glass with sand adhering to it. The sand is composed of quartz, feldspar, and a small amount of clay. The index of refraction of the glass is near 1.48. A drawing of a section through one of these tubes is reproduced as figure 85-B. In addition to these tubes, many pieces ranging in shape from spheres to teardrops are present, some of which have no sand adhering to them. Most of these pieces have a filament of glass extending from them. Another type in this collection was produced by glass flowing along the surface of the ground, forming rivulets and puddles. Fragments of this type have a flat base with adhering sand and a rounded, smooth upper surface without adhering sand. A drawing of a section cut from a fragment of this type is reproduced as figure 85-A. This drawing shows the flow structure as depicted by variation in the composition, color, and refractive index of the glass. Lechatelierite particles are present in this glass, but clear-cut photographs could not be obtained of them. The strain is irregularly distributed, exactly as it is in the bediasites and not as is postulated by Wright (240) for a small molten body cooled rapidly, namely, an outside zone of dilatational strain and an inside zone of compressional strain separated by a zone of no strain.

In consideration of the resemblance of tektites in composition to fused shaly sediments, a noncalcareous shale from the Carrizo formation was fused in the carbon arc. The glass which resulted had an index of refraction of 1.51. Lechatelierite particles are also present in this glass. Bubbles are rather abundant but not more so than in some of the more vesicular bediasites. The color of this glass is similar to that of the bediasites.

Volcanic ash beds are present in Grimes County in the general vicinity of the tektite area. Some of this ash was fused in the carbon arc. The resulting glass is nearly colorless, with a slightly greenish cast. It has an index of refraction ranging from slightly below to slightly above 1.485. Quartz and some feldspar are present in the ash as impurities. After fusion, the only impurities seen are lechatelierite particles.

Glassy material formed about a burning and cratered petroleum and gas well was examined. This material was collected and thin-sectioned by Dr. Duncan McConnell, of the geology department, The University of Texas, who very kindly allowed the writer to examine the slides. This material is mostly glassy, with some recognizable detrital grains scattered throughout. Some recrystallization has taken place with the development of crystallites somewhat resembling augite. The index of refraction of this glass is 1.55, which is much higher than that of the bediasites. This glassy material was re-fused in the carbon arc. The resulting glass has an index of refraction slightly under 1.54, is a clear, deep bottle-green, and has lechatelierite particles in it.

Another glass examined was formed in the fire box of a boiler at a creosoting plant. This glass contains no detrital minerals, has an index of refraction near 1.55, and has commenced to recrystallize. Recrystallization has taken place in that portion of the glass first deposited, while that last deposited is entirely glassy. The crystals in this glass also closely resemble augite.

Two australites were obtained from Mr. A. R. Allen of Trinidad, Colorado. These australites are from Charlotte Waters, Central Australia. The following data were obtained:

	Specific Gravity	N_{Na}	Weight
Australite No. 1	2.457	1.512	3.6 gm.
Australite No. 2	2.430	1.506	3.9 gm.

A thin plate was cut from australite No. 1. The color is somewhat darker than that of the average bediasite, and the flow structure, while present, is not so readily seen. The point of greatest interest is the presence of lechatelierite particles in this australite exactly similar to those found in bediasites and in the various fusion materials just described. Two sections cut from indochinites kindly furnished by Professor Lacroix contain lechatelierite particles. One of these is from Dalat, State of Annam, and the other specimen is from Muong Nong, lower Laos. Figures 1 and 2 of Plate 25 are photographs of lechatelierite particles in the former section.

A section was cut from a Bohemian moldavite kindly furnished by Mr. F. W. Cassirer of Paris. This section furnishes the most spectacular display of lechatelierite particles of any section yet examined. The size and shape of these particles vary greatly (fig. 86).

Particles are present of similar shape to those figured in indochinites and bediasites. In addition, there are many which have been drawn out into fantastic shapes such as corkscrew-like spirals and serpentine-like threads (fig. 87). Some are like ribbon fluted in such a manner that with certain orientations they entirely resemble the segmented appearance of a tapeworm. These much drawn-out types grade into the more nearly equal dimensional forms and are undoubtedly of the same material. The process by which these particles became so contorted is difficult to visualize. The melt must have been very hot and very fluid, and possibly escaping gas produced an effect similar to that of boiling, causing a turbulency sufficient to deform these drawn-out particles. They might also be explained by a piling up or jamming effect following the development of the flow structure.

Rutley (155) in 1885 described the microscopic structure of bouteillenstein and noted these inclusions. His description follows: "The section contains numerous spherical gas-bubbles and irregular enclosures, some fusiform, others club-shaped, while many assume bulbous and slightly branching forms. They are frequently produced into, or from, delicate capillary rods, and they usually contain several, sometimes a dozen or more, spherical gas-bubbles. The surfaces of these glass enclosures do not, as a rule, appear to be smooth and cylindrical, but they more resemble those of flint cores from which flakes have been artifically struck, or the surfaces of prisms of starch. They depolarize strongly from strain, and the glass immediately surrounding them is similarly affected, so that each glass enclosure is bordered by a nimbus which is traversed by dark brushes. The similarity of these glass enclosures to the one figured from the fulgurite of the Dom du Gouté is very striking." Rutley further states: "That bouteillenstein is an obsidian is denied by Makowsky; but (apart from its banded structure) its glass enclosures and numerous gas-bubbles and its almost perfect freedom from any products of crystallization, render its comparison with fulgurite not merely admissible, but possibly instructive."

Suess (194) in his 1900 work mentions these particles. The following is Suess' original statement:

> In Dünnschliffen von allen Vorkommnissen finden sich in wechselnder menge und Grösse, längliche oder rundliche Linsen von schwächer lichtbrechendem Glase eingeschlossen; ihre Streckung fällt stets mit der Richtung der Fluidalstructur Zusammen und tritt am stärksten hevor bei dem böhmischen Stücken. Doch besitzen sie auch in dem australischen Schliffe, wo sie sich etwas spärlicher vorfinden, oft eine ausgesprochene, beiderseitig zugeschärfte Lancetform; daneben finden sich ober auch rundliche oder unregelmässig umgrenzte Blättchen; hie und da liegen sich shief im Schliffe, ohne denselben der ganzen Dicke nach zu durchsetzen.

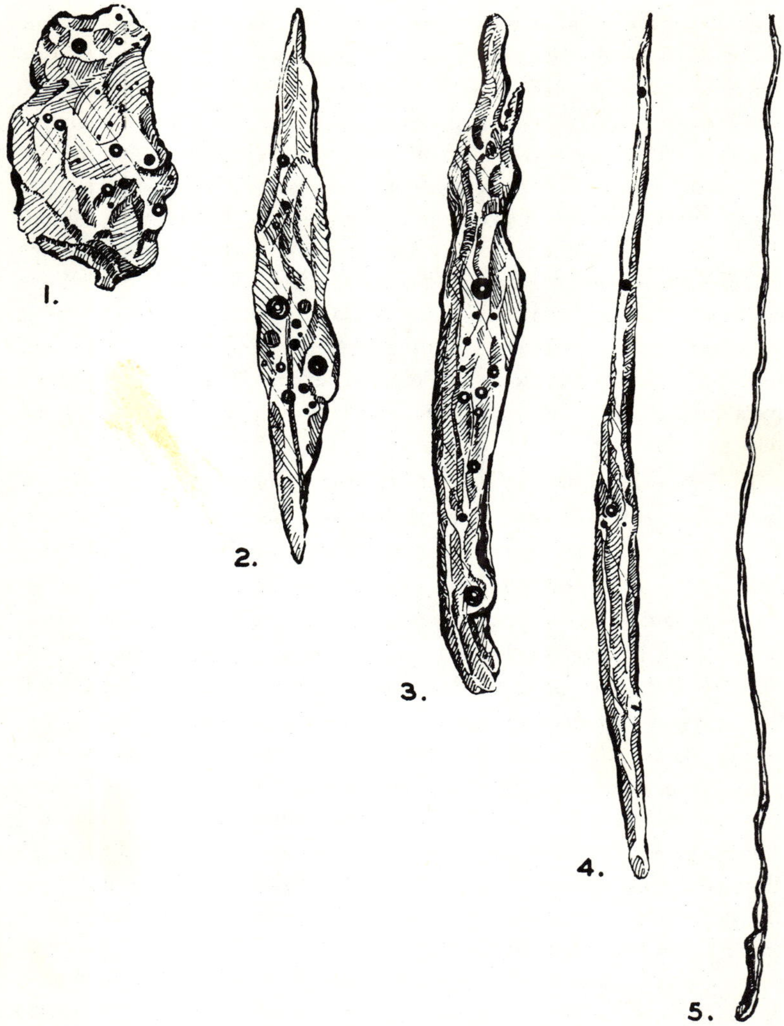

Fig. 86. Drawings of lechatelierite particles in a moldavite illustrating elongation during the development of flow structure. Assuming that the particles were originally about equidimensional, then the following ratio of elongation is shown: (1) 1–1.5; (2) 1–7; (3) 1–8; (4) 1–20; and (5) about 1–75. x100.

Lacroix (116) has figured in a photomicrograph an object which has very much the appearance of a lechatelierite particle. He states:

Les fragments de tectites, à forme quelconque, de Séan-Tô (Hai-nan) renferment souvent non seulement des bulles à forme distincte, mais encore de petites ponctuations d'un blanc de lait, ne dépassant pas les dimensions d'une tête d'épingle: au premier abord, elles pourraient être prises pour des inclusions d'une minéral étranger, mais l'examen microscopique, sous diverses

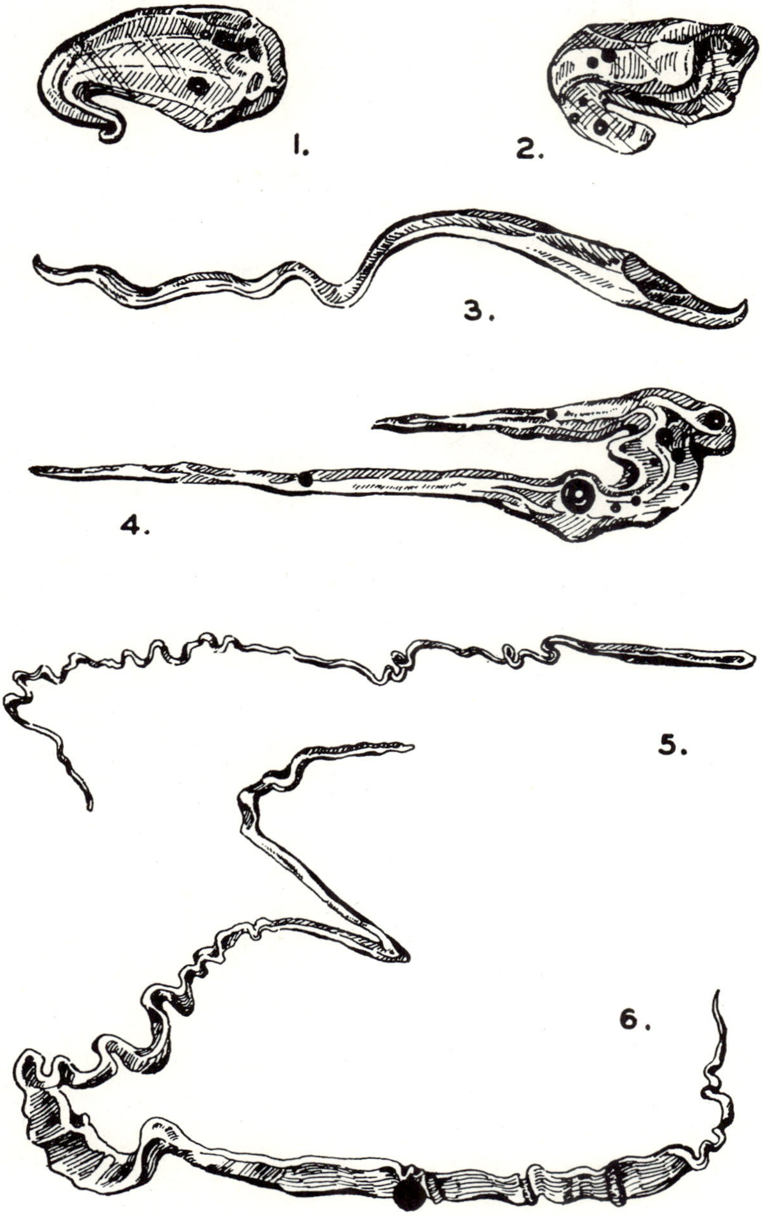

Fig. 87. Drawings of distorted lechatelierite particles in a moldavite illustrating turbulency during or jamming following the development of flow structure. x120.

incidences, montre qu' il s'agit d'agglomération de tres petites bulles gazeuses, conduisant sur quelques dixiémes ou même centièmes de millimétre à une ebauche de texture ponceuse.

An examination of Plate 25 will be sufficient to convince one that these particles are not groups of gas bubbles producing a pumiceous texture.

It is apparent from an examination of these various fusion products that glass, even of the composition of the bediasites, would recrystallize to some extent if maintained near the fusion point for a considerable time, as would be the case if they were fused by burning petroleum, gas, or coal. On the other hand, quick fusion and cooling can be produced naturally by electricity (lightning); by meteorite impact, as exhibited by the glasses about the Wabar, Henbury, and Canyon Diablo craters; and by passage of meteorites through the air.

* * * * * * *

From the present study, certain relevant facts have been ascertained. Tektites contain lechatelierite particles, which are nothing more nor less than fused silica (quartz). Glasses formed by burning petroleum and gas wells where heat is maintained for a considerable time show some recrystallization as well as some unfused sand grains. Lechatelierite particles were found in glass formed by a broken power line which arced through the soil, and no recrystallization was seen. Lechatelierite particles were found in glass fused from shale in the electric arc. The presence of lechatelierite particles suggests that fusion was rapid, that the heat was very intense, and that cooling was rapid. Only three natural modes of rapid and intense heating combined with rapid cooling can be mentioned. These are rapid heating caused by meteorites passing through the earth's atmosphere, rapid heating due to large meteorites fusing the soil upon striking the earth, and rapid heating caused by lightning striking the earth. The first mode is incapable of being demonstrated and rests upon an insecure footing of postulates, many of which have been proven incorrect. The second mode will undoubtedly be found to account for certain of the tektites, but contributing evidence will support this origin such as the presence of nickel, association with meteorites, presence of meteorite craters, and presence of detrital fragments in some of the border phase glasses.

The last mode of formation postulated, namely, that of lightning fusing the soil, has many points in its favor and also has several points that are difficult to explain. The chief advantages of this origin is that there is no necessity for a basis of nebulous postulates for either the chemical composition or the mode of fusion. The material necessary for the formation of tektites is present on earth, and the mode of fusion can readily be seen on any unsettled day. The chief disadvantage to this origin is self evident. So far as the writer is aware, glass masses formed by lightning have not been found. All lightning-produced glasses so far recognized, except some of the glazed surfaces such as are reported on peaks of the Alps, are found in sand and are fragile, thin-walled tubes with detrital grains on the exterior.

Editors' Comments on Papers 4 and 5

These two papers give some of the results of the authors' research in the indochinite strewn-field, primarily relating the Muong Nong-type and splash form-type indochinites to each other and to the soil and surficial rock in the area in which they occur. Although Lacroix (1935)* stated that "tectites sans formes figurées" (Muong Nong-type tektites) are related to splash-form types, for some time skeptics excluded such glass, as well as Libyan Desert glass, from the tektite category.

Specimens analyzed in this study of both Muong Nong- and splash form-type indochinites were found to be identical in composition. The way in which the Muong Nong-type indochinites occur indicates that they accumulated as puddles of melted surficial materials. Because of lateritization, the soil in which they lay is of different composition now than when the tektites formed. However, Hedge et al. (1965) found that the Sr^{87}/Sr^{86} ratios for the tektites and the soil are identical, indicating that the tektites were formed from the soil. In any theory of tektite origin, the manner in which the Muong Nong-type tektites were formed must be considered.

In the second paper the occurrence of mineral grains in Muong Nong-type indochinites is described. Glasses formed by meteorite impact commonly contain mineral grains. The absence of such grains in tektites had been used as evidence against the impact origin of tektites. The discovery of abundant relict mineral grains in some layers of Muong Nong-type indochinites eliminates one objection to the impact origin of tektites.

Dr. Kaset Pitakpaivan is Chief of the Geological Survey Division, Department of Mineral Resources, Bangkok, Thailand.

References

Hedge, C. E., Barnes, V. E., and Peterman, Z. E. (1965). Soil: source material for tektites. *Trans. Amer. Geophys. Union,* **46,** 545.

*Lacroix, Alfred (1935). Les tectites sans formes figurées de l'Indochine. *C. R. Acad. Sci., Paris.* **200**, 2129–2132.

4

Reprinted from the Proceedings of the National Academy of Sciences
Vol. 48, No. 6, pp. 947–955. June, 1962.

ORIGIN OF INDOCHINITE TEKTITES

Virgil E. Barnes and Kaset Pitakpaivan

THE UNIVERSITY OF TEXAS, AUSTIN, TEXAS AND GEOLOGICAL SURVEY DIVISION, ROYAL DEPARTMENT OF MINES, BANGKOK, THAILAND

Communicated by Preston E. Cloud, Jr., April 20, 1962

The origin of the small, curiously shaped glassy objects named tektites[1] has eluded scientists for nearly two hundred years. Because these objects appeared to be unrelated to the rocks and pebbles with which they were associated, many workers considered them to be meteorites. Evidence has been uncovered in Southeast Asia in connection with a study of the indochinite group of tektites, and specifically with the variety of tektites known as the Muong Nong-type, which shows that the indochinites are terrestrial in origin.

The Muong Nong-type of tektite glass was first described by Lacroix[2] from a locality a few kilometers south of Muong Nong, Laos, where the tektites occurred at a depth of about 1 meter. In an area of about 100 square meters, 67.5 kg of tektites were recovered, the largest piece weighing 3.2 kg. The blocky chunks at this locality are unlike the splash forms common in other tektite groups; yet Lacroix found this glass to have the other properties of tektites.

One of the chief aims of this investigation on tektites, supported by the National Science Foundation, was to visit the type area of the Muong Nong-type tektites. By the time the project was under way, Laos was closed to travel, but fortunately the Geological Survey Division, Royal Department of Mines, Thailand, knew of deposits of both normal and Muong Nong-type tektites in northeastern Thailand and were amenable to a cooperative project to study their field relations. In addition to the three Muong Nong-type deposits previously known by the Geological Survey Division, the authors found a fourth locality. We found also that Muong Nong-type tektites in some localities are scattered over the surface with normal-type tektites. A fifth locality of Muong Nong-type tektites was found at Dalat, and, as in Thailand, scattered fragments of the Muong Nong-type were found with normal indochinites. In Cambodia, Muong Nong-type indochinites were found scattered with normal indochinites in the vicinity of Kratié. Such tektites were also identified in collections from the Philippines and Texas.

A petrographic study of this material and of type specimens from Laos reveals: (1) a plane to folded internal lamination shown by variation from layer to layer in color and content of bubbles and lechatelierite, (2) a shimmery appearing linear structure within the laminae apparently produced by elongation of the melt products of mineral grains, and (3) lack of over-all internal tensional strain.

The occurrence in place, in areas a yard or two across, of as much as 20 kg of chunky tektite glass also differs from the manner in which normal tektites occur. Both types are of almost identical chemical composition and both are related chemically to the lateritic soil from which the Muong Nong-type tektite was excavated. Petrographic and other evidence is advanced in favor of an interpretation that the central part of the strewn-field was hotter than its margins, but no impact crater has been recognized. From this, it is possible to deduce that the indochinites may be the result of a head-on impact with a diffuse object such as a comet, as suggested by Urey,[3] or that they originated through some nuclear or electrical phenomenon not yet visualized.

Mode of Occurrence.—One of the deposits examined, near Phang Daeng, Thailand, was in laterite confined to an area of but a few square meters, beneath about 0.3 m of soil. The chunky tektite material revealed by excavation was roughly lens-shaped, and although the material was broken, with laterite separating the pieces, it was generally grouped together as a mass such as might have congealed from a puddle of molten materials. Ten kg or so of tektites were excavated from the site and perhaps as many more had been recovered previously and sent to the Royal Department of Mines. Test pits were dug along lines intersecting at right angles on the deposit, but no additional tektites were found. However, within 100 m to a kilometer of the deposit several dozen ordinary tektites and a few fragments of the Muong Nong-type were found on a thin covering of soil formed on Triassic sandstone, which is the dominant rock of the area.

Another Muong Nong-type deposit was visited near Kan Luang Dong, Thailand. Here tektites dug from a depth of 2 m in a well were from a mound-like deposit described as resembling a low white-ant hill. A portion of this mound is reported to remain in the side of the well but circumstances prevented removing the water and mud, therefore the tektites could not be studied in place. However, the 10 kg or so of tektites sorted from the sandy clay on the dump and obtained from the villagers are coated by sandy clay and are free of adhering laterite, showing that this occurrence is not associated with laterite.

Of the two remaining Muong Nong-type tektite localities in Thailand, one, at Nong Sapong, was revealed by a track worn down 0.7 m to the laterite layer where the tektites repose in a small area; the other, 81 km west of Nakhon Sakhon, was in laterite 0.7 m below the surface, exposed as a result of highway construction. In Thailand, therefore, the Muong Nong-type tektites in three cases out of four were associated with laterite formed in place.

The tektites are chemically similar to the laterite if the changes that take place during laterization are considered. This is taken as evidence that the tektites were present before laterization. Laterite, a mixture of oxides of iron and alumina with other substances which have a wide range in quantity, occurs commonly as porous, reddish concretionary crusts, residual from weathering in tropical and subtropical latitudes under conditions of good drainage and marked seasonal rainfall. The present chemical composition of both laterite and tektites is consistent with derivation of the laterite from an original surface material having a composition identical with that of the Muong Nong-type tektites. Changes in the composition of the surficial material since the tektites formed can be determined by comparison with the composition of the tektites (Table 1). We can therefore conclude that the climate in Southeast Asia when the indochinites formed was relatively dry and that the surficial materials from which tektites formed were not leached. After their fusion to a glassy state, some were transported and buried; others—especially the Muong Nong-type—were buried in place beneath surficial materials. The climate later became seasonally much more humid, resulting in leaching of surficial materials soluble in tropical ground waters, downward movement and precipitation of iron in the insoluble oxidized state, and leaching and removal of more soluble constituents. Before or during laterization, the Muong Nong-type tektite glass became fractured, and etching took place, accompanied by deposition of laterite along fractures and in etch pits.

TABLE 1

CHEMICAL ANALYSES OF MUONG NONG-TYPE TEKTITES AND ASSOCIATED LATERITE AND SOIL, PHANG DAENG, THAILAND

	Muong Nong-Type Tektites Lightest colored	Darkest colored	Laterite with tektites (depth 9–11 in.)	Laterite (depth 30 in.)	Soil (depth 4–6 in.)
Sample No. →	158177	158172	158178	158174	158175
SiO_2	76.9	80.4	45.8	50.9	84.6
Al_2O_3	10.6	9.5	12.2	12.8	5.4
TiO_2	0.67	0.62	0.54	0.59	0.45
Fe_2O_3	0.60	0.45	29.6	25.3	4.7
FeO	3.3	3.0	0.04	0.12	0.14
MnO	0.08	0.08	0.66	0.06	0.01
MgO	1.6	1.3	0.44	0.46	0.35
CaO	1.6	1.1	0.06	0.08	0.12
Na_2O	1.4	1.2	0.08	0.10	0.04
K_2O	2.3	2.2	0.86	1.2	0.60
H_2O^-	—	—	—	—	—
H_2O^+	0.24	0.14	9.3	8.2	3.1
P_2O_5	0.12	0.10	0.17	0.10	0.06
Total	99	100	100	100	100

Analyzed by Paul Elmore, Ivan Barlow, Samuel Botts, and Gillison Chloe, of the U.S. Geological Survey, using analytical methods similar to those described in the U.S. Geological Survey Bulletin 1036-C.

The Muong Nong-type indochinite is now known from 48 km west of Khon Kaen, Thailand, southeastward for a distance of 830 km to Dalat, Viet Nam, and in a southwest-northeast direction for more than 500 km from Kompong Speu, Cambodia, to Muong Nong, Laos. If systematic collecting were to be done, the area of occurrence would surely be extended.

Chemistry.—If, as postulated, the Muong Nong-type tektites are the result of local fusion of the ground to puddles of melted glass, then chemical analyses of the tektites should compare with analyses of the material around them, provided no change has taken place in these materials since the tektites formed. But laterization has taken place since they formed, which implies chemical change of the material since the tektites formed. During laterization, iron was leached and precipitated downward, alumina remained the same, hydration took place, and other constituents were leached and carried away.

Although the two analyses of Muong Nong-type tektites in Table 1 have little resemblance to the analyses of the laterite associated with them or the laterite 20 inches deeper, by using the Muong Nong-type tektite as a standard of comparison the changes which are thought to have taken place can be shown. In Table 2, the

TABLE 2

CHEMICAL CHANGE INVOLVED TO PRODUCE LATERITE IF THE SOURCE MATERIAL WAS OF THE SAME COMPOSITION AS THE MUONG NONG-TYPE TEKTITES AT PHANG DAENG, THAILAND

	A Laterite associated with tektites, analysis 158178	B Laterite associated with tektites, recalculated moisture free	C Average of tektite analyses 158177 and 158172	D Original amount present on basis of no change in Al_2O_3 ($C \times 1.34$)	E Gains and losses in per cent of original amount present ($B \times 100/D$)
SiO_2	45.8	50.6	78.65	105.4	48
Al_2O_3	12.2	13.5	10.05	13.5	100
TiO_2	0.54	0.60	0.64	0.86	70
Fe_2O_3	29.6	32.75	0.52	0.70	4,600
FeO	0.04	0.04	3.15	4.24	1
MnO	0.66	0.73	0.08	0.10	730
MgO	0.44	0.49	1.45	1.95	25
CaO	0.06	0.07	1.35	1.82	4
Na_2O	0.08	0.09	1.3	1.75	5
K_2O	0.86	0.95	2.25	3.03	31
H_2O	9.3	—	0.19	—	—
P_2O_5	0.17	0.19	0.11	0.15	127

laterite associated with the tektites (column A) is calculated to a dry basis (column B) and the average of the two tektite analyses from Table 1 is given in column C. Since Al_2O_3 is neither lost nor gained in the process of laterization and since it is more abundant in the laterite than in the tektites, an over-all loss of constituents by a factor of 1.34 must have taken place. If the amount of each constituent in column C is multiplied by the factor 1.34 (recorded in column D) and is compared with column B, the gains and losses can be shown in per cent of the original amount present (column E).

Fe_2O_3 has been enriched 46-fold, MnO 7-fold, P_2O_5 slightly, Al_2O_3 is constant, and all other constituents have been depleted in various amounts from slightly for TiO_2 to about half for SiO_2, three-quarters for K_2O and MgO to one-twentieth or less for Na_2O and CaO, and one-hundredth for FeO. These are exactly the kinds of changes that take place during laterization, and these results favor the laterite's forming from material of the composition of the Muong Nong-type tektites. The tektites survived even though half the silica was dissolved from the source materials of the laterite. The reason for this is that the source materials are fine grained and offer many times more surface area for attack than do the relatively large chunks of tektite.

The same comparison was made between the Muong Nong-type tektite and the laterite 20 in. beneath. It was found that iron is less abundant, manganese is unchanged, and leaching is less. The soil overlying the tektites shows no chemical similarity either to the laterite or to the Muong Nong-type tektites, suggesting that it is a recent accumulation of sandy material.

TABLE 3

CHEMICAL CHANGE INVOLVED TO PRODUCE LATERITE IF THE SOURCE MATERIAL WAS OF THE SAME COMPOSITION AS THE MUONG NONG-TYPE TEKTITE AT NONG SAPONG, THAILAND

	A Laterite associated with tektites, analysis 158179	B Laterite associated with tektites, recalculated moisture free	C Muong Nong-type tektite, analysis 158176	D Original amount present on basis of no change in Al_2O_3 (C × 0.97)	E Gains and losses in per cent of original amount present—(B × 100/D))
SiO_2	37.5	41.2	71.4	69.4	58
Al_2O_3	12.7	13.9	14.3	13.9	100
Fe_2O_3	39.3	43.1	0.50	0.49	8,800
FeO	0.07	0.08	4.5	4.4	2
MgO	0.05	0.05	2.1	2.0	2
CaO	0.05	0.05	1.9	1.8	3
Na_2O	0.08	0.09	1.5	1.5	6
K_2O	0.60	0.66	2.7	2.6	25
H_2O	8.9	—	—	—	—
TiO_2	0.58	0.64	0.84	0.82	78
P_2O_5	0.15	0.16	0.16	0.16	100
MnO	0.05	0.05	0.10	0.10	50

Analyses 158176 and 158179 by Paul Elmore, Ivan Barlow, Samuel Botts, and Gillison Chloe, of the U.S. Geological Survey, using analytical methods similar to those described in the U.S. Geological Survey Bulletin 1036-C.

TABLE 4

CHEMICAL CHANGE INVOLVED TO PRODUCE LATERITIC CLAY AT DEPTH OF 6 FEET IN WELL IF THE ORIGINAL MATERIAL WAS OF THE SAME COMPOSITION AS THE MUONG NONG-TYPE TEKTITE AT NONG SAPONG, THAILAND

	A Clay beneath tektites, analysis 158173	B Clay beneath tektites, recalculated moisture free	C Muong Nong-type tektite, analysis 158176	D Original amount present on basis of no change in Al_2O_3 (C × 1.15)	E Gains and losses in per cent of original amount present (B × 100/D)
SiO_2	63.9	69.2	71.4	82.4	84
Al_2O_3	15.2	16.5	14.3	16.5	100
Fe_2O_3	10.5	11.4	0.50	0.58	2,300
FeO	0.14	0.15	4.5	5.2	3
MgO	0.28	0.30	2.1	2.4	12
CaO	0.08	0.09	1.9	2.2	4
Na_2O	0.13	0.14	1.5	1.7	8
K_2O	1.2	1.3	2.7	3.1	42
H_2O	7.1	—	—	—	—
TiO_2	0.81	0.88	0.84	0.97	91
P_2O_5	0.06	0.06	0.16	0.18	33
MnO	0.02	0.02	0.10	0.12	17

Analyses 158173 and 158176 by Paul Elmore, Ivan Barlow, Samuel Botts, and Gillison Chloe, of the U.S Geological Survey, using analytical methods similar to those described in the U.S. Geological Survey Bulletin 1036-C.

The same comparisons have been made in Table 3 of Muong Nong-type tektites rom Nong Sapong, Thailand, and associated laterite, and in Table 4 of Muong Nong-type tektites and sandy clay from a depth of 6 feet in a nearby well. The results obtained are about the same as found at Phang Daeng, except that there is a lesser amount of leaching of the underlying sample at Nong Sapong.

The manner of occurrence of the Muong Nong-type indochinites, their internal properties, and their chemical properties as related to the material in which they lie, all point to their formation by fusion of the ground. The decisive evidence of the terrestrial origin of indochinites is furnished by the analyses in Table 5. In this

TABLE 5

CHEMICAL ANALYSES OF MUONG NONG-TYPE AND NORMAL-TYPE INDOCHINITES, THAILAND

	Normal tektite 48 km west of Khon Kaen	Muong Nong-type tektite Phang Daeng	Normal tektite Phang Daeng	Muong Nong-type tektite Phang Daeng
Number →	61-1034	61-1031	61-1032	61-1033
SiO_2	72.77	72.44	73.86	73.50
Al_2O_3	13.37	13.34	12.57	12.59
TiO_2	0.78	0.76	0.71	0.74
Fe_2O_3	0.37	0.29	0.23	0.41
FeO	4.34	4.29	4.00	4.03
MnO	0.10	0.10	0.076	0.094
MgO	2.00	2.00	1.98	1.93
CaO	2.01	2.16	2.24	2.16
Na_2O	1.48	1.61	1.62	1.58
K_2O	2.40	2.48	2.32	2.45
H_2O^-	0.03	0.01	0.00	0.00
H_2O^+	0.07	0.11	0.11	0.10
P_2O_5	0.04	0.11	0.08	0.15
Total	99.76	99.70	99.80	99.73
Ni, ppm	25	31	24	32

Analyzed by C. O. Ingamells, Mineral Constitution Laboratories, College of Mineral Industries, The Pennsylvania State University, University Park, Pennsylvania.

table, two analyses of Muong Nong-type indochinites and two of normal-type indochinites are paired. A glance shows that each pair is almost identical, and that both tektite types probably came from the same material.

Other Features of the Indochinite Strewn-Field.—Given such strong presumptive evidence that the Muong Nong-type indochinites formed by fusion of the ground and their identity in composition to normal-type indochinites, other features of the strewn-field deserve examination for their bearing on the origin of the indochinites.

Lechatelierite particles were found to vary in amount from locality to locality within the indochinite strewn-field, being much more abundant in the peripheral areas. This difference may be explained by greater heat and a longer period of heating in the central area (northeast Thailand and adjacent Cambodia and Laos), causing relatively more lechatelierite to diffuse into the surrounding glass and disappear. That this may be true is supported by the relatively smaller size of the lechatelierite particles from the central area, its smoother appearance, and the lesser abundance of bubbly lechatelierite.

It is thought that the bubbles associated with lechatelierite are formed from the gas and liquid-filled vacuoles which are so common in quartz, and that a study of these bubbles may give additional information on the relative temperature of formation of the various tektite groups. In the case of bubble-free lechatelierite, the viscosity would have been so low that all gas escaped. In the case of the lechatelierite with associated bubbles, it seems likely that the fluidity was just sufficient for the gas of the vacuoles to coalesce but that the viscosity was too great for the bubbles so formed to escape. In the case of the frothy lechatelierite, it seems likely that the viscosity was still higher and that the gas in the vacuoles expanded but did not coalesce.

The possibility is not ruled out that the lechatelierite content of the indochinites is related to the local soil and rock on which they lie rather than to the strewn-field as a whole, but if this were true, it would not explain the low content of lechatelierite of the tektites found on the broad outcrop of Triassic sandstone in the central area, or other properties of the lechatelierite indicating high temperatures in the central area.

"Fingers" in tektites were recently described by Barnes[4] for the first time and illustrated.[5] These "fingers," mostly more silicic (rarely less silicic) than the enclosing tektite glass, or schlieren with some characteristics of "fingers" have now been found in every tektite group except the javaites, billitonites, and Ivory Coast tektites where relatively few specimens have been available for study. "Fingers" were first seen in australites as light-colored, narrowly elongated inward extensions up to several millimeters long which are represented on the surface by protuberances. Their resistance to etching and the low index of refraction indicate that they are more siliceous than the surrounding glass. Moreover, when viewed while being rotated on a universal stage, they are seen to be coated with tiny particles that appear to be segregated lechatelierite. These particles are mostly at the intersections of very low but fairly sharp ridges with their crests toward the tektite glass. Some of these particles are so tiny that it is uncertain whether they are actually lechatelierite or are instead an optical effect caused by the intersecting ridges. It is thought that these "fingers" could have been formed only at high temperatures by volatilization of other constituents concentrating the silica.

Iron has a boiling point of 3,000°C, and its presence has a marked positive effect on the refractive index of a glass of which it is a part. Tektites contain a few per cent of iron, and since the "fingers" are markedly lower in refractive index than is the remainder of the tektite, it appears that iron volatilized along with other constituents as the silica was concentrated. Tektites with this kind of "finger" probably formed at temperatures in excess of 3,000°C. "Fingers," and schlieren with some of the properties of "fingers," in addition to their abundance in australites are rather common in indochinites and philippinites and were found in one moldavite.

Recently, "fingers" of a different character have been found which are darker than the surrounding glass, have a higher refractive index, have smooth surfaces without lechatelierite, and end in pits at the surface of the tektite. These "fingers" probably were produced by volatilization at temperatures lower than the boiling point of iron, causing iron to be enriched. Several were seen in indochinites from China and one is present in a bediasite.

"Fingers" provide evidence of temperature variations within the indochinite strewn-field. The indochinites from China near the margin of the strewn-field contain dark "fingers" indicative of temperatures of formation below 3,000°C, whereas those in the central part of the strewn-field contain light-colored "fingers" indicative of temperatures of formation above 3,000°C. Thus, evidence derived from variations in the content and appearance of lechatelierite, of bubble and froth content, and of the presence and nature of "fingers" all leads to the same conclusion: whatever the causal agent may have been, temperatures produced were higher in a broad area in the central part of the strewn-field, falling off toward its margin.

From the shapes of bubbles some deductions also can be made as to the cooling history of tektites and related glasses. Spherical bubbles show that the glass was not being deformed while it cooled, whereas elongate bubbles show that deformation was taking place. Bubbles in the Muong Nong-type tektite are predominantly elliptical, probably as a result of slow flowage of the glass as it cooled. Specimens with spherical and elongate bubbles are about equally numerous in normal indochinites. Many of the normal indochinites are flattened forms (plates and disks), and these may have fallen to the ground while still plastic enough to flatten, stretching the bubbles as they did so.

Conclusions.—A study of the internal features of the Muong Nong-type indochinites and of their chemical composition, as compared with the lateritic materials with which they are associated, furnishes strong evidence that they are the result of fusion of the ground before laterization. That the Muong Nong-type and normal-type tektites were formed by the same event is supported by the identical chemical composition of the two. Evidence from regional variation in nature and number of lechatelierite particles, bubbles, and "fingers" in normal-type indochinites indicates that a large area in the central part of the strewn-field was hotter than the margins, and information from bubble shapes indicates that many splash forms fell back to earth while still plastic.

None of this evidence is compatible with an extra-terrestrial origin for the indochinites, and little of it is consistent with their origin from the impact of a large meteorite or asteroid. The occurrence of two types of tektites so dissimilar in internal structure and outward appearance yet so similar in chemical composition could hardly represent a single extra-terrestrial shower, especially when one type is concentrated in the middle portion of their field of scatter. No evidence of an impact crater to which they might be attributed has been found, and if one had been found it would still be difficult to visualize the manner in which the Muong Nong-type glass could have segregated out according to temperature from the splash of a point impact.

Although impact glasses and tektites appear to form a continuous series, different members of the group may have been formed in different ways. Those showing evidence of two-stage melting such as could result from transit through the atmosphere (e.g., australites and javaites) are easier to explain by assigning an extra-terrestrial origin to them. Some, such as those from the San Francisco del Monte site in Manila, thought to be weathering out of tuff, may be the result of dense body impact of such force as to trigger a volcanic eruption[6] and produce a great caldera such as that of Taal. And others, such as the bediasites, moldavites, and perhaps the bulk of the philippinites, may have an origin similar to that of the indochinites.

The evidence is against an extra-terrestrial source or origin by dense body impact for the indochinite strewn-field. It could be explained, however, by the head-on impact of a diffuse object such as a comet[3] is thought to be or by some yet unvisualized and therefore uninvestigated nuclear or electrical phenomenon.

Grateful acknowledgment is made to the National Science Foundation for support of tektite research under grant NSF G-10236 administered by The University of Texas; to the Royal Department of Mines, Thailand, for their support of the research in Thailand, to Charles Mann and

Gaylord Walker of the United States Operations Mission to Cambodia, and to Edmond Saurin and Adrien Milliés-Lacroix, University of Saigon, for cooperation in Cambodia and Viet Nam, and to others. too numerous to name here, who made this research possible.

[1] Suess, F. E., *K. geol. Reich. Wien, Jb.*, **50**, 2, 193 (1900).

[2] Lacroix, A., *C.R. Acad. Sci.* (*Paris*), **200**, 2129 (1935).

[3] Urey, H. C., *Nature* (*London*), 179, 556 (1957).

[4] Barnes, V. E., *CERN Conference on Fission and Spallation Phenomena and Their Application to Cosmic Rays, September 26–29, 1961*, abstract 37.

[5] Barnes, V. E., *Scientific American*, **205**, 5, 58 (1961).

[6] Urey, H. C., these PROCEEDINGS, **41**, 27 (1955).

5

Reprinted from Science, December 27, 1963, Vol. 142, No. 3600, pages 1651-1652

Detrital Mineral Grains in Tektites

Abstract. *Abundant detrital crystalline mineral grains have been found in layered Muong Nong-type indochinite tektites from Nong Sapong, northeastern Thailand. These grains are an integral part of some tektite layers, and their presence furnishes strong presumptive evidence that indochinites, as well as other tektite groups in which layered specimens occur, formed from surficial earth materials.*

Most investigators now agree that tektites are the result of some kind of impact. The question is whether the impact took place here on earth, on the moon, or possibly on some other celestial body. If the impact took place on earth, tektites containing incompletely melted earth material should exist; but until now such material has not been found. Layered tektites of the Muong Nong type provided the logical place to search for included minerals because they show chemical and mineralogical evidence of having formed at lower temperatures than the normal sphere-, teardrop-, dumbbell-, and other splash-form types of tektites.

A petrographic microscopic examination of plates 1-m thick, cut from 15 specimens of layered indochinites from Nong Sapong, showed scarce to abundant detrital mineral grains in five of the plates (Fig. 1). Most of the mineral grains range in size between 0.01 and 0.1 mm. Several tested samples have birefringence and uniaxial-positive interference figures typical of quartz, and two grains isolated from crushed tektite have the refractive index of quartz. It is likely, therefore, that most of the included detrital mineral grains are quartz.

That the mineral grains were incorporated into the tektites when the melt was still very hot is apparent from their crackled appearance and hazy birefringence. The grains occur in bubbly layers rich in tiny lechatelierite particles and frothy lechatelierite (*1*), and these layers alternate with layers which are free of mineral grains and contain scant lechatelierite and bubbles. Thus the grains appear to have entered the melt while it was in the process of flowing, cooling, and becoming layered. The angularity of one of the larger grains (Fig. 2) indicates that at the time it was incorporated the melt had cooled below the point at which rounding of grain edges could take place.

Previously I described detrital mineral grains along faults in layered indochinites from Kan Luang Dong, Thailand, about 40 km south of Nong Sapong (*2*). These grains, as well as those included in tektite layers at Nong Sapong, are all the size of silt and finest sand. The country rock in the vicinity of these tektite localities is silty and sandy clay, and angular quartz grains up to 0.25 mm (rarely as much as 0.4 mm) are very abundant in the sandy clay and overlying soil. Such a correspondence in size and angularity between grains in layered tektites and grains in the material on which they lie strongly implies that layered tektites are locally derived.

By whatever manner the mineral grains of silt and fine-sand size were introduced along the faults, it is likely that dust of similar size was thrown into the air at the time of the faulting. If the dust settled on flowing melt it would have been incorporated, and the association of silica particles which had been subjected to different degrees of heating would thus be explained. The quartz particles which settled while the melt was hottest would have become lechatelierite, those which fell while the melt was at an intermediate temperature would have become frothy lechatelierite, and those particles incorporated after the temperature dropped below the melting point of quartz would have retained their crystallinity.

The supposed absence of transitional phases between tektites and earth materials has been considered as evidence against the terrestrial origin of tektites. This discovery of abundant detrital mineral grains in some layers of Muong Nong-type indochinites indicates the presence of this type of transitional phase (*3*).

VIRGIL E. BARNES
Bureau of Economic Geology, University of Texas, Austin

References and Notes

1. V. E. Barnes and K. Pitakpaivan, *Proc. Natl. Acad. Sci. U.S.* **48**, 947 (1962).
2. V. E. Barnes, *Trans. Am. Geophys. Union* **44**, 93 (1963).
3. This work was supported by grants from the National Science Foundation (NSF G-10236 and GP-408).

4 November 1963

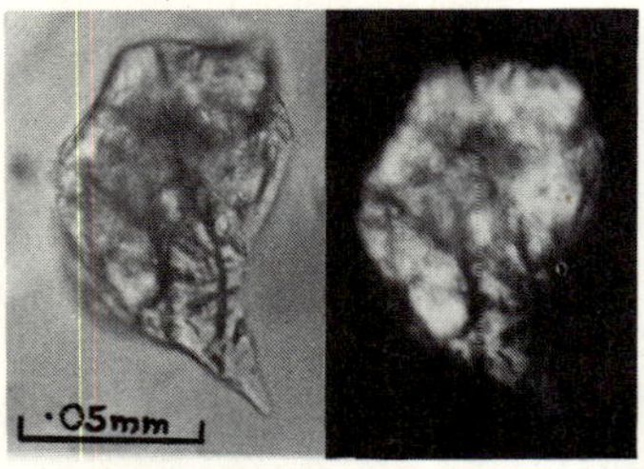

Fig. 1. (Left) Mineral grain in layered Muong Nong-type indochinite from Nong Sapong, Thailand, in plane-polarized light. (Right) The same mineral grain between crossed-nicols remains light, showing that it is crystalline. The streaks which remain dark between crossed nicol prisms may be fractures filled by isotropic glass.

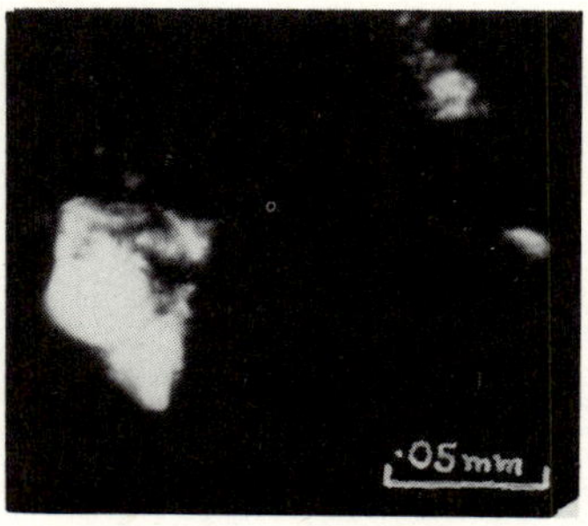

Fig. 2. Crystalline mineral grains in layered Muong Nong-type indochinite from Nong Sapong, Thailand, between crossed nicol prisms. The angularity of the larger grains shows that the melt was relatively cool when the grain was incorporated.

Editors' Comments on Paper 6

In this paper King* summarizes the geological information concerning Georgia tektites. Although the Georgia tektites occur in Pliocene or Pleistocene deposits, because of a 34-million-year K^{40}–Ar^{40} age King* suggests the possibility that the tektites have weathered out of older units farther inland and have been transported coastward.

Zahringer (1963) related Georgia tektites to bediasites using the potassium–argon method of age determination, and later Gentner et al. (1969)* found similar ages for the two groups using the fission-track method. This age is Oligocene rather than Eocene (40–60 million years) as the geologic age indicates. The bediasites in their northeastern occurrence have been found in place in the lower part of the Jackson Formation, of Eocene age (King, 1968). In the southwestern area of their occurrence the bediasites rest on the uppermost unit of the Jackson, which is transitional from Oligocene to Eocene (Barnes, 1960), and these bediasites could be coming out of Oligocene strata in this area. The bediasites in this southwestern area are more translucent than bediasites in general and are more nearly like the Georgia tektites in this respect, and the two are alike in composition. Although geological evidence is suggestive of two stratigraphic horizons of tektite occurrence, with the Georgia tektites belonging to the younger horizon, potassium–argon and fission-track dating, as noted above, give identical ages for Georgia tektites and bediasites. Further discussion of potassium–argon, fission-track, and geological dating will be found in Part III of this volume.

Dr. Elbert A. King, Jr. is now Professor and Chairman, Department of Geology, University of Houston, Houston, Texas, 77004.

References

Barnes, V. E. (1960). Significance of inhomogeneity in tektites. *Rept. 21st Intern. Geol. Congress, Copenhagen (Norden), Part 13,* pp. 328–338.

*Gentner, W., Storzer, D., and Wagner, G. A. (1969). New fission-track ages of tektites and related glasses. *Geochim. Cosmochim. Acta,* **33,** 1075–1081.

King, E. A., Jr. (1968). Stratigraphic occurrence of bediasites. Geol. Soc. Amer., Program 1968 Annual Meeting, pp. 160–161.

Zahringer, J. (1963). Isotopes in tektites. *In* "Tektites" (J. A. O'Keefe, ed.), pp. 137–149. Univ. Chicago Press, Chicago, Illinois.

6

Geochimica et Cosmochimica Acta 1964, Vol. 28, pp. 915 to 919. Pergamon Press Ltd. Printed in Northern Ireland

New data on Georgia tektites

ELBERT A. KING JR.*
Space Environment Division, Manned Spacecraft Center, National Aeronautics and Space Administration, Houston 1, Texas

Abstract—Geologic field work in Dodge County, Georgia, has demonstrated that all of the Georgia tektites have been found on a sand and gravel stratigraphic unit of probable Pliocene or Pleistocene age that unconformably overlies older Tertiary rocks. Because of the older potassium–argon ages of the tektites, it is probable that the tektites have been reworked and transported from some older stratigraphic unit to the north or northwest. It is recommended that the term "strewn field" be abandoned in reference to the Georgia tektite occurrence. The results of a chemical analysis of a new specimen (DGA-1) are similar to those of the only previous analysis of a Georgia tektite (USNM 1396) reported by CLARKE and CARRON (1961).

PREVIOUS WORK

THE FIRST tektites from Georgia were identified by Dr. E. P. HENDERSON of the Smithsonian Institution in 1938. STAIR (1955a, 1955b, 1956) published the absorption spectrum of a Georgia tektite, a photomicrograph of the same piece, and a photograph of another specimen from Georgia. BARNES and BRUCE (1959) visited the area, examined several specimens, and confirmed the Georgia tektite occurrence. BRUCE (1959) published photographs of several specimens from Georgia as well as a discussion of the finds. COHEN (1959) compared the Georgia tektites to the moldavites and bediasites and presented spectrochemical data and physical properties. SENFTLE and THORPE (1959) presented data on the magnetic susceptibility and magnetization of a Georgia tektite. REYNOLDS (1960), GENTNER and ZÄHRINGER (1960), ZÄHRINGER (1961) and ZÄHRINGER (1963) measured the potassium–argon ages of tektites from numerous localities. The most recent data indicate that the Georgia tektites are about 34 m.y. old. CLARKE and CARRON (1961) published a comprehensive physical and analytical study of a Georgia tektite (USNM 1396), including the only chemical analysis available until now. CLARKE and CARRON also discussed the possibility that the Georgia tektites might be artificial glass and compared the Georgia tektites with the single specimen from Martha's Vineyard Massachusetts. CLARKE and HENDERSON (1961) published photographs of several Georgia tektites, discussed the problems of their field occurrence and summarized most of the previous information. FURCRON (1961) discussed the geological age of the Georgia tektite shower and concluded that there was sufficient evidence to suggest a Pliocene–Pleistocene age for the fall. PINSON and SCHNETZLER (1962) analysed a fragment of a Georgia tektite for strontium and rubidium. KING (1962) reported finds of three new Georgia tektites and published photographs and brief descriptions of the specimens. KING also discussed the stratigraphy of the Georgia tektite occurrence and concluded that the stratigraphic evidence was insufficient to date the fall. TAYLOR (1963) analysed a fragment of a Georgia tektite (DGA-1) for oxygen isotopic composition.

STRATIGRAPHY

During late summer of 1962, the geology of Dodge County was mapped in the field at the scale of four inches per mile on aerial photographs furnished by the

Georgia Geological Survey. No stratigrapher has previously worked specifically with Dodge County, although Dodge County was included in larger reconnaissance mapping projects and regional studies. FURCRON (1961) summarized the previous stratigraphic work.

Dodge County contains two mappable stratigraphic units in the areas of the tektite occurrences. The upper unit, upon which all of the Georgia tektites have been found, is sand, clayey sand and gravel. The upper unit unconformably overlies a lower unit composed of clay and sandy clay. These relationships hold true wherever these units are found throughout Dodge County. The same upper unit unconformably overlies Oligocene rocks in Pulaski County, Georgia and marine Eocene rocks in Twiggs County. The lower unit is unfossiliferous at the surface, but is reported to

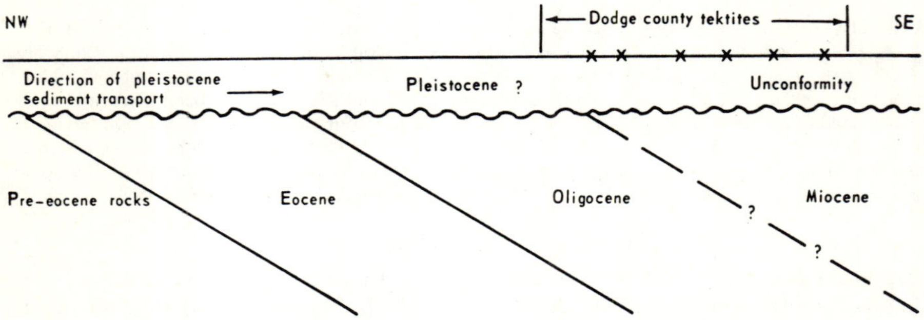

Fig. 1. Schematic cross section of a portion of the Georgia Coastal Plain.

grade into limestone containing Middle Miocene foraminifera in the down-dip subsurface. The age of the lower unit is thus believed to be Miocene. The first fossils found in well cuttings from Dodge County are Middle Oligocene foraminifera that occur at 100 to 150 feet below the surface. The upper unit is also unfossiliferous. The upper unit is believed to be Pleistocene because of its unconformable relationship to older rocks, coarse grain size, non-marine environment of deposition and general similarity to units of known Pleistocene age that are found elsewhere on the Gulf Coastal Plain. The occurrence of 34 m.y. old tektites on Pleistocene rocks does not present any problem. It is probable that the tektites have been eroded from older Early Oligocene or Late Eocene rocks to the north or northwest and were transported 10–15 miles to their present localities in Dodge County along with the Pleistocene sand and gravel. There are no tektite-surface morphological criteria by which transport of the tektites over such a short distance can be established or disproven. The transportation of the tektites from older stratigraphic units is clearly possible (Fig. 1), and the potassium–argon age of the tektites indicates that this has occurred. If tektites originated outside of the Earth–Moon system, they should contain measurable amounts of cosmic-ray-induced radioactivity; but this activity has not been found in any group of tektites. If the tektites did originate within the Earth–Moon system, they should be about the same age as the sedimentary rocks into which they fall. The only stratigraphic way to demonstrate conclusively that the reworking of the tektites from older units has occurred is to find a tektite *in situ* in one of the older stratigraphic units. Because of the rarity of specimens and the small number of good exposures of older units, such a tektite is not likely to be found. The find of a tektite *in situ* in the upper sand and gravel unit would not prove the age of the fall. The tektites are probably weathering out of the sand and gravel unit upon which they are found.

Analytical Data

King (1962) reported that three new specimens were found during field work and another tektite was obtained from a local resident. Plaster casts of these four specimens were given to the permanent collections of the Georgia Geological Survey.

Table 1. Chemical analysis of Georgia tektite DGA-1*

SiO_2	81·84 wt. %	
Al_2O_3	10·40	
Fe_2O_3	none	
FeO	2·30	
CaO	0·58	
MgO	0·67	Spectrographically detected
MnO	0·03	elements:
Na_2O	0·97	Ba, V, Zr—0·00X %
K_2O	2·15	Cu, Sr, Y, Yb—trace.
H_2O^+	0·32	
H_2O^-	0·22	
TiO_2	0·41	
P_2O_5	trace	
Total	99·89	

* Analyst, Jun Ito; weight of analyzed fragment, 4·442 g; weight of complete specimen, 25·945 g; analysis reported, 29 December 1962.

A fragment of one of the specimens (DGA-1) was analysed and the results of the analysis are presented in Table 1.

This analysis is very similar to the analysis of USNM 1396 reported by Clarke and Carron (1961) with slight exceptions. No ferric iron was detected in the original analysis of DGA-1. Ito confirmed this analysis by using an additional fragment weighing 0·75 g from the same specimen. Ferric iron content was determined by direct titration with titanous chloride solution in sulfuric acid after dissolution of the specimen in hydrofluoric and sulfuric acids in which methylene blue was used as the endpoint indicator. A trace of ferric iron was detected by this method; however, it is very doubtful that this trace was present in the tektite in this oxidation state. These wet chemical methods are liable to oxidation errors during the procedures of sample preparation, dissolution and transfer of the solution into another vessel for titration. The "essential" water value is greater than those values reported for most tekites and may be in error. The silica content, which is 1·3 per cent greater than USNM 1396, makes this the most siliceous North American tektite yet analysed.

On the basis of the only two analyses now available as well as previously reported physical properties, it appears that the Georgia tektites may have a much more restricted range of composition than the bediasites. However, it is unlikely that the small number of specimens (approximately 24) is representative of the total variation within the group.

An absorption spectrum* of a slice of DGA-1 is virtually identical to that of USNM 1396 presented by CLARKE and CARRON (1961), when corrected for specimen thickness.

TAYLOR (1963) analysed a fragment of DGA-1 for oxygen isotopic composition and obtained value of $\delta\, O^{18}/O^{16} = +9{\cdot}8‰$. The value was obtained by the fluorine extraction method and reported relative to the sea water standard.

Several types of artificial glass found with surface gravels in Dodge County were examined by microscopy and by X-ray fluorescence analysis. None of these glasses is similar to the Georgia tektites in composition, inclusions, or morphology. There is no reason to suspect an artificial origin of the tektites, especially in view of their 34 m.y. old potassium–argon ages.

There is no evidence of two periods of melting on the 15 specimens that the writer has examined, although several of the tektites have their primary shapes well preserved.

CONCLUSIONS

The Georgia tektites are found on rocks that are too young to be concordant with their older potassium–argon age. Because it is probable that the tektites have been reworked and transported from one of the older stratigraphic units to the north or northwest, it is recommended that the term "strewn-field" (implying that the tektites have fallen where they are now found) be abandoned in reference to the Georgia tektites.

None of the man-made glasses examined from Dodge County is similar to the Georgia tektites.

Acknowledgements—The Georgia Geological Survey supported the field work and Dr. A. S. FURCRON, Chief Geologist, was most helpful. Dr. S. M. HERRICK of the U.S. Geological Survey identified the microfossils. Mr. WILL SELLERS of Helena, Georgia, generously donated his time as a field companion. The Harvard University Department of Geological Sciences furnished the funds for the chemical analysis and a Summer Scholarship that was used in part for expenses in connection with the field work. This paper was prepared during the tenure of a National Science Foundation Summer Fellowship.

REFERENCES

BARNES V. E. and BRUCE G. A. (1959) Tektites in Georgia. *GeoTimes* **3** (7) 18.

BRUCE G. A. (1959) Tektites in Georgia. *Gems amd Minerals* No. 264, 22–23, 65–69.

CLARKE R. S. JR. and CARRON M. K. (1961) Comparison of tektite specimens from Empire, Georgia and Martha's Vineyard, Massachusetts. *Smithson. Misc. Colls.* **143,** 18.

CLARKE R. S. JR. and HENDERSON E. P. (1961) Georgia tektites and related glasses. *Georgia Min. Newsl.* **XIV,** 90–114.

COHEN A. J. (1959) Moldavites and similar tektites from Georgia, U.S.A. *Geochim. et Cosmochim. Acta* **17,** 150–153.

FURCRON A. S. (1961) Geologic age of the tektite shower and its associated rocks of the Georgia Coastal Plain. *Georgia Min. Newsl.* **XIV,** 115–119.

GENTNER W. and ZÄHRINGER J. (1960) The potassium–argon ages of tektites. *J. Geophys. Res.* **65,** 2492.

* Spectrum obtained by Dr. Carl Pitha, Crystal Growth Section, Air Force Cambridge Research Laboratories.

KING E. A. JR. (1962) Field investigation of Georgia tektites and description of new specimens. *Georgia Min. Newsl.* **XV,** 84–89.

PINSON W. H. JR. and SCHNETZLER C. C. (1962) Rubidium–strontium correlation of three tektites and their supposed sedimentary matrices. *Nature, Lond.* **193,** 233–234.

REYNOLDS J. H. (1960) Rare gases in tektites. *Geochim. et Cosmochim. Acta* **20,** 101–114.

SENFTLE F. E. and THORPE A. N. (1959) Magnetic susceptibility of tektites and some other glasses. *Geochim. et Cosmochim. Acta* **17,** 234–247.

STAIR R. (1955a) The spectral-transmissive properties of some tektites. *Geochim et Cosmochim. Acta* **7,** 43–50.

STAIR R. (1955b) Tektites and the lost planet. *Ann. Rep. Smithson Inst.* 1954, pp. 217–230.

STAIR R. (1956) Tektites and the lost planet. *Sci. Mon.* **83,** 3–12.

TAYLOR H. P. JR. (1963) Personal communication.

ZÄHRINGER J. (1961) Personal communication to Clarke and Henderson.

ZÄHRINGER J. (1963) *K–Ar Measurements of Tektites.* Second International Symposium on Tektites, Program, p. 12 (abstract).

Editors' Comments on Papers 7 and 8

Although Spencer (1933) and Chao et al. (1962) had reported metallic spherules in tektites, Chao, Dwornik, and Littler* were the first to show that some spherules in philippinites were nickel–iron having the mineralogy of meteoritic iron. This discovery forms another link in the body of evidence supporting the impact origin of tektites. Because of space limitations and the fact that this article is in a readily available international journal, the sections on "Mineralogy and Petrography" and "Chemical Composition" have been omitted. The reader is advised that an additional six plates in this article show exquisite photographs of the microstructure of spherules.

Spencer (1933), who first suggested the meteorite impact origin of tektites, has a strong advocate in Cohen (1959; 1960a, b; 1961; 1963). The experiment of Reid, Park, and Cohen* helps to confirm the conclusion of Chao, Dwornik, and Littler* that the natural spherules in philippinites resulted from admixture of meteoritic material, and this supports an impact origin for tektites. Although metallic spherules can be formed by melting philippinites under reducing conditions, such spherules are free of nickel and are distinct from the natural spherules.

Dr. E. C. T. Chao and co-authors are with the United States Geological Survey, Washington, D.C., 20042.

Dr. Arch M. Reid is now with the National Aeronautics and Space Administration, Manned Spacecraft Center, Houston, Texas, 77058. Dr. Alvin J. Cohen is now Professor in the Department of Earth and Planetary Sciences, University of Pittsburgh, Pittsburgh, Pennsylvania, 15213. Dr. Frederick R. Park is now in the Geosciences Department, Indiana University of Pennsylvania, Indiana, Pennsylvania, 15701.

References

Chao, E. C. T., Adler, I., Dwornik, E. J., and Littler, J. (1962). Metallic spherules in tektites from Isabela, Philippine Islands. *Science,* **135,** 97–98.

Cohen, A. J. (1959). Origin of Libyan Desert silica-glass. *Nature,* **183,** 1548–1549.

Cohen, A. J. (1960a). Germanium content of tektites and other natural glasses; implications concerning the origin of tektites. Rept 21st Intern. Geol. Congress, Copenhagen (Norden). Part 1, pp. 30–39.

Cohen, A. J. (1960b). Trace element relationships and terrestrial origin of tektites. *Nature,* **188,** 653–654.

Cohen, A. J. (1961). A semi-quantitative hypothesis of tektite origin by asteroid impact (abstract). *J. Geophys. Res.,* **66,** 2521.

Cohen, A. J. (1963). Asteroid- or comet-impact hypothesis of tektite origin: the moldavite strewn-field. *In* "Tektites" (J. A. O'Keefe, ed.), pp. 189–211. Univ. Chicago Press, Chicago, Illinois.

Spencer, L. J. (1933). Origin of tektites. *Nature,* **131,** 117–118, 876.

Geochimica et Cosmochimica Acta 1964, Vol. 28, pp. 971 to 980. Pergamon Press Ltd. Printed in Northern Ireland

New data on the nickel–iron spherules from Southeast Asian tektites and their implications

E. C. T. CHAO, EDWARD J. DWORNIK and JANET LITTLER*

Abstract—This paper presents new occurrences and mineralogic, petrographic, and chemical data on metallic spherules found in Philippine tektites from the Ortigas site of Mandaluyong near Manila and in indochinites from Dalat, South Viet Nam. Most of the metallic spherules contain kamacite, schreibersite and troilite. Schreibersite and troilite occur interstitially or in blebs among round or elongate fine-grained matrix kamacite. Where abundant, schreibersite coalesces and forms a network throughout the entire spherule. Troilite generally occurs in small round droplets within the schreibersite. The amount of schreibersite varies from less than 5 to about 35 per cent by volume and the troilite may constitute as much as 5 per cent of the spherules.

The composition of the phases present in the spherules was determined by the electron probe. In 9 analysed spherules from philippinites of the Ortigas site, the kamacite ranges from 1·6 to 4·5 per cent nickel. The average nickel content of kamacite in 3 analysed spherules from indochinites ranges from 4·7 to 12·9 per cent. The average nickel content in 4 schreibersite grains analysed from a single spherule ranges from 12·1 to 15·8 per cent.

The spherules in the tektites are nearly identical with the meteoritic spheroids from Meteor Crater, Arizona, in texture, mineral assemblage and nickel content as shown by the new data. We conclude that these spherules were formed as molten droplets of an impacting meteoritic body which was instrumental in producing the tektite glass.

INTRODUCTION

THIS report describes new occurrences and mineralogic, petrographic, and chemical data on metallic spherules discovered in philippinites from the Ortigas site in the Philippine Islands and in indochinites of Dalat, South Viet Nam, collected in 1962 by CHAO.

A preliminary report on the occurrence and gross composition of metallic spherules discovered in the Santiago Isabela philippinites of Luzon District, Phillippine Islands, has been published previously (CHAO *et al.*, 1962). Because of low nickel content in the Isabela tektite spherules, questions were raised regarding their meteoritic origin. The new material has afforded an opportunity to examine these questions in more detail.

The geological occurrence of the tektites was studied by CHAO during a visit to the Philippines and South Viet Nam in September and October 1962. Mineralogic and petrographic data on the spherules were obtained by CHAO and LITTLER by X-ray diffraction methods and the study of polished sections. The chemical compositions of the various mineral phases in the metallic spherules were obtained by DWORNIK with the use of the electron microprobe.

---

* Publication is authorized by the Director, U.S. Geological Survey.

GEOLOGICAL OCCURRENCES

Ortigas site, Philippine Islands

Several thousand tektites from the Ortigas site, Mandaluyong, about 5 miles east of Manila city limit, Luzon, Philippine Islands, were obtained with the help of Darwin Rossman of the U.S. Geological Survey. These tektites were contributed by Mr. Rafael Ortigas Jr. and his brother, Francisco Ortigas, to the Geological Survey for detailed laboratory studies. Among this large number of tektites are about 330 greenish pale brown, translucent, light colored tektites which contain metallic spherules.

The Ortigas site which is about 170 miles south of Santiago Isabela was discovered in the course of construction work on a drainage canal in the spring of 1962. The site is situated on a gently sloping plain or terrace about 130 ft above sea level and 80–100 ft higher than the present alluvial valley. As seen from the air, the site shows clear markings of an abandoned, terraced rice field. It is underlain by nearly flat-lying tuffaceous sandstone, conglomerate and pyroclastic rocks of the Guadalupe Formation of Pliocene age. The tuffaceous rocks are overlain by a blanket of tektite-bearing alluvial terrace gravel. Tektites were found in this blanket of gravel by local residents and members of the construction crew.

The tektite-bearing gravel consists chiefly of rounded pebbles and cobbles up to 5 in. in diameter composed of andesitic rocks, tuffaceous sandstone and silicified tuff. The pebbles are white, dark brown and purplish in color. A few large cobbles of dark gray basalt are also present. The large basalt cobbles are generally absent in the underlying tuffaceous conglomerates. Numerous manganese and iron oxide nodules, mostly about 1/4 in. in diameter, occur in the upper part of the gravel. They are particularly abundant near the surface. The matrix of the gravel is predominantly gray to grayish white plastic clay. In some places a layer of pebbly clay 5–6 in. thick, which probably accumulated as a result of rice cultivation, occurs on top of the gravel. The thickness of the gravel varies with the topography of the bedrock surface. In general, the gravel is less than 5 ft thick, and along fissures or enlarged joints it reaches a maximum thickness of about 6 ft. The age of the gravel is Quaternary.

Predominantly rounded shapes and the presence of crescent-shaped percussion marks on the tektites indicate that they are water-worn pebbles and have been transported and concentrated in the gravel. In order to estimate the abundance of tektites, approximately one cubic yard of the gravel in an undisturbed area at this site was excavated and the gravel was washed and sieved to recover tektites. A total of 184 tektites, mostly small fragments, were recovered; of these 37 are rounded pebble shaped. The largest specimen recovered is $1\frac{1}{2}$ in. in diameter. Two of the tektites are of the greenish pale brown, translucent, spherule-bearing type.

The spherule-bearing Ortigas site tektites range in color from greenish pale brown to yellowish pale brown. They are lighter in color and more translucent than typical philippinites. They can be easily distinguished when viewed moistened and in front of a strong light source. Some show a streaky marbleized texture. They range in size from less than 1/4 in. to nearly 2 in. in length. Most are subrounded broken fragments, but others are well rounded and pebble-shaped. The bulk specific gravity of this group of about 330 philippinites ranges from less than 2·34 to 2·49. The measured index of refraction on 10 selected polished wafers ranges from 1·5136 to 1·5189. The measured index of refraction on 15 selected, powdered samples using 0·002 step immersion oil and Na light ranges from 1·509 to 1·521 ±0·001. Polished thin wafers of the greenish pale brown philippinites have a striking streaky marbleized appearance. Bands which appear milky in hand specimens are dark in wafers. The darker color may be due to a difference in the valence state of iron in the tektite. X-ray diffraction examination of the milky type of glass revealed no microlites of any kind. Lechatelierite-type inclusions are generally abundant in these tektites.

The metallic spherules were discovered in wafers cut from the greenish pale brown philippinites. Every specimen of this type from the Ortigas site examined to date contains metallic spherules ranging from a few microns to over 800 μ in diameter.

Dalat, South Viet Nam

Metallic spherules have been found in two indochinites from two localities within the city limits of Dalat, South Viet Nam. One locality is at the reactor site of the Atomic Reactor Center of Viet Nam and the other locality is a road cut, 2/3 km north of Khu Van Kiep in the northern part of the city of Dalat. The reactor site is located on a knoll (elevation 1504 m), underlain by deeply weathered yellowish to reddish brown mica schist of the Dalat Series, believed to be of early Paleozoic age. The deeply weathered schists are cut by quartz veins some of which are tourmaline-bearing. The bedrock is overlain by a soil with three distinct layers. In ascending order these are (1) a yellowish silt composed of the deeply weathered micaceous schist, (2) a layer 2 to 7 in. thick containing subangular fragments of milky quartz and quartz-tourmaline rocks, and (3) a chocolate-brown top soil one to two feet thick. The tektites occur in the stony soil layer and on the surface. The erosional surface on which the tektites are found is younger than a basalt of Pliocene–Pleistocene age. The spherule-bearing tektite (specimen IDC-1-62) is one of about 1500 tektites collected at the reactor site and was found on the surface. It is a small curved specimen 2 cm by 2 cm in lateral dimensions and about 4 mm thick. It has typical flow structure, is rather vesicular and contains abundant lechatelierite-type inclusions.

A second spherule-bearing tektite (specimen IDC-8-62) is one of several tektites found along a vertical bank of a road cut in the northern part of the city of Dalat and about 2 km northwest of the reactor site. The soil profile here consists of the chocolate brown top soil underlain by nearly one foot of brick-red clay above the weathered schist. Tektites were found scattered in the red clayey layer, which is locally exposed along the vertical bank. The spherule-bearing tektite is a small irregular fragment 7 × 7 × 5 mm across.

Mineralogy and Petrography

Metallic spherules from the Ortigas site philippinites

Spherules occur at random within tektite specimens from the Ortigas site in both clear and dark bands of the tektite glass (Fig. 1(a)). Several were found protruding into spherical vesicles and one was found almost completely inside a vesicle

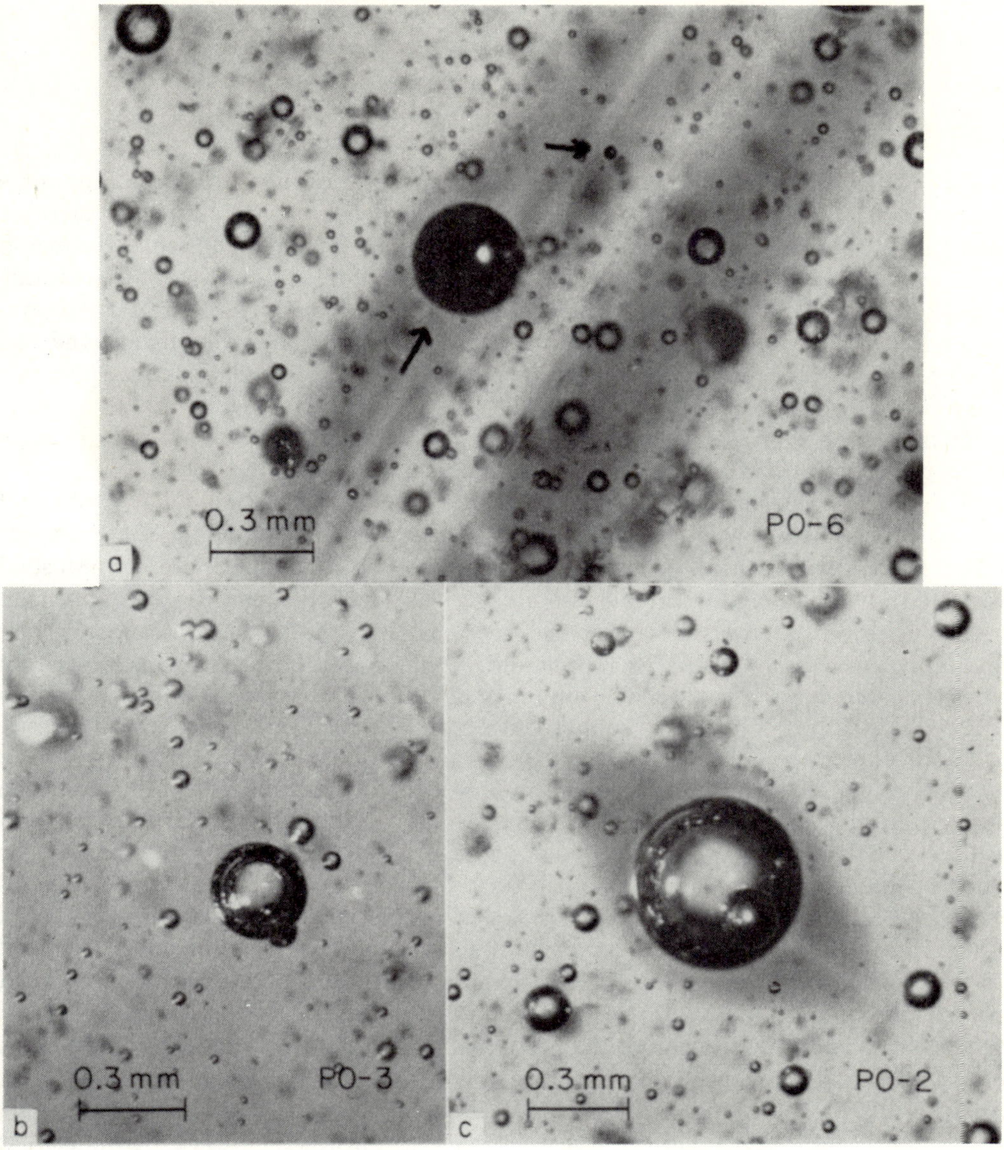

Fig. 1(a). Photograph of philippinite wafer from the Ortigas site showing the occurrence of one large and one small nickel–iron spherule (see arrows). Circles are vesicles. Specimen PO-6, transmitted and reflected light.

Fig. 1(b). Photograph showing the occurrence of a nickel–iron spherule inside a vesicle. Specimen PO-3, transmitted and reflected light.

Fig. 1(c). Photograph showing the occurrence of a nickel–iron spherule above a large vesicle. The vesicle is surrounded by a dark ring. Specimen PO-2, transmitted and reflected light.

(Fig. 1(b)). Some of the spherules are surrounded by a halo of slightly darkened glass. Others are located near a vesicle ringed by glass darker than the matrix glass (Fig. 1(c)). In this latter case the spherule bears no relationship to the darker glass. The dark bands of glass are believed to be higher in iron content and perhaps in trivalent iron.

ORIGIN OF THE SPHERULES

In the discussion of the origin of spherules from the Santiago Isabela philippinites (CHAO and others, 1962), we compared the mineral composition of these metallic spherules in tektites with spherules from deep sea sediments, with spherules recovered from Antarctic ice and those collected at mountain tops, as well as those metallic spherules which occur in impactite glass associated with terrestrial meteorite craters. At that time, because of the low nickel content of the kamacite and the lack of information regarding the other phases, the origin of the tektite spherules was left somewhat in doubt.

It is now appropriate to re-examine the origin of the tektite spherules in view of the new data on texture, mineral assemblage and chemical composition.

The nickel–iron spherules from tektites are without exception solid, fresh, and free of oxide phases, whereas many of the alleged cosmic magnetic spherules are hollow and consist essentially of an iron oxide phase, probably magnetite (unpublished work). In texture and mineral assemblage, the tektite spherules resemble the fresh meteoritic spheroids from Meteor Crater, Arizona and the nickel-iron spherules found in the Wabar impactite. Spheroids from Meteor Crater, Arizona, consist chiefly of kamacite with variable amounts of interstitial schreibersite and troilite (Figs. 6(a) and 6(b)). The fine, randomly orientated kamacite grains are characteristic of rapid quenching.

It is conceivable that the nickel–iron spherules in the tektites could have been derived from elements that were present in the pre-existing parent material and were concentrated by diffusion under reducing conditions, very much like the smelting process of iron ore. For purposes of comparison, while studying Thailand tektites, CHAO collected some slag samples in a locality about 50 kilometers west of Khon Kaen. This slag contains many iron spherules. Two of these spherules were studied in the same manner as the tektite spherules. The slag spherules consist of two mineral phases, α-iron and cementite in perlitic intergrowth (Fig. 7). The X-rayed and polished specimen was analysed by our colleague Cynthia Mead, who found iron with no detectable nickel (the lower limit of detection is approximately 0·2 per cent nickel). It is not uncommon, however, to find an association of α-iron with some impurities of phosphorus and sulfur as represented by schreibersite (Fe_3P) and troilite (FeS), in iron and steel manufacture.

If a criterion of the meteoritic or terrestrial origin of the spherules is to be found in their chemical composition, apparently it must be based on the nickel content of the spherules or of the kamacite phase alone. The occurrence of spherules with as much as 10·1 and 12·9 per cent nickel in the kamacite phase (Table 2(a)) is highly suggestive of meteoritic origin. Where the schreibersite also contains as much as 15 per cent nickel (Table 2(b)), this evidence is considerably strengthened. In view of the high silica content of tektites, the possibility that the spherules were derived from native nickel–iron in the parent rock which was converted into tektite glass is exceedingly remote. Accordingly we re-affirm our hypothesis that the nickel–iron spherules were produced during meteorite impact and the impacting body became dispersed in a melt composed of the target rocks formed by shock-induced fusion (CHAO, 1962; 1963).

Other Implications

Accepting the meteorite impact hypothesis as the mechanism for the formation of tektites, it is reasonable to expect that only a small amount of glass formed in proximity to the impacting body would contain molten projectile material. This may account for the rarity of spherules even in large collections of tektites.

The presence of the nickel–iron spherules with either schreibersite or troilite or both does not reflect completely whether the impacting body was of iron or stony meteorite composition. If one prefers the impacting body to be a comet (Urey, 1963; Barnes and Pitakpaivan, 1962), then the comet must contain, in addition to the hypothetical ice particles, particles of nickel iron in excess of 800 μ in diameter.

If one follows the lunar impact origin for tektites (Chapman and Larson, 1963; Chao, 1963), then tektites arrived as cold bodies before plunging into the Earth's atmosphere. Fusion-stripping by ablation of a parent body in the atmosphere such as that favored by O'Keefe (1963) will tend to destroy the nickel–iron spherules by converting the metallic iron to ferrous iron or wüstite and perhaps ferric iron (Marvin, 1963) if the oxygen pressures are high. It is difficult to see how the phosphide and sulfide in the spherules could survive under conditions of fusion stripping in the atmosphere.

The spherules are believed to have condensed either from a vapor or a molten liquid. The close association of the sulfide and phosphide within the iron and the lack of mixing with vaporized silicate phases perhaps favor the derivation of the spherules from a molten liquid. The impact or shock-induced temperature of the glass and the projectile must have exceeded 1535°C, the melting point of iron. The random process of condensation perhaps also explains in part the wide variation of nickel content in kamacite and the wide variation of sulphur and phosphorus in the spherules.

Acknowledgement—The authors are greatly indebted to Mr. Rafael Ortigas Jr. and his brother, Mr. Francisco Ortigas for the large, generous gift of philippinites from which abundant spherules were obtained. We wish also to thank our colleague, David B. Stewart who carried out the least squares refinement of the cell dimensions of the schreibersite.

References

Barnes V. E. and Pitakpaivan Kaset (1962) Origin of indochinite tektites. *Proc. Nat. Acad. Sci.* **48**, (6) 947–955.

Chao E. C. T., Adler I., Dwornik E. J. and Littler Janet (1962) Metallic spherules in tektites from Isabela, the Philippine Islands. *Science* **135**, (3498) 97–98.

Chao E. C. T. and Littler Janet (1962) The petrography of impactites and tektites with special reference to a dense impactite glass from the Ries Crater, Germany, (abs.). *J. Geophys. Res.* **67**, (9) 3548–3549.

Chao E. C. T. (1963) The petrographic and chemical characteristics of tektites. Chapter 3, p. 51–94, in *Tektites* (Ed. John A. O'Keefe). Univ. of Chicago Press, Chicago.

Chapman Dean R. and Larson Howard K. (1963) On the lunar origin of tektites. *J. Geophys. Res.* **68** (14) 4305–4358.

Clark S. P. Jr. (1961) Iron meteorites. *Ann. Rep. Geophys. Lab.*, p. 183–185. Carnegie Inst. of Washington, 1960–1961.

Groeneveld D. (1959) A new iron meteorite from Bellsbank Barkly West District. *Trans. Geol. Soc. S. Africa* **62**, 75–79.

Hägg G. (1928) Röntgenographische Studien über die binär Systeme von Eisen mit Phosphor, Arsen, Antimon, und Wismut. *Z. Krist.* **68**, 470–472.

Henderson E. P. and Perry S. H. (1958) Studies of seven siderites. *Proc. U. S. Nat. Mus.* **107**, 339–403.

Marvin U. B. (1963) Mineralogy of the oxidation products of the Sputnik 4 fragments and of iron meteorites. *J. Geophys. Res.* **68**, (17) 5059–5068.

O'Keefe John A. (1963) The origin of tektites. Chapter 8, pp. 167–188, *Tektites* (Ed. John A. O'Keefe). University of Chicago Press, Chicago.

Palache C., Berman H. and Frondel C. (1944) *The System of Mineralogy of J. D. Dana, and F. S. Dana, Yale Univ. 1837–1892.* Seventh edition Vol. I, pp. 124–125.

Urey H. C. (1963) Cometary collisions and tektites. *Nature, Lond.* **197**, (4864) 228–230.

8

Geochimica et Cosmochimica Acta 1964, Vol. 28, pp. 1009 to 1010. Pergamon Press Ltd. Printed in Northern Ireland

Synthetic metallic spherules in a Philippine tektite

ARCH M. REID,* FREDERICK R. PARK* and ALVIN J. COHEN†

Abstract—Metallic spherules have been synthesized by melting a Philippine tektite under reducing conditions and quenching to a glass. The spherules are smaller than those which occur naturally in philippinites and are metallic iron with no detectable nickel, sulphur or phosphorus.

WHEN iron-bearing rocks are melted in a reducing environment, iron may appear in the quenched product as a separate metallic phase. In the experiment described here a Philippine Islands tektite was melted under reducing conditions to determine (a) whether metallic spherules would form and (b) whether such spherules bear any resemblance to naturally-occurring spherules in Isabela tektites (CHAO *et al.*, 1962).

Slices cut from a tektite from Santiago, Isabela, Philippine Islands, were polished and examined by reflected light microscopy until we were convinced that no metallic spherules were present (at least within the limits of resolution of the microscope up to magnifications of 1000 ×). The specimens were partly ground and melted in closed high-purity graphite crucibles in an induction furnace, which was held at 1650–1700°C for 10 min. Sample C260a formed a glass button in the bottom of the crucible (Fig. 1(a)). No metallic spherules were found in this glass. Sample C260b was mixed with powdered graphite before heating. The end product of heating C260b was a vesicular glass cylinder which during the run pushed up the lid of the crucible so that the upper portion (light colored part in Fig. 1(b)) was exposed to the air during part of the run. The metallic spherules were found in the lower, darker (due to dispersed graphite) portion of sample C260b.

Two other samples were melted under reducing conditions without producing spherules. One (Fig. 2(a)) is an obsidian from Glass Mountain, California. Melting the dark obsidian along with powdered graphite produced a white, highly vesicular glass. A moldavite from Dukovany in Moravia was also melted in a graphite crucible (Fig. 2(b)) without producing spherules.

Sample C260b, the Isabela tektite, after heating, was cut and polished. Most of the spherules are less than one micron in diameter and display circular sections in polished slices of the glass (Fig. 3). The few larger grains are ovoid or irregular. The largest spherule examined has a mean diameter of 11 μ. Under the microscope the spherules appear to consist predominantly of a single phase. To confirm the identification of this phase as metallic iron, an X-ray powder diffraction pattern of the glass was obtained. The spherules constitute only a small fraction of the total

* Mellon Institute, Pittsburgh, Pennsylvania.

† Department of Earth and Planetary Sciences, University of Pittsburgh, Pittsburgh, Pennsylvania.

sample, but the most intense lines of both graphite and α-iron were observed on the film.

The spherules were examined with the ARL electron probe microanalyser at Mellon Institute. No nickel was detected in the iron which indicates that the nickel content is below 0·3 weight per cent. No phosphorus or sulfur was detected. Besides iron, traces of Si, Al, Ca, K and Ti are present within the spherule. In a traverse made across the 11 μ spherule the iron content of the surrounding glass was found to be very low, close to the limits of detection. At a distance of 4 μ from the spherule the iron content of the glass increases and there is a gradient in increasing iron content as the beam from the probe approaches the spherule. The other elements detected are distributed uniformly throughout the spherule. The electron microprobe cannot detect elements of atomic number less than 12 so that carbon, which is probably present in these spherules, was not detected.

Melting an Isabela tektite under reducing conditions resulted in the separation of part of the iron from the melt in the form of small metallic spheres. Since the solubility of nickel in glasses formed under reducing conditions is very low (SMITH and COHEN, 1963), nickel probably also has segregated into the metallic phase. However, Isabela tektites contain very little nickel and the iron to nickel ratio of the glass is around 1200 to 1.

Naturally-occurring tektite spherules differ from the metallic spherules described above in that they contain appreciable amounts of nickel, phosphorus and sulphur. Thus their composition suggests a meteoritic origin (CHAO *et al.*, 1962) and is not consistent with an origin by segregation from the melt. If the tektites were molten under reducing conditions, the composition of the meteoritic spherules may have been modified by the addition of iron from the melt.

Acknowledgements—The Philippine Islands tektites used in this work were obtained as exchange specimens from E. P. HENDERSON of the U.S. National Museum. The moldavite sample was presented to us by Professor R. ROST of Charles University, Prague. We thank Dr. L. F. VASSAMILLET and A. SCOTT for examining our material with the electron probe microanalyser and J. ANANIA who performed the iron and nickel analyses. One of the authors (F. R. P.) thanks Dr. T. B. MASSALSKI for helpful discussion. This work was supported by grant NsG-57-60 Supplement 1-62 from the National Aeronautics and Space Administration.

REFERENCES

CHAO E. C. T., ADLER I., DWORNIK E. J. and LITTLER J. (1962) Metallic spherules in tektites from Isabela, Philippine Islands. *Science* **135,** 97–98.

SMITH H. L. and COHEN A. J. (1963) Absorption spectra of cations in alkali-silicate glasses of high ultraviolet transmission. *Phys. Chem. Glasses* **4,** 173–187.

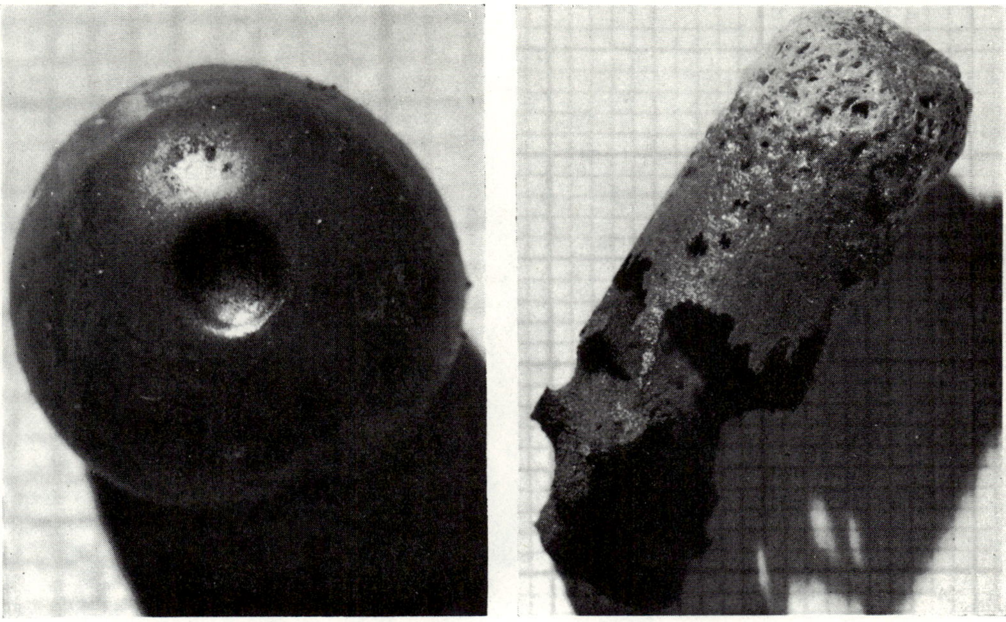

(a) (b)

Fig. 1. (a) Glass button formed by melting fragments of an Isabela tektite. (b) Glass cylinder formed by melting fragments of an Isabela tektite with powdered graphite. Scale in mm.

(a) (b)

Fig. 2. (a) Obsidian, Glass Mountain, California. Center piece is a section of the glass cylinder formed by melting obsidian fragments with powdered graphite. (b) Glass button formed by melting fragments of a moldavite from Dukovany in Moravia. Scale in mm.

1010

Fig. 3. Reflected light photomicrograph of spherules in a glass formed by melting fragments of an Isabela tektite. Magnification ×1645.

Editors' Comments on Papers 9, 10, and 11

These three papers concern the kinds of mineral inclusions so far found in tektites. The Muong Nong-type indochinites in which coesite was found are the same specimens in which quartz grains had been reported previously (Barnes, 1963)*.

The discovery of coesite by Walter* gives another constraint to any acceptable theory of tektite origin. Shock was somehow involved in the event that formed the Muong Nong-type tektites. Walter* shows that coesite is unstable at high temperature, and he disagrees with the suggested origin for the Muong Nong tektites, namely, melting of terrestrial rock with the subsequent collection of the melt in puddles, because he believes the length of time at high temperature would be too long for coesite to survive.

However, if Muong Nong-type indochinites formed as puddles, it is possible that the coesite and associated quartz, partly changed to lechatelierite, were incorporated as particles settling out of the air on to the melt as it flowed. Particles of this size would have cooled almost instantaneously and by the time they were incorporated in the melt it, too, may have been cool enough for the coesite to survive (Barnes and Russell, 1966)*. That some mineral grains were incorporated in Muong Nong-type indochinites after the melt had collected in puddles is shown by the presence of detrital mineral grains along a fault. That the melt was still fairly hot is shown by the formation of crystallites along the same fault, probably nucleated by dust (Barnes, 1964)*.

As suggested by Walter*, further efforts to determine the occurrence of coesite in tektites is needed. Subsequently Glass* found weak evidence of coesite in a Muong Nong-type tektite, but more work should be done on samples from other localities, and on Libyan Desert glass.

In the second paper Clarke and Wosinski* give additional evidence relating splash-form and Muong Nong-type tektites and relating tektites to impactites. This finding removed one criterion that had been used to identify splash-form tektites, namely the absence of crystalline minerals other than magnetic spherules. Because of space limitations and the fact that this article is in a readily available international journal, the sections on optical examination, electron microscopy, electron microprobe studies, and morphology have been omitted. An additional five plates in this article are devoted to excellent photographs of microstructure of the baddeleyite in the Martha's Vineyard tektite and of various analogs, as well as of electron microprobe scanning beam photographs of the Martha's Vineyard tektite and its baddeleyite inclusion.

In the third of these papers, Glass* identified zircon and chromite together in one specimen of Muong Nong-type tektite, and concluded that the association of these two minerals indicated a sedimentary source material. The source might well be broadened to include soils from such deposits.

Both the chromite and zircon described by Glass* show evidence of shock, but no work has been done to show the shock pressure involved to produce the effect found. A better understanding is needed of the pressure–temperature relations involved in shock and the role of shock in the formation of tektites. When parameters for these relationships have been defined, the use of minerals such as zircon and lechatelierite in tektite thermometry will be more nearly precise.

Dr. L. S. Walter is now Head of the Planetology Branch, Goddard Space Flight Center.

Dr. Billy P. Glass is now Professor, Department of Earth and Planetary Sciences, University of Delaware, Newark, Delaware, 19711.

References

*Barnes, V. E. (1963). Detrital mineral grains in tektites. *Science,* **142,** 1651–1652.

Barnes, V. E. (1964). Terrestrial implication of layering, bubble shape and minerals along faults in tektite origin. *Geochim. Cosmochim. Acta,* **28,** 1267–1271.

Barnes, V. E., and Russell, R. V. (1966). Devitrification of glass around collapsed bubbles in tektites. *Geochim. Cosmochim. Acta,* **30,** 143–152.

Reprinted from *Science*, **147**, 1029–1032 (1965)

9

Coesite Discovered in Tektites

Abstract. *Coesite has been identified by x-ray diffraction and electron microprobe chemical analysis as a constituent of inclusions in Muong Nong type tektites from Phaeng Dang, Thailand. The fine coesite grains are mixed with coarse quartz in the core of the inclusions, and the core is surrounded by frothy lechatelierite. The mixture of SiO_2 phases indicates that these tektites have been quenched from high temperatures and that modifications in texture and chemical composition from the original parent material have been minimal.*

The number of investigations on tektites has increased enormously because there are indications that tektites may come from the moon. Many of these investigations are prompted by the belief that some knowledge of the nature and origin of the moon's surface can be obtained before man's landing there.

Barnes (*1*) has pointed out the peculiarities of the Muong Nong type tektites, and in describing their chemistry and petrology has indicated that this tektite type is rich in silica (generally about 80 weight percent SiO_2), has a layered appearance, and relatively abundant inclusions of silica glass (lechatelierite) which is often bubbly or frothy. He has further described (*2, 3*) the occurrence of crystalline inclusions as well as peculiar nonspheroidal bubbles. Muong Nong type tektites are found in the area of Thailand.

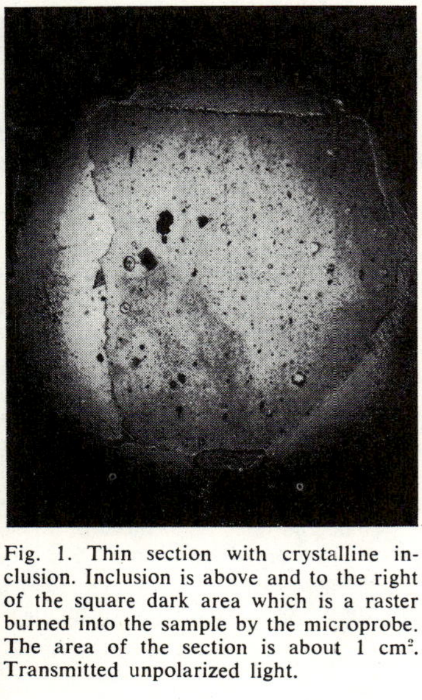

Fig. 1. Thin section with crystalline inclusion. Inclusion is above and to the right of the square dark area which is a raster burned into the sample by the microprobe. The area of the section is about 1 cm^2. Transmitted unpolarized light.

I have examined an unusual inclusion in a thin section of a Muong Nong type tektite from Phaeng Dang, Thailand (*3*). The very small (less than a millimeter) inclusions have cores of brownish material with a birefringence which was barely discernible. In each case the core was surrounded by a rim of frothy lechatelierite (Figs. 1 and 2). The abundance of these inclusions in this material is such that one inclusion occurs in about every other thin section. Since the core generally has a diameter of 0.1 mm or less, the abundance of this material in these specimens is probably less than 100 parts per million.

Electron microprobe analysis showed that the inclusions are approximately 100 percent SiO_2, quartz being the reference standard. In addition, the microprobe was used to check the bulk composition of a tektite specimen in which an inclusion was found even though the composition of the Muong Nong tektites from Phaeng Dang has already been presented by Barnes (*1*). The method for this analysis was similar to a random point-count method used in petrographic analysis. One hundred points around the entire 1-cm^2 area of the tektite section were analyzed for Si, Fe, and Al. The percentage (by weight) values were obtained from x-ray intensities by comparison with a series of standard synthetic glasses of tektite compositions, the necessity of corrections thus being obviated.

The results of the microprobe analysis are shown in the histograms in Fig. 3. Since the points analyzed did not happen to include any lechatelierite (which is of minor abundance in the section) the silica value indicated by the histogram is probably low by about 1 percent (by weight). Nevertheless, the values obtained from Fig. 3, namely, 79.5 percent SiO_2, 3.5 percent Fe oxide (*4*), and 9.8 percent Al_2O_3, agree very well with the values for Muong Nong type tektites of Barnes (*1*).

The first x-ray diffraction pattern was obtained with a Chesley-type pinhole camera. In this instrument the x-rays are collimated by the fine bore (0.076 cm) of a length (1 cm) of lead barometer glass. The sample, in the form of a fragment of a thin section, is attached to sticky mylar tape and placed on the brass sample holder. The sample holder is then clamped into a bracket which is adjustable so that the spot to be analyzed can be positioned over the bore of the collimater tube. The x-rays penetrate the sample, and the diffraction pattern appears on the film which is clamped to the pedestal in the other half of the camera.

The accuracy of the *d*-values (interplanar spaces) obtained with this type of camera is greatly limited because of its small size. The distance from sample to film must be accurately known, and this was measured by calibration from the strong (104) reflection of a calcite standard which was run independently since the small size of the sample did not permit the use of internal standards. Measurement of the diffraction lines themselves was the major source of error. For example, with CuKα radiation, the major line, if measured as 0.555 cm resulted in the value *d* being 3.111, but if determined as 0.560 cm resulted in *d* being 3.088.

In Table 1 the results of several diffraction analyses of one inclusion are compared with the values obtained for

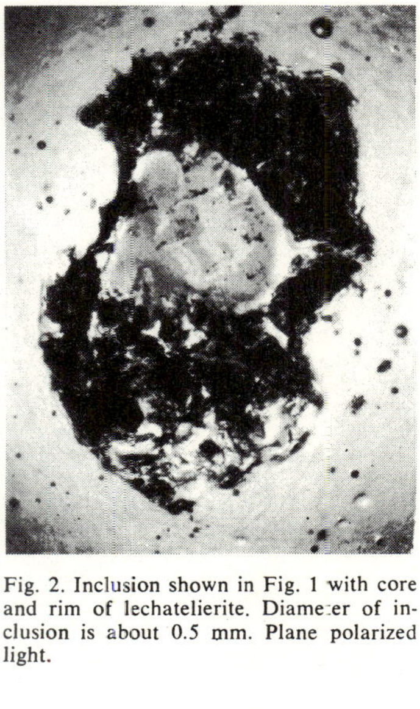

Fig. 2. Inclusion shown in Fig. 1 with core and rim of lechatelierite. Diameter of inclusion is about 0.5 mm. Plane polarized light.

Table 1. X-ray diffraction pattern data for standard coesite and tektite.

Coesite stand.*		ASTM† stand.			Sec. 1, Exp. 4*		Sec. 1, Exp. 12*		Sec. 1, Exp. 13‡	
I	*d* (Å)	*I*	*d* (Å)	hk*l*	*I*	*d* (Å)	*I*	*d* (Å)	*I*	*d* (Å)
		5	6.22	020						
		5	4.40	021						
8	3.424	50	3.432	130, 111	5	3.447	5	3.380	5	3.426
10	3.088	100	3.098	002, 040	10	3.102	10	3.077	10	3.099
5	2.750	15	2.77	220, (041)	1	2.75	5	2.742	3	2.755
5	2.694	15	2.68	131					3	2.705
2	2.484	5	2.350	201, $\bar{2}41$			5	2.491		
1	2.288	10	2.303	112, 150			5	2.297	1	2.311
1	2.183	10	2.195	240, $22\bar{3}$			3	2.254	1	2.267
2	2.031	10	2.034	151, 310‡			1	2.041	3	2.032
1	1.839	10	1.846	330			2	1.861		
2	1.784	10	1.789	$\bar{2}61$			2	1.794	1	1.795
3	1.706	15	1.716	260, 222‡			4	1.716	2	1.716
		10	1.711	113, $\bar{3}52$‡			4	1.708	2	1.703

* Average of two measurements. † American Society of Testing Materials. ‡Average of three measurements.

a coesite standard, the material for which was donated by J. Fahey of the U.S. Geological Survey. The measurement error can be appreciated when the values obtained for coesite are compared with the accepted values of Dachille and Roy (*5*).

In view of the fact that microprobe analysis showed the composition of the inclusions to be SiO_2, the measurement error and the fact that the values obtained for many measurements centered about the accepted values for coesite, the identification of coesite as a constituent of the inclusions is considered quite reliable. In Fig. 4 the diffraction patterns obtained for one sample and a coesite standard may be compared directly. Two exposures of the sample are used so that the darkened inner lines as well as the weak outer lines can be compared with the standard. Although the holes in the centers of the films are offset owing to movement of the films, the centers of the two sets of diffraction rings are coincident.

Since coesite seldom produced discrete spots in the diffraction pattern, the number of individual grains in the sample volume must be quite large—somewhat over a hundred with random orientation. The average diameter of these grains must be on the order of 1 micron in order to permit continuous diffraction lines from a sample volume of about 10^{-9} cm^3. In addition, there seems to be little or no preferred orientation of either the quartz or the coesite.

Quartz was identified in another inclusion (Table 2). Diffraction patterns at several locations of this inclusion showed a slight increase in the ratio of coesite to quartz in the center as opposed to that at the periphery of the brownish core.

The total diameter of the inclusions was about 0.5 mm. In every case, however, the brownish, fine-grained material, which contained the coesite, was rimmed by frothy lechatelierite and made determinations of index of refraction on the aggregate impossible. Probably the lechatelierite was formed by melting of the crystalline silica aggregate. The petrographic relationship of the aggregate to the lechatelierite is shown at high magnification in Fig. 5.

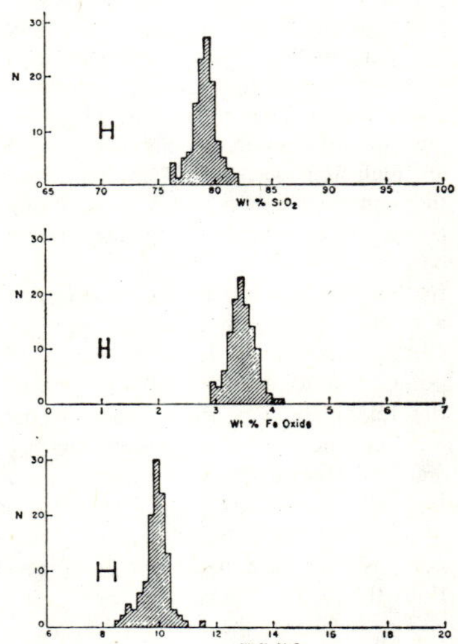

Fig. 3. Results of microprobe analysis of 100 random points in the section shown in Fig. 1. Brackets in each histogram show the limits of the 4-σ statistical variation to be expected at a 95-percent confidence level from the number of quanta counted in the analyses.

The index of refraction of the bulk glass of the tektite was 1.492.

A small (1 g) sample of a Phaeng Dang tektite was cleaned and coarsely crushed. Portions were placed in a shallow dish containing an oil of refractive index close to that of the bulk tektite glass. From this material about seven fragments which contained inclusions were selected, mounted on a glass rod, and exposed to x-rays in a Debye-Scherrer camera.

The diffraction pattern obtained by a 12-hour exposure of this material is compared with the pattern of a coesite standard (Fig. 6); only two coesite lines are apparent in the pattern of the sample. Weaker lines are obscured because of the abundance of spots produced by the quartz grains, scattering by glass in the sample, and the long exposure time. The smooth, continuous appearance of the coesite lines, especially when compared with the broad spots produced by the quartz, is an indication that the quartz in this sample occurs in several large grains whereas the coesite occurs as many, very fine grains.

The patterns also indicate that the abundance of quartz is much greater than that of coesite. Thus the abundance of coesite in the tektites is probably much less than the 100 ppm abundance of the inclusions themselves. This emphasizes the value of the pinhole camera in these determinations since without the selectivity of the collimator, the coesite lines are considerably obscured by those of quartz.

Discovery of coesite in nature was first made by Chao *et al.* (*6*) in samples of (shocked) Coconino sandstone from Meteor Crater, Arizona. Other occurrences have since been reported, and in each case the origin of the coesite was attributed to impact metamorphism. Thus the coesite in the tektite inclusions probably has a similar origin.

The relative sizes of the quartz and coesite grains in the inclusions, as determined from the x-ray patterns, tend to indicate that the reaction of quartz to form coesite has been incomplete.

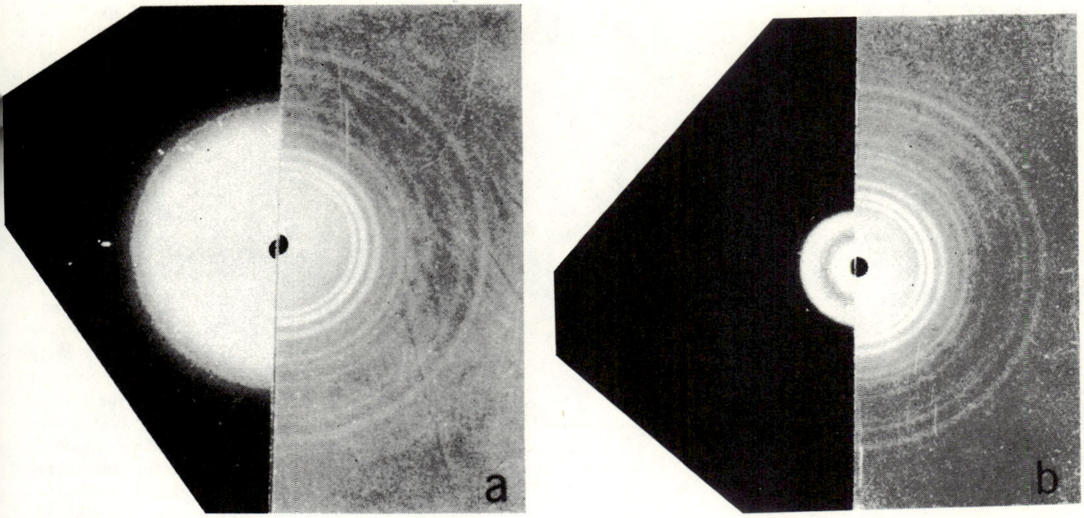

Fig. 4. Transmission diffraction patterns taken with Chesley camera cut and matched to enable comparison of the coesite pattern (on the right) with coesite in the inclusion (on the left). *a*, Developed to show light outer rings; *b*, developed to show dark inner rings in the pattern of the inclusion.

However, the observation (*7*) that coesite, when exposed to temperatures from 1100° to 1350°C for short periods, decomposed to metastable quartz prompted investigation of this decomposition at higher temperatures. Because of the presence of lechatelierite which is obviously derived from the crystalline silica, the temperatures to which the inclusions were exposed probably exceeded 1700°C for periods of the order of seconds (enough time to permit bubbles to form and to allow for excess heating and subsequent cooling but not enough to allow homogenization of the silicate glasses). In runs on coesite which lasted 6 minutes and 1 hour at both 1500° and 1600°C, the run products consisted of cristobalite in the 1-hour runs and cristobalite plus a small amount of coesite in the 6-minute runs. Therefore, at higher temperatures, cristobalite is to be expected as the decomposition product of coesite, an indication that the quartz of the inclusions is probably primary. Furthermore, all the diffraction lines produced by the inclusion in one section could be attributed to coesite; this indicates that quartz, if present at all, was a very minor constituent.

Table 2. Diffraction analysis of second section.

Mineral*	ASTM			Sec. 2 Exp. 10		Sec. 2 Exp. 18	
	hk*l*	*d*	*I*	*d*	*I*	*d*	*I*
Q	100	4.26	35	4.25	8	4.22	5
C	130	3.43	50	3.50	3	3.50	7
Q	101	3.34	100	3.34	10	3.29	10
C	002	3.098	100	3.076	5	3.12	3
C	220	2.77	15	2.73	1	2.74	2
C	131	2.68	15			2.65	3
Q	110	2.46	12	2.45	4	2.45	2
Q	102	2.28	12	2.28	2	2.23	3
C	240	2.195	10			2.198	3
Q	200	2.128	9	2.115	2	2.129	3
Q	201	1.980	6	1.964	4	1.962	2
Q	112	1.817	17	1.808	5	1.820	5
C	$\bar{2}61$	1.789	10			1.779	1

* C, coesite; Q, quartz.

An extremely important question is the condition of formation of the frothy lechatelierite. DeCarli and Jamieson (*8*) have shown that crystalline quartz can be made amorphous by shock at temperatures no greater than 1100°C during the shock and 700°C immediately after the shock. These temperatures would be insufficient to mobilize the glass to the extent that bubbles could form, and the pressure effects of the shock would certainly prevent the formation of bubbles. Cooling from this temperature must have been quite rapid in order to prevent the formation of cristobalite from the coesite.

This consideration imposes fairly stringent conditions on theories of the origin of tektites. Calculations by Greenland and Lovering (*9*) indicate that tektite spheres of 1-cm radius (approximately the size of the Phaeng Dang fragments) would cool from 2000°K to 1000°K in about 35 seconds. According to this, then, tektites might cool from 1700°C to the temperature at which the coesite decomposition is slowed considerably in an interval of about 10 seconds. In view of the amount of cristobalite experimentally formed in 6 minutes, 10 seconds is probably about the maximum quench time permitted. As reported (*9*), the time required for a tektite of 5-cm radius to cool through the same temperature limits is about 175 seconds. In this period, a detectable amount of cristobalite would probably have formed. This limitation on the size is

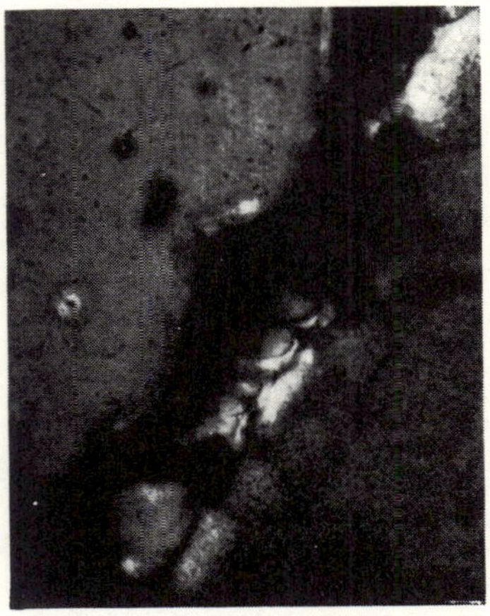

Fig. 5. Photomicrograph of tektite glass (upper left), frothy lechatelierite (center) and crystalline inclusion (lower right) in transmitted, plane polarized light. Width of field of view approximately 0.1 mm. Note filaments of lechatelierite which support inclusion in place and lack of apparent contact between inclusion and tektite glass.

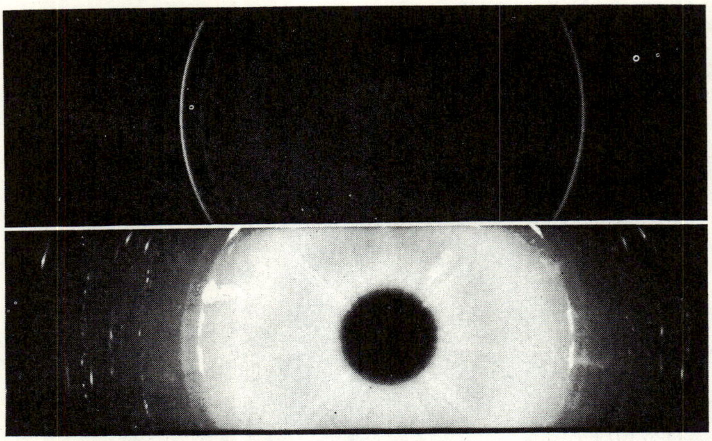

Fig. 6. Debye-Scherrer patterns of coesite (above) and inclusion-glass mixture (below). Weak broad line can be attributed to coesite as well as one weaker line toward the center of the pattern. Other intense spots are due to coarse grains of quartz.

in disagreement with the origin for the Muong Nong tektites proposed by Barnes (*10*) who suggested that they have formed by melting of terrestrial rock and the subsequent collection of this melt into pools.

There is a strong possibility that heating effects due to and immediately after the shock resulted in extreme heating of the crystalline material of the inclusions while the less dense tektite glass remained at relatively low temperatures. This would permit formation of coesite to be quenched quickly through the use of the body of the material as a heat sink.

Thus the rapid heating and cooling cycle to which the tektites were subjected would not permit a great deal of movement and homogenization of any separate phases in the original tektite material. Thorough mixing of the constituents to the degree shown by the histograms of Fig. 3 would certainly take many minutes at even moderately high temperatures. The material from which the tektites were formed was either glass or extremely fine-grained before the quartz was transformed to coesite by impact metamorphism. This limitation on the source material places a severe restriction on the terrestrial origin of tektites. Certainly the Phaeng Dang tektites could not have formed through mixing of the melts of various sedimentary rocks as proposed by Taylor (*11*), and this casts a doubt of such a source material for other tektites. It also seems unlikely that the comets or meteorites which formed the several tektite-strewn fields would be able, in each case, to select already homogenized material from which to make tektites.

The occurrence of the coesite-lechatelierite mixture could conform with either the Chapman (*12*) lunar-direct entry hypothesis in which debris from a lunar impact is ejected to the earth or to the O'Keefe (*13*) lunar-parent body hypothesis in which the coesite would be formed in a large body with sharp thermal gradients around the inclusions with subsequent removal of the fragments by erosion from the parent body rather than by ablation as liquid drops. Alternately the frothy lechatelierite could be formed as liquid drops by ablationary heating of the parent body.

The bulk composition of the Muong Nong type tektites and the analysis of the Phaeng Dang specimen reported here serve to relate these tektites to other tektites in the Far Eastern strewn field since, as Barnes has already pointed out (*1*), these tektites are end-members being richer in SiO_2 and correspondingly depleted in other constituents. It may, of course, be coincidence that those tektites which show the results of a low degree of heating and a rapid quench happen to be those with the greatest silica content. On the other hand, it may be reasonable to infer that the other tektites were derived from material similar in composition as well as texture to the Muong Nong type tektites perhaps by fractional volatilization.

The discovery of coesite in tektites is consistent with a lunar origin of this material. Furthermore, the primary nature of the texture which is indicated by the lack of decomposition products supports O'Keefe and Cameron's (*14*) suggestion that the original material was an acidic ash flow or welded tuff. The relatively high silica content of the source material may be due to several factors affecting the differentiation of lunar rocks which are discussed by Walter (*15*).

Note added in proof: Several analyses of less siliceous Muong Nong type tektites reported by Barnes (*1*) were overlooked. Nevertheless, the most siliceous indochinites are of the Muong Nong type and thus far coesite has been found only in those rich in silica. Further effort to determine the occurrence of coesite in tektites is certainly indicated. Since the preparation of this report, inclusions similar to those described here have been identified visually in specimens from Kan Long Dong.

L. S. Walter

Theoretical Division, Goddard Space Flight Center, Greenbelt, Maryland

References and Notes

1. V. E. Barnes, *Geochim. Cosmochim. Acta* **28**, 893 (1964).
2. ———, *Science* **142**, 1651 (1963); *Geochim. Cosmochim. Acta* **28**, 1267 (1964).
3. This inclusion was first noticed by Dr. M. Ross of the U.S. Geological Survey. He pointed it out to Dr. J. A. O'Keefe of NASA who in turn brought it to my attention. Several more inclusions were then found in other sections of the same and other Phaeng Dang specimens which had been obtained from V. E. Barnes of the University of Texas.
4. For the purpose of analysis, Fe oxide is defined as iron oxide in the approximate ratio generally found in tektites, $Fe_2O_3 : FeO :: 1 : 10$.
5. F. Dachille and R. Roy, *Z. Krist.* **111**, 451 (1959).
6. E. C. T. Chao, E. M. Shoemaker, B. M. Madsen, *Science* **132**, 220 (1960).
7. F. Dachille, R. Zeto, R. Roy, *ibid.* **140**, 991 (1963).
8. P. S. DeCarli and J. C. Jamieson, *J. Chem. Phys.* **31**, 1675 (1959).
9. L. P. Greenland and J. F. Lovering, *Nature* **192**, 1195 (1962).
10. V. E. Barnes and K. Pitakpaivan, *Proc. Natl. Acad. Sci. U.S.* **48**, 947 (1962).
11. R. D. Cherry and S. R. Taylor, *Geochim. Cosmochim. Acta* **22**, 164 (1961).
12. D. R. Chapman and H. K. Larson, *J. Geophys. Res.* **68**, 4305 (1963).
13. J. A. O'Keefe, in *Tektites* (Univ. of Chicago Press, Chicago, 1963), p. 228.
14. ——— and W. Cameron, *Icarus* **1**, 271 (1962).
15. L. S. Walter, *Trans. Am. Geophys. Union* **43**, 464 (1962); *Ann. N.Y. Acad. Sci.*, in press.
16. I thank M. Ross of the USGS and J. A. O'Keefe of NASA for drawing attention to these inclusions; F. Wood and J. Bishop for aid in the analytical and experimental work; K. Fredriksson and G. Arrhenius for use of the Chesley camera; J. A. O'Keefe and P. DeCarli for helpful discussion; and V. E. Barnes, who collected and donated the tektite material.

8 February 1965

10

Geochimica et Cosmochimica Acta 1967, Vol. 31, pp. 397 to 406. Pergamon Press Ltd. Printed in Northern Ireland

Baddeleyite inclusion in the Martha's Vineyard tektite

ROY S. CLARKE JR.
Smithsonian Institution, Washington, D.C.
and
JOHN F. WOSINSKI
Corning Glass Works, Corning, New York

(*Received* 20 *December* 1965)

Abstract—The Martha's Vineyard tektite is an enigmatic specimen that is unique in two ways. It is the only specimen from the island of Martha's Vineyard, and it contains an inclusion of baddeleyite, monoclinic ZrO_2. Baddeleyite defects are well known in artificial glass and have recently been observed in impactite. This is its first occurrence in a tektite.

The baddeleyite inclusion was identified by the use of reflected light microscopy, electron microscopy, and electron microprobe studies. This data is sufficient to establish the identification, even though it was not possible to study the inclusion in thin-section or to obtain an X-ray diffraction pattern. The morphology of the inclusion is discussed and shown to be different than that for baddeleyite inclusions reported in impactite. The presence of baddeleyite in the Martha's Vineyard tektite suggests another similarity between tektites and impactites.

INTRODUCTION

THE WORD provocative could well be used to describe the Martha's Vineyard tektite. This enigmatic specimen has presented a challenge, and consequently it has been the object of detailed study by a number of workers over the last five years. The specimen is unique both from consideration of its point of origin, the island of Martha's Vineyard, and by the fact it contains an inclusion of baddeleyite, monoclinic ZrO_2. Although a great deal of evidence has been accumulated to support the idea that the Martha's Vineyard specimen is a true tektite, questions remain concerning its relationship to other tektites and impactites that will probably not be answered satisfactorily until similar specimens are found and made available for study. This paper is concerned primarily with descriptive studies of the baddeleyite inclusion.

PREVIOUS WORK

The circumstances of the find of the Martha's Vineyard tektite in 1959 in an erosional gully overlooking the beach, the general geologic setting of that area, and the initial description of the specimen were reported by KAYE, SCHNETZLER and CHASE (1961). They pointed out that until other specimens were found in the area it would remain a question as to whether the island of Martha's Vineyard is a real tektite locality. Although a number of individuals have searched no other specimens have been reported from this area. CLARKE and CARRON (1961) gave additional data on the Martha's Vineyard tektite and pointed out its remarkable similarity to the first Georgia tektite to receive detailed study. They suggested that one possible explanation of this unexpected similarity of two specimens found separated by hundreds of miles in distance and over 20 years in time, was an artificial origin for the Martha's Vineyard and Georgia specimens. Somewhat later CLARKE and HENDERSON (1961) reported the presence of a zirconium-rich inclusion in the Martha's Vineyard specimen, and this remains a unique observation for tektites. Similar inclusions were not known at that time in other natural glasses. However, it was well known that zirconium-rich inclusions are common crystalline defects called "stones", in artificial glasses.

During this period a K–Ar age of 33·7 ± 2·2 m.y. was obtained on the Martha's Vineyard tektite (ZÄHRINGER, 1963; ZÄHRINGER and GENTNER, 1963). This was in agreement with ages of a Georgia and a Texas tektite, and with the ages previously reported on one of the same Georgia tektites and on a Texas tektite by REYNOLDS (1960). Later, FLEISCHER, PRICE and WALKER (1965) using the fission track dating method confirmed the K–Ar age for the Martha's Vineyard specimen, their value being 35·5 ± 8·3 m.y. However, fission track measurements on 6 Georgia tektites did not agree with the accepted K–Ar values. Two Georgia tektites, USNM 1396 and USNM 2136, gave fission track ages of <0·003 and 0·57 (±0·10) m.y., while the K–Ar values on these same specimens are 33·9 ± 1·7 and 35·0 ± 2·6 m.y. respectively (ZÄHRINGER, 1963).

SCHNETZLER and PINSON (1964) applied the Sr–Rb age method to many tektites. They found that the Martha's Vineyard tektite fits in the group of other North American tektites on a 400 m.y. isochron indicated for tektites in general. The isochron indicates in this case the time of differentiation of the tektite parent material and not that of formation of specimens. Later work, however, indicates that the isocron relationship may not be valid (SCHNETZLER, oral communication).

The zirconium-rich inclusion was intensively studied and identified as baddeleyite, monoclinic ZrO_2, by CLARKE and WOSINSKI (1963). The data supporting this identification are treated below in detail. Baddeleyite was unknown as an inclusion in other natural glasses until EL GORESY (1965) reported its presence in impact glasses from the Ries Kessel structure in Germany and from Aouelloul crater, Africa. He observed zircon partially decomposed to baddeleyite in Ries glass, and zircon crystals that had undergone complete breakdown to baddeleyite in Aouelloul glass. In the latter case the original zircon crystal outlines were preserved in some of the specimens.

APPEARANCE OF INCLUSION

An opaque-appearing inclusion was first noticed in the Martha's Vineyard tektite when the specimen was immersed in mineral oil and examined in strong transmitted light. The index of refraction of the oil was 1·47 and that of the glass 1·48, minimizing refraction differences. Under these illumination conditions the inclusion became obvious. It was located 1 cm in from the irregular broken edge of the specimen, which was 8 mm thick at that point. The inclusion lay 3 mm from one surface and 5 mm from the other. Figure 1 is a photograph of the inclusion taken through the surface of the specimen using transmitted light. The opaque-appearing core, or nucleus, was 0·26 mm long and 0·12 mm wide. The surrounding dark translucent halo, suggestive of a reaction zone, was 1·0 mm long and 0·52 mm wide.

The inclusion was prepared for study by carefully cutting it out of the tektite specimen with a thin diamond saw. This piece of glass was mounted in a bakelite block in such a way that the long axis of the inclusion (Fig. 1) was perpendicular to the surface of the block. The mounted specimen was then ground until the interior of the core of the inclusion was exposed. Less than half the inclusion was consumed in the operation. Figure 2 is a photograph at 44 times magnification of the mounted and polished specimen. The light colored oval area is the core of the inclusion, the reaction halo is thin and flat, and conforms to the flow structure of the enclosing glass. The original top and bottom surfaces of the glass specimen are well to the left and right of the photograph. The line across the top of the picture is a sawed edge, the material above it being bakelite. Slight irregularity below this edge was due to chipping of the glass during preparation of the mount. The polished inclusion surfaces was 0·26 mm long from the edge of the core to the end of the reaction zone, and the core was 0·14 mm long and 0·10 mm wide.

Identification of Inclusion

The inclusion has been studied by a number of techniques, all of which give data indicating that it is baddeleyite, monoclinic ZrO_2. Due to the inclusion's small size and the fact that it is the only known specimen of its kind, it has not been possible as yet to obtain an X-ray diffraction pattern or to study it petrographically in thin-section. However, the data when considered in its entirety lead to an unambiguous identification. The results of studies using reflected-light microscopy, electron microscopy, and detailed electron microprobe analysis are discussed in detail below.

* * * * * * *

Discussion

The Martha's Vineyard tektite remains an enigmatic specimen despite intensive study over a number of years. No other individual tektite specimen has been studied by so many different techniques and by such a number of investigators. Still, the circumstances of its find remain in doubt. Evidence in the form of more tektite specimens from Martha's Vineyard has not been produced to establish that this is a true tektite locality and not just an adventitious occurrence.

The tektite is also unique in another way. It contains an inclusion which has never been observed before in tektites and only recently observed in other natural glass systems. The data presented here demonstrate that the inclusion is baddeleyite, monoclinic ZrO_2.

The general morphology of the Martha's Vineyard baddeleyite inclusion is also of significance. It differs from that reported for impact glass and suggests a slow reaction rather than an instantaneous one associated with impact phenomena.

Laboratory refractory-corrosion tests of artificial tektite glasses and glasses in the alumino-silicate family indicate zircon crystals are resistant to zirconia breakdown at temperatures as high as 1550°C over periods of several days (Clarke and Wosinski, 1963). However, monclinic zirconia breakdown and recrystallization in these same glass systems can take place at lower temperatures (1450–1500°C)

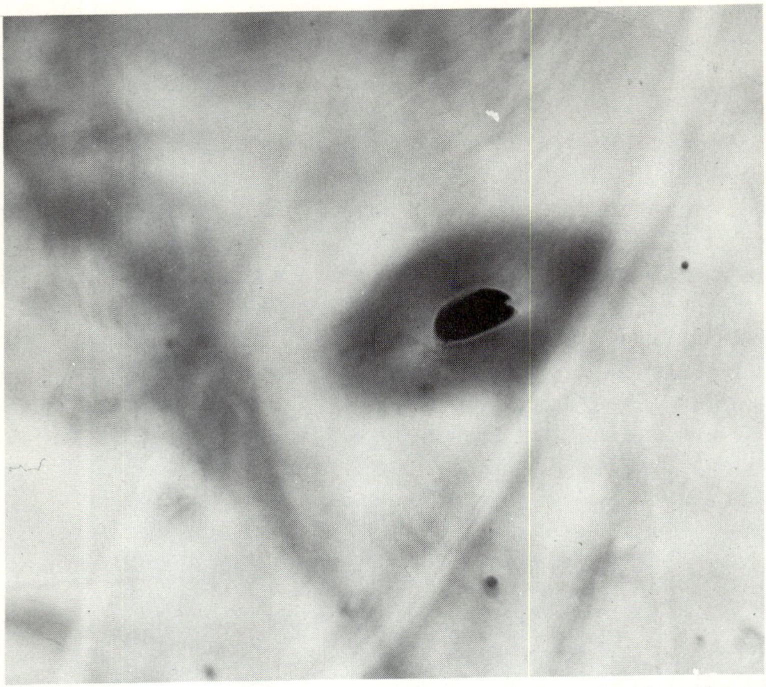

Fig. 1. Inclusion in Martha's Vineyard tektite. Photograph taken through the specimen with transmitted light, showing dark central area and lighter halo, 40 ×.

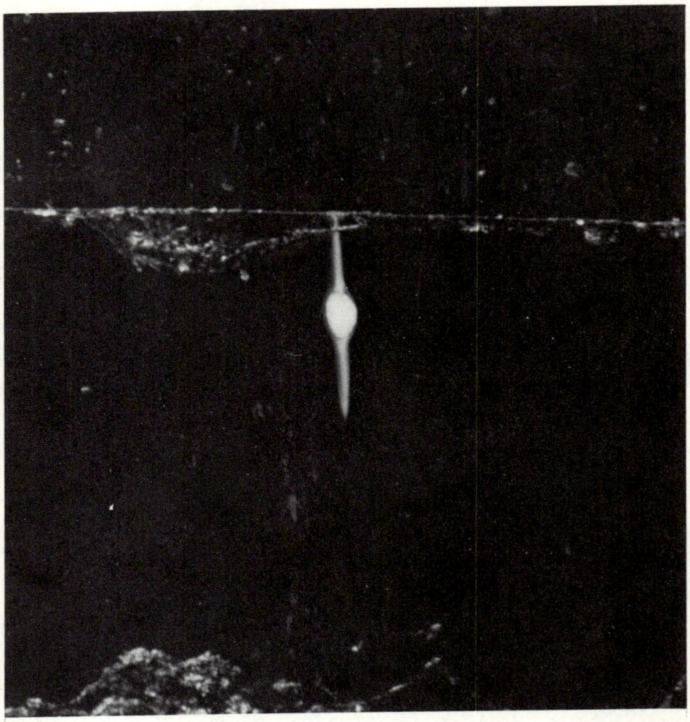

Fig. 2. Photograph of Martha's Vineyard tektite inclusion after having been cut out of specimen, mounted in Bakelite, and a polished surface exposing the interior of the inclusion prepared. Line across top of photograph is interface between specimen and Bakelite. Halo shown in Fig. 1 extends up and down and is shown to be flat, 44 ×.

and in short periods of time (2–8 hr). The tests performed dealt only with zircon, zirconia and the glasses mentioned. Glass batch materials which would include fluxes, whether they be natural or artificial, would certainly influence for either zircon or zirconia the temperature required for breakdown and the rate at which breakdown is produced. Practical experience in glass melting technology has shown that in these systems baddeleyite defects can be derived from either zircon or zirconia. The possible mineral source of the Martha's Vineyard inclusion is therefore left to speculation and chance.

The accumulation of dating and other data over the last several years has strengthened the case for the Martha's Vineyard tektite being a true tektite. It is a unique specimen due to its locality and its baddeleyite inclusion, a common mineral found as a defect in artifical glasses. If the assumption is made that the Martha's Vineyard specimen is a legitimate tektite, its baddeleyite inclusion becomes significant. This observation combined with those of El Goresy (1965) on terrestrial impact glass establishes another similarity between these two types of natural glass.

Note added in proof—King (1966) has identified an inclusion of baddeleyite after zircon in a Georgia tektite.

Acknowledgments—A number of individuals have cooperated in this work. Cynthia W. Mead of the U.S. Geological Survey did an electron microprobe study that was a great help in the early stages of this work. Kurt Fredriksson and Charles Fiori of the Smithsonian Institution and Raymond Fitzgerald of the University of California, San Diego, did much of the electron microprobe work reported here. The electron micrographs were prepared by Gerald B. Carrier of the Corning Glass Works. Grover Moreland of the Smithsonian Institution gave valuable assistance in a number of critical sample preparation operations. This work was supported in part by National Aeronautics and Space Administration grants NsG-71-60 and NsG-688.

References

Clarke R. S. Jr. and Henderson E. P. (1961) Georgia tektites and related glasses. *Georgia Min. Newsl.* **14,** (4) 90–114.

Clarke R. S. Jr. and Carron M. K. (1961) Comparison of tektite specimens from Empire, Georgia and Martha's Vineyard, Massachusetts. *Smithson. Misc. Coll.* **143** (4).

Clarke R. S. Jr. and Wosinski J. F. (1963) The ZrO_2 inclusion in the Martha's Vineyard tektite. Oral presentation at the *Second International Symposium on Tektites, Pittsburgh,* 5–7 September.

El Goresy A. (1965) Baddeleyite and its significance in impact glasses. *J. Geophys. Res.* **70,** 3453–3456.

Fleischer R. L., Price P. B. and Walker R. M. (1965) On the simultaneous origin of tektites and other natural glasses. *Geochim. Cosmochim. Acta* **29,** 161–166.

Kaye C. A., Schnetzler C. C. and Chase J. N. (1961) Tektite from Martha's Vineyard, Massachusetts. *Geol. Soc. Am. Bull.* **72,** 339–340.

King E. A., Jr. (1966) Baddeleyite inclusion in a Georgia tektite. *Trans. Am. Geophys. Union* **47,** 145.

Ramdohr P. (1963) Oral communication. Mineralogical Institute, University of Heidelberg, Heidelberg, Germany.

Reynolds J. H. (1960) Rare gases in tektites. *Geochim. Cosmochim. Acta* **20,** 101–114.

Schnetzler C. C. and Pinson W. H. Jr. (1964) Variation of strontium isotropes in tektites. *Geochim. Cosmochim. Acta* **28,** 953–969.

Setser J. L. and Ehmann W. D. (1965) Zirconium and hafnium abundances in meteorites, tektites and terrestrial materials. *Geochim. Cosmochim. Acta* **28,** 769–782.

Zähringer J. (1963) K-Ar measurements of tektites. *Radioactive Dating*, 289–305. International Atomic Energy Agency.

Zähringer J. and Gentner W. (1963) Radiogenic and atmospheric argon content of tektites. *Nature* **199**, 583.

11

eprinted from
1 August 1970, Volume 169, pp. 766-769

SCIENCE

Zircon and Chromite Crystals in a Muong Nong–Type Tektite

Abstract. *Chromite, zircon, and quartz crystals (identified by x-ray diffraction) have been recovered from a 2.07-gram sample of Muong Nong–type tektite. The absence of eskolaite (Cr_2O_3) and baddeleyite (ZrO_2) supports a previous conclusion that Muong Nong–type tektites were not heated as intensely as other tektite groups. X-ray asterism studies indicate that the crystals are shocked, which supports an impact origin. The presence of chromite and zircon together suggests that the Muong Nong–type tektite was produced from sedimentary material.*

During the investigation of the chemical variation within a Muong Nong–type (layered) tektite from Indochina, several crystalline inclusions, identified as chromite and zircon, were discovered. The discovery was not entirely unexpected, however, since previous work had indicated that the Muong Nong–type tektites were formed at lower temperatures than the splash-form types were and thus provide the logical place to search for minerals included in tektites (*1*). The minerals described in this report are believed to be relict and are therefore of obvious importance in indicating the nature of the parent material from which and the process by which the tektite was formed.

The chromite and zircon inclusions were discovered by crushing and sieving a 5.4-g sample of Muong Nong–type tektite. The fraction consisting of particles 110 to 149 μm in size (2.07 g) was then separated into 14 fractions based on specific gravity by use of heavy liquids (Bromoform and *N,N*-dimethyl formamide). Microscopic examination of each fraction revealed that many of the grains in the heavier fractions (specific gravity, 2.42 to 2.75) contain crystalline inclusions that were identified as zircon and chromite by x-ray diffraction. During further investigations, quartz and possibly coesite were identified by x-ray diffraction in frothy grains from the lightest fractions.

Microprobe analysis of several grains from each specific gravity fraction showed that the tektite has a wide range in composition (approximately 64 to 80 percent SiO_2), but its average composition is similar to that of other previously analyzed Muong Nong–type tektites (*2*). A thin section from the tektite shows highly contorted flow lines and numerous frothy and subangular lechatelierite particles, as well as a layered structure that is also evident in the hand specimen.

All of the mineral inclusions were completely surrounded by glass. Two chromite crystals were recovered; they are opaque and have a bronze color and submetallic luster. One is an octahedral crystal about 75 μm in width (Fig. 1); the other is an irregular grain (56 by 87 μm). These crystals were identified from x-ray diffraction patterns obtained with a Debye-Scherrer camera. Fourteen lines were present and all match lines for chromite. No eskolaite (Cr_2O_3) lines were observed. The composition of one of the chromite crystals was determined by electron microprobe analysis (Table 1). The grain containing the chromite crystal was mounted in epoxy, ground down to expose the crystal, and then polished. Nine chromites of known compositions were used for standards. The silica content of the chromite inclusion is rather high (average, ~ 2 percent) and variable (increasing toward the edges); otherwise the grain was fairly homogeneous, and the chromite-glass interface seemed to be rather sharp. Unfortunately, the grain was lost in an attempt to polish down to the maximum diameter of the chromite crystal; therefore no spectrometer traces or counts across the interface are available.

Approximately 15 zircon crystals were recovered. The largest dimension for each crystal is between 60 and 100 μm. They are generally white and opaque—although one euhedral crystal is partially translucent with an opalescent appearance. They vary in shape from euhedral, doubly terminated crystals to rounded grains and to irregular but rounded grains (Fig. 2). X-ray diffraction patterns were obtained for five different grains. Approximately 25 lines were observed for each pattern;

all of them match zircon lines. No lines for baddeleyite (ZrO_2) were observed.

Elemental scans of two zircon crystals indicated the presence of large amounts of Zr and Si, small amounts of Al, Fe, and Ca, and trace amounts of Ti and Hf. Semiquantitative analyses of three zircons (one euhedral crystal, one rounded grain, and one irregular rounded grain) were performed with the electron microprobe (Table 1). Quartz and zirconia were used as standards to determine the amount of Si and Zr, respectively. Glass standards were used to determine the Al, Fe, and Ca content. The zircons were fairly homogeneous and all three have approximately the same composition.

Scanning electron beam photographs of the zircon crystals also show that the crystals are homogeneous with regard to Zr, Si, Al, and Hf and that the boundaries are sharp (Fig. 2). Spectrometer traces of Zr and Si show that the zircons are fairly homogeneous and that the zircon-glass interface is sharp. Neither the scanning electron beam photographs nor the spectrometer traces show the presence of ZrO_2 (baddeleyite).

One transparent, weakly birefringent area in the thin section was shown by microprobe analysis to be nearly pure SiO_2 and may be a relict quartz grain. Quartz was identified from an x-ray diffraction pattern obtained from some frothy glassy particles separated from the fraction with a specific gravity of less than 2.198. Eleven lines were observed; ten of these lines match those given for quartz (American Society for Testing and Materials 5-0490). However, one line, with a *d*-spacing of 2.76 Å, did not match any line for quartz but does closely match the third most intense line for coesite (2.77 Å). During further examination of the x-ray pattern, a very faint line with a *d*-spacing of 3.08 Å was observed. This is close to the most intense line for coesite (3.098 Å). The second most intense line for coesite (3.432 Å), if present, is masked by the most intense line for quartz (3.43 Å). Therefore, coesite is probably present in the sample, but its identification is not well established. However, coesite has been previously identified in similar particles from other Muong Nong–type tektites (*3*).

Crystalline inclusions in tektites are extremely rare. Metallic spherules [kamacite and iron phosphide (?) of probable meteoritic origin] have been reported in some Australasian tektites (*4*). Magnetite spherules (with associated iron and wüstite) that probably exsolved from the tektite glass, have been reported by Kleinmann (*5*). Coesite has been described by Walter (*3*) in a Muong Nong–type tektite, and baddeleyite has been found in the Martha's Vineyard tektite and a Georgia tektite (*6*). However, prior to this discovery the only apparently relict mineral reported in a tektite was quartz, which was identified in Muong Nong–type tektites by Barnes (*7*) and Walter (*3*).

Clarke and Wosinski (*6*) and King (*6*) consider the presence of baddeleyite in tektites as additional evidence of the impact origin of tektites because of El Goresy's (*8*) work which showed that zircon tends to decompose into baddeleyite and silica as a result of impact. El Goresy states that the presence of baddeleyite in impact glasses is indicative of a very high temperature of fusion of the original rock containing zircon. Experimental work on the system ZrO_2-SiO_2 indicates that zircon breaks down at 1775°C under atmospheric pressure (*9*). Kirby (*10*) reports 1900°C as the breakdown temperature for zircon.

Chromite is a refractory mineral; however, Trojer has shown that chromite can break down to eskolaite (Cr_2O_3) in temperatures as low as 1200°C if heated in a highly acidic siliceous glass medium (see *11*, p. 540).

Table 1. Chemical composition of chromite and zircon inclusions as determined by electron microprobe analysis; T, trace; N.D., not detected. Values are expressed in percentages.

Chromite		Zircon*	
Cr_2O_3	57.6	SiO_2	32
FeO	22.5	ZrO_2	59
Al_2O_3	7.0	TiO_2	T
MgO	10.0	Al_2O_3	2
SiO_2	2	FeO	1
TiO_2	0.5	MnO	N.D.
MnO	0.28	MgO	N.D.
		CaO	0.3
		ThO_2	N.D.
		P_2O_5	N.D.
		HfO_2	T
		Y_2O_3	N.D.

* Average of three analyses.

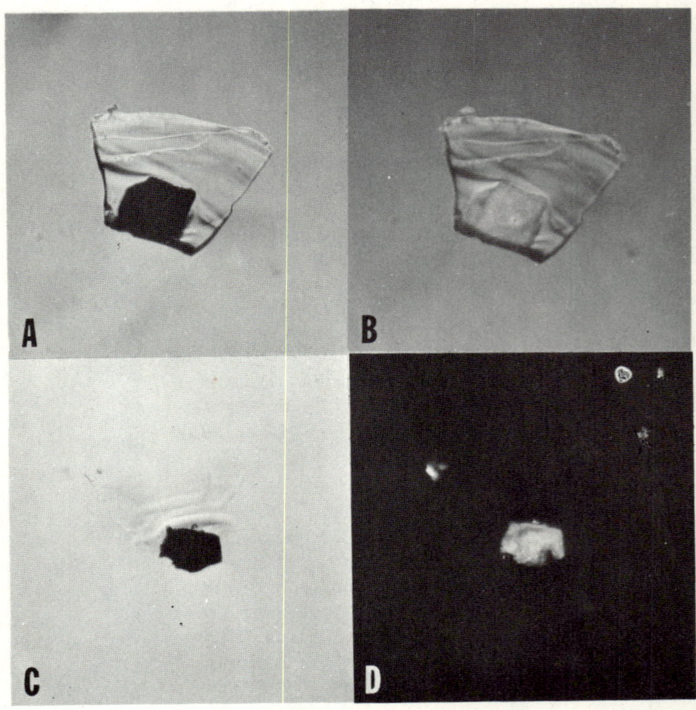

Fig. 1. Photomicrographs of chromite crystals. (A) Euhedral chromite crystal surrounded by tektite glass, in transmitted light. Note flow structure around the chromite crystal (opaque). (B) The same crystal as in (A) in transmitted and reflected light under crossed Nicols. Note octahedral shape of the crystal. (C) Second chromite crystal in transmitted light. (D) The same crystal as in (C) in reflected light and crossed Nicols. Note that a portion of the crystal near the lower right corner is apparently isotropic. The scale is the same for all the photomicrographs.

If baddeleyite described in the Martha's Vineyard and Georgia tektites is a decomposition product of zircon produced during the formation of the tektites, then the presence of zircon and absence of baddeleyite in the Muong Nong–type tektite indicates that this tektite was not heated as intensely as the Martha's Vineyard and Georgia tektites. This supports previous conclusions that the Muong Nong–type tektites were formed at lower temperatures than other tektite groups (*12*).

Nearly all investigators now agree that tektites were produced by impact. X-ray asterism studies of the chromite and zircon grains from the Muong Nong–type tektite support this conclusion. Shock damage can be detected by rotation of single-crystal grains in an x-ray beam (*13*). Shocked crystals produce diffraction patterns that range from spots to streaks to arcs, depending on the degree of shock. Of the mineral inclusions studied, all except one euhedral zircon grain produce continuous diffraction lines similar to those produced by a powder. The euhedral zircon crystal produced a diffraction pattern consisting of elongate spots. Unfortunately, no work has been done which would indicate the shock pressure needed to produce this degree of asterism in chromite and zircon crystals.

The chemical compositions of tektites have been compared to various terrestrial materials. Some investigators maintain that they most closely resemble acid igneous rocks, others note the similarities to sedimentary rocks and soils (*14*), while a third group feels that tektites are not similar to any terrestrial rock type (*15*). Several investigators have suggested a lunar impact origin for tektites (*16*).

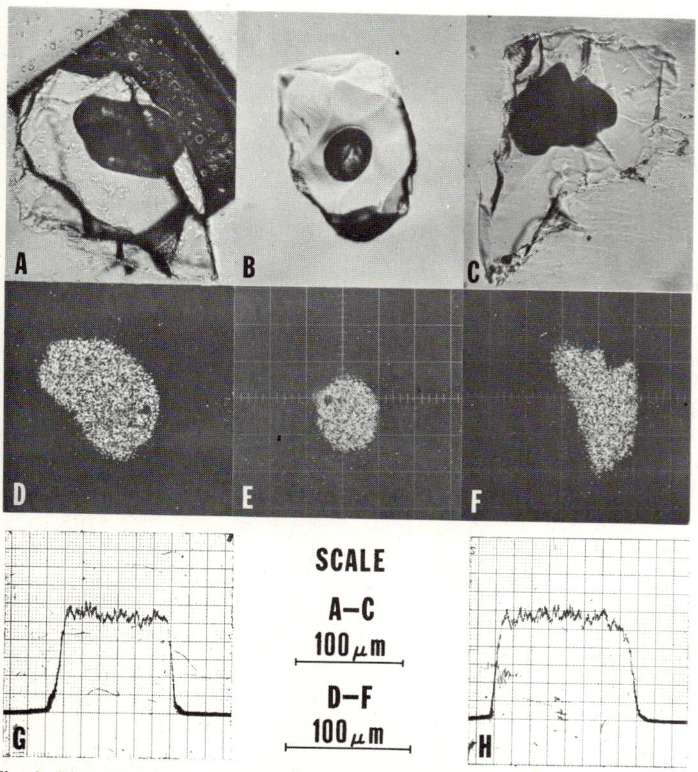

Fig. 2. Photomicrographs (A–C), electron microprobe scanning photographs for Zr (D–F), and Zr spectrometer traces (G and H) of three zircon crystals. (A) Photomicrograph of polished section in transmitted light showing outline of euhedral zircon crystal. The black area outside of the grain is epoxy burned by the electron beam. (B) Muong Nong–type tektite fragment containing a rounded zircon crystal. Transmitted light. (C) Polished section showing an irregular zircon grain. Transmitted light. Zirconium scanning photographs of (D) the euhedral zircon crystal shown in (A); (E) the rounded zircon crystal shown in (B); and (F) the irregular zircon grain shown in (C). Zirconium spectrometer trace (G) across the euhedral crystal shown in (A) and (H) across the irregular grain shown in (C).

The presence of chromite and zircon together indicates a sedimentary source material for the Muong Nong–type tektite. Chromite usually occurs in ultrabasic rocks, such as peridotites and serpentines derived from them. Although zircon has been described in basalts, it is more commonly found as an accessory of acidic igneous rocks such as granite, syenite, diorite, and others (*17*). Thus it seems unlikely that chromite and zircon would form together in a single igneous rock. On the other hand, zircon and chromite are resistant to chemical and mechanical erosion and are stable in sedimentary deposits. (It should be pointed out, however, that chromite does not commonly occur as a constituent of detrital sedimentary deposits.) Therefore, it seems likely that the source material of the Muong Nong–type tektite was a sedimentary material and not an igneous rock. This does not exclude a lunar origin for the tektite, however, since processes forming the lunar regolith tend to mix components from various rock types.

The subangular to rounded shapes of many of the inclusions, especially the zircon crystals shown in Fig. 2B, are also suggestive of a detrital sedimentary deposit. It may also be significant that both the zircon and chromite crystals are approximately the same size (Fig. 1) and have similar specific gravities (4.68 and 4.6, respectively), suggesting that the parent material was gravitationally sorted. This again points to a sedimentary material as the source material for the Muong Nong–type tektites.

A comparison of the chemical composition of the Muong Nong chromite with terrestrial, meteoritic, and lunar (from Apollo 12) chromites shows that it is more similar to terrestrial chromites. The Muong Nong chromite is in fact quite similar in composition to terrestrial chromites but has a higher MgO content than most meteoritic chromites. Although chromite was not found in Apollo 11 samples, it has been described in Apollo 12 samples (*18*). All the Apollo 12 chromites were found to have TiO_2 contents greater than 4 percent, whereas the Muong Nong chromite has a TiO_2 content of approximately 0.5 percent. Additional samples will be required, however, before it can be determined whether the Apollo 12 chromites are typical of lunar chromites.

The discovery of chromite and zircon crystals in a Muong Nong–type tektite increases the number of relict minerals now known to occur in tektites. The absence of any signs of decomposition of zircon or chromite to baddeleyite or eskolaite, respectively, supports previous observations which indicate that the Muong Nong–type tektites were less intensely heated than other tektite groups. X-ray asterism studies of the chromite and zircon crystals support an impact origin for the Muong Nong–type tektite. The presence of chromite and zircon together and the rounding of the crystals, along with evidence of sorting, all indicate that a sedimentary deposit is the source material for the Muong Nong–type tektite.

Note added in proof: After the above investigation was completed, corundum, rutile, and monazite, as well as quartz and zircon, have been tentatively identified by x-ray diffraction in another specimen of Muong Nong-type tektite from Indochina. These inclusions will be discussed in detail in a later publication.

BILLY P. GLASS*
Goddard Space Flight Center, Greenbelt, Maryland 20771

References and Notes

1. V. E. Barnes, *Science* **142**, 1651 (1963).
2. The average composition of the tektite determined by electron microprobe analysis is (in percent): SiO_2, 76; Al_2O_3, 11.5; Fe as FeO, 4.1; MgO, 1.5; CaO, 1.1; K_2O, 2.6; and Na_2O, 1.1. This compares well with Muong Nong–type tektite analyses given by D. R. Chapman and L. C. Scheiber [*J. Geophys. Res.* **74**, 6742 (1969)]. For example, compare the above analysis with Chapman and Scheiber's sample No. 34.
3. L. S. Walter, *Science* **147**, 1029 (1965).
4. E. C. T. Chao, in *Tektites*, J. A. O'Keefe, Ed. (Univ. of Chicago Press, Chicago, 1963), pp. 51–94.
5. B. Kleinmann, in *Abstracts* (3rd International Tektite Symposium, Corning, New York, 1969), p. 26.
6. R. S. Clarke, Jr., and J. F. Wosinski, *Geochim. Cosmochim. Acta* **31**, 397 (1967); E. A. King, *Trans. Amer. Geophys. Union* **47**, 145 (1966).
7. V. E. Barnes, *Geochim. Cosmochim. Acta* **28**, 1267 (1964).
8. A. El Goresy, *J. Geophys. Res.* **70**, 3453 (1965).
9. F. P. Hall and H. Linsley, *Phase Diagrams for Ceramists* (American Ceramics Society, Columbus, Ohio, 1947), p. 157.
10. D. Kirby, *Metallurgia* **30**, 65 (1944).
11. A. El Goresy, H. Fechtig, J. Otteman, in *Shock Metamorphism of Natural Material*, B. M. French and N. M. Short, Eds. (Mono, Baltimore, 1968), pp. 531–553.
12. V. E. Barnes, in *Tektites*, J. A. O'Keefe, Ed. (Univ. of Chicago Press, Chicago, 1963), pp. 25–50.
13. F. Dachille, P. Gigl, P. Y. Simons, in *Shock Metamorphism of Natural Materials*, B. M. French and N. M. Short, Eds. (Mono, Baltimore, 1968), p. 555.
14. V. E. Barnes, *Geochim. Cosmochim. Acta* **33**, 1121 (1969).
15. C. C. Schnetzler and W. H. Pinson, Jr., in *Tektites*, J. A. O'Keefe, Ed. (Univ. of Chicago Press, Chicago, 1963), pp. 95–129.
16. H. H. Nininger, *Sky Telesc.* **2**, 8 (1943); J. A. O'Keefe, in *Space Research*, H. K. Bijl, Ed. (Proceedings International Space Science Symposium, 1st, Nice, 1960) (North-Holland, Amsterdam, 1960); D. R. Chapman and H. K. Larson, *J. Geophys. Res.* **68**, 4305 (1963).
17. E. S. Dana and W. E. Ford, *Dana's Textbook of Mineralogy* (Wiley, New York, 1963), pp. 493 and 611.
18. S. E. Haggerty and H. O. A. Meyer, in preparation.
19. I thank Dr. J. A. O'Keefe for helpful suggestions and encouragement. C. W. Kouns and Chris Inman prepared the samples and helped with the photography. F. Wood, Jr., and W. Riffle helped with electron microprobe and x-ray analyses, respectively. Discussions with Drs. J. A. O'Keefe, H. O. A. Meyer, S. K. Saxena, and L. S. Walter have been helpful. Prof. V. E. Barnes, Drs. B. M. French, R. S. Clarke, Jr., and P. D. Lowman critically reviewed the manuscript. Judith A. Glass helped prepare the manuscript. The research was done while B.P.G. was on active duty with the U.S. Army Corps of Engineers assigned to Goddard Space Flight Center.
* National Research Council postdoctoral resident research associate at the Goddard Space Flight Center.

30 April 1970

Editors' Comments on Paper 12

One of the bits of evidence that frequently has been cited for an extraterrestrial origin of tektites is the presence of near-perfect vacuums in tektite bubbles, which theoretically could not have formed at the earth's surface under atmospheric pressure. Unknowingly at the time, another theory of the formation of such bubbles had its beginning in Manila in January 1961. Dr. Beyer, philippinite expert and archeologist, gave us three specimens that had been heated in the Manila fire. The specimens were among the last of the philippinites to be sliced for study, and it was with great interest that we observed the collapsed bubbles with devitrification around them. The evidence that the bubbles probably were formed by water vapor, later absorbed into the glass around the bubbles and leaving almost perfect vacuums, shows that tektites need not have formed under vacuum conditions.

In this paper we also suggest, from experimentation with solar furnace melts, that tektites may have been formed from porous materials through the unidirectional application of very intense heat, so that the volatile substances were driven before the heat front. A dozen of the solar furnace melts were sent to Dr. Frank E. Senftle of the U.S. Geological Survey for water determinations. The results of this work were reported by Gilchrist et al. (1969), who found that the solar furnace glasses contained several times as much water as tektites contain and concluded: "If the process suggested by Barnes and Russell is an important one in the formation of tektites, it seems that the process could not have taken place under atmospheric conditions but occurred in at least a partial vacuum. The present water content of glasses prepared in the atmosphere reflects the water content of the parent materials plus the water absorbed from the atmosphere when remelted." However, if these melts had been made under conditions of low humidity rather than in Austin, Texas, where humidity is high, and/or at higher temperatures, perhaps the water content would have matched that found in tektites.

The original purpose of the solar furnace experimentation was to learn more about lechatelierite formation and the behavior of other minerals in high-temperature melts. The results of these experiments are mentioned by Barnes (1971)* but are not related to previous findings. Before the solar furnace experimentation was completed, Barnes (1964) described and speculated on the origin of rayed bubbles that he found in moldavites and indochinites. The results of the solar furnace runs with muscovite and biotite mixed with powdered synthetic tektite glass indicate that the explosive disintegration of a hydrous mineral, suggested in that paper, is the correct interpretation for the origin of such rays.

Other minerals melted with synthetic tektite glass included apatite, asbestos, augite, beryl, fluorite, garnet, gypsum, hornblende, microcline, quartz, olivine, plagioclase, serpentine, and zircon. All minerals promptly disappeared except quartz, which changed to lechatelierite. Zircon made an opaque glass and its behavior is discussed by Barnes (1971)*.

Dr. Richard V. Russell, formerly a graduate student at The University of Texas at Austin, is now at the Kungl. Universitet i Stockholm, Geologiska Institutionen, Mineralogiska avdelningen, Box 6801, 113 86 Stockholm, Sweden.

References

Barnes, V. E. (1964). Rayed bubbles in tektites. *Geochim. Cosmochim. Acta,* **28,** 1373–1375.

*Barnes, V. E. (1971). Description and origin of large tektite from Thailand. *Chem. Erde,* **30,** 13–19.

Gilchrist, J., Thorpe, A. N., and Senftle, F. E. (1969). Infrared analysis of water in tektites and other glasses. *J. Geophys. Res.,* **74,** 1475–1483.

12

Geochimica et Cosmochimica Acta, 1966, Vol. 30, pp. 143 to 152. Pergamon Press Ltd. Printed in Northern Ireland

Devitrification of glass around collapsed bubbles in tektites

VIRGIL E. BARNES and RICHARD V. RUSSELL*
Bureau of Economic Geology, The University of Texas, Austin, Texas

(*Received in revised form* 30 *August* 1965)

Abstract—Devitrification was first observed around partly collapsed bubbles (vacuoles) in Santa Mesa philippinites heated by the fire which destroyed Manila near the end of World War II. This localization of devitrification furnishes independent evidence that bubbles in normal tektites were formed by water vapor and that the very low pressure in normal tektite bubbles resulted from the absorption in or reaction of the water vapor with the surrounding tektite glass. Although tektites are essentially anhydrous they do contain a trace of water, in fact more than enough to form all the bubbles found in tektites. Experiments show that a similar amount of devitrification and partial collapse of bubbles takes place in philippinites when heated about 4 days at a temperature of 825°C. Under identical conditions of time and temperature australites, indochinites and javanites showed equivalent amounts of change in shorter times. Bediasites showed no change in 7 days. Neither collapse nor devitrification took place around bubbles in Muong Nong-type tektites, indicating that water vapor was not an important constituent of the gas forming these bubbles and that the pressure in the bubbles approaches equilibrium with atmospheric pressure.

INTRODUCTION

THREE Santa Mesa philippinites were given to the writer by Dr. H. Otley Beyer from a collection stored in a house that burned when Manila was destroyed during the retreat of the Japanese in 1945. Because of the storage of a large volume of inflammable material, the house burned for three days. When the ashes cooled, Dr. BEYER (1961) recovered many tektites ranging from specimens showing little heating effect to some which were changed in color and shape. Those showing heating effects also show devitrification around partially collapsed bubbles.

Heating experiments on fresh specimens reported here give more precise information on the range of time and temperature needed to produce the amount of devitrification found in the philippinites that passed through the Manila fire.

DATA

Tektites heated in Manila fire

Two of the three heated specimens are brown, whereas unheated philippinites are black. They softened slightly, as shown by imprints of objects against which they rested. The third specimen developed a slight brownish cast on one side, became slightly iridescent, but did not change shape. A petrographic examination of millimeter-thick plates prepared from these specimens shows that the two which softened are devitrified around partly collapsed bubbles and to a lesser extent around their peripheries, whereas the third specimen shows no devitrification.

* Present address: Mineralogiska Institutionen, Kungstensgatan 45, Stockholm VA, Sweden.

Table 1. Comparison of selected properties of unheated Santa Mesa philippinites with the properties of Santa Mesa tektites which passed through the Manila fire

	Unheated	Heated		
	Average of 49	A	B	C
Exterior color	Black	Black	Grayish black	Brownish black
Deformation	—	None	Distinct	Distinct
Devitrification*				
Around bubbles	0	0	4	5
Around periphery	0	0	1	3
Refractive index	1·5137	1·5103	1·5162	1·5162
Maximum size in mm				
Lechatelierite	0·22 × 0·59	0·03 × 0·06	0·05 × 0·15	0·2 × 0·5
Spherical bubbles	1·93	0·13	0·04	0·05
Elongate bubbles	1·1 × 2·2	None	0·11 × 0·21	0·2 × 0·3
Abundance*				
Lechatelierite	3·7	1	3	2
Bubbles	5·4	2	1	4
Strength*				
Flow structure	5·0	2	2	3
Flow structure strain	3·8	4	2	3
Overall strain	2·5	1	1	1

* Devitrification, abundance and strength are stated in relative terms from none, 0, to the maximum amount seen in any tektite, 7.

These and various other properties of the three tektites are compared in Table 1 with the properties of average unheated Santa Mesa tektites.

Although the heated specimens show weaker flow structure, flow structure strain and overall strain and a lesser abundance and smaller size of lechatelierite particles and bubbles than unheated specimens, these properties for the heated tektites are all within the range for unheated Santa Mesa tektites. Heating experiments described in the next section show that the lower values for strength of flow structure, flow structure strain and overall strain probably resulted in part from heating.

Bubbles around which devitrification has taken place are unlike bubbles previously seen in any tektite. The bubbles are irregular in shape; the irregularity appears to be greater for large bubbles than for small ones. In addition to the stronger distortion of large bubbles, the glass-bubble interface is papillose and has a minute pattern of wrinkling. The irregularities are of a type that could have resulted from the partial collapse of a strong vacuum bubble causing the bubble-glass interface to wrinkle as the tektite glass softened. The crests of the wrinkles are toward the glass.

Bubble irregularities are illustrated in Fig. 1. The portion of the bubble shown in Fig. 1(a) has a sharp keel, and some of the wrinkles contain grinding sludge which photographs dark. Because of the Lakeside cement filling, this bubble appears light, whereas bubbles entirely encased in glass appear dark (Fig. 1b).

Because all other bubbles of similar volume in the specimen are spheres, it is concluded that the tiny elongate bubble (Fig. 1b) was squeezed off and migrated

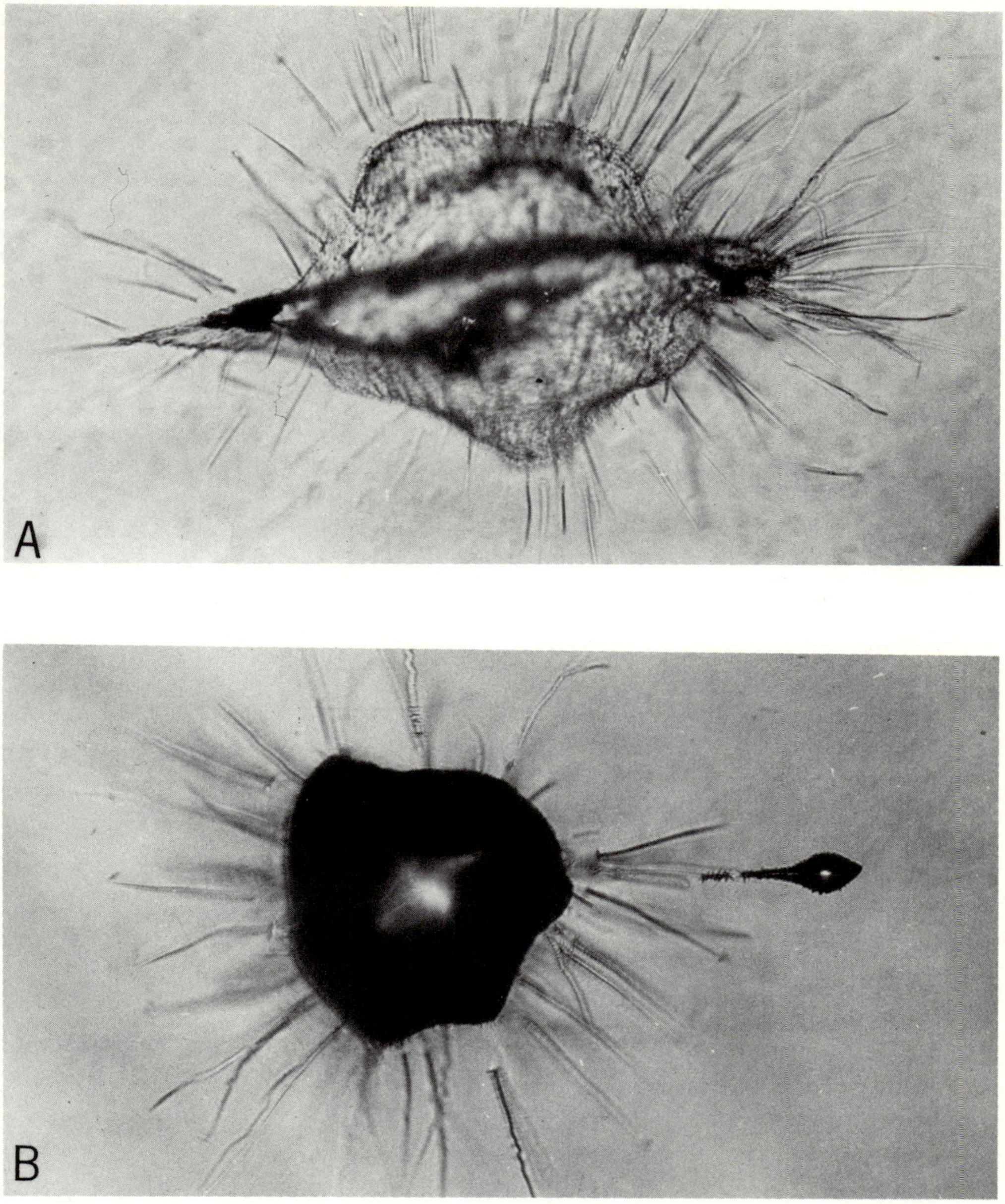

Fig. 1. Devitrification around collapsed bubbles in Santa Mesa philippinite heated in Manila fire, SM-C. Hair- or lath-like crystallites have faint cross-structure.
(a) Irregularly shaped bubble filled by Lakeside cement appears light because the cement enables light from below the tektite plate to be transmitted through the bubble (×220). (b) Bubbles entirely within tektite glass appear black because light from below the tektite plate is reflected by the bubble surfaces. During collapse the small spear-head-shaped bubble at right appears to have separated from the main bubble and migrated (×240).

144

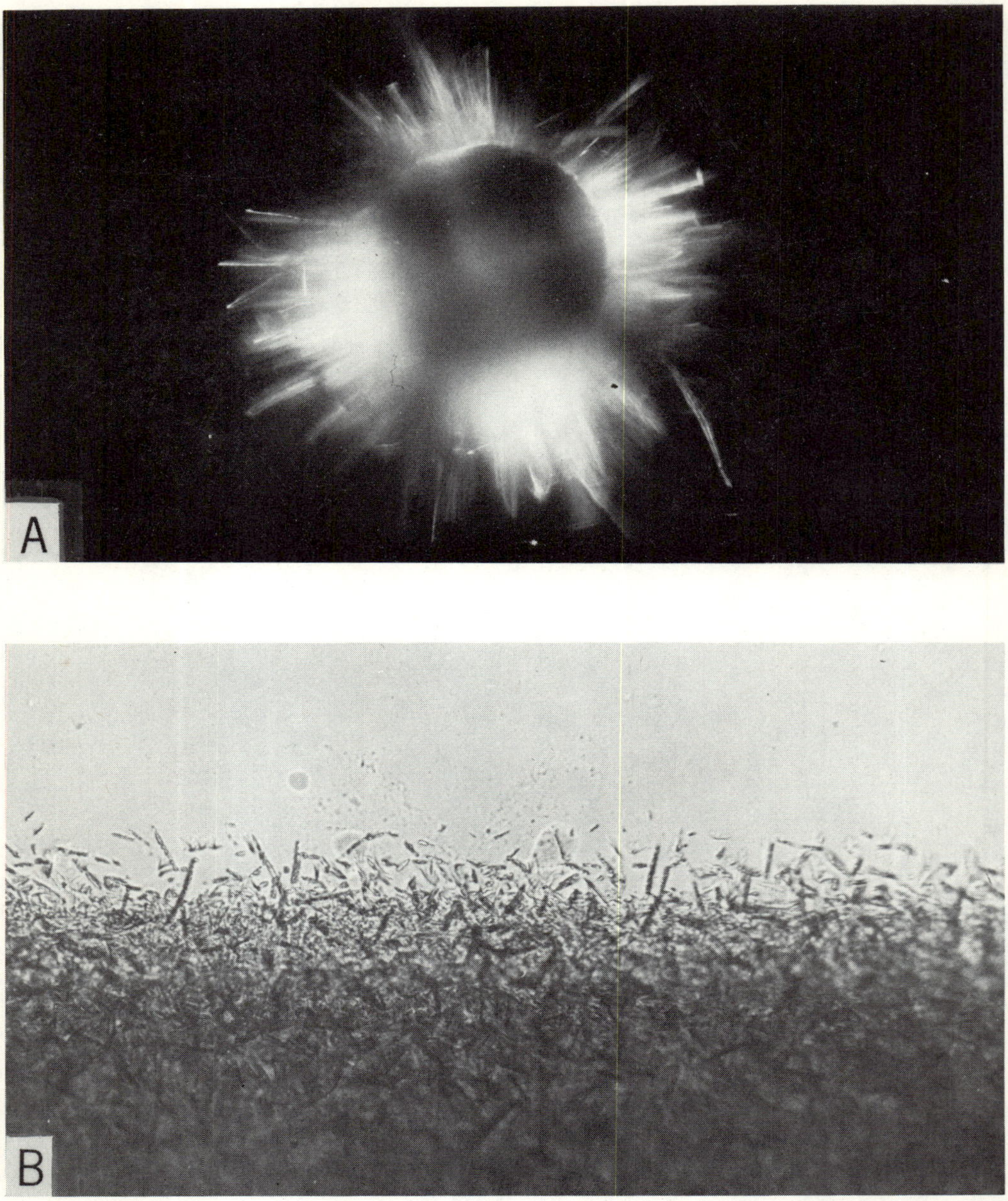

Fig. 2. Devitrification in Santa Mesa philippinite heated in Manila fire, SM-C. (a) Birefringence and extinction cross photographed between cross-nicols was produced by hair- or lath-like crystallites in 0·1 mm zone around slightly collapsed bubble (×200). (b) Devitrified zone 0·02 mm thick at periphery of specimen slopes steeply towards bottom of photograph (×300).

from the main bubble during its collapse. The glass in the vicinity of the bubble was probably fairly soft for migration to have been possible.

The hair-like crystals, in zones about 0·1 mm thick around bubbles, range from almost straight to curved and many have an indistinct structure which gives them a feathery appearance. Most of the hair-like crystals terminate at the bubble-glass interface; however, some appear to terminate away from the interface (Fig. 1b). The hair-like crystals have moderate birefringence and are of higher refractive index than the enclosing glass. Where they are sufficiently numerous an extinction cross appears between crossed-nicols (Fig. 2a). The mineral has not been identified.

A discontinuous devitrification zone about 0·02 mm thick on external surfaces, but not present on cut surfaces, is one-fifth the thickness of the devitrification zone around bubbles. The crystals within the peripheral zone are mostly hair-like and perpendicular to the surface; some, such as those shown in Fig. 2b, appear to be variously oriented prismatic crystals, which may be hypersthene similar to that described by BEALL, MACDOWELL and WOSINSKI (1965).

Tektites heated under controlled conditions

Experiments were of two types, 1. to determine the approximate temperature at which bubbles collapse, and 2. to determine the length of time at this temperature needed to achieve an equivalent amount of collapse and devitrification as found to be present in the Santa Mesa tektites which passed through the Manila fire.

In these experiments, although temperature and time were precisely controlled, it was impossible to be sure that samples were entirely comparable. Even if the same specimen is apportioned to the various runs, individual samples may behave differently upon heating because tektites are not completely homogeneous. In the case of the australites and Santa Mesa philippinites, the specimens were small and individual specimens were used for each run. Although trends were apparent, individual specimens showed evidence of variation in their rates of devitrification.

For the first series of experiments, to determine the approximate temperature at which bubbles collapse, two Santa Mesa Philippine tektites and two layered Muong Nong-type tektites from Muong Nong, Laos were used. The 1200°C temperature used for the first run, based on the experiments by LACROIX (1932) in which he suspended tektites at various temperatures and recorded the amount they sagged, turned out to be much too high. Other runs were made at 1000°C and 900°C, and in both cases the temperature was still so high that the specimens softened and elliptical bubbles became spherical through the action of surface tension. A fourth run at 750°C produced no visible effect except a slight darkening of color. The final run made at 825°C produced crystallization around collapsed bubbles in Santa Mesa tektites comparable to that observed in the Santa Mesa specimens which passed through the Manila fire. The Muong Nong specimens showed no visible effect.

The experimental data are tabulated in Tables 2 and 3. An important fact brought out in Table 2 is that, with increments of temperature, devitrification in Santa Mesa tektites takes place around bubbles at lower temperatures than it does in the body of the tektite glass. This shows that the glass around these bubbles is different and more reactive perhaps from the presence of water. In the

Table 2. Tektite heating data using temperatures of 750, 825, 900, 1000 and 1200° C for 1, 2, 3, 1 and 2 days respectively

Specimen	Nd	Relative abundance: Bubbles	Relative abundance: Lechatel-ierite	Temp., first bubble defor-mation	Temperature, first devitrification: Around bubbles	Temperature, first devitrification: Around lechatel-ierite	Temperature, first devitrification: In body of glass	Degree of devitrifi-cation
Philippinites								
SM—SS	1·5068	6	4	825*	825			0
SM—TT	1·5066	6E†	6	825*	825	900	1000	1
Indochinites								
MN—C	1·5093	6E	1	*		900	900	1
$30850B_1$	1·5110	6E	1	*		900	900	1

Relative abundance and degree of vitrification are stated in relative terms from none, 0, to the maximum amount seen in any tektite, 7. The blank spaces show that no change occurred in 7 days. The philippinites are from Santa Mesa and the indochinites are from near Muong Nong, Laos, the type locality of Muong Nong-type tektites.

* During the run at 900° C all elliptical and deformed bubbles returned to spherical shape because of the action of surface tension as the tektites softened.

† Elliptical bubbles are identified by the letter E.

devitrification process higher water content around bubbles would have acted only as a catalytic agent, the final devitrification character of the glass would have been a function of the bulk composition.

Other important facts shown in Table 2 are: (1) that there is no devitrification around bubbles in Muong Nong tektites, and (2) that bubbles in Muong Nong tektites show no evidence of wrinkling at 825°C, whereas the elliptical bubbles in Santa Mesa tektite SM-TT did wrinkle at this temperature. Bubbles became perfect spheres in both Santa Mesa and Muong Nong tektites at 900°C. The lack of wrinkling at 825°C indicates that bubbles in Muong Nong tektites contain an appreciable amount of gas and the lack of devitrification indicates that the glass around these bubbles is not reactive.

At constant temperature with increments of time, as shown in Table 3, devitrification takes place in glass around bubbles sooner than it does in the body of the glass, and thus the conclusion is supported that the glass around bubbles is more reactive. The data in Table 3 also indicate that somewhat lower temperatures would have been more favorable for the study of javanites and Corning glass X-191-BUZ, and that somewhat higher temperatures will be necessary for the study of bediasites. At a temperature of 850°C and 7 days heating, bediasites were unaffected, and at 825°C and 2 days heating, bubbles in javanites were so encased in crystallites that the bubble-glass interfaces were largely masked from view.

Although it would be informative to know the appropriate temperatures for the study of bediasites and javanites and the results of a series of runs at these temperatures, such studies would be unlikely to provide additional evidence to explain why bubbles in tektites behave the way they do upon heating. However,

Table 3. Tektite heating data using temperature of 825° and 850° C for periods of 2, 3, 6, and 7 days

Specimen	Nd	Relative abundance: Bubbles	Relative abundance: Lechatel-ierite	Day of first bubble defor-mation	Day of first devitrification: Around bubbles	Day of first devitrification: Around lechatel-ierite	Day of first devitrification: In body of glass	Degree of devitrifi-cation
Corning								
X-191-BUZ*	1·5050	4	0				2	7
Javanite								
J-DD	1·5145	2	2	2	2		2	6
Indochinites								
Dalat-Z	1·5057	5	6	2	2	3	3?	5
PD-EE*	1·4910	5E†	2			3	3?	4
Australites								
LW-1	1.5143	6	6					
LW-2	1·5135	3	1	2	2	3?	3	3
LW-3	1·5125	4	4					
LW-4	1·5148	3	4					
Philippinites								
SM-UU	1·5110	3	2					
SM-VV	1·5153	3	6	2	2	6	6	1
SM-WW	1·5128	4	3					
SM-XX	1·5128	4	2					
31212	1·5110	3	3	2	2	3	6?	1
Bediasites								
B-8	1·4922	2	1					0
30763	1·4870	1	0					0
30764–9*	1·5063	6	7					0

Relative abundance and degree of devitrification are stated in relative terms from none, 0, to the maximum amount seen in any tektite, 7. The blank spaces show that no change occurred in 7 days. Specimen J-DD is from Sangiran Dome; Dalat-Z from Dalat, South Viet Nam; PD-EE Muong Nong-type tektite from Phang Daeng, Thailand; LW-1 to 4 from Lake Wilson; SM-UU to SM-XX from Santa Mesa; 31212 from Pugad Babuy; bediasites are from Grimes County, Texas.

* Specimens heated at 850° C.

† Elliptical bubbles are identified by the letter E.

such studies would be a relatively easy way to learn whether or not normal tektites of the various tektite groups contain abnormally high vacuums.

Incidental observations of possible interest to workers making tektite heating experiments are listed as follows:

1. Crystals similar to those described in Muong Nong tektites by Barnes (1964a) are present in runs at 1200°C, specimen $30850B_1$, and at 1000 and 1200°C, specimen MN-C. In the crystal-bearing specimen originally described from Muong Nong, crystals occurred in only a few layers of the many composing the specimen. It is possible, but highly improbable, that out of 10 samples heated, three of the samples heated to the highest temperatures would by chance have contained minerals before they were heated.

2. Faults are present in two samples of the specimen of Muong Nong-type

tektite from Phang Daeng, PD–EE. In the sample heated 3 days at 850°C, the fault is marked by a layer of tiny bubbles. In the other sample, heated for 7 days, fault-breccia blocks are along a portion of the fault and a few crystals of tridymite(?) and abundant spindles of quartz(?) are along the boundaries of the breccia blocks. It is likely that these crystals were present before the sample was heated.

3. Thin stony- or pebbly-appearing films of devitrification enveloped some lechatelierite particles in heated samples. In most, crystallinity was not detected: in a few, definite clear crystals can be seen projecting from the surface of the lechatelierite. Clear unaltered lechatelierite particles are commonly present in samples containing stony-appearing particles. Stony-films first appeared in samples either heated to 900°C or for 3 days except that in one sample stony-films first appeared in the 6-day run. The films serve to show that lechatelierite particles are sharply bounded and not gradational into the surrounding glass.

4. A brown zone around bubbles in Lake Wilson australites heated for 2 days is composed of extremely fine matted brown(?) crystallites in brown glass. With longer heating all of the glass of the tektite turned brown. In tektites from other strewn-fields the zone of devitrification around bubbles is either the same color, lighter colored from white crystallites, or rarely a bleached halo.

After this manuscript was completed, Dr. Dean Chapman, Ames Research Center, Moffett Field, California, sent a plate cut from a Lake Wilson australite which showed devitrification around papillose bubbles. The bubbles are in an initial stage of collapse, devitrification products are sparse, and the plate is without overall strain. It is likely that this specimen was heated in an intense brush fire or camp fire under circumstances which allowed the specimen to cool slowly.

5. Devitrification did not take place around bubbles in lechatelierite particles in any of the experiments. If devitrification took place where bubbles were partly within lechatelierite and partly within the surrounding glass, it occurred only around that portion of the bubble in the surrounding glass.

6. Samples taken from the furnace and allowed to cool rapidly in air were under about the same strain as unheated samples, while samples allowed to cool slowly in the furnace were under much weaker strain than unheated samples, thus confirming that the amount of strain depends mostly on the cooling history.

DISCUSSION

The very low pressure found in tektite bubbles was interpreted by SUESS (1951), and accepted by many other tektite workers, as evidence that tektites solidified where pressure was only a small part of atmospheric pressure on earth. Because it was believed that bubbles with such low pressure could not be formed at atmospheric pressure, it was concluded that tektites are extraterrestrial in origin.

However, although he did not think it likely, Suess mentioned another way in which low pressure might have been developed in tektite bubbles, namely, that the bubbles may have been produced by water vapor which later dissolved in the surrounding glass, leaving a vacuum. The heating experiments described here show that this is the more likely explanation for low pressure in bubbles.

The work of FRIEDMAN (1958) is in agreement with this conclusion. Friedman showed that tektites containing as little as 2 ppm of water will vesiculate slightly upon rapid heating, and concluded that the bubbles in some tektites could have been formed by water vapor which subsequently dissolved in the glass.

Earth materials of composition similar to that of tektites contain water and gaseous constituents which occur interstitially, adsorbed on grains, within vacuoles in minerals, and as constituent parts of minerals. Upon heating, the loosely held interstitial and adsorbed water and gas would be the first to escape, followed by the contents of vacuoles and the volatile material, mostly water, forming constituent parts of minerals. Bubbles formed at the end of the heating cycle therefore would have been chiefly filled by water vapor. As the melt cooled, the water vapor in the bubbles would have been absorbed in or would have reacted with the glass immediately surrounding the bubbles.

The presence of water in glass is known to lower its viscosity and to promote crystallization. Therefore, upon reheating, the water-bearing glass in the vicinity of the bubbles would have softened, thus allowing devitrification to take place. The presence of a thin devitrification zone next to weathered tektite surfaces also indicates that water may be the ingredient promoting devitrification. Tektites were exposed to weathering for many centuries, and it is likely that a thin hydrated zone formed during this time.

The argument might be posed that devitrification has begun around bubbles because of the well known fact that surfaces encourage nucleation. While it cannot be denied that this is a factor, the observation that crystals are absent around Muong Nong-type tektite bubbles and on fresh cut surfaces allows us to conclude that surface nucleation must be coupled with another factor, namely, a higher water content in the glass immediately around the bubble.

In Muong Nong-type tektites the lack of deformation of bubbles during heating indicates that the gas pressure in the bubbles is about in equilibrium with atmospheric pressure, and the lack of devitrification around the bubbles indicates that water vapor probably was not an important constituent of the gas. These conclusions are in agreement with the observations by ZÄHRINGER (1963) who measured a pressure of 40 mm of mercury in the bubbles of a Muong Nong tektite and determined that the gas in the bubbles has a similar composition to the gas in the atmosphere.

If, as chemical evidence strongly suggests, the Muong Nong-type and normal indochinites originated from the same material (BARNES and PITAKPAIVAN, 1962) how is it possible in one case for bubbles to be exceedingly good vacuums and in the other for bubbles to contain normal air approaching equilibrium with the atmosphere? The possibility that these two distinctly different types of tektites formed at different times or in different places is vanishingly remote. Most likely both were formed as part of the same process, and the Muong Nong-type represent unfined or little fined melt while normal tektites represent fined melt.

BARNES (1964b) showed that puddle-like deposits of Muong Nong-type indochinites were faulted, and postulated that at the time of faulting the more fluid melt was thrown into the air as splash forms. In a unidirectional application of intense heat as occurred during the formation of tektites, the temperature would have decreased rapidly with depth below the surface of the substance being melted.

The surficial melt would have been the most fluid and consequently most highly fined, and as fluidity became less with depth, fining also would have decreased in amount. During solar furnace melting experiments this is exactly what happened. The surface material melted and became very fluid the instant the hot spot touched it. Melting progressed with relatively little bubbling, and most of the volatile substances escaped through the porous material in the opposite direction from the heat source as the front of intense melting advanced.

O'KEEFE (1964) demonstrated that the low water content of tektites could not be accounted for by the removal of water in bubbles and postulated that tektites were formed from a dry material. The present work does not prove that solar furnace glass has as low a water content as tektite glass but does demonstrate a mechanism for removing water from a melt without the aid of bubbles. Positive proof would be furnished that this is true if analyses showed that the water content of solar furnace glass is within the range of that for tektite glass.

If tektite melts are formed from porous materials such as soils through unidirectional application of very intense heat the volatile substances, just as with the solar furnace, would be driven before the rapidly advancing front of melting material instead of bubbling through it as happens in the glassmaker's pot. If tektite melts were formed in this manner bubbles would have been of little importance in removing volatile substances, and dry rock as a source material for tektites is unnecessary. It is also unnecessary to postulate that ignimbrites on the moon, if present, are significantly drier than those on earth.

Although superficially resembling the shardy structure of ignimbrites, the shimmery structure of Muong Nong-type indochinites is distinctly different. Ignimbrites are composed of glassy shards, that is, fragments of explosively disrupted glassy bubbles. These shards, mixed with hot gases including water vapor, formed fiery density currents which carried the shards to their place of deposition. The shards settled and became welded to form the rock called ignimbrite. In thin section, the shards of ignimbrites are seen to retain their distinctive shapes. The shimmery structure in Muong Nong-type tektites has a completely different appearance, and the very minute wavy parallel threads are thought to be drawn-out melts of tiny mineral grains.

In the solar furnace melting experiments, no structure equivalent to that of Muong Nong-type structure was formed either with soil or with mixtures of minerals and Corning synthetic tektite glass. In a matter of 10 to 30 sec a glass comparable to normal tektite glass was produced ranging in color from the light green of moldavites to the brown of indochinites. Bubbles were scarce, lechatelierite mostly without bubbles was common, and flow structure was strong.

The Muong Nong-type structure is a peculiar structure and possibly in its appearance gives a clue to its origin. The small amount of lechatelierite present in Muong Nong-type indochinites compared with the amount of quartz present in the soils of southeast Asia is a puzzling feature of these tektites. Is it possible that shock pulverized the quartz grains and that the tiny particles when melted were elongated by subsequent flowage?

That shock did occur is shown by the finding of coesite by WALTER (1965) in a Muong Nong-type tektite from Phang Daeng, Thailand, collected by BARNES and

PITAKPAIVAN (1962). In view of the probably high temperatures involved in the formation of the Muong Nong-type indochinites, the scarcity of lechatelierite in them, and the instability of coesite at high temperatures, it is probable that the coesite-bearing grains were introduced during the accumulation of the melt in puddles after the temperature was considerably lower than at the time the tektites formed. That detrital mineral grains were introduced after the melt formed was postulated previously by BARNES (1963).

Muong Nong-type tektites occur in localized deposits weighing many tens of kilograms. Individual blocks in these deposits weigh up to several kilograms. It seems reasonable that fragments in such deposits originally were parts of individual masses or puddles. The absence of shard structure is evidence that these blocks were not blocks of ignimbrite ejected from the moon by meteorite impact. The chemical composition of the gas and its pressure in the bubbles is evidence that these blocks were not derived from other types of fused moon material, but that they were derived from a terrestrial source.

CONCLUSION

When Santa Mesa tektites passed through the Manila fire bubbles collapsed and glass around the bubbles became partially devitrified. This evidence supported by evidence from heating experiments shows that normal tektite bubbles probably resulted from water vapor, and that high vacuums in the bubbles probably resulted from the absorption in or reaction of the water vapor with the glass surrounding the bubbles. Although this evidence shows that tektites could have formed at atmospheric pressure on earth it is not proof that tektites formed on earth. However, confirmation of the fact that bubbles containing very low pressures could have formed on earth eliminates one of the objections voiced against the terrestrial origin of tektites.

Acknowledgments—Grateful acknowledgment is made to the National Science Foundation for support of this tektite research under grants NSF G-10236 and GP-408 administered by The University of Texas; and to Dr. DEAN CHAPMAN, Ames Research Center, Moffett Field, California, for the Lake Wilson australites used in this research.

REFERENCES

BARNES V. E. (1963) Detrital mineral grains in tektites. *Science* **142,** 1651–1652.

BARNES V. E. (1964a) Petrography of tektites from near Muong Nong, Laos. *Trans. Amer. Geophys. Un.* **45,** 82.

BARNES V. E. (1964b) Terrestrial implication of layering, bubble shape and minerals along faults in tektite origin. *Geochim. et Cosmochim. Acta* **28,** 1267–1271.

BARNES V. E. and PITAKPAIVAN K. (1962) Origin of indochinite tektites. *Proc. Nat. Acad. Sci. U.S.* **48,** 947–955.

BEALL G. H., MACDOWELL J. F. and WOSINSKI J. F. (1965) Devitrification of tektite glass. *Trans. Amer. Geophys. Un.* **46,** 117–118.

BEYER H. O. (1961) Final note on war damage, and summary of the surviving collection, in *Philippine Tektites*, vol. 1, pt. 2, pp. 186–188. Univ. Philippines, Quezon City and Manila.

Friedman I. (1958) The water, deuterium, gas and uranium content of tektites. *Geochim. et Cosmochim. Acta* **14**, 316–322.
Lacroix A. (1932) Les tectites de l'Indochine. *Arch. museum natl. hist. nat. Paris* 6, **8**, 139–240.
O'Keefe J. A. (1964) Water in tektite glass. *J. Geophys. Res.* **69**, 3701–3707.
Suess H. E. (1951) Gas content and age of tektites. *Geochim. et Cosmochim. Acta* **2**, 76–79.
Walter L. S. (1965) Coesite discovered in tektites. *Science* **147**, 1029–1032.
Zähringer J. (1963) K–Ar measurements of tektites, in *Radioactive Dating*, pp. 289–305. IAEA, Vienna.

Physical and Chemical Data for Tektites

II

Editors' Comments on Papers 13 and 14

Because of space limitations, papers on physical properties included in this Part will be limited to three, two of which, concerning the influence of high temperatures in tektite origin, are included in this discussion. Most properties have been listed by Barnes (1967)* with a brief statement and bibliographic references given for each.

Friedman, Thorpe, and Senftle* concluded from magnetic susceptibility data that the tektites studied could not have formed from the materials with which they are now associated. This is not surprising, as the moldavites and bediasites, the two groups discussed, have been extensively reworked into deposits unrelated to their primary sites of deposition.

One of the important questions concerning tektites is whether or not they reflect the composition of the material from which they came. The presence of lechatelierite, schlieren, and other inhomogeneities suggests that tektites did not remain in a molten state sufficiently long for more than minimal changes in composition to have taken place through volatilization. This study supports the conclusion that tektites are little changed in composition over that of their source material.

In opposition to this view Walter (1967) attributes variation in tektite composition to vapor fractionation. He started with material of a Muong Nong-type tektite composition having a silica content of about 82 per cent. It is not invariably true that Muong Nong-type tektites are high in silica; in fact, refractive index studies show that in two out of three deposits of Muong Nong-type indochinites examined, the range of silica content is greater than that of splash-form indochinites in the same area (Barnes, 1971)*.

In the second paper, the viscosity data of Friedman and Parker* give a basis for understanding the bubble-free character of some specimens of Libyan Desert glass. Tektite glass varies in bubble content from specimens that are forthy to others free of bubbles. One of the problems connected with Libyan Desert glass is how some specimens could have been fined to such a high degree. The minute wavy fluidal structure of the glass shows that it was not melted long, yet there are chunks of this glass weighing several pounds which are essentially free of bubbles. Their data, when extrapolated to 2000°C, give 2 min for the time for removal of bubbles larger than 0.2 mm in diameter, which is a reasonably short time for the fining of the glass and retention of fluidal structure.

Dr. Irving Friedman is now with the United States Geological Survey in Denver, Colorado, 80225.

References

Barnes, V. E. (1967). Tektites. *In* International Dictionary of Geophysics, K. Runcorn, ed., vol. 2, pp. 1507–1518, Pergamon Press, Oxford.

*Barnes, V. E. (1971). Description and origin of large tektite from Thailand. *Chem. Erde,* **30,** 13–19.

Walter, L. S. (1967). Tektite compositional trends and experimental vapor fractionation of silicates. *Geochim. Cosmochim. Acta,* **31,** 2043–2063.

13

(Reprinted from Nature, Vol. 187, No. 4743, pp. 1089–1092, September 24, 1960)

COMPARISON OF THE CHEMICAL COMPOSITION AND MAGNETIC PROPERTIES OF TEKTITES AND GLASSES FORMED BY FUSION OF TERRESTRIAL ROCKS

By

Dr. IRVING FRIEDMAN, Dr. ARTHUR THORPE and Dr. FRANK E. SENFTLE

U.S. Geological Survey, Washington, D.C.

IT has been suggested by some investigators[1] that tektites are of terrestrial origin and hence were formed in an atmosphere of composition similar to that surrounding the Earth to-day. In particular, Hawkins[2] has pursued the hypothesis that tektites may be a special form of fulgurite. In order to shed more light on this problem an attempt has been made to determine the changes in chemical composition that silicate rocks would undergo when heated to 2,000–3,000° C., in air in a solar furnace. The fused materials were analysed chemically and magnetically before and after fusion.

The rock samples were heated by placing them at the focal spot of the solar furnace at the U.S. National Bureau of Standards. This unit consists of an 8-ft. square mirror controlled photoelectrically by a heliostat and adjusted to reflect the Sun's light into a 60-in. diameter parabolic mirror of 24-in. focal length. A relatively large specimen is fastened in a holder that can be moved independently around three mutually perpendicular azimuths by means of individual electric motor drives. The operator can view the sample through a small hole in the centre of the parabolic mirror. The hot focus can thus be moved to different parts of the specimen by the motor drives on the sample holder in order to melt an area larger than the focal area. The construction of the furnace makes it necessary to have the sample mounted vertically, and any molten material produced tends to drop off after a relatively short time. This limits the time of heating for those rocks the melts of which are very fluid, such as basalt. In the case of the bentonite and basalt glass, the material dropped off after a few minutes of heating at temperatures estimated to be more than 2,500° C. The more siliceous melts, on the other hand, were far more viscous, and the molten area could be kept hot for periods of 1/2 hr. or more before they dropped off.

The temperature measurements were crude estimates, at best. The melts were far from black bodies, and tended to reflect the Sun into the optical pyrometer, which tended to make the temperature readings high. The value of $2,500 \pm 300°$ C. is thus a rough estimate corrected for reflexion.

Total chemical analyses by the method of Shapiro and Brannock[3] were made on all the unheated and fused specimens. As magnetic measurements have been used to show the state and quantity of iron in glasses (see Senftle and Thorpe[4]), the magnetic susceptibility and saturation magnetization were also measured on all samples before and after heating. The results listed in Table 1 have been reduced to a water-free basis.

In tektites the monovalent alkali oxides (K_2O, Na_2O) are generally lower than the divalent oxides (FeO, MgO, and CaO). It was expected, therefore, that due to loss of volatiles, the chemical composition of the fused rocks might approach the average composition of some of the tektites as shown in Table 2. The loss of silica and alkalis and the concentration of alumina are apparent in most of the samples. However, the changes are small, and one can conclude that much higher temperatures and/or longer periods of heating would be necessary to change the composition of even an obsidian to that of an indochinite, two materials the silica and alumina composition of which are fairly close even before heating.

Another interesting observation is that the ferric–ferrous iron ratio in the fused samples tends to approach the average value of 0·4. The iron in the basalt and rhyolite glasses oxidized whereas that in the obsidian and other more oxidized materials was reduced. The reduction of ferric to ferrous iron in air begins at some temperature between 1,300° C. and 1,400° C. primarily due to the decomposition of Fe_2O_3. Abou-el-azm[5] has examined the change in composition of many silicate glasses with temperature, and has shown that the ferrous iron increases rapidly above 1,400° C. By extrapolation of his results it can be shown that in the region of 1,700° C. the conversion of Fe^{3+} to Fe^{2+} reaches a limiting value. He has also found that if the ferric–ferrous ratio is low (high ferrous iron) initially, part of the ferrous iron tends to be oxidized due to contact of the glass with the air at the surface of the melt, or by oxygen contributed by the other constituents.

2

His work indicates that the melt approaches an equilibrium between Fe^{3+} and Fe^{2+} depending on the available oxygen and the temperature of the melt. Thus, it can be expected that as the temperature is raised above 1,700° C., the ferric–ferrous ratio of fused rock will approach a limiting value for a given oxidizing atmosphere.

In the experimental work reported here the fused rock glass was in contact with 1/5 atmosphere of oxygen at approximately 2,500° C. At this oxygen pressure and high temperature the ferric–ferrous iron ratio appears to approach an equilibrium value of about 0·4. In comparison, the same ratio for tektites is somewhat smaller as shown in Table 2. This difference in the ferric–ferrous ratio between fused rocks and tektites suggests that the tektites were probably heated in a gas phase the partial pressure of oxygen of which was less than 1/5 atmosphere.

If the last heating of the tektites occurred as they passed through the atmosphere, it is possible that the major portion of the heating occurred in the upper atmosphere at oxygen pressures many orders of magnitude less than at the surface—perhaps as low as 10^{-4} to 10^{-6} atmosphere. As the air density increased, the objects were slowed down and the heat produced was lost more efficiently. This is in accordance with the observation that stony meteorites that are heated sufficiently in their passage through the upper atmosphere to form a fusion crust are cool enough to handle by the time they strike the Earth.

The saturation magnetization, σ, of the molten rocks is similar to that reported for tektites. The data in Table 1 show that, after melting, the magnetization dropped to zero (with the exception of the sand specimen from Big Stone Gap, Virginia) in accordance with similar measurements on tektites reported by Senftle and Thorpe[4]. Zero magnetization indicates complete solution of the iron, and one therefore concludes that complete solution of the iron has taken place in these melts within the heating time. The above authors have previously shown evidence that the minimum heating time necessary to dissolve particulate iron oxide is probably of the order of 1 min. for temperatures of the order of 2,000° C. and maybe slightly less for higher temperatures. For the experimental temperatures and chemical composition, the heating time shown here can be considered ample to produce tektite-like glass. The heating period of tektites is probably greater than one minute, but not necessarily larger than fif-

Table 1. Chemical Composition of Rock Melts before and after Heating*

	Sand, Big Stone Gap, Va. 3104–2, 3		Bentonite, Phoenix, Ariz. 3104–6, 7		Basalt glass, Hawaii 67–3, 4617–4		Obsidian, Iceland 3024–1 4617–1		Rhyolite, Thomas Mts., Utah, 4617–3*A*, 3*B*		Rhyolite, New Mexico, 2–2 3104–4, 5	
SiO_2	89·3	87·0	68·7	68·2	51·8	51·8	73·8	74·7	77·3	76·6	76·7	76·7
Al_2O_3	4·5	6·8	18·6	19·0	15·0	15·0	13·7	14·0	12·6	13·2	13·0	13·1
Fe_2O_3	2·5	1·0	1·3	0·5	0·9	2·4	0·7	0·4	0·9	0·3	0·3	0·3
FeO	1·4	2·5	0·00	0·72	10·3	8·8	1·7	1·9	0·05	0·68	0·79	0·68
MgO	0·39	0·59	6·9	7·1	7·4	7·4	0·12	0·13	0·08	1·0	0·04	0·05
CaO	0·77	1·0	1·2	1·1	11·8	11·7	1·0	1·0	0·75	1·5	0·58	0·51
Na_2O	0·21	0·20	2·9	2·8	2·4	2·4	4·8	4·3	3·4	2·9	4·0	4·0
K_2O	1·1	0·92	0·49	0·53	0·56	0·55	4·0	3·5	5·1	4·7	4·6	4·6
$Na_2O + K_2O$	1·31	1·12	3·39	3·33	2·96	2·95	8·8	7·8	8·5	7·6	8·6	8·6
Fe^{3+}/Fe^{2+}	1·63	0·36	0·91	0·62	0·08	0·24	0·37	0·19	16·2	0·40	0·34	0·40
Magnetic susceptibility χ (10^{-6} e.m.u./gm. units)	41	32	1·33	1·23	17·1	16·0	4·75	3·31	0·2	1·1	11·6	1·41
Magnetization, σ (10^{-4} e.m.u./gm. units)	2,640	660	1·20	0	0	0	50·6	0	19·6	0	107	0
Heating time	45 min.		3 min.		5 min.		15–20 min.		15–20 min.		5 min.	

* Analyses performed by Rapid Rock Analysis Group, U.S. Geological Survey.

Table 2. AVERAGE COMPOSITION OF TEKTITES (COMPUTED FROM DATA GIVEN BY BARNES, REF. 6)

	Philippine tektites	Texas tektites	Moldávites	Indochinites	Billitonite and assoc. tektites (not Java)	Australites	Java tektites
SiO_2	71·21	75·64	79·60	73·14	70·50	73·33	73·73
Al_2O_3	12·57	14·59	11·02	12·48	12·53	12·75	11·38
Fe_2O_3	1·47	0·41	0·15	0·19	0·74	0·50	0·83
FeO	4·19	4·00	2·24	4·98	5·27	4·25	4·46
MgO	2·90	1·28	1·30	2·00	3·07	1·98	2·39
CaO	3·19	0·05	1·92	2·51	3·14	3·01	2·49
Na_2O	1·52	1·36	0·51	1·45	1·75	1·25	1·15
K_2O	1·93	1·85	3·00	2·40	2·28	2·05	2·32
H_2O^+	0·37	0·05	0·07	0·28	0·45	0·19	0·25
H_2O^-	—		0·04	0·10	0·09	0·09	0·06
TiO_2	0·89	0·81	0·80	0·94	0·85	0·67	0·87
MnO	0·11	0·01	0·11	0·13	0·16	0·16	0·11
	100·35	100·05	101·39	100·60	100·83	100·23	99·99
Fe^{3+}/Fe^{2+}	0·31	0·09	0·06	0·03	0·13	0·11	0·17
Average magnetic susceptibility χ (10^{-6} e.m.u./gm. units)*	$7{\cdot}0_4$	$5{\cdot}2_9$	$2{\cdot}6_5$	$6{\cdot}3_3$	$7{\cdot}5_0$	$6{\cdot}1_6$	$7{\cdot}3_0$

* See Senftle, F., and Thorpe, A. (ref. 4).

teen minutes, except perhaps for specimens containing a very high silica content, for example, Libyan Desert glass (this glass is not considered to be a true tektite by some investigators).

The change in the value of the magnetic susceptibility is not considered too important from the point of view of the formation of tektites. Its magnitude simply reflects the total iron content, and the observed changes can be explained by the solution of the iron in the glass. Thus, the high susceptibility for the sandstone is due to incomplete solution of the iron in the glass, whereas the high value for basalt glass is due to the high iron content. The rest of the rocks contain less iron than most tektites, and hence their corresponding magnetic susceptibilities are lower than those of tektites.

Table 3 shows the results of the magnetic measurements on the unheated and fused rock in which tektites have been found. The values of the magnetic susceptibility and saturation magnetization are not too unlike that for tektites in general, but the results show that the magnitude of the susceptibility is quite different from that of the tektites found in the area. It indicates that the moldavite contains less iron than the enclosing sand, whereas bediasites contain more iron than the local rock. One cannot help but conclude that tektites are not formed from the local rock in which they are found, for example, by local surface heating or effects of lightning.

In summary, the chemical and magnetic measurements of several types of rocks have been made before and after fusing at 2,500° C. in air at atmospheric pressure. The changes in composition indicate that tektites were either (*a*) melted at temperatures in excess of 2,500° C., or (*b*) heated at this temperature for periods in excess of 20 min. Thus, the small change

Table 3. Comparison of the Magnetic Properties of Tektites and the Glass Made from the Enclosing Rock Formation

Sample	Magnetic susceptibility χ (e.m.u./gm.)	Saturation magnetization (e.m.u./gm.)
Sand from Nechov, Bohemia*	$3{\cdot}35 \times 10^{-6}$	$31{\cdot}5 \times 10^{-4}$
Glass made from sand	$6{\cdot}45 \times 10^{-6}$	0
Moldavite from the same site*	$2{\cdot}04 \times 10^{-6}$	0
Sandstone from Wellborn sandstone of the Jackson group, Texas†	$2{\cdot}46 \times 10^{-6}$	$1{\cdot}2 \times 10^{-4}$
Glass made from sand	$2{\cdot}41 \times 10^{-6}$	0
Bediasites from Texas	$4{\cdot}2$–$5{\cdot}4 \times 10^{-6}$	0

* Specimens collected and contributed by A. J. Cohen, Mellon Institute, Pittsburgh, Pa.

† Specimen collected and contributed by G. S. Hawkins, Boston University Observatory, Boston, Mass.

6

noted in the composition suggests that the observed composition of tektites probably reflects the composition of the material from which they came. The larger ferric–ferrous ratio of the fused rocks compared to that of tektites suggests that the tektites were melted in an oxygen atmosphere the pressure of which was less than that on the surface of the Earth. The changes observed in the magnetic properties indicate that a heating period of 15–20 min. at 2,500° C. is sufficient to produce magnetic properties similar to those of tektites, but does not preclude the possibility of longer periods of heating. Comparison of the magnetic susceptibility of tektites and the melted rock from the same location show that tektites could not be formed from the rocks in which they are found. The above data suggest that tektites were not formed by heating of surface material *in situ* or by lightning.

We are indebted to Dr. Van Valkenburg of the U.S. National Bureau of Standards for use of the solar furnace. We also thank our colleagues who contributed specimens or helped us by their critical discussions.

Publication of this article is authorized by the Director of the U.S. Geological Survey.

[1] Symposium on Tektites, *Geochim. Cosmochim. Acta*, **14** (1958).

[2] Hawkins, G. S., *Nature*, **185**, 300 (1960).

[3] Shapiro, L., and Brannock, W., U.S. Geol. Survey Bull., 1036-*C* (1956).

[4] Senftle, F. E., and Thorpe, A., *Geochim. Cosmochim. Acta*, **17**, 234 (1959).

[5] Abou-el-azm, A., *Trans. Soc. Glass Tech.*, **38**, 101 (1954).

[6] Barnes, V. E., Univ. of Texas Pub., 3945, 477 (1940).

Printed in Great Britain by Fisher. Knight & Co., Ltd., St. Albans.

14

JOURNAL OF GEOPHYSICAL RESEARCH VOL. 74, NO. 27, DECEMBER 15, 1969

Libyan Desert Glass: Its Viscosity and Some Comments on Its Origin[1]

IRVING FRIEDMAN

U. S. Geological Survey, Denver, Colorado 80225

C. J. PARKER

Corning Glass Works, Corning, New York 14830

The viscosity of Libyan desert glass has been determined over the temperature range 975°C to 1090°C. Its viscosity of 10^{13} poises at 1000°C and 10^{10} at 1250°C is approximately 2 orders of magnitude less than that of fused silica, but it is 3½ orders of magnitude greater than that of moldavites and 6½ orders of magnitude greater than that of obsidian. The viscosity has been used to calculate the time-temperature relation necessary to 'fine' the Libyan desert glass and to remove the larger gas bubbles. For the six samples studied, it would take 47 days at 1600°C to remove the bubbles larger than 0.2 mm diameter, the largest diameter found in our samples. The time needed to remove the larger bubbles would be 0.5 day at 1800°C and 2 min at 2000°C.

Clayton and Spencer [1934] published a description of a natural glass found scattered over at least 2000 km² of the Libyan desert. This glass has an unusual chemical composition; an analysis by M. H. Hey and quoted by the above authors is as follows:

SiO_2 = 97.58%, TiO_2 = 0.21%, Al_2O_3 = 1.54%, Fe_2O_3 = 0.11%

FeO = 0.23%, CaO = 0.38%, Na_2O = 0.34%, H_2O (above 110°C) = 0.05%

A water analysis by *Friedman* [1958] gave a value of approximately 0.064%, which is in reasonable agreement with the analysis of Hey.

The origin of this glass has not been determined. On the basis of chemistry, one might infer that the quartz sand, or Nubian orthoquartzite, upon which the glass is found might indeed be the source material from which the glass was formed by fusion.

Fusing material of the composition of Libyan desert glass would have required very high temperatures. Even if a high-temperature source were available to fuse the hundreds of kilograms of material that is present, a further problem remains—that of 'fining' the glass, or removing the trapped gas bubbles.

Manufacturing a large piece (pieces up to 7.4 kg have been collected) of almost pure silica glass practically free of bubbles by direct fusion is difficult, and vacuum fusion of quartz at ~1750°C is usually employed in making clear fused silica bodies.

At the suggestion of members of the staff of the Corning Research Laboratory, the actual viscosity of Libyan desert glass was measured (Figure 1).

For comparison Figure 1 also shows the viscosity versus temperature curves for pure silica glass and for obsidian containing various amounts of water. (Normal-flow obsidian would contain about 0.1–0.2% water by weight.)

Using Stokes' law and the viscosities given in Figure 1, we can calculate the time necessary for bubbles of various sizes to rise to the surface of a mass of 'fluid' glass kept at various temperatures.

$$v = \tfrac{2}{9} r_b^2 \rho g / \eta$$

where v is the velocity (cm/sec) of the rising bubble.

r_b = radius (cm) of the bubble.

ρ = liquid density minus gas density in the

[1] Based on a talk presented at the Third International Tektite Symposium, Corning, New York, April 1969. Publication authorized by the Director, U. S. Geological Survey.

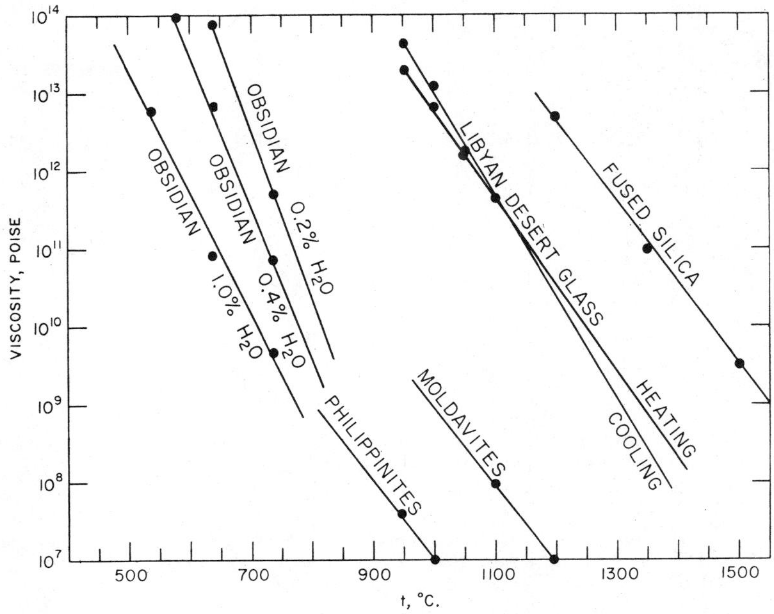

Fig. 1. Viscosity in poises versus temperature in degrees Centigrade for Libyan desert glass, fused silica, moldavites and philippinites (tektites), and obsidian. Data for Libyan desert glass, this paper; fused silica, *Volarovich and Leont'eva* [1936]; moldavites and philippinites, *Hoyte et al.* [1965]; obsidian, *Friedman et al.* [1963].

bubble. (For our calculation we will use $\rho = 2.2$.)

g = acceleration of gravity.

η = viscosity in poises.

If we assume a maximum distance of bubble travel equal to the smallest dimension of the largest piece of Libyan desert glass examined (approximately 2 cm), we can solve for the time necessary to remove bubbles of various sizes.

These times are minimum because the pieces initially were probably larger and the smallest dimension of the pieces examined may not have been parallel to the direction of bubble travel.

Assuming a bubble size of 0.02 cm diameter and a viscosity of 10^7 poises,

$$v = \frac{2/9(10^{-2})^2 2.2(980)}{10^7}$$

$$= 4.8 \times 10^{-9} \text{ cm/sec}$$

$$4 \times 10^8 \text{ sec/2 cm} = \text{approx. 13 yr}$$

Figure 2 is a photograph of a cut and polished surface of a piece of Libyan desert glass. The specimen is about 1 cm thick.

In the six pieces of Libyan desert glass that we examined, the largest bubbles remaining are about 0.02 cm in diameter. If the glass was formed on the surface of the earth by fusion of desert sand or sandstone, and if bubbles of initially larger size than 0.02 cm diameter were present, it would take about 47 days at 1600°C ($\eta = 10^5$ poises) to remove bubbles of diameters greater than 0.02 cm. Extrapolating the viscosity data to even higher temperature, we find that at 1800°C about 0.5 day would be required to remove bubbles larger than 0.02 cm diameter, and at 2000°C, about 2 min.

If the similarity in chemistry between the Libyan desert glass and the desert sands and rocks in the vicinity of the finds (A. Cohen, personal communication) is more than an unusual coincidence, we are faced with the problem of melting and fining large quantities of the sand or rock. The fining problem gives some limits to the time-temperature relations that probably are obtained.

Keeping a mass of glass at, say, 1800°C for days would require either a continuous heat source or initially heated material of sufficient

Fig. 2. Polished slab of Libyan desert glass photographed in oblique light. Note high concentration of small bubbles along left edge but relative absence of such fine bubbles elsewhere in specimen. Sample shown in approximately 1 cm thick.

volume that the interior would have cooled slowly.

The most satisfying picture of the melting process would be to have a very high temperature act for a short time. This would tend to give a shallow zone of melting and fining and a high temperature gradient, resulting in less intermediately fined and partially fused rock than if we assume a lower temperature acting for a longer time. It might be significant in this connection that no partially fused or sintered material was collected during the Clayton expedition in 1932.

Many of the pieces of Libyan desert glasses are elongated fragments, and the bubbles are aligned and often elongated parallel to the long axis of the fragments, indicating linear flowage. In contrast, the flow lines or schlieren observed in most tektites are curvilinear, and the flowage is related, in general, to the external aerodynamic shapes of the tektite specimens.

The small size of the bubbles might be attributed to the small pore size of the original sandstone and to the very low water content of the sandstone before fusion. However, the small number of bubbles in most of the volume of the specimens, of even small size (<0.01 cm), rules against such an explanation. The appearance and distribution of bubbles in the specimens examined (see Figure 2) suggest that the small bubbles were able to merge into larger bubbles and that many of these large bubbles must have risen beyond the limits of the present eroded sample.

O'Keefe [1964] considered the loss of water from tektites. The shape of many tektites indicates aerodynamic shaping, probably in the earth's atmosphere. If we assume that the tektites were formed by impact of an extraterrestrial body on the earth, the fluidity of the tektite is limited, since if it had been very fluid, surface tension would have been inadequate to hold it together against the disruptive accelerating forces as it was thrown from the impact crater. On the other hand, as O'Keefe pointed out, the lack of large bubbles in tektites indicates a low viscosity. This dilemma does not apply to the Libyan desert glass because the lack of aerodynamic shaping and flow banding indicates an origin where rapid travel through a gas (atmosphere) did not occur.

References

Clayton, P. A., and L. J. Spencer, Silica-glass from the Libyan desert, *Mineral. Mag., 23,* 501–508, 1934.

Friedman, I., The water, deuterium, gas and uranium content of tektites, *Geochim. Cosmochim. Acta, 14,* 316–322, 1958.

Friedman, I., William Long, and R. L. Smith, Viscosity and water content of rhyolitic glass, *J. Geophys. Res., 68,* 6523–6525, 1963.

Hoyte, A. F., P. Wirtz, and F. E. Senftle, Electrical resistivity and viscosity of tektite glass, *J. Geophys. Res., 70,* 1985–1994, 1965.

O'Keefe, J. A., Water in tektite glass, *J. Geophys. Res., 69,* 3701–3707, 1964.

Volarovich, M. P., and A. A. Leont'eva, *J. Soc. Glass Technol., 20,* 139, 1936.

(Received June 9, 1969.)

Editors' Comments on Paper 15

A prodigious amount of data on tektites has been accumulated by scientists at Ames Research Center, Moffett Field, California. In addition to the 6000 specific-gravity determinations made for the report reproduced here, another 41,000 specific-gravity measurements have been made from which 530 tektites were selected and analyzed.

Because of space limitations, of the fact that this article is in an available international journal, and because it is digested in the article by Chapman (1971)*, the last thirteen pages exclusive of References have been omitted. Chapman, Larson, and Scheiber* explain, in the material omitted, the simplicity of the specific-gravity method for learning about specific-gravity variation, as well as about variations in Si, Al, and Fe in tektites from locality to locality within a strewn-field and from strewn-field to strewn-field. They conclude that the number of porous specimens present are too few to significantly alter the shape of population polygons (histograms) depicting specific-gravity variations. Refractive index measurements can be used for the same purpose but are much more time-consuming and require mutilation of specimens.

In addition to the data compiled and analyzed in this paper, Dean Chapman and his co-workers have contributed many other data concerning tektites, including one of the most ambitious chemical investigations of tektites ever undertaken. Some of the papers giving results of their experimentation and data gathering are listed with the references to this commentary.

Dr. Dean R. Chapman is now Chief, Thermo- and Gas-Dynamics Division, Ames Research Center, Moffett Field, California, 94035.

References

Centolanzi, F. J., and Chapman, D. R. (1966). Vapor pressure of tektite glass and its bearing on tektite trajectories determined from aerodynamic analysis. *J. Geophys. Res.,* **71,** 1735–1749.

Chapman, D. R. (1960). Recent re-entry research and the cosmic origin of tektites. *Nature,* **188,** 353–355.

Chapman, D. R. (1964). On the unity and origin of the Australasian tektites. *Geochim. Cosmochim. Acta,* **28,** 841–880.

*Chapman, D. R. (1971). Australasian tektite geographic pattern, crater and ray of origin, and theory of tektite events. *J. Geophys. Res.,* **76,** 6309–6338.

Chapman, D. R., Keil, K., and Annell, C. (1967). Comparison of Macedon and Darwin glass. *Geochim. Cosmochim. Acta,* **31,** 1595–1603.

Chapman, D. R., and Larson, H. K. (1963). On the lunar origin of tektites. *J. Geophys. Res.,* **68,** 4305–4358.

Chapman, D. R., and Scheiber, L. C. (1969). Chemical investigation of Australasian tektites. *J. Geophys. Res.,* **74,** 6737–6776.

15

Geochimica et Cosmochimica Acta 1964, Vol. 28, pp. 821 to 839. Pergamon Press Ltd. Printed in Northern Ireland

Population polygons of tektite specific gravity for various localities in Australasia*

DEAN R. CHAPMAN, HOWARD K. LARSON and LEROY C. SCHEIBER

N.A.S.A., Ames Research Center, Moffett Field, California

Abstract—Measurements of specific gravity by the method of liquid flotation have been made on about 6000 tektites from 18 different localities in Australasia, from 1 locality in Texas and 2 in Czechoslovakia. Comparison of specific-gravity population polygons for various localities has led to the unanticipated conclusion that the australite population in southwest Australia is essentially the same as the philippinite population, rather than the population elsewhere in Australia. The javanite population appears closely related to certain populations in Australia. In several localities the presence of two superimposed populations is demonstrated. The specific-gravity evidence indicates that the Australasian tektites represent a single event comprising many distinguishable clusters, some of which have partially overlapped.

INTRODUCTION

AT the turn of this century, the Austrian geologist SUESS (1900) gave the collective name of "tektites" to three widely separated groups of natural glass objects then known to be scattered over parts of Australia, Czechoslovakia and the island of Billiton. In the intervening decades, multitudes of tektites also have been discovered in numerous other areas: in southeast Asia, the Philippines and Indonesia, as well as on the continents of North America and Africa. This gradual revelation that tektites are widely spread has been accompanied by a most curious circumstance. Less difference sometimes is revealed between chemical analyses of two tektites found in different countries several thousand kilometers apart than between two tektites found in the same country only a few kilometers apart. Such circumstances have complicated the problem of ascertaining which of the widely dispersed tektite groups originate from the same event and which originate from different events.

During a field trip in 1962 to the major tektite collections around the world, two of us (DRC and HKL) routinely made measurements of the density of numerous tektites at each locality visited. The specific gravity of a total of 3600 specimens representing 14 different Australasian tektite localities were determined systematically during this trip. Subsequent analysis revealed some interesting patterns interconnecting certain of these tektite groups and some tantalizing gaps in these patterns. In an effort to fill some of the gaps, additional measurements were made in the laboratory on about 2400 tektites, including specimens from four additional Australasian sites as well as sites from Texas and Czechoslovakia. These latter specimens were obtained on loan from various museums and collectors listed subsequently. Thus, the total number of specimens measured is about 6000, and the total number of localities 21.

The chemical similarities between various Australasian tektites are quite

* The loosely used geographical term "Australasia" is employed herein for purposes of brevity to designate the particular global expanse illustrated in Fig. 1.

remarkable, as emphasized in the recent paper of Schnetzler and Pinson (1963). The family-like chemistry of Australasian tektites fortunately is such that an accurate measurement of the specific gravity provides a good measure of the distribution of at least the three principal chemical elements, Si, Al, Fe. An increase in density, for example, is accompanied by a linear decrease in Si and a corresponding linear increase in Al Fe, and the alkalis (Barnes, 1939; Chao, 1963b). When a large number of specimens is considered, the presence of internal bubble cavities introduces some scatter in the correlated data of major element chemistry versus bulk specific gravity, but does not upset the general correlation. It does not follow, however, that the tektite population at two different localities necessarily would be the same if their average density were the same; for one locality may comprise a relatively homogeneous cluster with a narrow range of specific gravity, while the other a very heterogeneous cluster with a wide range, but with the same average value. If, however, the population polygon is the same at two localities, this means that the entire spectrum of number distribution within each band range of specific gravity is the same; hence, the specific-gravity populations would be the same and the statistical distribution of major element chemistry might be expected to be essentially the same.

As emphasized by Chao (1963b), tektites were formed by a rapid mixing of materials: the turmoil of internal striae, the wide compositional variations between different glass inclusions, the iron–nickel spherule inclusions, and certain glass inclusions indicating splatter during penetration, all point toward a picture of tektite formation by violent mixing of more than one type of parent material during an impact process. A population polygon, therefore, is regarded as a reflection of the statistical intricacies of this natural mixing process insofar as concerns the principal rock or mineral components being mixed. Variations in minor chemical components, of course, could not be detected in a population polygon because of their small effect on specific gravity.

The most compelling evidence to date for associating the Australasian tektites as a unit has been provided by potassium–argon age determinations. The results of such measurements have been given by Reynolds (1960), Gentner and Zähringer (1960), and, most recently by Zähringer (1963). In spite of some differences between the numerical values obtained for tektite age in these investigations, it is nevertheless certain that there are at least three distinct age groups: an old group in the United States, a middle-aged group in Czechoslovakia and a relatively young group in Australia, Indonesia, southwest Asia, South China, the Philippines and the Ivory Coast of Africa. The precise subgroupings, if any, of the young tektites are not entirely clear. Thus, the most recent K–A measurements of Zähringer suggest to him that the young tektites may, in turn, comprise two separate divisions: an Ivory Coast group (1·3 m.y.) and an Australasian group (0·7 m.y.). The K–A age may provide only an upper limit on the age of formation since degassing might not have been 100 per cent complete for all tektites during their formation.

The uncertainty in geological age datings of tektites does not enable fine distinctions to be made between Australasian tektite groups. Barnes (1961) has expressed his opinion that the Australasian group itself comprises three or four different events, each of a different geological age. On the other hand, the geological observations of CHAO (1963a) indicate to him that the Australasian tektites possibly are of a single age. It is clear that additional observations are desirable to help resolve existing uncertainties about the ages and groupings of tektites. The principal objective of this investigation is to provide new observational data pertaining to the question of whether the Australasian tektites represent a single or multiple events.

EXPERIMENTAL METHOD AND INTERPRETATION OF MEASUREMENTS

The salient feature of the method employed for specific gravity measurement is that it is especially suitable for making precise relative comparisons between various tektite groups. All measurements were made with one instrument and a single technique, namely that of liquid flotation. This technique provides a high precision essentially independent of the tektite mass. The same balance weights, flotation weights, supporting platinum wire and nulling counterweight were employed throughout all measurements of this investigation, both in the field abroad and in our laboratory. Crystals of zinc iodide were dissolved in tap water and titrated to provide a liquid of density variable in small increments anywhere from 2·6, in which all tektites floated, down to 1·0, in which all sank. By successively immersing each tektite in solutions of incrementally increasing concentration of the zinc iodide (the density of which is measured at each change) until the specimen changed from a sinker to a floater, the specific gravity of each tektite was determined to fall within certain narrow limits. The flotation technique is sufficiently sensitive that variation of $\pm 0{\cdot}001$ g/cm^3 can readily change a floater to a sinker, or vice versa, and the balance employed for measuring the liquid density is sufficiently sensitive that variations of $\pm 0{\cdot}0003$ are readily measured. Hence, for comparison purposes the relative accuracy is about $\pm 0{\cdot}001$. Due to systematic errors, the absolute accuracy would not be as great (being about $\pm 0{\cdot}003$ as determined from calibration by a precision-ground glass sphere of known mass and known diameter); but absolute accuracy is not relevant to our objective of making sensitive relative comparisons—with the same instrument—between the population polygons of tektites from different areas.

The population spectrum of tektite specific gravity at each locality was determined basically as a histogram, but is characterized herein by a normalized population polygon. In most cases, the density of several hundred tektites was measured for each locality; in all cases the number of specimens having density between regular band-width increments of 0·01 g/cm^3 were determined and this number divided by the total for the locality yielded the percentage of specimens falling within each 0·01 incremental range. The data uniformly are plotted as a function of density in such a manner that the ordinate of a point plotted at a density of 2·415, for example, represents the per cent of the total population having density within the 0·01 range between 2·410 and 2·420, while that plotted at 2·425 represents the per cent falling between 2·420 and 2·430 etc. It was also found that the correct incremental density range corresponding to the mode, and a reasonable measure of the percentage corresponding to modal density, could be determined with fair accuracy from as little as 30 specimens randomly selected in areas where the polygon of specific gravity happens to correspond to a relatively uniform population.

Population polygons as reflections of variations in chemistry rather than cavities

A change in tektite specific gravity can be produced by variations either in chemistry or in internal bubble cavities. A population polygon reflects the combined effect of both variations. It has been found that for almost all tektite groups, however, variations in chemistry dominate those of cavities in determining the shape of a polygon. BAKER and FORSTER (1943) came to this conclusion from their studies of australite specific gravity. We have come to the same conclusion from study of other Australasian tektites, with few exceptions, for the reasons outlined in the remainder of this section.

Sizeable cavities appear as irregularities at the lower end of the density range of a given population polygon. This low range corresponds to a relatively small per cent of the total sample, as will be seen. In order to estimate the relative number of tektites containing cavities sufficient in number and size to affect the normalized population polygon, 10 to 20 thin sections about 1 mm thick were sliced from a suite of specimens encompassing the entire density range encountered in each of the localities SA1 (Thailand), P1 (Philippines), and I1 (Java). From observations in transmitted light, an obviously solid portion without bubbles was selected from each slice so that its true glass density could be measured. In this way the minimum glass density, $\rho_{\min}$, was determined at each of these localities. Specimens with bulk density ρ_B less than $\rho_{\min}$ would indicate the presence of sizeable internal cavities, rather than an unusually low glass density. The relative number of specimens with $\rho_B < \rho_{\min}$ can be determined from the population polygon. The results are as follows:

Locality	$\rho_{\min}$, g/cm^3	Per cent of specimens with $\rho_B < \rho_{\min}$
SA1, Thailand	2·399	4·1
P1, Philippines	2·422	5·6
I1, Java	2·378	0·8

In a large sample of tektites from a given locality, the presence of only 5 per cent, or less, with $\rho_B < \rho_{\min}$ would not affect the population polygon by a significant amount compared to the much larger amounts by which the population polygons differ among different tektite localities.

From independent observations, it has been deduced that these circumstances also apply to the australites. In the collection of Dr. GEORGE BAKER (locality A7a) a count was made of the number of fragment specimens exhibiting sizeable bubble cavities visible to the unaided eye on the fractured surface, relative to the number not exhibiting such cavities. (Microscopic vesicles do not affect appreciably the bulk specific gravity, so these need not be considered.) About 4 per cent of these fragments exhibited such cavities; a percentage of the same order as deduced above for the northern localities of the Australasian strewn-field. Consequently, it is to be expected that variations in internal cavities within tektites can have only a minor influence on the shape of the population polygon relative to the chemical variations in silica, alumina and iron; and a comparison between two population polygons is interpreted primarily as a comparison between the statistical variations in chemistry between two tektite clusters.

One exception to these circumstances is represented by a certain class of tektites from Manila Bay. Here a "spongy" type of specimen of irregular shape is found among the normal rizalites. This type comprises about 15 per cent of the total population (BEYER, 1961–1962) and, if included in the population polygon, would broaden considerably the lower range of bulk specific gravity while lowering by somewhat less than 15 per cent the polygon ordinates over the remaining range of specific gravity. This spongy type later is treated separately as a component population with its own population polygon reflecting variations in bubble cavities more than variations in chemistry. It is possible that the large chunky (or Muong Nong type) tektites found in southeast Asia, and the small core type tektites found in Java, may constitute additional exceptions analogous to the spongy variety found near Manila, but we have not investigated this point.

* * * * * * *

REFERENCES

BAKER GEORGE and FORSTER H. C. (1943) The specific gravity relationships of australites. *Amer. J. Sci.* **241,** 377–406.

BAKER GEORGE (1956) Nirranda strewn-field australites, southeast of Warrnambool, Western Victoria. *Mem. Nat. Mus. Vict.* No. 20, 59–172.

BARNES VIRGIL E. (1939) *North American Tektites.* Univ. of Texas Publication, Contributions to Geology, Pt. 2, pp. 477–582.

BARNES VIRGIL E. (1961) Tektites. *Sci. Amer.* **205,** (5) 58–65.

BEYER H. OTLEY (1961–1962) *Philippine Tektites.* University of Philippines Pub., Vol. 1, Pts. I and II, Manila.

CHAO E. C. T. (1963a) *Geological Occurrences of Some Southeast Asian and Australian Tektites.* Paper presented at 44th Annual Meeting, American Geophysical Union, Washington, D.C. April 17–20, 1963, abstract in *Trans. Amer. Geophys. Un.* **44,** (1) 93.

CHAO E. C. T. (1963b) The petrographic and chemical characteristics of tektites, in *Tektites* (Ed. J. O'Keefe), pp. 51–94. University of Chicago Press, Chicago.

CHAPMAN DEAN R., LARSON HOWARD K. and ANDERSON LEWIS A. (1962) *Aerodynamic Evidence Pertaining to the Entry of Tektites into the Earth's Atmosphere.* NASA TR R-134.

CHAPMAN DEAN R. and LARSON HOWARD K. (1963) On the lunar origin of tektites. *J. Geophys. Res.* **68,** 4305–4358.

CHAPMAN DEAN R. (1963) On the unity and origin of the Australasian tektites. *Geochim. et Cosmochim. Acta* **28,** 841–880.

COHEN ALVIN J. (1963) Asteroid- or comet-impact hypothesis of tektite origin: the moldavite strewn-fields, in *Tektites* (Ed. J. O'Keefe), pp. 189–211. University of Chicago Press, Chicago.

GENTNER W. and ZÄHRINGER J. (1960) Das Kalium-Argon-Alter von Tektites. *Z. Naturf.* **15a,** (2) 93–99.

VON KOENIGSWALD G. H. R. (1963) Tektite studies V: strain in tektite glass. *Kon. Ned. Akad. Wet. Proc. Ser. B* **66,** (3) 92–97.

REYNOLDS J. H. (1960) Rare gases in tektites. *Geochim. et Cosmochim. Acta* **20,** 101–114.

SCHNETZLER C. C. and PINSON W. H. JR. (1963) The chemical composition of tektites, in *Tektites* (Ed. J. O'Keefe), pp. 95–129. University of Chicago Press, Chicago.

SUESS FRANZ E. (1900) Die Herkunft der Moldavite und Verwandter Gläser: *Jb. Geol. Reichsansk,* **50,** (2) 193–382.

ZÄHRINGER J. (1963) *K–A-measurements of Tektites.* Paper presented at Second International Symposium on Tektites, Pittsburgh, Penn., Sept. 5–7, 1963.

Editors' Comments on Paper 16

Most of the basic data on tektites available through 1963 are summarized in this paper and a plethora of data published since are briefly reviewed or mentioned in the following pages of commentary. For ease of relating the following comments to the paper, the same headings are used. However, many additional properties not previously investigated are included under new headings.

Size

A Muong Nong-type indochinite weighing 12.8 kg (28 lb) reported from northeastern Thailand (Barnes, 1971)* is probably from Laos (Pitakpaivan, 1972).

Specific Gravity and Refractive Index

Chapman et al. (1964)* measured specific gravities on about 6000 tektites and compared histograms of specific-gravity values, finding distinctive histograms for various areas of tektite occurrence. Refractive index values also give distinctive histgrams and in addition show the unity of the indochinite–philippinite (Barnes, 1963) and the moldavite (Barnes, 1969) strewn-fields.

Magnetic Properties

Booker and Harrison (1966) note that the intensity of magnetization in most tektites is less than 10^{-7} emu/g. After heating in a field of 0.5 oersted, tektites show a remarkable increase in magnetization, which suggests that they were formed in a magnetic field considerably smaller in strength than the earth's field.

Ostertag et al. (1969) concluded from magnetic susceptibility studies that tektites were formed above 1500°C and that glasses which simulate the oxidation state of iron in tektites can be formed by conventional glass-melting procedures in air.

Senftle et al. (1969) found: "With the exception of one low value, the magnetic susceptibility of the normal microtektites measured ranged from 4.7×10^{-6} to 11.7×10^{-6} emu/g; the bottle-green specimens ranged from 5.1×10^{-6} to 14.3×10^{-6} emu/g. In comparison, the Australasian tektites range from 5.6×10^{-6} to 11.2×10^{-6} emu/g."

Fusion Temperature, Cooling Rate, Viscosity

In the last sentence in this section in the reproduced paper, the reference to "Long" should have been "Friedman and Long (1962);" this reference in Friedman (1963) is listed as "Personal communication."

Hoyte et al. (1965), from measurement of the electrical resistivity of tektites from 330°K to 450°K, and with application of an empirical formula developed by Littleton (1933), found that a viscosity curve can be constructed for temperatures between 1370° and 1770°K sufficiently accurate for aerodynamic calculations.

Friedman and Parker (1969)* found that the viscosity of Libyan Desert glass is 10^{13} poises at 1000°C and 10^{10} poises at 1250°C.

Coefficient of Thermal Expansion

Hubbard et al. (1956) found the coefficient of thermal expansion for an australite to be 4.2×10^{6} in the range 100–400°C. In addition to expansion characteristics of australites, the chemical durability, hygroscopicity, and spectral transmittance of australites were examined, as well as the pH response of australite glass electrodes.

Elasticity and Sound-Wave Velocities

Levengood (1966), in determining the internal elastic energy variations in tektites, measured flaw parameters and critical stress values for various tektites using a technique based on surface defect formation. He concluded that these properties "may be related to their geologic age" and that "tektites were produced in a fairly stable environment."

Seeger (1966), in studying sound-wave velocities in tektites and other natural glasses, found that compressional sound-wave velocities vary only by about 6 per cent among all tektite groups, thus giving an additional indication of the unity of origin of the tektites.

Soga and Anderson (1967) used the resonant-sphere technique in measuring the elastic properties of a moldavite and an indochinite. They concluded "that, when these tektites were solidified, they were quenched from temperatures substantially higher than their annealing points."

Included Minerals

Chantret et al. (1967) questioned that the grains of quartz are detrital and suggested instead that they are relicts of primary quartz. Baddeleyite has been identified in a Georgia tektite by King (1966b) and in the Martha's Vineyard tektite by Clarke and Wosinski (1967)*. Glass (1970)* described zircon and chromite in a Muong Nong-type indochinite and listed in addition corundum, rutile, and monazite as being present in another specimen.

Metallic Spherules

Reid et al. (1964)* synthesized nickel-free metallic spherules by melting a Philippine tektite under reducing conditions. Park and Reid (1964) concluded that the presence of a nickel–iron alloy is probably the most valid single criterion of an extraterrestrial origin for small magnetic particles and spherules. Senftle et al. (1964) favored the injection of part of the meteorite body to form the spherules. Brett (1967) found high nickel contents in some spherules of impactites, caused by diffusion of iron from the spherule into the surrounding glass. In tektites the glass around the spherules is not enriched in iron and Brett concluded that the tektites were formed in an atmosphere with very low oxygen pressure. Kleinmann (1969) described magnetite-bearing spherules in tektites.

Major Elements

More than 200 major element analyses of tektites have been published since 1963. Barnes (1967)* and the following publications include a large proportion of these analyses: Bouška and Povondra (1964), Schnetzler and Pinson (1964a), Taylor (1964), Taylor and Sachs (1964), Taylor and Solomon (1964), Chao et al. (1966), King (1966a), Philpotts and Pinson (1966), Taylor (1966), Cuttita et al. (1967), Chapman, Keil, and Annell (1967), Konta and Mraz (1969), and Chapman and Scheiber (1969). Altogether Chapman and Scheiber analyzed 530 Australasian tektites, of which 59 are included in their 1969 paper and the rest are to be published later.

Minor Elements

Numerous minor and trace element analyses of tektites have been published since 1963. Some of the publications listed above, and the following publications, include a large proportion of these data: Annell et al. (1964), Lovering and Morgan (1964)*, Setser and Ehmann (1964), Baedecker and Ehmann (1965)*, Ehmann (1965), Greenland and Lovering (1965), Schnetzler et al. (1966*, 1967), Tanner and Ehmann (1967), Norman and Haskin (1968), Morgan (1969), Pinson and Griswold (1969), Rybach and Adams (1969)*, Schmidt (1970), Ehmann and Rebagay (1970), and Ehmann and Showalter (1971).

Isotopes

Zahringer (1963) reviewed isotopes in tektites. Isotope studies published since include the following: Schnetzler and Pinson (1964b), Tilles (1964), Pinson et al. (1965), Lippolt and Wasserburg (1966), Taylor and Epstein (1966), Taylor et al. (1967), De Laeter and Jeffery (1967), Schnetzler et al. (1967, 1969), Taylor and Epstein (1969), Compston and Chapman (1969), and Wampler et al. (1969)*. Other references on isotopes will be found in the section on age determination.

Water in Tektites

Water has been determined in tektites and other glasses by Gilchrist et al. (1969) using infrared analysis. They found an average water content of 0.012 ± 0.004 wt. per cent, and that glass made in the solar furnace is almost four times as high in water, indicating that tektites were made in a partial to high vacuum. O'Keefe et al. (1964), by electrodeless discharge through a bubble in a tektite, found the spectrum of hydrogen, which they suggested might be from water in the bubbles decomposed by the discharge. O'Keefe (1964) concluded that water contents as low as found in tektites cannot be produced from "ordinary terrestrial materials by the impact of a meteorite or comet."

Gas in Tektites

O'Keefe et al. (1962) found neon, helium, and oxygen in a large tektite bubble and concluded that the source was the atmosphere. Jessberger and Gentner (1971) found that the gas in Muong Nong-type indochinites and Libyan Desert glass is atmospheric in composition and that the pressure in the bubbles is in the 100 mm range.

Earth and Lunar Materials

Possible source materials for tektites have been analyzed chemically, including rocks and glasses in the vicinity of impact features, terrestrial rocks, and lunar rocks and soils (Taylor, 1965, 1966, 1967; Taylor and Kolbe, 1965; Kolbe et al., 1967; Engelhardt, 1967; Compston and Taylor, 1969; Showalter et al., 1972; as well as in some of the references given above).

References

Annell, C., Cuttitta, F., Chao, E. C. T., and Fletcher, J. D. (1964). Minor elements in selected Australasian tektites (abstract). *Trans. Amer. Geophys. Union,* **45,** 81.

*Baedecker, P. A., and Ehmann, W. D. (1965). The distribution of some noble metals in meteorites and natural materials. *Geochim. Cosmochim. Acta,* **29,** 329–342.

*Barnes, V. E. (1967). Tektites. *In* "International Dictionary of Geophysics" (K. Runcorn, ed.) Vol. 2, pp. 1507–1518. Pergamon Press, Oxford.

Barnes, V. E. (1968). Petrography of new tektite localities in Thailand and the Philippines. Rept. 23rd Intern. Geol. Congress, Prague, Vol. 13, pp. 27–36.

Barnes, V. E. (1969). Petrology of moldavites. *Geochim. Cosmochim. Acta,* **33,** 1121-1134.

*Barnes, V. E. (1971). Description and origin of large tektite from Thailand. *Chem. Erde,* **30,** 13–19.

Booker, J. R., and Harrison, C. G. A. (1966). Magnetic properties of tektites (abstract). *Trans. Amer. Geophys. Union,* **47,** 144–145.

Bouška, V., and Povondra, P. (1964). Correlation of some physical and chemical properties of moldavites. *Geochim. Cosmochim. Acta,* **28,** 783–791.

Brett, R. (1967). Metallic spherules in impactite and tektite glasses. *Amer. Mineral,* **52,** 721–733.

Chantret, F., Marion, C., and Pellas, P. (1967). Expanded quartz inclusions in Muong-Nong tektites. 30th Ann. Meeting Meteoritical Society, October 25–27, Ames Research Center, Moffett Field, California.

Chao, E. C. T., Cuttitta, F., Carron, M. K., Annell, C., and Mount, P. (1966). New data on some Ivory Coast tektites (abstract). U.S. Geol. Survey. P. P. 550-A, p. A127.

Chapman, D. R., Keil, K., and Annell, C. (1967). Comparison of Macedon and Darwin glass. *Geochim. Cosmochim. Acta,* **31,** 1595–1603.

*Chapman, D. R., Larson, H. K., and Scheiber, L. C. (1964). Population polygons of tektite specific gravity for various localities in Australasia. *Geochim. Cosmochim. Acta,* **28,** 821–839.

Chapman, D. R., and Scheiber, L. C. (1969). Chemical investigation of Australasian tektites. *J. Geophys. Res.,* **74,** 6737–6776.

*Clarke, R. S., Jr., and Wosinski, J. F. (1967). Baddeleyite inclusion in the Martha's Vineyard tektite. *Geochim. Cosmochim. Acta,* **31,** 397–406.

Compston, W., and Chapman, D. R. (1969). Sr isotope patterns within the southeast Australasian strewn-field. *Geochim. Cosmochim. Acta,* **33,** 1023–1036.

Compston, W., and Taylor, S. R. (1969). Rb–Sr study of impact glass and country rocks from the Henbury meteorite crater field. *Geochim. Cosmochim. Acta,* **33,** 1037–1043.

Cuttitta, F., Clarke, R. S., Jr., Carron, M. K., and Annell, C. S. (1967). Martha's Vineyard and selected Georgia tektites: new chemical data. *J. Geophys. Res.,* **72,** 1343–1349.

De Laeter, J. R., and Jeffery, P. M. (1967). Tin: its isotopic and elemental abundance. *Geochim. Cosmochim. Acta,* **31,** 969–985.

Ehmann, W. D. (1965). On some tantalum abundances in meteorites and tektites. *Geochim. Cosmochim. Acta,* **29,** 43–48.

Ehmann, W. D., and Rebagay, T. V. (1970). Zirconium and hafnium in meteorites by activation analysis. *Geochim. Cosmochim. Acta,* **34,** 649–658.

Ehmann, W. D., and Showalter, D. L. (1971). Elemental abundance trends in the Australite strewn field by non-destructive neutron activation. Proc. NATO Advanced Study Institute "Activation Analysis in Geochemistry and Cosmochemistry," Kjeller, Norway, Sept. 7–12, 1970. Universitetsforlaget, Oslo, Norway.

Engelhardt, W. (1967). Chemical composition of Ries glass bombs. *Geochim. Cosmochim. Acta,* **31,** 1677–1689.

Friedman, I. (1963). The physical properties and gas content of tektites. *In* "Tektites" (J. A. O'Keefe, ed.), pp. 130–136. Univ. Chicago Press, Chicago, Illinois.

*Friedman, I., and Parker, C. J. (1969). Libyan Desert glass: its viscosity and some comments on its origin. *J. Geophys. Res.,* **74,** 6777–6779.

Gilchrist, J., Thorpe, A. N., and Senftle, F. E. (1969). Infrared analysis of water in tektites and other glasses. *J. Geophys. Res.,* **74,** 1475–1483.

*Glass, B. P. (1970). Zircon and chromite crystals in a Muong Nong-type tektite. *Science,* **169,** 766–769.

Greenland, L., and Lovering, J. F. (1965). The variation of iron and manganese in tektites. *Geochim. Cosmochim. Acta,* **29,** 563–567.

Hoyte, A., Senftle, F., and Wirtz, P. (1965). Electrical resistivity and viscosity of tektite glass. *J. Geophys. Res.,* **70,** 1985–1994.

Hubbard, D., Krumrine, E. M., and Stair, R. (1956). Australite (meteoric) glass. *Trans. Amer. Geophys. Union,* **37,** 767–778.

Jessberger, E., and Gentner, W. (1971). Mass spectrometric analyses of gas inclusions in Muong Nong glass and Libyan Desert glass (abstract). *E⊕S (Trans. Amer. Geophys. Union),* **52,** 366.

King, E. A., Jr. (1966a). Major element composition of Georgia tektites. *Nature,* **210,** 828–829.

King, E. A., Jr. (1966b). Baddeleyite inclusion in a Georgia tektite. *Trans. Amer. Geophys. Union,* **47,** 145.

Kleinmann, B. (1969). Magnetite bearing spherules in tektites. *Geochim. Cosmochim. Acta,* **33,** 1113–1120.

Kolbe, P., Pinson, W. H., Jr., Saul, J. M., and Miller, E. W. (1967). Rb–Sr study on country rocks of the Bosumtwi crater, Ghana. *Geochim. Cosmochim. Acta,* **31,** 896–875.

Konta, J., and Mraz, L. (1969). Chemical composition and bulk density of moldavites. *Geochim. Cosmochim. Acta,* **33,** 1103–1111.

Levengood, W. C. (1966). Internal elastic energy variations in tektites. *J. Geophys. Res.,* **71,** 613–618.

Lippolt, H. J., and Wasserburg, G. J. (1966). Rubidium-Strontium-Messungen an Gläsern vom Bosumtwi-Krater und an Elfenbeinküsten-Tektiten. *Z. Naturforsch.,* **21a,** 226–231.

Littleton, J. T. (1933). Critical temperatures in silicate glasses. *Ind. Eng. Chem.,* **25,** 748–755.

*Lovering, J. F., and Morgan, J. W. (1964). Rhenium and osmium abundances in tektites. *Geochim. Cosmochim. Acta,* **28,** 761–768.

Morgan, J. W. (1969). Uranium and thorium in tektites. *Earth Planet. Sci. Lett.,* **7,** 53–63.

Norman, J. C., and Haskin, L. A. (1968). The geochemistry of Sc: a comparison to the rare earths and Fe. *Geochim. Cosmochim. Acts,* **32,** 93–108.

O'Keefe, J. A. (1964). Water in tektite glass. *J. Geophys. Res.,* **69,** 3701–3707.

O'Keefe, J. A., Lowman, P. D., Jr., and Dunning, K. L. (1962). Gases in tektite bubbles. *Science,* **137,** 228.

O'Keefe, J. A., Walter, L. S., and Wood, F. M., Jr. (1964). Hydrogen in a tektite vesicle. *Science,* **143,** 39–40.

Ostertag, W., Erickson, A. A., and Williams, J. P. (1969). Magnetic susceptibility of some synthetic and natural tektites. *J. Geophys. Res.,* **74,** 6805–6810.

Park, F. R., and Reid, A. M. (1964). A comparative study of some metallic spherules. *In* "Cosmic dust." *Ann. N.Y. Acad. Sci.,* **119,** 250–281.

Philpotts, J. A., and Pinson, W. H., Jr. (1966). New data on the chemical composition and origin of moldavites. *Geochim. Cosmochim. Acta,* **30,** 253–266.

Pinson, W. H., Jr., and Griswold, T. B. (1969). The relationship of nickel and chromium in tektites: new data on the Ivory Coast tektites. *J. Geophys. Res.,* **74,** 6811–6815.

Pinson, W. H., Jr., Philpotts, J. A., and Schnetzler, C. C. (1965). K/Rb ratios in tektites. *J. Geophys. Res.,* **70,** 2889–2894.

Pitakpaivan, K. (1972). Personal communication.

*Reid, A. M., Park, F. R., and Cohen, A. J. (1964). Synthetic metallic spherules in a Philippine tektite. *Geochim. Cosmochim. Acta,* **28,** 1009–1010.

*Rybach, L., and Adams, J. A. S. (1969). The radioactivity of the Ivory Coast tektites and the formation of the Bosumtwi crater (Ghana). *Geochim. Cosmochim. Acta,* **33,** 1101–1102.

Schmidt, D. (1970). Zinnspurengehalte in Gesteinen, Mineralen, Tektiten, Einschlagsgläsern Chrondriten und Achrondriten (abstract). *In* "Jahresbericht, Max-Planck-Institut für Kernphysik, Heidelberg," pp. 217–218.

Schnetzler, C. C., Philpotts, J. A., and Pinson, W. H., Jr. (1969). Rubidium–strontium correlation study of moldavites and Ries Crater material. *Geochim. Cosmochim. Acta,* **33,** 1015–1021.

Schnetzler, C. C., and Pinson, W. H., Jr. (1964a). A report on some recent major element analyses of tektites. *Geochim. Cosmochim. Acta,* **28,** 793–806.

Schnetzler, C. C., and Pinson, W. H., Jr. (1964b). Variation of strontium isotopes in tekties. *Geochim. Cosmochim. Acta,* **28,** 953–969.

*Schnetzler, C. C., Pinson, W. H., Jr., and Hurley, P. M. (1966). Rubidiium–strontium age of the Bosumtwi crater area, Ghana, compared with the age of the Ivory Coast tektites. *Science,* **151,** 817–819.

Schnetzler, C. C., Philpotts, J. A., and Thomas H. H. (1967). Rare-earth and barium abundances in Ivory Coast tektites and rocks from the Bosumtwi Crater area, Ghana. *Geochim. Cosmochim. Acta,* **31,** 1987–1993.

Seeger, C. R. (1966). Sound-wave velocities in some tektites and natural glasses. *Trans. Amer. Geophys. Union,* **47,** 144.

Senftle, F. E., Thorpe, A. N., and Lewis, R. R. (1964). Magnetic properties of nickel–iron spherules in tektites from Isabela, Philippine Islands. *J. Geophys. Res.,* **69,** 317–324.

Senftle, F. E., Thorpe, A. N., and Sullivan, S. (1969). Magnetic properties of microtektites. *J. Geophys. Res.,* **74,** 6825–6833.

Setser, J. L., and Ehmann, W. D. (1964). Zirconium and hafnium abundances in meteorites, tektites and terrestrial materials. *Geochim. Cosmochim. Acta,* **28**, 769–782.

Showalter, D. L., Wakita, H., Smith, R. H., Schmitt, R. A., Gillum, D. E., and Ehmann, W. D. (1972). Chemical composition of sawdust from lunar rock 12013 and comparison of a Java tektite with the rock. *Science,* **175,** 170–172.

Soga, N., and Anderson, O. L. (1967). Elastic properties of tektites measured by resonant sphere technique. *J. Geophys. Res.,* **72**, 1733–1739.

Tanner, J. T., and Ehmann, W. D. (1967). The abundance of antimony in meteorites, tektites and rocks by neutron activation analysis. *Geochim. Cosmochim. Acta,* **31,** 2007–2026.

Taylor, H. P., Jr., and Epstein, S. (1966). Oxygen isotope studies of Ivory Coast tektites and impactite glass from the Bosumtwi Crater, Ghana. *Science,* **153,** 173–175.

*Taylor, H. P., Jr., and Epstein, S. (1969). Correlations between O^{18}/O^{16} ratios and chemical compositions of tektites. *J. Geophys. Res.,* **74,** 6834–6844.

Taylor, S. R. (1964). Nickel-rich tektites from Australia. *Nature,* **201,** 281–282.

Taylor, S. R. (1965). Similarity in composition between Henbury impact glass and australites. *Geochim. Cosmochim. Acta,* **29,** 599–601.

Taylor, S. R. (1966). Australites, Henbury impact glass and subgreywacke: a comparison of the abundances of 51 elements. *Geochim. Cosmochim. Acta,* **30,** 1121–1136.

Taylor, S. R. (1967). Composition of meteorite impact glass across the Henbury strewn field. *Geochim. Cosmochim. Acta,* **31,** 961–968.

Taylor, S. R., Erlank, A. J., and Gurney, J. J. (1967). K/Rb ratios in australites. *Geochim. Cosmochim. Acta,* **31,** 953–960.

Taylor, S. R., and Kolbe, P. (1965). Geochemistry of Henbury impact glass. *Geochim. Cosmochim.*

Acta, **29,** 741–754.

Taylor, S. R., and Sachs, M. (1964). Geochemical evidence for the origin of australites. *Geochim. Cosmochim. Acta,* **28,** 235–264.

Taylor, S. R., and Solomon, M. (1964). The geochemistry of Darwin glass. *Geochim. Cosmochim. Acta,* **28,** 471–494.

Tilles, D. (1964). Stable silicon isotope ratios in tektites. *Geochim. Cosmochim. Acta,* **28,** 1015–1017.

*Wampler, J. M., Smith, D. H., and Cameron, A. E. (1969). Isotopic comparison of lead in tektites with lead in earth materials. *Geochim. Cosmochim. Acta,* **33,** 1045–1055.

Zahringer, J. (1963). Isotopes in tektites. *In* "Tektites" (J. A. O'Keefe, ed.), pp. 137–149. Univ. Chicago Press, Chicago, Illinois.

16

TEKTITES. *Introduction.* Tektites are natural glasses whose origin is still in question. Authorities now generally agree that they are the result of some kind of impact, but opinions differ as to whether the impact took place on the Earth, on the Moon, or on another planet. Tektites are found widely distributed in definite areas (strewn-fields) in various parts of the world and came to rest in groups at different times. Most occurrences have been given distinctive names, such as moldavites for those found in Czechoslovakia and bediasites for those found in Texas. Others are named after the place or country in which they were found; for example, australites, javanites, billitonites, indochinites, and philippinites. A few groups, such as the Ivory Coast tektites, Libyan Desert glass, Georgia tektites, and the lone specimen from Martha's Vineyard, are not specifically named. Macusanite, formerly called americanite, is obsidian and not tektite as originally thought.

The various groups of tektites have $^{40}K - ^{40}A$ ages ranging from 34 m.y. (million years) in the case of the Texas bediasites of Eocene age and about 15 m.y. for those of Miocene age from Czechoslovakia to 0·7 m.y. for the Pleistocene tektites of Australia and southeast Asia. From geological considerations the Australian tektites appear to be much younger than 0·7 m.y.

Libyan Desert glass and Darwin glass are here treated with the tektites because they have characteristics in common with tektites and there is no clear way to exclude the two glasses. However, some workers exclude both glasses from the tektites and include them with the impactites.

True impactites are associated with craters presumably produced by meteorite impact, and usually meteorites are present near the craters. An exception is the Aouelloul crater around which impactite glass is found, but around which, so far, no meteorites have been found. Impactites associated with meteorites are found around al Hadida (Wabar) craters, Saudi Arabia; Meteorite Crater, Arizona; and Henbury craters, Australia.

Physical Characteristics of Tektites

Physical characteristics discussed herein include size, colour, hardness, specific gravity, refractive index, magnetic properties, absorption spectra, fusion temperatures, cooling rate, viscosity, and internal features best observed with a petrographic microscope. The internal features furnish clues to the character of the material from which tektites formed, as well as information about the manner in which they formed (Barnes 1940, 1958, 1960, 1962a, 1963b, 1963c; Chao 1963). Some of the more characteristic internal features are glassy particles, bubbles, flow structure, layering, strain, and "fingers". Less commonly seen features include rayed bubbles, faults, included detrital minerals, metallic spherules, a zirconium-rich inclusion, and minerals which have grown *in situ.*

Size. Tektites are most often about the size of a walnut, but they range from considerably smaller to as large as a grapefruit. They may be in the form of spheres, tear-drops, dumb-bells, rods, disks, flanged buttons, or may be deeply furrowed and pitted from corrosion (Baker 1963). Wind-abraded masses of Libyan Desert glass weigh as much as 7·4 kg, and some of the fragmented, layered puddles of indochinites originally may have weighed several hundred kg. Layered tektites of this type were first described by Lacroix (1935) from near Muong Nong, Laos. Such layered tektites are referred to as Muong Nong-type tektites.

Colour. Most tektites appear to be black. When held up to the light their translucency is mostly visible on thin edges, and they are generally yellowish brown. Those from Bohemia, however, are a clear light green; those from Moravia are greyish green; and some from the Libyan Desert are a clear straw yellow.

Hardness. Tektite hardness ranges between 6 and 7 on Mohs' scale (Baker 1959).

Specific gravity and refractive index. The two most commonly determined physical characteristics of tektites are specific gravity and refractive index. The compilation in Table 1 combines data tabulated by Chao (1963) and unpublished data in the writer's files. Specific gravity and refractive index show a direct relationship; that is, the higher the refractive index, the higher the specific gravity. Both of these properties show an inverse relationship to silica; that is, the higher the specific gravity and refractive index, the lower the silica. Most of the specific gravity measurements are bulk measurements and may not represent the true specific gravity of the tektite glass if bubbles are present. For this reason refractive index, which is not influenced by bubble content, indirectly gives the most accurate view of variation of silica content in tektites.

Philippinites have the highest average refractive index of all tektite groups, followed closely by australites and javanites and at more widely spaced intervals by normal indochinites, bediasites, moldavites, Darwin glass, and Libyan Desert glass.

Magnetic properties. The magnetic susceptibility of tektites as reported by Sentfle and Thorpe (1959) and Thorpe *et al.* (1963) is summarized in Table 2. Most obsidian shows measurable intensity of magnetization, whereas tektites do not. By heating obsidian to 1450°C, the intensity of magnetization is reduced to zero, indicating by analogy that tektites were heated to a temperature higher than this.

Friedman (1963) summarized the work of Olte and Siegel (1961) on the electromagnetic constants of tektites. Australites, indochinites, philippinites, and moldavites gave ε^1/μ values between 42×10^{-6} and 52×10^{-6}. Dielectric constants for these tektites were found to be between 6 and 7·4; the Libyan Desert glass gave a distinctly different value of 4·2.

Absorption spectra. Friedman (1963) in reviewing absorption spectra reported that the spectral range has been investigated from 0·3 microns in the ultra-violet through the visible to 26 microns in the infra-red.

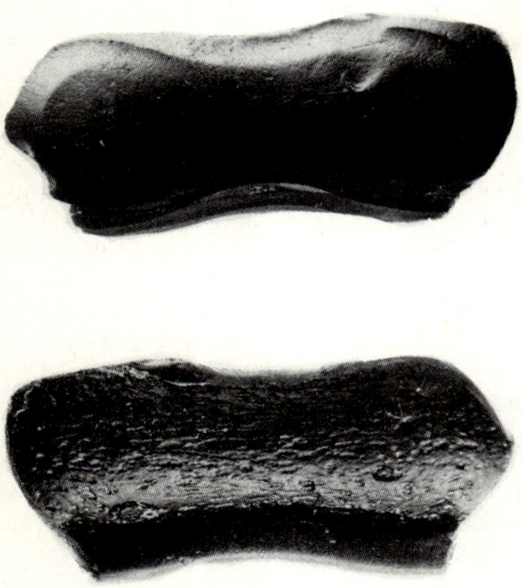

A

B

Fig. 1. Very slightly etched australite dumb-bell shows effects of ablation (X 1·7).
A. Ring waves on forward side formed during flight of tektite through atmosphere. Pit (upper right) resulted as softened glass collapsed into a bubble cavity.
B. On posterior side original rough surface of tektite contrasts sharply with smooth secondarily melted glass in flange.

Table 1. Specific gravity and refractive index of tektites

	Specific gravity			Refractive index		
	Range	Ave.	No of specimens	Range	Ave.	No. of specimens
Australites	2·31–2·51	2·412	1,141	1·4970–1·5190	1·5117	119
Bediasites	2·329–2·433	2·377	469	1·4828–1·5150	1·4954	138
Darwin glass	2·275–2·296	2·288	4	1·4740–1·4915	1·4806	10
Indochinites						
*Chao (1963)	2·396–2·461	2·436	289	1·4972–1.5148	1·5076	58
Normal				1·4958–1·5148	1·5052	189
Muong Nong-type				1·4850–1·5110	1·4979	99
Javanites	2·396–2·495	2·437	68	1·5000–1·5168	1·5091	45
Libyan Desert glass		2·208	1	1·4600–1·4642	1·4616	7
Moldavites	2·303–2·385	2·347	73	1·4798–1·4987	1·4903	117
Philippinites	2·390–2·467	2·444	520	1·5038–1·5204	1·5126	251

* Chao's tabulation does not distinguish between normal and Muong Nong-type indochinites, and for this reason is shown separately.

Table 2. Magnetic susceptibility of tektite, impactite, and igneous glasses

	No. of specimens	Magnetic susceptibility e.m.u./g ($\times 10^{-6}$)		Intensity of magnetization e.m.u./g ($\times 10^{-4}$)
Australites	7	5·8–6·9	6·1	0
Bediasites	26	3·7–7·9	5·6	0
Georgia tektite	1		3·6	0
Indochinites	10	5·1–7·8	6·6	0
Ivory Coast	1		8·6	
Javanites*	10	6·7–7·8	7·3	0
Libyan Desert glass	1		−0·20	0
Moldavites	2	2·3–3·0	2·6	0
Philippinites	18	6·2–8·6	7·0	0
Aouelloul (impactite)	1		3·8	0
Macusanite (obsidian)	1		0·41	3·8

* Includes two billitonites.

Fusion temperature, cooling rate, viscosity. Fusion temperatures of tektites as tabulated by Baker (1959) range from 1055°C for billitonites to 1450°C for Darwin glass. Tektites are resistant to thermal shock.

Hammond (1950) found upon examining the strain characteristics of two moldavites that they cooled through the range between 700°C and 600°C at a rate exceeding 50°C/min.

The viscosity of an indochinite was determined by Volarovich and Leontieva (1939) to be $1{\cdot}06 \times 10^9$ poise at 950°C and $1{\cdot}22 \times 10^5$ poise at 1410°C. A moldavite gave values of $1{\cdot}7 \times 10^9$ poise at 908°C and $2{\cdot}3 \times 10^5$ poise at 1410°C. Friedman (1963) and Long reported that a bediasite has a viscosity about two orders of magnitude greater than rhyolite obsidian at the same temperature.

Fig. 2. Moderately etched platy indochinite from Dalat, South Viet Nam (X 1·2).

Glassy particles. Glassy particles isolated from bediasites were determined to be lechatelierite on the basis of their refractive index (Barnes 1940). Similar particles have been found in tektites from all other groups, as well as in glass produced by known sources of sudden and intense heat and rapid cooling, such as impactites, glass produced by atomic bomb explosion, glass produced by the arcing of power lines through the soil, and in the thin tubes and films of glass produced when lightning strikes the Earth. Lechatelierite is not present in obsidian.

Other glassy particles of refractive index intermediate to that of tektite glass and lechatelierite, and mostly larger than lechatelierite particles, are also found in a few tektites. Such glassy particles could be readily explained if tektites containing such particles were formed from sandy shales containing aggregated clumps of quartz grains (sandstone intraclasts).

The abundance, size, and shape of lechatelierite particles, and the abundance of bubbles associated with the lechatelierite, are of importance in interpreting the history of a tektite glass. These characteristics of tektites are dependent on the temperature of the melt and the length of time it remained molten; the hotter the melt and the longer it remained molten, the fewer and smaller are the lechatelierite particles, and the less bubbly the lechatelierite.

In Table 3 the abundance of lechatelierite is stated in a relative manner, assigning a numerical value to the abundance of particles from 0, if lechatelierite is absent, to 7 for the specimens containing the most particles. Lechatelierite particles are least abundant in australites and in moldavites from Moravia and most abundant in Aouelloul crater glass, Darwin glass, and moldavites from Bohemia. The indochinites as a group are rich in lechatelierite and the bediasites, javanites, and philippinites have an intermediate amount.

Barnes and Pitakpaivan (1962) postulated that the bubbles associated with lechatelierite are formed from the

*Table 3. Relative abundance of lechatelierite and bubbles, and relative strength of flow structure, flow structure strain, and over-all strain in tektites and impactite glasses**

		Relative abundance		Relative strength		
	No. of specimens	Lechatelierite particles	Bubbles	Flow structure	Flow structure strain	Over-all strain
Australites	80	1·9	1·8	4·7	4·7	4·8
Bediasites	64	3·0	2·4	4·9	4·9	1·7
Darwin glass	7	5·9	6·4	3·7	4·4	0·1
Indochinites						
Normal type	158	3·9	3·6	5·0	4·3	4·0
Muong Nong-type	29	2·2	5·6	3·5	3·5	0·0
Javanites	29	2·7	3·3	5·0	4·3	1·7
Libyan Desert glass	5	7·0	4·2	3·8	3·8	0·0
Moldavites	53	4·8	4·2	4·7	5·7	3·5
Bohemia	48	5·1	4·4	4·9	5·7	3·4
Moravia	5	1·4	2·4	2·8	5·2	4·8
Philippinites	93	3·2	4·4	3·9	3·6	3·7
Aouelloul (impactite)	7	6·1	6·3	6·4	4·6	0·0

* The higher the number, the greater the abundance or strength.

gas- and liquid-filled vacuoles which are so common in quartz, and that a study of these bubbles may give additional information on the relative temperature of formation of the various tektite groups. Experiments with solar furnace melts of quartz and synthetic tektite glass in air at atmospheric pressure and temperatures up to about 2000°C show that lechatelierite free of bubbles formed at the highest temperatures, bubbly lechatelierite formed at intermediate temperatures, and frothy lechatelierite in which the gas in the vacuoles expanded but did not coalesce formed at the lowest temperatures.

The bubble characteristics of lechatelierite are given in Table 4. On the basis of frothy lechatelierite content, the Muong Nong-type glass formed at the lowest temperatures, followed by Aouelloul crater glass and Darwin glass at slightly higher temperatures. Only a trace of froth has been found in normal indochinites, philippinites, and bediasites, and none has been found in australites, javanites, or moldavites.

Bubbly lechatelierite, because of its wider distribution, is an even better indicator than frothy lechatelierite of the relative temperature of formation of the various groups. Very few specimens of australites and bediasites contain bubbly lechatelierite, showing that they formed at the hotter end of the range. About one-third of the philippinites, moldavites, and javanites contain bubbly lechatelierite, showing that they formed at intermediate temperatures. Toward the lower end of the temperature range, most of the lechatelierite in indochinites, Aouelloul crater glass, and Darwin glass is bubbly. At still lower temperatures, the lechatelierite of the Muong Nong-type glass became frothy but was not hot enough for the froth to coalesce to form bubbles.

Fig. 3. Deeply etched bediasite from Grimes County, Texas (X 1).

Bubbles. Bubbles are found in all tektite groups and are of two categories: those associated with lechatelierite particles (discussed in the previous section) and those distributed throughout the body of the glass. The abundance, size, and shape of bubbles distributed throughout the tektite glass are also of importance in interpreting the history of the glass. As a melt becomes hotter it becomes more fluid and more of the vaporization products escape, resulting in retention of fewer and smaller bubbles. Also, the hotter a fluid, the more likely that the bubbles will be spherical. Irregular bubbles indicate that the melt did not become sufficiently fluid for surface tension to draw the bubbles into spheres; elliptical bubbles indicate that the melt flowed as it cooled.

The abundance of bubbles is stated in Table 3 in a relative manner. Bubbles are least abundant in australites, bediasites, and Moravian moldavites, most abundant in Darwin glass, Aouelloul crater glass, and Muong Nong-type indochinites, and of intermediate abundance in philippinites, Bohemian moldavites, Libyan Desert glass, normal indochinites, and javanites. Those bediasites which are free of bubbles are also free or nearly free of lechatelierite, are highly translucent, and are found mostly in the southeastern area of their distribution.

The relative number of specimens containing elliptical bubbles and those in which all bubbles are spherical are shown in Table 4. In bediasites, all bubbles are spheres; in Libyan Desert glass, all are elongate forms. The bubbles in australites, javanites, and philippinites are mostly spherical, and those in Muong Nong-type tektites and moldavites are mostly elliptical, whereas in Darwin glass and in normal indochinites the two kinds of bubbles are fairly evenly distributed. The irregularly-shaped pointed bubbles found in some layered indochinites and in Libyan Desert glass represent residual pore space in material that barely melted.

Flow structure. The relative strength of flow structure is given in Table 3. Philippinites, Darwin glass, and Muong Nong-type glass have somewhat weaker than average flow structure and that in the Moravian tektites is weaker still. Flow structure in Aouelloul crater glass is very strong.

Flow structure strain. Flow structure strain in tektites results from the difference in coefficient of expansion of adjacent layers of tektite glass. Strain developed as the glass cooled. The relative strength of flow-structure strain, as observed in mm-thick plates between crossed nicols, is given in Table 3. Moldavites have above average flow-structure strain; philippinites and Muong Nong-type glass are below average.

Over-all strain. The over-all internal strain present in most normal tektites is exactly the kind of strain a small molten body would be expected to have upon rapid cooling. Because the heat flows outward, the first part of a molten body to solidify is the surface. The inside is still liquid and expanded. As it cools it shrinks, and since the outside is rigid, tensional strain develops.

The relative strength of over-all internal strain is given in Table 3. Over-all strain is absent in the chunky, layered Muong Nong-type tektites, Aouelloul crater glass, and Darwin glass. Both bediasites and javanites have weak over-all strain.

"Fingers." "Fingers" in tektites were first seen in australites and described by the author (1961). They are light-coloured, finger-like inward extensions up to several millimetres in length and are discernible as protuberances on the posterior surfaces of occasional australites. They are of lower refractive index and more highly siliceous than the enclosing tektite glass. "Fingers", and schlieren with some of the properties of "fingers", are relatively abundant in australites, common in indochinites and philippinites, and so far have been found in

Table 4. Shape of bubbles and bubble characteristics of lechatelierite in tektite and impactite glasses

	No. of specimens	Bubbles		Lechatelierite	
		Spherical (per cent)	Elliptical or irregular (per cent)	Bubbly (per cent)	Frothy (per cent)
Australites	80	99	1	6	0
Bediasites	63	100	0	6	0
Darwin glass	7	43	57	71	57
Indochinites					
Normal type	149	40	60	87	4
Cambodia	31	58	42	45	10
China	5	60	40	100	0
Thailand	93	35	65	68	1?
Vietnam	20	30	70	90	10
Javanites	29	97	3	28	0
Libyan Desert glass	5	0	100	0	0
Moldavites	53	19	81	34	0
Bohemia	48	15	85	38	0
Moravia	5	60	40	0	0
Philippinites	93	82	18	37	3
Muong Nong-type*	39	8	92	28	90
Aouelloul (impactite)	7	14	86	86	71

* Includes 4 from Cambodia, 5 from Vietnam, 29 from Thailand, and 1 bediasite.

Table 5. Mean composition and range in composition of major constituents in tektites

	Libyan Desert glass (1)	Darwin glass (4)		Martha's Vineyard tektite (1)	Georgia tektite (1)	Moldavites (14)		Bediasites (25)		Indochinites (52)	
		Mean	Range			Mean	Range	Mean	Range	Mean	Range
SiO_2	97·58	87·98	86·34–89·81	80·6	80·54	79·01	74·91–82·68	76·56	71·89–81·31	73·68	70·58–81·36
Al_2O_3	1·54	7·04	6·13–8·00	11·3	11·21	11·09	9·44–13·80	13·69	10·96–17·56	12·47	8·87–14·39
Fe_2O_3	0·11	0·36	0·19–0·63	0·4	0·33	0·30	0·14–2·05	0·23	0·00–0·77	0·32	0·03–0·82
FeO	0·23	1·53	0·89–2·08	2·2	2·40	2·15	1·13–3·36	3·74	2·29–5·26	4·52	2·81–5·63
MgO	tr	0·76	0·58–0·92	0·7	0·65	1·49	0·22–2·20	0·63	0·37–0·95	2·05	1·11–2·93
CaO	0·38	0·07	nil–0·17	0·7	0·61	2·08	1·16–3·74	0·64	0·41–0·96	2·17	1·00–3·48
Na_2O	0·34	0·11	0·01–0·15	1·1	1·16	0·52	0·26–0·78	1·52	1·20–1·84	1·42	0·90–2·00
K_2O	none	1·07	0·87–1·36	2·4	2·38	3·04	2·20–3·76	2·07	1·60–2·43	2·39	1·84–3·16
TiO_2	0·21	0·78	0·51–1·24	0·5	0·43	0·69	0·31–1·40	0·76	0·53–1·05	0·82	0·47–1·03
MnO	tr	0·00	tr–0·01	0·05	0·05	0·08	0·02–0·18	0·04	0·01–0·07	0·11	0·06–0·32
Total	100·49	99·70		99·9	99·78	100·45		99·88		99·95	

	Australites (32)		Billitor ites and javanites (14)		Ivory Coast (3)		Philippinites (21)		All tektites (168)
	Mean	Range	Mean	Range	Mean	Range	Mean	Range	Range
SiO_2	73·06	68·91–79·51	71·32	68·30–76·40	71·05	68·00–76·56	70·87	68·90–72·10	68·00–97·58
Al_2O_3	12·23	9·36–15·42	12·04	9·86–13·50	14·60	11·54–16·46	13·48	12·08–15·23	1·54–17·56
Fe_2O_3	0·60	0·23–1·48	0·77	0·06–2·25	0·18	0·17–0·18	0·79	0·50–2·03	0·00–2·25
FeO	4·14	3·11–5·30	5·10	3·17–6·81	5·51	3·99–6·46	4·30	3·03–5·32	0·23–6·81
MgO	2·04	1·35–2·49	3·18	2·38–4·96	3·29	2·88–3·60	2·67	2·23–3·65	tr–4·96
CaO	3·38	1·48–5·10	2·95	2·22–3·92	1·67	1·40–2·00	3·14	2·50–3·97	nil–5·10
Na_2O	1·27	0·91–1·84	1·55	0·77–2·46	1·71	1·32–2·35	1·41	1·18–1·76	0·01–2·46
K_2O	2·20	1·25–2·56	2·19	1·57–2·76	1·53	0·82–1·92	2·31	1·69–2·56	0·00–3·76
TiO_2	0·68	0·08–0·90	0·78	tr–1·10	0·73	0·60–0·80	0·83	0·63–1·04	tr–1·40
MnO	0·12	tr–0·42	0·14	0·08–0·32	0·08	0·06–0·09	0·09	0·08–0·16	tr–0·42
Total	99·72		100·02		100·35		99·89		

one moldavite. Dark "fingers", higher in refractive index than the enclosing glass, recently have been found in indochinites and in one bediasite. These "fingers" mostly end in pits at the surface of the tektites.

Rayed bubbles. Rays are patchily distributed around bubbles in a few indochinites and in one moldavite (Barnes 1962b). The rays mostly have a refractive index in the range 1·49 to 1·50, are up to about 2 mm in length, and were shot outward from the bubbles into the surrounding glass, as is shown by distortion of lechatelierite particles near which they pass.

Faults. Layers are offset along faults in Muong Nong-type tektites from both Kan Luang Dong, Thailand (Barnes 1963a) and Muong Nong, Laos (Barnes 1964a). Layering is probably the result of the accumulation of melt in puddles. Evidence that the faulting took place while the melt was very hot is: 1. perfect welding of the fault-fractures with detrital grains in the Kan Luang Dong specimens equally embedded in the glass on either side of the welds, 2. growth of minerals along the faults in the Kan Luang Dong specimens, and 3. warping of the fault planes in specimens from both localities by movement of the melt following faulting.

Included detrital minerals. In addition to detrital mineral grains along faults, abundant detrital mineral grains not associated with faults have been found in layered tektites from Nong Sapong, Thailand (Barnes 1963c). The detrital grains are quartz and occur in bubbly layers rich in tiny lechatelierite particles and frothy lechatelierite, alternating with layers which are free of mineral grains and contain scant lechatelierite and bubbles. Thus, the grains appear to have entered the melt while it was in the process of flowing, cooling, and becoming layered. Some of the detrital grains have been changed to coesite, a high-pressure form of silica (Walter 1965). The quartz grains in the associated soil and rock are of about the same grain size and angularity as mineral grains in the tektite layers and along faults.

Metallic spherules. The presence of metallic spherules in Darwin glass, australites, and indochinites was reported by Spencer (1933). Chao *et al.* (1962) discovered metallic spherules in several philippinites from the Isabela district. The spherules range in size from less than 0·1 to 0·5 mm in diameter, and the major phase, kamacite, contains unidentified inclusions which may be iron phosphide.

Zirconium-rich inclusion. Clarke and Henderson (1961) identified a bubbly inclusion in the Martha's Vineyard tektite as being rich in zirconium, probably derived from a zirconium mineral such as zircon. If this mineral were originally zircon, its present state indicates that the Martha's Vineyard tektite formed at a temperature in excess of 2550°C.

Minerals formed in situ. The first minerals found which grew *in situ* in tektites were along faults in layered tektites from Kan Luang Dong, Thailand (Barnes 1963a). More recently, abundant mineral grains have been found which grew in layered tektites from Muong Nong, Laos (Barnes 1964a). At least three minerals are present, none of which has been positively identified.

Chemical Characteristics of Tektites

The literature on chemical composition of tektites has been reviewed by Schnetzler and Pinson (1963), who summarized chemical data available through August 1962.

Major elements. Schnetzler and Pinson tabulated analyses for major elements in tektites not previously listed by Barnes (1940). In the same publication, Chao (1963) published 21 new analyses of bediasites. All published analyses on major element content of tektites, as well as unpublished analyses for 16 indochinites (Barnes 1964b) and 1 Borneo tektite (Wilford and Barnes 1964), are summarized in Table 5. Mean values in order of decreasing silica content are shown for the different groups. For each group, the range of composition is given to show the wide overlap from group to group, with the exceptions of Libyan Desert glass and Darwin glass.

In using this table, it should be remembered that new analyses are averaged with older analyses in which alumina was underestimated and ferrous iron overestimated by about 1 per cent, potash slightly underestimated, and lime and magnesia slightly overestimated (Barnes 1964b). The older analyses also show a much wider scatter on variation diagrams than the newer ones, indicating that in general they may be less reliable. Of the various groups, analytical data are probably most representative for the bediasites because careful selection was made of material for analysis, and modern and accurate analytical methods were used to obtain the data for this tektite group. Similar careful procedures are needed for all other tektite groups.

Minor elements. In the Schnetzler and Pinson (1963) tabulation of data for trace elements (except those in the gaseous state) in tektites, values are converted to parts per million by weight for all constituents. These data and data by Haskin and Gehl (1963) are summarized in Table 6 giving the number of specimens analysed and the range in amount of each element in each strewn-field. Schnetzler and Pinson in addition gave mean values, method of determination, and publication reference for data they tabulated.

Neon, helium, and oxygen were identified by O'Keefe *et al.* (1962) in a bubble in a bediasite by spectrographically analysing light produced in the gas by exposure in a 96-megacycle radio frequency field. Friedman (1963) tabulated analyses of gas in tektites, and this information is summarized in Table 7. Friedman found that the CO, CO_2, and H_2 reported probably resulted from reaction of carbon impurities in the apparatus with water in the tektites.

Isotopes. Isotopes in tektites and the age dating of tektites through isotopic variations have been reviewed by Zähringer (1963b). The similarity of lead isotope ratios for tektites and average modern galena, compiled by Zähringer from Tilton (1958) and Bate (1955), is shown in Table 8.

Strontium isotope ratios for 18 tektites compiled by Pinson (1962) are summarized in Table 9.

Deuterium in 18 tektites determined by Friedman (1958) are shown in Table 10.

Table 6. Trace element content of tektites

Element	Australites		Darwin glass		Indochinites		Ivory Coast tektites		Javanites*		Libyan Desert glass		Moldavites		North American tektites		Philippinites	
	No.	Range	No.	Range	No.	Range	No.	Range	No.	Range	No.	Range	No.	Range	No.	Range	No.	Range
Barium	24	540–800			8	300–2000							10	2600–7700			1	420
Beryllium	?	4							?	4			?	2				
Boron	10	6–22							5	10–19			16	3–16				
Caesium	24	1·1–4·2																
Chromium	34	60–165			15	70–220			6	300–425			27	20–330	2	30–40	1	170
Cobalt	25	10–20			6	3·6–14							2	8	2	10		
Copper	38	2·9–18			8	7·4–29			8	3·6–12·1			4	1·6–11·4	2	2·4–5·0	3	3·8–6·2
Gallium	25	0·9–11·4	1	2·5	8	8·2–100	1	9·8	3	4·6–8·2	3	1·3–1·8	11	3·1–20			1	7·1
Germanium	2	0·04–0·06	1	0·52	4	0·09–0·19	1	0·42	3	0·40–0·62	3	0·70–0·80	2	0·15–0·16	2	0·13–0·23	2	0·40–0·41
Lead	1	2·90			3	2·4–8·6					2	4·7–5	3	2·4–6			2	1·4–5
Lithium	25	37–650											3	14–650	2	65–100		
Nickel	38	17–49	2	162–310	19	8–220	1	105	8	15–350	1	3·3	45	8–27	3	14–20	4	22–32
Rare earths																		
Yttrium	1	27							1	8			9	2	1	31	1	31
Lanthanum	1	38			1	47							1	29	1	42	1	33
Praseodymium	1	12·6													1	14·8	1	10·2
Neodymium	1	34·5													1	35	1	31
Samarium	1	5·2	1	11·3											1	4·4	1	6·85
Europium	1	1·5	2	1·4–1·9	1	1·5							1	0·83	1	2·3	1	1·4
Gadolinium	1	7·7													1	8·2	1	5·6
Terbium	1	0·59													1	0·61	1	0·95
Dysprosium					1	5·3							1	2·8				
Holmium	1	1·15													1	1·36	1	0·94
Erbium	1	4·0													1	3·6	1	3·6
Thulium	1	0·40													1	0·55	1	0·57
Ytterbium	1	2·56													1	4·01	1	3·4
Lutecium	1	3·6													1	0·47		
Rubidium	31	62–800			13	70–131			1	98			4	130–1000	6	66–600	10	70–141
Scandium	24	8·9–16			2	<1			1	3			9	3			1	<1
Silver	1	<1											2	<1	2	<1		
Strontium	31	141–550			19	85–550			1	153			17	80–1000	6	152–550	10	100–190
Thorium	3	5·0–18·4							1	9·6			4	10·8–18·6	3	3·3–4·7		
Uranium	9	1·6–2·7			9	1·1–3·0			3	1·4–2·7	2	0·8–1·1	9	1·5–3·0	6	1·1–2·4	7	1·5–2·7
Vanadium	25	50–99			6	54–100							12	18–41	2	30–40		
Zinc	21	1·7–19			2	3·2–12·4			9	1·3–19·7			5	3·2–10·0	2	13·4–16·0	3	6·0–13·4
Zirconium	25	300–550			8	40–270							15	110–550	2	550	1	200
Water	2	50–70			4	50–100			2	20–40	1	640	2	3–5	2	20–50	5	20–140

* Includes billitonites and Borneo tektites.

Table 7. Gas content in tektites and obsidian in volume per cent

	Australite (cleaned)	Australite	Billitonite	Moldavite	Philippinite	Colombian obsidian
Total gas cm^3/g	0·14	0·33	0·23	0·66	0·11	0·83
CO_2	6	6	46	12·6	21	27·1
CO	92	7	47	33·1	56	24·4
H_2	1·5	0·5	6·5	41·1	6	35·3
H_2O	tr	86	—	—	15	—
CH_4	—	—	—	tr	—	2·2
SO_2	—	—	0·2	—	—	—
N_2	—	—	tr	—	—	—
O_2	—	—	—	0·6	—	1·1

Table 8. Isotopic composition of lead in tektites and galena

	$^{206}Pb/^{204}Pb$	$^{206}Pb/^{207}Pb$	$^{206}Pb/^{208}Pb$
Australite	18·81	1·206	0·486
Libyan Desert glass	19·11	1·213	0·488
Moldavite	18·55	1·185	0·483
Philippinite	18·83	1·199	0·482
Average modern galena	18·9	1·19	0·482

Table 9. Strontium isotope ratios in tektites

	Number of specimens	87/86	87/86 corrected	86/88	84/88
Australites	2	0·7202	0·7174	0·1185	0·0067
Bediasite	1	0·7207	0·7186	0·1187	0·0068
Georgia tektite	1	0·7154	0·7144	0·1189	0·0066
Indochinites	5	0·7204	0·7198	0·1192	0·0067
Javanite	1	0·7219	0·7194	0·1186	—
Martha's Vineyard tektite	1	0·7177	0·7156	0·1189	0·0071
Moldavites	2	0·7246	0·7246	0·1194	0·0068
Philippinites	5	0·7179	0·7181	0·1195	0·0067

Table 10. Water content and relative deuterium concentration in tektites

	Number of specimens	H_2O(%)	Mole %D
Australites	2	0·005–0·007	0·0143–0·0148
Bediasites	2	0·002–0·005	0·0146–0·0148
Billitonite	1	0·004	0·0151
Indochinites	4	0·007–0·01	0·0140–0·0166
Javanite	1	0·002	0·0148
Libyan Desert glass	1	0·064	0·0143
Moldavites	2	0·0003–0·0005	
Philippinites	5	0·002–0·014	0·0132–0·0157
Aouelloul (impactite)	1	0·02	0·0140
Macusanite (obsidian)	1	0·23	0·0127
Atom bomb glass*	1	0·05	

* Alamogordo, New Mexico

Oxygen isotopic analyses of tektites, impact glasses, soils, and macusanite by Taylor and Epstein (1962) and Silverman (1951) are summarized in Table 11.

Isotopes produced by cosmic rays, such as ^{3}He, ^{21}Ne, ^{26}Al and ^{10}Be, have been searched for but none have been found.

Argon-40 is important in tektite age determinations. The summarization of ^{40}A content, potassium content, and ages of tektites in Table 12 is from Zähringer (1963a). By use of fission track dating, Fleischer and Price (1963) arrived at the same ages for tektites as found by Zähringer, who used the potassium-argon method.

Table 11. Oxygen isotopic analyses of tektites, impactites, soils, and obsidian

	Number of specimens	δ $^{18}O/^{16}O$ per mil
Tektites		
Australites	2	9·7
Bediasites	8	8.9–10·8
Georgia tektite	1	9.8
Indochinites	4	9·7–10·3
Javanites (2) and billitonite	3	9·6–10·6
Libyan Desert glass	1	10·2
Moldavites	2	10·3–10·4
Philippinites	4	10·1–10·4
Impactites		
Henbury craters	1	14·5
Ries glass	2	7·9–9·0
al Hadida (Wabar) craters	1	12·9
Soils		
Elberton, Georgia (A-zone)	2	10·1–10·5
Phang Daeng, Thailand (laterite)	1	15·2
Obsidian (macusanite)		
Macusani, Peru	1	12·0

Origin of Tektites

Tektites were originally considered to be a glassy species of lava (Mayer 1788), or products of prehistoric glass manufacture (Lindaker 1792). Other postulated origins include their formation by lightning striking the Earth (Rutley 1885); obsidian shot from Moon volcanoes (Verbeek 1897); destruction of a planet similar to Earth (Linck 1928) with spurted material becoming comets which periodically brought tektites to Earth (Barnes 1957); residue of the final drying out of blobs of silica gel left behind by the evaporation of highly siliceous water (Van Lier 1933); ablation stripping of a parent body as it skipped through the Earth's atmosphere (Adams 1963); and grazing impact of a comet which dragged melted loess into the air, later to be dropped as tektites (Sun 1963).

The writer (1940) found the preponderance of evidence weighted in favour of tektites having formed from rocks and soils containing quartz grains, rapidly melted and rapidly cooled; much additional evidence that this is true has been found since that time. O'Keefe's (1963) review of tektite origin is based on his belief that tektites are from the Moon. Cohen (1963) adopted the terrestrial impact origin of tektites based on the comet-asteroid impact hypothesis of Urey (1955, 1957) and the meteorite impact hypothesis of Spencer (1933).

Tektite properties of probable significance in deciding between a terrestrial or lunar origin for tektites are tabulated in Table 13. Other properties which appear to be of little importance in deciding between a lunar and terrestrial origin include: splash form shapes, magnetic susceptibility, intensity of magnetization, bubbles, flow structure, flow structure strain, over-all strain, "fingers", faults, metallic spherules, and surface sculpture produced by chemical etching. O'Keefe (1963) argued that the type of sculpture on tektites attributed by most workers to chemical etching is a result of ablation. However, similarly etched surfaces are common on obsidian pebbles which have never left the ground.

Most of the objections to a lunar origin listed in Table 13 are based on the strong probability that the Moon was not differentiated, as the Earth has been, to give a siliceous crust (Barnes *et al.* 1958). If, however, the crust of the

Table 12. Abundance of argon 40 and potassium and ages of tektites

	Number of specimens	K(%)	Rad. ^{40}A (10^{-8} cm^3/gm)	Age (10^6yr)
Australites	6	1·76–1·99	5·1–5·9	0·68–0·76
Bediasites	6	1·46–1·71	204–282	33·7–35·0
Georgia tektites*	3	1·99–2·02	268–282	33·7–35·0
Indochinites	4	1·68–2·09	4·8–6·0	0·71–0·73
Ivory Coast tektite	1	1·30	6·6	1·3
Javanites**	3	1·63–2·10	4·7–6·0	0·72–0·73
Moldavites	8	2·07–2·72	119–157	14·4–14·9
Philippinites	5	2·00–2·29	5·4–6·6	0·68–0·73

* Includes Martha's Vineyard tektite.

** Includes one billitonite and a specimen from Borneo.

Moon proves to be similar in chemical composition to that of the Earth, a strong objection still remains to the origin of tektites from the Moon. This objection is the presence of detrital quartz grains in a few layered indochinites. These grains are similar in size and shape to detrital quartz grains in the soils and rocks on which the tektites rest. It is most unlikely that similar quartz-rich soils and rocks could have accumulated on the surface of the waterless and airless Moon.

The presence in layered indochinites of detrital quartz grains is strong presumptive evidence that this tektite strewn-field is of terrestrial origin. Since no crater is obvious, this strewn-field may be the result of the impact of a diffuse object, such as a comet (Urey 1957), or it may be the result of some nuclear or electrical phenomenon not yet fully understood. When evidence derived from a study of the indochinite strewn-field is applied to moldavites, bediasites, philippinites, Libyan Desert glass, and Darwin glass, it appears that they are probably of similar origin.

Table 13. Check list for terrestrial or lunar origin of tektites

	Favours terrestrial origin	Favours lunar origin
Bona fide ablation features (australites and javanites)	No	Yes
Absence of *bona fide* ablation features in other tektites	Yes	No
Specific gravity (2·303–2·495)	Yes	No
Refractive index (1·4798–1·5204)	Yes	No
Permittivity ε^1/μ	Yes	?
Presence of lechatelierite	Yes	No
Presence of rayed bubbles	Yes	No
Presence of included detrital mineral grains	Yes	No
Presence of zirconium-rich inclusion	Yes	No
Presence of minerals formed *in situ*	Yes	No
Bulk chemical composition	Yes	No
Trace element composition	Yes	No
^{18}O–^{16}O ratio	Yes	No
Presence of layering	Yes	No
Low gas pressure in bubbles	No	Yes

Australites and javanites show definite evidence of aerodynamic ablation and for this reason must be presumed to have entered or re-entered the Earth's atmosphere. Therefore, they are either of extraterrestrial origin (Chapman and Larsen 1963) or they were ejected beyond the atmosphere immediately following their formation from Earth materials.

Acknowledgments

Publications consulted freely in preparing this review include "Tektites", edited by John A. O'Keefe, The University of Chicago Press; "Tektites", November, 1961, *Scientific American*, and "Tektites", 1962, *ICSU Review*. Unpublished data incorporated in this review were obtained with support from National Science Foundation grants NSF G-10236 and GP-408, administered by The University of Texas.

Bibliography

Two outstanding tektite workers not specifically cited are F. E. Suess and Charles Fenner. Lists of their publications, and those of many other tektite workers not cited, will be found in bibliographies by Barnes (1940), Brown (1953), Baker (1959), and in "Tektites", edited by John A. O'Keefe, The University of Chicago Press, 1963. Except for Brown, these sources also furnish the chief comprehensive reviews of tektite research.

ADAMS E.W. (1963) in *Tektites*, Chicago: The University Press, 150.

BAKER G. (1959) *Mem. nat. Mus. Vict.*, **23,** 313.

BAKER G. (1963) in *Tektites*, Chicago: The University Press, 1.

BARNES V.E. (1940) University of Texas Pub. 3945, 477.

BARNES V.E. (1957) *Geo Times*, **1,** 12, 6.

BARNES V.E. (1958) *Geochim. et Cosmochim. Acta*, **14,** 267.

BARNES V.E. (1960) Rept. 21st Intern. Geol. Cong., Norden, **13,** 328.

BARNES V.E. (1961) *Scientific American*, **205,** 5, 58.

BARNES V.E. (1962a) *I.C.S.U. Review*, **4,** 75.

BARNES V.E. (1962b) *Trans. Am. Geophys. Union*, **43,** 466.

BARNES V.E. (1963a) *Trans. Am. Geophys. Union*, **44,** 93.

BARNES V.E. (1963b) in *Tektites*, Chicago: The University Press, 25.

BARNES V.E. (1963c) *Science*, **142,** 1651.

BARNES V.E. (1964a) *Trans. Am. Geophys. Union*, **45,** 82.

BARNES V.E. (1964b) *Geochim. et Cosmochim. Acta*, **28,** 893.

BARNES V.E., KOPAL Z. and UREY H.C. (1958) *Nature*, **181,** 1457.

BARNES V.E. and PITAKPAIVAN K. (1962) *Proc. Nat. Acad. Sci.*, U.S., **48,** 947.

BATE G.L. (1955) Ph.D. dissertation, Columbia University.

BROWN H. (1953) *A bibliography of meteorites*, Chicago: The University Press.

CHAO E.C.T., ADLER I., DWORNIK E.J. and LITTLER J. (1962) *Science*, **135,** 97.

CHAO E.C.T. (1963) in *Tektites*, Chicago: The University Press, 51.

CHAPMAN D.R. and LARSON H.K. (1963) *J. Geophys. Res.*, **68,** 4305.

CLARKE R.S. JR. and HENDERSON E.P. (1961) *Georgia Min. Newsletter*, **14,** 90.

COHEN A.J. (1963) in *Tektites*, Chicago: The University Press, 189.

FLEISCHER R.L. and PRICE P.B. (1963) *Gen. Elect. Res. Lab. Rpt.* 63-RL-3501M.

FRIEDMAN I. (1958) *Geochim. et Cosmochim. Acta*, **14,** 316.

FRIEDMAN I. (1963) in *Tektites*, Chicago: The University Press, 130.

HAMMOND C.R. (1950) *Popular Astron.*, **58,** 345.

HASKIN L. and GEHL M.A. (1963) *Science*, **139,** 1056.

LACROIX A. (1935) Acad. Sci. Paris C.R., **209,** 2129.

LINCK G. (1928) *Neues Jahrb.*, **57,** 223.

LINDAKER J.T. (1792) *Gesell. Bömischer Naturforsch.*, **2,** 272.

MAYER J. (1788) *Böhmische Gesell. Wiss.*, **1787,** 265.

O'KEEFE J.A. (1963) in *Tektites*, Chicago: The University Press, 167.

O'KEEFE J.A., LOWMAN P.D. JR. and DUNNING K.L. (1962) *Science*, **137,** 228.

OLTE A. and SIEGEL K.M. (1961) *Astrophys. J.*, **133,** 706.

PINSON W.H. JR. (1962) Mass. Inst. Tech. Contract No. AF 19(604)-6135, Final Report.

RUTLEY F. (1885) *Geol. Soc. London, Quart. J.*, **12,** 152.

SCHNETZLER C.C. and PINSON W.H. JR. (1963) in *Tektites*, Chicago: The University Press, 95.

SENTFLE F.E. and THORPE A. (1959) *Geochim. et Cosmochim. Acta*, **17,** 234.

SILVERMAN S.R. (1951) *Geochim. et Cosmochim. Acta*, **2,** 26.

SPENCER L.J. (1933) *Nature*, **132,** 571.

SUN M.S. (1963) *Trans. Am. Geophys. Union*, **44,** 93.

TAYLOR H.P. JR. and EPSTEIN S. (1962) *J. Geophys. Res.*, **67,** 4485.

THORPE A.N., SENTFLE F.E. and CUTTITTA F. (1963) *Nature*, **197,** 836.

TILTON G.R. (1958) *Geochim. et Cosmochim. Acta*, **14,** 323.

UREY H.C. (1955) *Proc. Nat. Acad. Sci.*, U.S., **41,** 27.

UREY H.C. (1957) *Nature*, **179,** 556.

VAN LIER R.J. (1933) *K. Akad. Wetensch. Amsterdam, Pr.*, **36,** 4, 454.

VERBEEK R.D.M. (1897) *K. Akad. Wetensch. Amsterdam, Verh.*, **5,** 421.

VOLAROVICH M.P. and LEONTIEVA A.A. (1939) *C.R. (Doklady) Acad. Sci. U.R.S.S.*, **22,** 589.

WALTER L.S. (1965) *Science*, **147,** 1049.

WILFORD G.E. and BARNES V.E. (1964) *Trans. Am. Geophys. Union*. **45,** 82.

ZÄHRINGER J. (1963a) in *Radioactive dating*, Internat. Atomic Energy Agency, Vienna, 289.

ZÄHRINGER J. (1963b) in *Tektites*, Chicago: The University Press, 137.

V.E. BARNES

Editors' Comments on Papers 17–20

Four papers have been chosen as illustrative of recent work on elemental abundances in tektites and how these abundances compare with those in terrestrial rocks. The first two papers relate tektites generally to terrestrial rocks, and the next two papers relate one tektite strewn-field to the kind of rock in which an impact crater was excavated.

Although under some conditions of heating osmium and rhenium are highly volatile, Lovering and Morgan* concluded, from the few data available, that terrestrial acid igneous or sedimentary rocks are possible parent material for tektites, providing these elements were not significantly depleted during the time tektites were formed.

Baedecker and Ehmann* found that iridium and platinum and the iridium:gold ratio for tektites is similar to that in sediments and in granitic rocks, and they note that Taylor (1962) and Ehmann (1965) pointed out the similarity of trace-element abundances in tektites and common terrestrial materials. In view of the finding of two distinct groups of australites with an order-of-magnitude difference in gold content, it would be interesting to know if one of these groups might correspond to the older australites reported by Fleischer et al. (1969)*. To conserve space the "Experimental" and "Results" sections of the paper by Baedecker and Ehmann* have been omitted.

Schnetzler, Pinson, and Hurley* concluded that the Ivory Coast tektites are most probably the fusion products of meteoritic impact in Birrimian basement rocks at the Bosumtwi crater site, and also point out that Birrimian rocks are widespread in the Ivory Coast.

Rybach and Adams*, in their study of uranium, thorium, and potassium abundances, also arrived at the conclusion that Ivory Coast tektites are related to the rocks in which Bosumtwi crater is excavated.

Dr. John F. Lovering is now Professor and Chairman, Department of Geology, University of Melbourne, Parkville, Victoria 3052, Australia. Dr. J. W. Morgan is now Research Associate, Enrico Fermi Institute and Department of Chemistry, University of Chicago, Chicago, Illinois, 60637.

Dr. Philip A. Baedecker, formerly a graduate student of Dr. William D. Ehmann at the University of Kentucky, is now Assistant Professor, Department of Chemistry, Institute of Geophysics and Planetary Physics, University of California, Los Angeles, California 90024.

Dr. L. Rybach is now Professor of Geophysics, Eidg. Technische Hochschule, Zurich, Switzerland.

References

Ehmann, W. D. (1965). On some tantalum abundances in meteorites and tektites. *Geochim. Cosmochim. Acta,* **29,** 43–48.

*Fleischer, R. L., Price, P. B., and Woods, R. T. (1969). A second tektite fall in Australia. *Earth Planet. Sci. Lett.,* **7,** 51–52.

Taylor, S. R. (1962). The chemical composition of australites. *Geochim. Cosmochim. Acta,* **26,** 685–722.

17

Geochimica et Cosmochimica Acta 1964, Vol. 28, pp. 761 to 768. Pergamon Press Ltd. Printed in Northern Ireland

Rhenium and osmium abundances in tektites

J. F. LOVERING* and J. W. MORGAN
Department of Geophsyics, Australian National University, Canberra, A.C.T. Australia

Abstract—Rhenium and osmium abundances have been determined in two australites, two javaites, one philippinite, two indochinites, two moldavites and one bediasite by a neutron activation technique. Rhenium abundances range from $0{\cdot}14 \times 10^{-10}$ to $1{\cdot}23 \times 10^{-10}$ g/g and osmium abundances from $2{\cdot}4 \times 10^{-10}$ to $13{\cdot}8 \times 10^{-10}$ g/g. Measurements of rhenium and osmium abundances in terrestrial acid igneous and sedimentary rocks suggest that either rock types could be suitable parent materials for tektites.

INTRODUCTION

RHENIUM and osmium are both among the rarest of the chemical elements and very little is known about their distribution in nature. What little data there is on the geochemistry of rhenium has been summarized by FLEISCHER (1959, 1960) and the only recent advances have been concerned with rhenium abundance determinations in 29 iron meteorites and one chondrite (HERR *et al.*, 1961) and a collection of twelve igneous rocks by MORRIS and FIFIELD (1961). Osmium geochemistry is very poorly understood although a recent study by BATE and HUIZENGA (1963) of osmium abundances in ten chondrites, two achondrites, four terrestrial rocks and one tektite together with meteorite data reported by HERR *et al.* (1961) considerably increased the number of measurements available up to the present.

Rhenium and osmium concentrations in tektites could provide important information on the way in which the tektites have formed and the nature of their parent material. For example, it has already been shown that zinc, and to a lesser extent, copper have apparently been selectively volatilized from the parent tektite material during the fusion process which formed the tektites (GREENLAND and LOVERING, 1963). In their most highly oxidized states, both Os and, to a generally lesser extent, Re are extremely volatile elements as oxides as shown by melting-point and boiling-point data shown in Table 1. If selective volatilization processes have in fact operated during the fusion of the tektites (GREENLAND and LOVERING, 1963) it might be predicted that Os and Re would be depleted in tektites relative to likely parent materials.

In the present study a neutron activation method was developed to simultaneously determine Re and Os abundances in each tektite sample analysed. The tektite samples analysed were aliquots of the samples previously prepared by GREENLAND and LOVERING (1963) and analysed by them for Zn and Cu. The numbers used to differentiate the samples analysed in Table 3 are the same as those used by GREENLAND and LOVERING.

ANALYTICAL METHOD

The neutron activation technique developed to determine both Re and Os in rocks and meteorites down to concentration level of the order of 10^{-11} g/g will be

* At present Visiting Professor, Department of Geology, The University of Rochester. Rochester, New York.

described in detail elsewhere (MORGAN and LOVERING, 1963). Briefly it consists in exposing samples and standards for one week to a high thermal neutron flux (around 9×10^{12} neutrons/cm²/sec), adding Re and Os carriers and fusing with Na_2O_2, separating first the Os and then the Re by means of anion exchange techniques, precipitating the Os as the osmium thio-urea chromium thiocyanate complex Os $(Thio)_6$ $Cr(SCN)_6$ and the Re as the tetraphenyl arsonium perrhenate As $(C_6H_5)_4$ Re O_4. The nuclides detected are Re^{186}($t\frac{1}{2}$ = 88·9 hours) and Os^{191} ($t\frac{1}{2}$ = 16 days) and activities were measured using a 2 in. end window β-counter, with a background of 30–40 cpm. The counting efficiency was about 20 per cent, and samples and standards were usually counted for at least $\times 10^4$ counts. No self absorption correction was necessary for the Re sources.

Preparation of standards

A standard solution containing 6 μg/g Re and 40 μg/g Os made up in 3N HCl was prepared from *Specpure* $KReO_4$ and $(NH_4)_2OsCl_6$ dried to constant weight at 150°C. Concentrations of the initial concentrated solutions were checked gravimetrically using the osmium thio-urea chromium thiocyanate complex and tetraphenyl arsonium perrhenate. Then 30 mg of *Specpure* SiO_2 were weighed into a silica ampoule, 50–60 mg of the standard solution added and the ampoule reweighed. The liquid was centrifuged down, absorbed by the silica and the standards were then dried overnight at 80°C.

Preparation of samples

Tektite samples previously ground and analysed for Cu and Zn by GREENLAND and LOVERING (1963) were used in the present work. Approximately 0·1 g samples of the powdered tektites were weighed into clean, dry silica ampoules which were then sealed.

RESULTS

The accuracy of the Re and Os analyses determined by the present method may be judged by comparing the results on the standard granite G-1 and the standard diabase W-1 with previously published data (Table 2). The results of the present work are in good general agreement with those neutron activation Re determinations by MORRIS and FIFIELD (1961) on the other hand the neutron activation Os determinations by BATE and HUIZENGA (1963) are consistently higher than those in the present study but at these very low Os abundance levels even these discrepancies are not serious.

MORRIS and FIFIELD also reported Re concentrations of 100 and 120×10^{-10} g/g in a sample of Johnson–Matthey and Co. *Specpure* silica. Another sample of *Specpure* silica (Batch No. 23453) was analysed in the present study and found to contain considerably higher amounts of Re (i.e. 416 and 435×10^{-10} g/g). MORGAN and LOVERING (1963b) had previously shown a similar variability in the thorium content of the same batch No. 23453 (i.e. 4·2 and $0{\cdot}9 \times 10^{-8}$ g/g). It would seem that this batch of *Specpure* silica had presumably been slightly contaminated in the production process and we would agree with the conclusion of MORRIS and FIFIELD (1961) that this illustrated the difficulty of obtaining reliable data for such rare elements in rocks.

Re and Os abundances in ten tektites from six of the recognized localities have been reported in Table 3. The precisions on the Re analyses varied from ±80 per cent to ±15 per cent as the analytical technique itself was improved. The Os sensitivity is less than the Re sensitivity and at the low Os abundance levels it was only possible to report the analyses with standard deviations of around ±50 per cent based on the very low count rates of the samples (of the order of 1 cpm) using a β-counter with a background 30 cpm.

Discussion

In their highest stable oxidation states, both Re and Os oxides are very volatile compounds (Table 1) and heating parent tektite material under oxidizing

Table 1. Melting point and boiling point data for Re and Os compounds*

	Melting point (°C)	Boiling point (°C)
Rhenium	3167 ± 60	—
ReO_3	—	decomposes 400
Re_2O_7	approx. 220	sublimes 450
Osmium	2700	5300
OsO_2	decomposes 650	—
OsO_4	39·5, 41	130

*Data from Handbook of Chemistry and Physics (Ed. C. D. Hodgeman, 43rd Edition 1961–1962). Chemical Rubber Publ., Ohio.

conditions might be expected to result in selective volatilization of both Re and Os and impoverishment of both elements in tektites. However, Thorpe *et al.* (1963) have recently studied ferric/ferrous ratios in all tektite families and found very low ratios (~0·05) which they concluded implies melting in an atmosphere with a very low partial pressure of oxygen. If the tektites have been melted under reducing conditions, then the relative behaviour of Re and Os would probably be very different than if the heating had taken place under oxidizing conditions. When heated under reducing conditions, osmium oxides are readily reduced to osmium metal (Mellor, 1936) which is an extremely refractory metal with a boiling point >5300°C (Table 1). On the other hand, rhenium oxides are much more stable when heated under reducing conditions (Sidgwick, 1950) so that under these circumstances Re may still behave as a volatile element and be impoverished in tektites while Os would behave as a relatively involatile element and might not be as highly depleted as Re. As a result of this behaviour, it might be expected that Os/Re ratios in tektites would be higher than the ratios in possible parent materials.

Os/Re ratios in tektites

Herr *et al.* (1961) have determined the Os/Re ratio in the silicate phase of the Mocs "ordinary" chondrite as 11·3, which is in good agreement with the "cosmic" ratio of 11·9 reported by Cameron (1959). Herr *et al.* also determined Os/Re ratios in 29 iron meteorites and found a range of 5·12–19·6. The only reliable data on Os/Re ratios in terrestrial igneous rocks may be calculated from the data in Table 2 in which Re and Os abundances in the "standard" granite sample G-1 and

Table 2. Neutron activation determinations of Re and Os abundances in some terrestrial materials

Rock	Locality	Re (10^{-10} g/g)	Os (10^{-10} g/g)	Os/Re	References
Biotite granite	Liruei, Nigeria	2·2	—	—	MORRIS and FIFIELD (1961)
Riebeckite-aegerine granite	Liruei, Nigeria	2·5	—	—	MORRIS and FIFIELD (1961)
Albite-riebeckite granite	Liruei, Nigeria	11·2	—	—	MORRIS and FIFIELD (1961)
Rhyolite	Liruei, Nigeria	11·4	—	—	MORRIS and FIFIELD (1961)
Standard granite G–1	Westerly, Rhode Island	(5·2–8·4)6·7	—	—	MORRIS and FIFIELD (1961)
		—	2·0	—	BATE and HUIZENGA (1963)
		7·89 ± 0·09 3·34 ± 0·04 } 5·61	0·65 ± 0·39 ≤ 0·48	0·08 ≤ 0·14	this work
Standard diabase W–1	Centerville, Virginia	(5·6–8·4)7·1	—	—	MORRIS and FIFIELD (1961)
		—	4·6	—	BATE and HUIZENGA (1963)
		4·41 ± 0·07 4·05 ± 0·04 } 4·23	2·26 ± 0·55 2·85 ± 0·43 } 2·56	0·51 0·70 } 0·60	this work

Table *cont.*

Specpure SiO_2	Unknown Batch No.	100 120		— —		— —		Morris and Fifield (1961)
	J. M. Batch No. 23453	416 ± 4 435 ± 4		82 ± 12 843 ± 16		— —		this work
Quartz-rich sandstone	Wilpena Pound, South Australia	0·27 ± 0·26 0·42 ± 0·14	0·34	≤ 0·65 0·55 ± 0·43		≤ 2·4 1·3		this work
Shale*	State Circle Shale, Etheridge Creek, A.C.T., Australia	0·54 ± 0·25 0·48 ± 0·26	0·51	5·56 ± 1·09 3·51 ± 1·06	4·54	10·3 7·4	8·9	this work
Dark shale with pyrite	Captain's Flat Railway Station, N.S.W. Australia	859 ± 4·8 792 ± 5·2	825	4·38 ± 0·58 2·52 ± 0·61	3·45	0·005 0·003	0·004	this work

* Specimen analysed has been described as sample 13 by Compston and Pidgeon (1962).

the "standard" diabase sample W-1 are reported. The Os/Re ratios are very low G-1: $\sim$0·1, W-1: $\sim$0·6) and indicate a considerable impoverishment of Os relative to Re in crystal igneous rocks—probably the greater part of the earth's osmium is concentrated in the earth's metallic core.

Of the three sedimentary rocks examined, the Wilpena Pound sandstone and the State Circle shale have Os/Re ratios rather higher than either G-1 or W-1 while the dark shale with pyrite from Captains Flat has an extremely low ratio around 0·004. These are the only data so far available on Re and Os abundances in sedimentary rocks.

The Os/Re ratios in the tektites vary from about 5 to $\leq$ 28 and are considerably higher than the ratios so far observed in the terrestrial rocks. However, it has already been shown that Os/Re ratios in tektites would be expected to be higher than the ratios of their parent materials, if the tektites have been formed by high temperature fusion under reducing conditions. Under those circumstances, and on the basis of known Os/Re ratio data, it can be concluded that terrestrial rocks (both sedimentary and igneous) are possible parent materials for tektites.

Absolute Re and Os abundances in tektites and possible tektite parent materials

It is no exaggeration to say that there are probably as many theories concerned with the origin of tektites as there are workers in the field. On the other hand, it is probably fair to say that most theories currently argued are concerned with two basic questions:

1. Whether terrestrial sedimentary or acid igneous rocks are suitable parent materials for tektites; and
2. Whether tektites are terrestrial or extra-terrestrial in origin. If they are extra-terrestrial, then it is now generally considered that the most likely origin is by asteroid or large meteorite impact on the lunar surface.

If sedimentary rocks can be shown to be the most suitable parent materials, then an extra-terrestrial origin for tektites would be excluded. On the other hand, if acid igneous rocks are suitable parent materials, then either a terrestrial or a lunar origin is possible, provided that the moon has undergone igneous differentiation at some time during its evolution with the formation of acid igneous rocks on its surface.

At the present time there is very little information available on Re and Os abundances in either sedimentary or igneous rocks and we are currently investigating abundances in a variety of geochemically important rock-types.

Rhenium. FLEISCHER (1959) has quoted work of BASITOVA who reports an average 60 $\times 10^{-10}$ g/g Re in acid igneous rocks. Neutron activation determinations of Re abundance in four acid igneous rocks by MORRIS and FIFIELD (1961) range from 2·2 to 11·4 $\times 10^{-10}$ g/g and in the granite G-1 are a little less than 10 $\times 10^{-10}$ g/g (Table 2). The Re abundances in acid igneous rocks are at the most an order of magnitude higher than tektite abundances so that if some depletion of Re has taken place during the formation of the tektites, acid igneous rocks would seem to be suitable parent materials for tektites.

The only Re abundances previously measured in sedimentary rocks were reported by PETERSEN *et al.* (1959) who studied siltstones and sandstones of the Shinarump member of the Triassic Chinle formation in Coconino County, Arizona

Table 3. Re and Os abundances in tektites

Tektite group	Specimen number	Rhenium 10^{-10} g/g		Osmium 10^{-10} g/g		Os/Re	
Australites	T 31	0·72 ± 0·10	0·98	3·9 ± 2·6	7·1	5·4	7·2
		1·23 ± 0·15		10·2 ± 3·4		8·3	
	T 50	0·47 ± 0·13	0·53	22·4 ± 1·9	3·0	5·1	5·7
		0·59 ± 0·12		3·5 ± 2·2		5·9	
Javaites	T 57	0·22 ± 0·12		2·8 ± 1·9		12·7	
		≤ 0·10		—		—	
	T 60	0·14 ± 0·09	0·19	2·8 ± 1·4	3·4	20·0	17·9
		0·24 ± 0·12		4·0 ± 1·5		16·7	
Philippinites	T 9	0·91 ± 0·06	0·76	9·4 ± 1·0	11·6	10·3	16·5
		0·61 ± 0·04		13·8 ± 5·7*		22·6	
Indochinites	T 4	0·50 ± 0·40		≤ 14		≤ 28	
	T 6	≤ 0·54		13·4 ± 7·3		≥ 25	
Moldavites	T 3	≤ 0·21		≤ 18·5		—	
		≤ 0·18		≤ 10·4		—	
	T 55	0·20 ± 0·03		2·9 ± 1·0	2·9	14·5	
		—		2·8 ± 1·2		—	
Bediasites	T 2	0·94 ± 0·34		≤ 10·5		≤ 11	
		≤ 0·34		≤ 7·2		—	

* Corrected for ruthenium contamination.

which were associated with basal conglomerates containing uraninite ore. They reported Re abundances so high (from 0·005 to 0·1 per cent) that these rocks essentially constitute Re ore deposits and it is unlikely that all sedimentary rocks have such high Re contents. In the course of the present work Re abundances were measured in a quartz-rich sandstone, a shale and a dark shale containing pyrite (Table 2). The sandstone and the ordinary shale have Re abundances almost identical with those of tektites so that if either or both of these rocks are the parent materials from which tektites have formed, then no significant volatilization of Re has occurred during the formation of the tektites. On the other hand, if the dark shale is considered as a possible tektite parent then Re must have been very greatly depleted during tektite formation as the dark shale contains nearly 10^4 times more Re than does the tektites.

Depending on the degree to which Re has or has not been selectively volatilized during the formation of tektites, then either acid igneous rocks or terrestrial sedimentary rocks could be suitable parent materials for tektites.

Osmium. The only acid igneous rock which has recently been analysed for Os is the standard granite G-1 and it has already been pointed out that some

2

uncertainty exists as to its Os content. Bate and Huizenga (1963) found 2×10^{-10} g/g Os while our results indicated only about $0{\cdot}65 \times 10^{-10}$ g/g Os (Table 2).

Of the three sedimentary rocks examined in the present study, both the clay shale and the dark shale contained about 4×10^{-10} g/g Os while the quartz-rich sandstone contained around $0{\cdot}55 \times 10^{-10}$ g/g Os.

The Os abundances in tektites range from about 2 to 13×10^{-10} g/g so that if Os is not significantly depleted in tektites, then either acid igneous rocks or sedimentary rocks may well be suitable parent materials for tektites.

Conclusions

As far as the limited data on terrestrial acid igneous or sedimentary rocks are concerned, the observed Re and Os abundances in tektites may be explained if either materials are taken as possible parent materials for tektites. Other studies (Lovering, 1960; Greenland and Lovering, 1963) have suggested that acid igneous rocks could be suitable parent materials for tektites and the Re–Os abundance studies reported here are not inconsistent with this suggestion.

References

Bate G. L. and Huizenga J. R. (1963) Abundance of ruthenium, osmium and uranium in some cosmic and terrestrial sources. *Geochim. et Cosmochim. Acta* **27,** 345–360.

Cameron A. G. W. (1959) A revised table of abundances of the elements. *Astrophys. J.* **129,** 676–699.

Compston W. and Pidgeon R. T. (1962) Rubidium-strontium dating of shales by the total-rock method, *J. Geophys. Res.* **67,** 3493–3502.

Fleischer M. (1959) The geochemistry of rhenium, with special reference to its occurrence in molybdenite. *Econ. Geol.* **54,** 1406–1413.

Fleischer M. (1960) The geochemistry of rhenium—addendum. *Econ. Geol.* **55,** 607–609.

Greenland L. and Lovering J. F. (1963) The evolution of tektites: elemental volatilization in tektites. *Geochim. et Cosmochim. Acta* **27,** 249–259.

Herr W., Hoffmeister W., Hirt B., Geiss J. and Houtermans F. G. (1961) Research on the dating of iron meteorites by the rhenium–osmium method. *Z. Naturf.* **16a,** 1053–1058.

Lovering J. F. (1960) High temperature fusion of possible parent materials for tektites. *Nature, Lond.* **186,** 1028–1030.

Mellor J. W. (1936) *A Comprehensive Treatise on Inorganic and Theoretical Chemistry*, pp. 702–713 (Volume XV). Longmans Green, London.

Morgan J. W. and Lovering J. F. (1963a) The determination of rhenium and osmium in rocks and meteorites by neutron activation analysis in preparation.

Morgan J. W. and Lovering J. F. (1963b) The determination of uranium and thorium in rocks by neutron activation analysis. *Analyt. Chim. Acta* **28,** 405–417.

Morris D. F. C. and Fifield F. W. (1961) Rhenium content of rocks. *Geochim. et Cosmochim. Acta* **25,** 232–233.

Petersen R. G., Hamilton J. C. and Myers A. T. (1959) An occurrence of rhenium, associated with uraninite, in Coconino Country, Arizona. *Econ. Geol.* **54,** 254–267.

Sidgwick N. V. (1950) *Chemical Elements and their Compounds* (Volume II). Clarendon Press, Oxford.

Thorpe A. N., Senftle F. E. and Cuttitta F. (1963) Magnetic and chemical investigations of iron in tektites. *Nature, Lond.* **197,** 836–840.

18

Geochimica et Cosmochimica Acta 1965, Vol. 29, pp. 329 to 342. Pergamon Press Ltd. Printed in Northern Ireland

The distribution of some noble metals in meteorites and natural materials*

P. A. BAEDECKER and W. D. EHMANN
Department of Chemistry, University of Kentucky
Lexington, Kentucky

(*Received* 15 *August* 1964)

Abstract—Abundances of gold, iridium and platinum have been determined in a wide range of meteoritic and terrestrial materials. The technique of thermal neutron activation analysis, as used in this study, couples the advantages of high sensitivity and freedom from reagent and laboratory contamination. Abundances of these elements in the chondrites are used to derive values for their "cosmic" atomic abundances. Implications of the data to theories of nucleosynthesis and the chemical history of meteorites and tektites are discussed.

INTRODUCTION

THE determination of noble metal abundances has always attracted the attention of analysts due to the value of these elements in commerce. In much of the early work on their abundances in natural materials only the most abundant elements, such as silver, were directly determined. The abundances of gold, platinum, iridium and other members of this group were then estimated by use of abundance ratios derived from analyses of the principal ores of these elements.

Most of the classical determinations of these elements were plagued with problems of reagent and laboratory contamination, due to the use of noble metals in the common coinage and some types of laboratory-ware. These analytical problems are discussed in some detail by GOLDSCHMIDT (1958).

One of the first attempts to use the technique of activation analysis for determination of a noble metal in meteorites was the work of GOLDBERG and BROWN (1949) on palladium abundances in siderites. Since that time, activation analysis has been applied to the direct determination of many of the noble metals in the sub-microgram abundance region characteristic of most natural materials. The technique combines the advantages of high sensitivity and inherent freedom from reagent and laboratory contamination.

* * * * * * *

DISCUSSION

Taking the chondritic abundances of Table 3 (averaged by type in Table 7) as fairly representative of primitive solar non-volatile material, "cosmic" atomic abundances may be calculated for these elements. These atomic abundances together with

Table 5. Abundances of gold and iridium in tektites

Specimens	Description	Au (ppm)	Ir* (ppm)
Australite	Cockburn site, S.A.	0·0015	—
Australite	Eucla site, W.A.	0·0024	—
Australite	Mt. Vernon site, W.A.	0·0132	—
Billitonite		0·0242	—
Georgia tektite	Specimen E.Y.-W.B. Gibbons School site	0·0067	—
Java tektite	Sangiran Dome site	0·0013	—
Libyan Desert glass	U.A.R.	0·0194	$\leq 5 \times 10^{-5}$
Moldavite	Locenice site	0·0039	—
Philippinite	Isabella site	0·0118	$\leq 1 \times 10^{-5}$
Texas tektite	Fayette County site	0·0003	—
Thailand tektite	Kan Luang Dong site (Unstrained type)	—	$\sim 8 \times 10^{-5}$
Thailand tektite	Kan Luang Dong site (Ordinary type)	0·0013	$\sim 7 \times 10^{-5}$
Soil	Found associated with the unstrained Thailand tektites	—	$\leq 1 \times 10^{-5}$

* Near the limit of sensitivity of the technique, as discussed in the text.

Table 6. Abundances of gold and iridium in terrestrial materials

Specimens	Description	Au (ppm)	Ir* (ppm)
Darwin Glass	Presumed impact glass from Tasmania	0·0031	$\sim 20 \times 10^{-5}$
Deep-sea sediment Argentine Basin.Core (Core V-18-35)	100 cm depth	0·0026	$\leq 140 \times 10^{-5}$
	355 cm depth	—	$\leq 1{\cdot}2 \times 10^{-5}$
	675 cm depth	0·0021	$\leq 3{\cdot}3 \times 10^{-5}$
	1045 cm depth	0·0030	$\leq 14 \times 10^{-5}$
Deep-sea sediment Mid. Pacific Ocean Core (Goldberg's core MP-38)	12–16 cm depth	—	$\leq 30 \times 10^{-5}$
	56–60 cm depth	0·0090	$\leq 73 \times 10^{-5}$
	214–218 cm depth	—	not detected
	369–373 cm depth	—	not detected
GR	Standard granite	0·010†	—
G–1	Standard granite	0·002†, 0·0020	$\leq 7 \times 10^{-5}$‡
W–1	Standard diabase	0·004†, 0·0048, 0·0049	$\leq 5 \times 10^{-5}$

* Near the limit of sensitivity for this technique, as discussed in the text.

† These data were derived from irradiated samples that were delayed in transit to our laboratories. Low activity levels suggest that they be regarded as valid only within a factor of two.

‡ The Ir abundance of 0·006 ppm reported for G-1 by Rushbrook and Ehmann (1962) should be discarded based on this later analysis, using a more sensitive counting technique.

several others found in the literature are tabulated in Table 8. It should be noted that the silicon abundances used in these calculations are actual experimental abundances in each specimen determined by VOGT and EHMANN (1965), using the technique of fast neutron activation analysis. These abundances would be slightly different if only the olivine–hypersthene (the L-group of UREY and CRAIG, 1953), or the carbonaceous chondrites were used as the basis of the cosmic abundance calculations, as suggested by several authors. Such group atomic abundances may be calculated from the data of Table 7, but the cosmic abundance values presented in Table 8 are based on the average for all chondrites analyzed in this work.

It is noted that the experimental atomic abundances obtained in this work substantiate rather well the theoretical calculations of CLAYTON and FOWLER (1961). The experimental data for platinum obtained in this work suggest the theoretical value is somewhat low, but definite conclusions are not possible since only four chondrite analyses were used in this comparison.

Previous analyses for rhodium, ruthenium, and osmium in chondrites suggest that siderophilic elements are more abundant in the H-group ordinary chondrites than in the L-group chondrites (UREY, 1964). Reference to Table 7 confirms that this is also true for the three siderophile elements gold, iridium and platinum.

It has been previously suggested by EHMANN *et al.* (1961) that the gross abundance of iridium in ordinary chondrites may be used to estimate the metal phase content in these meteorites due to the virtually 100 per cent siderophilic nature of this element. This suggestion was made on the basis of analyses of five (four of them from H-group), chondrites for which the metal phase abundances were well known. A strong correlation between metal phase content and iridium abundance was indicated, and an average metal phase concentration for iridium of 2·85 ppm was calculated on the basis of these five analyses. This is in excellent agreement with the value obtained experimentally by NICHIPORUK and BROWN (1962) of 2·8 ppm for the iridium concentration in chondritic metal phase. However, subsequent analyses of L-group chondrites have shown that the situation is not so simple. The concentration of iridium in these meteorites is variable with respect to the known metal phase content, and is in general higher than would be predicted by assuming a constant metal phase iridium abundance of 2·8 ppm. Also, the Murray carbonaceous chondrite has nearly normal abundances of iridium, gold, and platinum, in the absence of any appreciable metal phase.

A better correlation may be made between these metals and total iron content. The ratio of total iron in the H-group to total iron concentration in the L-group chondrites, according to UREY and CRAIG (1953) is 1·28, while the ratio of the average H-group iridium concentration to the average L-group concentration is 1·34. Similarly, the ratio for gold is 1·41. The ratio of the average metal phase content for the H-group chondrites to the average metal phase content for the L-group chondrites is 2·40 according to UREY and CRAIG (1953). The same ratio according to KEIL (1962) is 2·57.

The above observations seem to be consistent with the suggestion that the carbonaceous chondrites are primitive, relatively undifferentiated matter, from which the other meteorite types may have evolved (MASON, 1960). The evolutionary scheme

Table 7. Abundances of gold, iridium and platinum by chondrite type

Type	Average Au (ppm)	Average Ir (ppm)	Average Pt (ppm)	Abundance ratio Pt:Ir:Au
Olivine–bronzite (H-group)	0·24	0·51	1·6	6·7:2·1:1
Olivine–hypersthene (L-group)	0·17	0·38	1·4	8·2:2·2:1
Enstatite	0·37	0·083	1·3	3·5:0·22:1
Carbonaceous	0·17	0·38	1·5	8·8:2·2:1
All chondrites in Table 3	0·22	0·41	1·5*	6·8:1·9:1

* Based on four best Pt determinations only.

Table 8. Atomic abundances of gold, iridium and platinum relative to Si = 10^6 atoms

Reference (Source of data)	Atomic abundances, Si = 10^6 Au	Ir	Pt
Suess and Urey (1956) (Chondrite analyses and interpolations)	0·145	0·821	1·625
Cameron (1959) (Adjusted chondritic abundances)	0·145	0·494	1·28
Clayton and Fowler (1961) (Calculations from theories of nucleosynthesis)	0·13	0·39	0·80
This work (Chondrite analyses)*	0·18	0·35	1·28

* Calculations based on actual silicon abundances in the meteorites analyzed, as determined by Vogt and Ehmann (1965).

proposed follows the degree of iron reduction in the different classes of meteorites. The L-group chondrites would represent a later stage in which iron has been partially reduced. Elements more noble than iron would be concentrated in the metallic phase and follow a Prior's Law type of relationship with decreasing noble metal concentration in the metallic phase as total metallic phase abundance in the chondrite increases. The H-group chondrites would then represent a still later stage in which the metal phase concentration in the chondrite has been increased by further iron reduction with a corresponding dilution of the noble element concentration in the metallic phase. In the enstatite chondrites the iron is still more reduced and these meteorites lie still further along the evolutionary chain. The stony-irons, irons, and achondrites might then represent the final results of metal–silicate fractionation. Although this picture may present an oversimplification of the relationship between the various meteorite types, it seems consistent with the observed extent of chemical fractionation among the various classes.

The abundances of iridium and platinum appear to be anomalously low in the enstatite chondrite Abee, in spite of its high metal phase content of 23·8 per cent (DAWSON, MAXWELL and PARSONS, 1960). It is hoped that it will be possible to obtain other enstatite chondrites to determine if this anomaly is general for this type of chondrite. Until other enstatite chondrites are studied, no great significance should be attached to this apparent fractionation.

Analyses for these three elements in chondrites, siderites, meteoritic troilite, tektites and various terrestrial materials, as reported in Tables 3–6, confirm the suggestion that these elements are predominantly siderophilic in nature. In addition, significant chalcophilic tendencies are noted for gold and platinum on the basis of a single meteoritic troilite analysis. Iridium shows virtually no chalcophilic tendencies. All three elements are strongly depleted in silicate matrices as shown by the abundance data in metal-poor achondrites, meteoritic olivine, tektites and various terrestrial materials. Of these elements, iridium shows the greatest abundance variability. For example, the iridium abundance in the Sikhote–Alin siderite was found to be only 0·014 ppm in spite of near normal gold and platinum abundances, while iridium abundances as high as 2·5 ppm were found in other siderites.

If the ratios of Pt:Ir:Au are tabulated for various meteoritic and terrestrial materials as in Table 9, it becomes obvious that iridium is strongly depleted in the Earth's crust, in meteoritic troilite and in the tektites. While the ratios in Table 9 are certainly only approximate, the processes of differentiation operative in the formation of the Earth's crust appear to have discriminated against iridium to the extent of two orders of magnitude, if a chondritic starting material is assumed.

The extent of the depletion of iridium in the Earth's crust suggests that this element might serve as a sensitive indicator of an extraterrestrial origin for materials such as "cosmic" spherules which are found in sediments, glacial ice and upper atmosphere dust samples. Nickel, which has been commonly used for this purpose is a common ingredient in many commercial alloys and is not depleted as much in average terrestrial materials relative to the meteorites, as iridium. Iridium has the added advantage of being extremely easy to determine, using the technique of activation analysis and high flux reactors, at abundance levels as low as 10^{-11} g/g. We suggest that this iridium abundance test may be the best chemical indicator for meteoritic particles and we hope to check this suggestion in our laboratories.

The extremely low abundances of gold and iridium in the tektites are similar to those found in sediments and granitic rocks. Although the iridium analyses in these materials are good to a factor of two, at best, the Ir:Au ratio reported in Table 9 for the tektites is still clearly most similar to that for terrestrial rocks and sediments. Hence, these materials show the same sort of iridium depletion with respect to gold as exhibited by the crustal rocks of the Earth. The similarity between trace element abundances in tektites and common terrestrial materials has been pointed out by TAYLOR (1962) and has also been shown in analyses for various other trace elements in this laboratory (EHMANN, 1965).

Some suggestions of groupings are noted among the tektite occurrences on the basis of their gold contents. The australite from the Mt. Vernon site in Western Australia, the billitonite, the philippinite and the Libyan Desert glass all have similar high gold contents. Two more centrally located sites in Australia yielded specimens with gold abundances approximately an order of magnitude lower. Hence, these

Table 9. Abundance ratios for platinum, iridium and gold in various types of meteoritic and terrestrial materials

Type of material	Reference	Pt:Ir:Au (by weight)
Continental igneous rocks	VINOGRADOV (1956)	5:0·1:1
Siderites	NICHIPORUK and BROWN (1962); GOLDBERG, UCHIYAMA and BROWN (1951)	7·6:2·6:1
Chondrite metal phase	NICHIPORUK and BROWN (1962); VINCENT and CROCKET (1960b)	6·1:2·0:1
Chondrites	This work	6·8:1·9:1
Meteoritic troilite	This work	7:0·02:1
Tektites	This work	(–:0·006:1)*
Terrestrial rocks and sediments	This work	(–:0·02:1)*

* Estimated from Tables 5 and 6, using an average iridium abundance of approximately 5×10^{-5} ppm in both tektites and the terrestrial rocks and sediments.

tektite data would be consistent with suggestions of an overlap of two distinct distributions in the Australian strewn-field. Somewhat surprising is the low gold abundance in the bediasite analyzed, as compared to the other tektite occurrences. Other bediasites should be analyzed for gold to determine if this deficiency is general for the Texas occurrence.

Acknowledgements—The assistance of the following individuals and institutions that provided the specimens analyzed in this work is gratefully acknowledged: Dr. K. R. DAWSON (Geological Survey of Canada, Ottawa), Dr. R. E. FOLINSBEE (University of Alberta), Mr. E. P. HENDERSON (U.S. National Museum), Dr. E. L. KRINOV (Academy of Sciences, U.S.S.R.), Dr. C. B. MOORE (Arizona State University) and Dr. W. F. READ (Lawrence College, Appleton, Wisconsin) for the meteorite specimens; Dr. M. H. HEY (British Museum) for the Libyan Desert glass; Dr. V. E. BARNES (University of Texas) and Dr. B. DAILY (South Australian Museum, Adelaide) for some of the tektite specimens; Prof. S. W. CAREY (University of Tasmania) for the Darwin glass; Mr. W. BRASWELL, Mrs. E. YAWN (Eastman, Georgia) and Mr. T. E. ALLEN (Atlanta, Georgia) for the Georgia tektite; Dr. M. EWING (Lamont Geological Observatory) for the Argentine Basin cores; Dr. H. DE LA ROCHE (Centre de Recherches Petrographiques et Geochemiques, Nancy, France) for the sample of standard granite GR; and the U.S. Geological Survey for samples of standard rocks G–1 and W–1.

The authors wish to express their thanks to Mr. DONALD G. ALLIE for his assistance in sample preparation and in post-irradiation processing. The continuing support of the U.S. Atomic Energy Commission through contract AT-(40-1)-2670 and the University of Kentucky Faculty Research Fund is also gratefully acknowledged. One of the authors (P. A. B.) wishes to express his thanks to the University of Kentucky Graduate School for fellowship support received during a portion of this study.

REFERENCES

BAEDECKER P. A. (1964) Platinum, gold and mercury in meteorites and natural materials by neutron activation analysis. M.S. thesis, University of Kentucky, Lexington.

CAMERON A. G. W. (1959) A revised table of abundances of the elements. *Astrophys. J.* **129,** 676–699.

Clayton D. D. and Fowler W. A. (1961) Abundances of heavy nuclides. *Ann. Phys. (N.Y.)* **16,** 51–68.

Dawson K. R., Maxwell J. A. and Parsons D. E. (1960) A description of the meteorite which fell near Abee, Alberta, Canada. *Geochim. et Cosmochim. Acta* **21,** 127–144.

DeGrazia A. R. and Haskin L. (1964) On the gold contents of rocks. *Geochim. et. Cosmochim. Acta* **28,** 559–564.

Ehmann W. D. (1965) On some tantalum abundances in meteorites and tektites. *Geochim. et Cosmochim. Acta* **29,** 43–48.

Ehmann W. D., Amiruddin A., Rushbrook P. R. and Hurst M. E. (1961) Some trace element abundances in the Bruderheim meteorite. *J. Geophys. Res.* **66,** 3581.

Goldberg E. and Brown H. (1949) Neutron pile as a tool in quantitative analysis. *Science* **109,** 347–353.

Goldberg E., Uchiyama A. and Brown H. (1951) The distribution of nickel, cobalt, gallium, palladium, and gold in iron meteorites. *Geochim. et Cosmochim. Acta* **2,** 1–25.

Goldschmidt V. M. (1958) *Geochemistry.* Oxford University Press, London.

Hamaguchi H., Nakai T. and Kamenoto Y. (1961) Determination of platinum, iridium, and palladium in meteorites by neutron activation. *Nippon Kagaku Zasshi* **82,** 1489–1493 (*Chem. Abs.* **56,** 6650a).

Krinov E. L. (1960) *Principles of Meteoritics.* Pergamon Press, London.

Keil K. (1962) On the phase composition of meteorites. *J. Geophys. Res.* **67,** 4055–4061.

Mason B. (1960) The origin of meteorites. *J. Geophys. Res.* **65,** 2965–2970.

Nichiporuk W. and Brown H. (1962) Platinum and iridium abundances in meteorites. *Phys. Rev. Letters* **9,** 245–246.

Rushbrook P. R. and Ehmann W. D. (1962) Iridium in stone meteorites by activation analysis. *Geochim. et Cosmochim. Acta* **26,** 649–657.

Suess H. E. and Urey H. C. (1956) Abundanees of the elements. *Rev. Mod. Phys.* **28,** 53–74.

Taylor S. R. (1962) The chemical composition of australites. *Geochim. et Cosmochim. Acta* **26,** 685–722.

Urey H. C. (1964) A review of atomic abundances in chondrites and the origin of meteorites. *Rev. Geophys.* **2,** 1–34.

Urey H. C. and Craig H. (1953) The composition of the stony meteorites and the origin of the meteorites. *Geochim. et Cosmochim. Acta* **4,** 36–82.

Vincent E. A. and Crocket J. H. (1960a) Studies in the geochemistry of gold—I. The distribution of gold in rocks and minerals of the Skaergaard intrusion, East Greenland. *Geochim. et Cosmochim. Acta* **18,** 130–142.

Vincent E. A. and Crocket J. H. (1960b) Studies in the geochemistry of gold—II. The gold content of some basic and ultrabasic rocks and stone meteorites. *Geochim. et Cosmochim. Acta* **18,** 143–148.

Vinogradov A. P. (1956) Regularity of distribution of elements in the Earth's crust. *Geochemistry (USSR) (English Transl.)* **1,** 1–43.

Vogt J. R. and Ehmann W. D. (1965) Silicon abundances in stony meteorites by fast neutron activation analysis. *Geochim. et Cosmochim. Acta* (In press).

Reprinted from *Science*, **151**, 817–819 (1966)

19

Rubidium-Strontium Age of the Bosumtwi Crater Area, Ghana, Compared with the Age of the Ivory Coast Tektites

Abstract. *Rocks from the vicinity of Bosumtwi crater, Ghana, and a representative collection of Ivory Coast tektites have been analyzed mass spectrometrically for rubidium, strontium, and strontium isotopic composition. The data from the rocks of the crater area yield an age of 1.97 × 10⁹ years* (λ_{Rb} = *1.47 × 10⁻¹¹ year⁻¹). The data for the Ivory Coast tektites fall on this isochron. This identity of age values for the Ivory Coast tektites and the Birrimian basement rocks of West Africa strongly supports the hypothesis of terrestrial formation for these tektites. The evidence available at present suggests that the Ivory Coast tektites are most probably the fusion products of meteoritic impact at the Bosumtwi crater site.*

The purpose of this study was (i) to determine if the Ivory Coast tektites fall on the 400-million-year isochron which has been suggested (*1*) for tektites from the other three major geographic localities, namely, Australasia, Czechoslovakia, and North America; and (ii) to compare the rubidium, strontium, and strontium isotopic compositions of the Ivory Coast tektites with those of the rocks at the Bosumtwi crater in Ghana. Maclaren (*2*) first suggested, in 1931, that the Bosumtwi, or Ashanti, crater was formed by meteoritic impact. The discoveries of coesite (*3*) and shatter cones (*4*) at the crater support this theory. Cohen (*5*) suggested that the Ivory Coast tektites, which occur approximately 300 km to the west of the crater, might have been produced by the impact. Since then, K-Ar analyses of Ivory Coast tektites (*6*) and Bosumtwi glass (*7*) have yielded the same age, within rather large experimental uncertainty, of 1.3 (± 0.3) × 10^6 years. Recently, fission track dating (*8*) has supported this agreement between the time of melting of the crater glass and the Ivory Coast tektites.

All but two of the tektites analyzed, GSFC Nos. 3 and 4, were from a group of 75 samples recently obtained from the Ivory Coast. Details of this collection, such as specimen locality, weight, density, and so forth, are in preparation for later publication. The samples analyzed were chosen to cover the specific gravity range of the collection and to cover the range of Rb/Sr ratios, which were estimated from rough x-ray fluorescence measurements. All of the tektites, except GSFC No. 4, are from the approximately 45- by 80-km strewn-field area described by Lacroix (*9*). Tektite GSFC No. 4 was found in 1963 in a prospecting pit 2 km south of Adzopé; this is approximately 100 km to the south of the previously defined strewn-field.

The rubidium, strontium, and strontium isotopic compositions for the tektites are given in Table 1; those for the rocks of the Bosumtwi crater area, in Table 2. The rubidium and strontium determinations are by stable isotope dilution techniques, while the isotope ratio analyses are on unspiked strontium. The basic analytical procedures have been previously described (*1*). Precision of the rubidium and strontium isotope dilution analyses, as demonstrated by replicate analyses, is approximately ± 2 percent. The strontium isotopic ratios reported in Tables 1 and 2 were determined on four different mass spectrometers of similar design (12-inch radius of curvature, 60° sector, single focusing) at the Massachusetts Institute of Technology, the Goddard Space Flight Center, and the National Bureau of Standards. Most of the samples were analyzed on at least two of the machines; only the average values for each sample are reported in the tables. No systematic bias was observed between the machines; the average difference between $(Sr^{87}/Sr^{86})_N$ analyzed on the same sample was less than 0.1 percent for samples analyzed repeatedly on the same mass spectrometer and for samples analyzed on different machines.

The tektites in Table 1 are listed in order of increasing specific gravity. Thus they are probably listed in approximate order of decreasing SiO_2 content, as the inverse correlation between specific gravity and SiO_2 content has been demonstrated in several studies (*10*). The data in Table 1 suggest an inverse correlation between specific gravity and rubidium, but no correlation is evident between specific gravity and strontium.

The Ivory Coast tektites exhibit a rather restricted range in both rubidium and strontium, as has been observed for most elements in tektites. The strontium content in GSFC No. 4, however, is far outside the range of the other 12 samples. This value has been checked by two independent analyses, which agreed within ± 1 percent. The values for rubidium and strontium in Table 1 are in excellent agreement with the single reported analyses in an Ivory Coast tektite of 64 ppm Rb and 316 ppm Sr (by x-ray flourescence techniques) by Gentner *et al.* (*7*). They also report rubidium and strontium in a Bosumtwi glass from the same general locality as glass No. 10 in Table 2. Their results (of 73 ppm Rb and 358 ppm Sr) are again in excellent agreement with our values.

The Bosumtwi area is underlain by rocks of the Birrimian System (Pre-Cambrian). Except for a small amount of recent clays and agglomerates, the rocks in the immediate vicinity of the crater are metamorphosed sediments of Birrimian age (predominately phyllites with interbedded graywackes and quartzites) and intrusive granites (*11*). We are unaware of any reported determinations of the absolute age of these rocks from the Bosumtwi area. However, numerous determinations have been made on very similar Birrimian rocks from the Ivory Coast. Approximately 80 mineral and whole-rock Rb-Sr age determinations have been made on the Birrimian rocks from the Ivory Coast (*12*), and they indicate an age for this system of between 1.9 and 2.0 × 10^9 years.

The analyses of the granite reported in Table 2 confirm the identification of the rocks of the Bosumtwi area as Birrimian. With the same decay constant for Rb^{87} as was used in the Ivory Coast Birrimian calculations (λ_{Rb} = 1.47 × 10^{-11} yr^{-1}), the granite data yield an age of 1.97 × 10^9 years.

The data in Tables 1 and 2 are plotted in Fig. 1. The phyllite and graywacke (samples Nos. 12 and 13) have Rb/Sr ratios that are so low they show almost no enrichment of Sr^{87} even in 2 × 10^9 years. The glass and suevite (samples Nos. 10 and 8) lie on the 1.97 × 10^9 year isochron, despite the fact that the K-Ar analyses indicate they were formed approximately 1.3 × 10^6 years ago (*7*). The most logical interpretation is that they were produced by the mixing and fusion of the country rock of Birrimian age during the event that produced the crater, and merely reflect the Rb-Sr Birrimian age of their parent material. Calculations based on the Rb, Sr, and Sr isotopic composition of the phyllite, granite, and Bosumtwi glass indicate the glass could be a mixture of approximately 75 percent phyllite and 25 percent granite.

The Czechoslovakian, Australasian, and North American tektites are also plotted in Fig. 1, based on previously reported data (*1*). The 400-million-year isochron, through the average of each of these three groups, is shown. The Ivory Coast tektites definitely do not lie on this isochron, and thus the data presented here on the Ivory Coast tektites do not clarify the significance of this isochron.

However, the plotted points for the Ivory Coast tektites lie essentially on the Birrimian-age isochron of 1.97×10^9 years. Only three of the tektite points fall exactly on the isochron; the others all occur, at varying distances, slightly to the left of the line. This selective scatter to the left of the line can be ascribed to fractional volatilization of Rb relative to Sr during the fusion process which formed the tektites. Several recent studies suggest that selective volatilization of Rb relative to Sr took place during the fusion process that produced tektites (*13*) and, further, that the SiO_2 content decreases (hence the density increases) with increasing volatilization (*14*). The tektites listed in Table 1 (in order of increasing specific gravity) could be considered, therefore, as being listed roughly in order of increasing degree of volatilization and increasing loss of rubidium relative to strontium. There is an inverse correlation of rubidium with density, and a direct correlation of density with the distance of the points to the left of the line. That is, the samples at the top of the table, which according to the referenced studies should have suffered the least fractionation, lie on the line, while samples with increasing density (that is, less SiO_2 and more volatilization) fall, in general, further to the left of the isochron.

The age relationships presented here for the Ivory Coast tektites and the Birrimian rocks support the hypothesis of terrestrial origin for these tektites. The probability that these tektites, if extraterrestrial, would have just the right combination of rubidium, strontium, and strontium isotopic composition to lie on the isochron of the country rock on which they fell seems rather remote. Because much of the basement of West Africa is approximately 2×10^9 years old, this age value does not point unequivocally to the Bosumtwi crater as their place of formation. However, chemical similari-

Table 1. Rb, Sr, and Sr isotopic composition of Ivory Coast tektites.

Sample No.*	Rb (ppm)	Sr (ppm)	Sr^{86}/Sr^{88}	Sr^{87}/Sr^{86}	$(Sr^{87}/Sr^{86})_N$†	No. of analyses
T5702	75.3	289	0.1197	0.7224	0.7231	2
T5658	72.8	288	.1201	.7205	.7226	1
T5708	68.1	288	.1195	.7228	.7230	1
GSFC-4	63.6	399	.1197	.7178	.7187	3
T5704	66.5	303				
T5707	72.1	275				
T5709	65.4	325	.1195	.7221	.7224	2
T5663	67.6	277				
T5646	67.6	289	.1197	.7219	.7228	2
T5699	55.0	318				
T5673	67.5	287	.1201	.7196	.7217	2
GSFC-3	64.8	304	.1192	.7227	.7232	1
T5659	59.0	313	.1201	.7205	.7226	2

*Samples listed in order of increasing density. †Measured Sr^{87}/Sr^{86} ratio normalized to a constant Sr^{86}/Sr^{88} ratio of 0.1194.

Table 2. Rb, Sr, and Sr isotopic composition of rocks and glass from the Bosumtwi Crater area, Ghana.

Sample No. and description	Rb (ppm)	Sr (ppm)	Sr^{86}/Sr^{88}	Sr^{87}/Sr^{86}	$(Sr^{87}/Sr^{86})_N$*
No. 12. Phyllite from Obuom mine, south of Kumasi.	10†	500†	0.1181	0.7075	0.7036
No. 13. Graywacke, interbedded with sample No. 12.	35.5	625	.1194	.7044	.7044
No. 14 W.R. Granite, ½ mile from Ijisu, N.E. of crater. Whole rock fraction.	301	98.9	.1192	.9616	.9610
No. 14 C.F. Same granite as No. 14 W.R.; coarse fraction (+100 mesh).	344	85.9	.1187	1.0440	1.0409
No. 8. Suevite from Ata stream bed, Bosumtwi crater.	53.8	181	.1198	0.7255	0.7266
No. 10. Dark glass separated from suevite from Ata stream bed.	64.1	351	.1194	.7177	.7179

*See second footnote to Table 1. †Estimates by x-ray fluorescence.

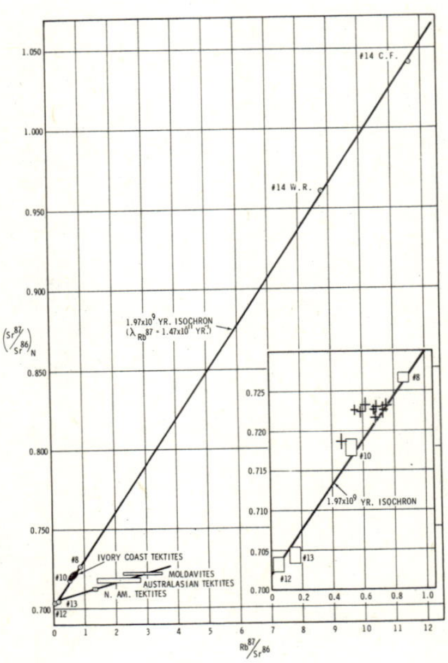

Fig. 1. Plot of Sr^{87}/Sr^{86} versus Rb^{87}/Sr^{86} for Bosumtwi crater rocks and Ivory Coast tektites. On the full plot the Bosumtwi rocks are shown as circles, while the field of Ivory Coast tektites is shown as a cross-hatched area. The Bosumtwi sample numbers correspond to the numbers in Table 2. The values covered by the other three groups of tektites (*1, 13*) are also shown on the full plot. In the large-scale inset, which refers to the lower left portion of the large diagram, the errors are shown, the Bosumtwi area rocks by boxes, and the individual Ivory Coast tektites by crosses.

ties between the Ivory Coast tektites and glass from the Bosumtwi crater support this hypothesis of common origin. Gentner *et al.* (*7*) report the close similarity between the Bosumtwi crater glass and the Ivory Coast tektites for seven trace elements. Our Rb, Sr, and strontium isotopic analyses support this close similarity between the two types of materials. In addition, preliminary chemical analyses of these samples show a close similarity for the major elements examined: Fe, Ti, Ca, Mg, K, Na, and Mn (*15*). The concentrations of all these elements, except Ca, in the one piece of Bosumtwi glass available, lie within the rather restricted range of the 11 analyzed Ivory Coast tektites.

These three lines of evidence—the Rb-Sr age correlation (age of parent material), the K-Ar age correlation (age of fusion), and the chemical correlation—strongly support the hypothesis that the Ivory Coast tektites are related to the phenomenon that formed the Bosumtwi crater.

Note added in proof. An independent Rb-Sr isotopic study of three Ivory Coast tektites and five Bosumtwi crater glasses has recently been completed by Lippolt and Wasserburg (*Z. Naturforsch.*, in press). Their data and conclusions are essentially in agreement with this paper.

C. C. SCHNETZLER
Goddard Space Flight Center, Greenbelt, Maryland

W. H. PINSON
P. M. HURLEY
Department of Geology and Geophysics, Massachusetts Institute of Technology, Cambridge

References and Notes

1. C. C. Schnetzler and W. H. Pinson, *Geochim. Cosmochim. Acta* **28**, 953 (1964).
2. M. Maclaren, *Geog. J. London* **78**, 270 (1931).
3. J. Littler, J. J. Fahey, R S. Deitz, E. C. T. Chao, *Geol. Soc. Am. Spec. Papers* **68**, 218 (1961).
4. H. P. T. Rohleder, *Centralbl. Mineral. Geol.* **316** (1934).
5. A. J. Cohen, *J. Geophys. Res.* **66**, 2521 (1961).
6. J. Zahringer, *Radioactive Dating* (International Atomic Energy Agency, Vienna, 1963), p. 289.
7. W. Gentner, H. J. Lippolt, O. Müller, *Z. Naturforsch.* **19a**, 150 (1964).
8. R. L. Fleischer, P. B. Price, R. M. Walker, *Geochim. Cosmochim. Acta* **29**, 161 (1965).
9. A. Lacroix, *Compt. Rend. Acad. Sci. Paris* **199**, 1539 (1934).
10. V. E. Barnes, *Texas Univ. Publ.* No. **3945**, 477 (1939); E. C. T. Chao, in *Tektites*, J. O'Keefe, Ed. (Univ. of Chicago Press, Chicago, 1963), p. 51.
11. N. R. Junner, *Gold Coast Geol. Surv. Bull.* **8**, 5 (1937); H. P. T. Rohleder, *Geog. J. London* **87**, 51 (1936); John Saul, personal communication.
12. M. Bonhomme, *Ann. Fac. Sci. Clermont* **5**, Geol. Mineral. (1962); M. Vachette, *Comp. Rend. Acad. Sci. Paris* **258**, 1569 (1964).
13. J. A. Philpotts and W. H. Pinson, *Geochim. Cosmochim Acta*, in press; W. H. Pinson, C. C. Schnetzler, J. A. Philpotts, H. W. Fairbairn, *Trans. Amer. Geophys. Union* **46**, 118 (1965). Preliminary work on heating of tektites to 1800° to 2000°C, in both air and vacuum, indicates loss of Rb of between 15 and 35 percent, while the Sr remains essentially constant.
14. J. A. Philpotts and W. H. Pinson, *Geochim. Cosmochim. Acta*, in press; L. S. Walter and M. K. Carron, *ibid.* **28**, 937 (1964); L. S. Walter, *Science* **147**, 1029 (1965).
15. W. H. Pinson, J. A. Philpotts, J. Hwang, unpublished data on analyses by x-ray fluorescence and atomic absorption.
16. We are indebted to the following for donating samples: John Saul and Elliot Miller, of Saul Miller, Consulting Geologists; E. C. T. Chao of the U.S.G.S.; R. M. Walker and R. L. Fleischer of the General Electric Research Laboratory; and the Société de Développement Minéral de la Côte d'Ivoire. Financial aid for this research was furnished in part by NASA through the Center for Space Research at the Massachusetts Institute of Technology and in part by the U.S. Atomic Energy Commission. Field work for the collection of the tektites in the Ivory Coast by John Saul and Elliott Miller was sponsored by the National Geographic Society. We also thank W. R. Shields of the National Bureau of Standards for the use of mass spectrometers, H. H. Thomas, M. L. Bottino, and P. D. Fullagar for aid in analytical work, and members of the GSFC Geochemistry Laboratory for helpful discussions.

18 October 1965

20

Geochimica et Cosmochimica Acta, 1969, Vol. 33, pp. 1101 to 1102. Pergamon Press. Printed in Northern Ireland

The radioactivity of the Ivory Coast tektites and the formation of the Bosumtwi Crater (Ghana)

L. RYBACH
Institute for Crystallography and Petrography, Swiss Federal Institute of Technology
8006 Zurich, Switzerland

and

J. A. S. ADAMS
Department of Geology, Rice University, Houston, Texas 77001

(*Received* 5 *March* 1969; *accepted* 31 *March* 1969)

Abstract—Further evidence for the impact origin of the Ivory Coast tektites and the Bosumtwi Crater is presented: uranium, thorium and potassium abundances show very similar patterns in tektites and crater rocks.

U, Th and K have been determined in 20 Ivory Coast tektites by nondestructive gamma-ray spectrometry. Five total rock samples from the Bosumtwi Crater have been analyzed by the same method. The experimental procedures were the same as described in BALACEK and ADAMS (1965) and in ADAMS (1964). The concentrations have been calculated from the measured gamma-ray spectra by the method given in RYBACH *et al.* (1966). The K and U concentrations in the tektites have been checked by electron microprobe and fluorimetric methods.

As compared to other tektite groups, the Ivory Coast tektites are noteworthy in having an average of 1·9 ppm Th and a mean Th/U ratio of 2·6, both of which are significantly lower than values in any single specimen or group of tektites reported thus far (O'KEEFE, 1963; CHERRY and ADAMS, 1963; BALACEK and ADAMS, 1965). The mean K content (1·55% K) is also significantly lower than values for other tektite groups. The mean K/U ratio ($1{\cdot}32 \times 10^4$) is clearly "terrestrial" (cf. WASSERBURG *et al.*, 1964).

The old ($2 \cdot 10^9$ yr) Precambrian rocks from the Bosumtwi Crater area show signs of intensive shock deformation which was studied in thin sections; in samples from the inner slope of the crater quartz with cleavage planes, diaplectic quartz-glass, plagioclase glass (partly recrystallized) and coesite aggregates have been found. STÖFFLER (1966) describes similar features in the Ries suevite.

The mean U, Th and K concentrations of the Crater rocks are compared to the tektite values below; a very close correspondence is evident.

	U (ppm)	Th (ppm)	K (%)	$(K/U) . 10^4$	Th/U
Ivory Coast tektites (mean of 20)	1·10	2·9	1·55	1·32	2·6
Bosumtwi Crater rocks (mean of 5)	1·05	2·6	1·50	1·43	2·5

The results reported here are consistent with the conclusions of Gentner *et al.* (1964), Schnetzler *et al.* (1966) and Taylor and Epstein (1966)—based on K/Ar age determinations, strontium and oxygen isotopic analysis—that the Crater rocks were the source of the tektites, and that the Bosumtwi Crater was formed by meteorite impact (cf. Maclaren, 1931).

Acknowledgments—The loan of rock samples and/or tektite specimens by the Department of Geology and Geophysics, Massachusetts Institute of Technology, Cambridge, USA, and by the Max-Planck-Institut für Kernphysik, Heidelberg, W. Germany, is kindly acknowledged.

References

Adams J. A. S. (1964) Laboratory gamma-ray spectrometer for geochemical studies. In *The Natural Radiation Environment*, (editors J. A. S. Adams and W. Lowder), pp. 485–497. University of Chicago Press.

Balacek K. J. and Adams J. A. S. (1965) Radioactivity in tektites. *Trans. Amer. Geophys. Union* **46,** 545.

Cherry R. D. and Adams J. A. S. (1963) Gamma-spectrometric determination of thorium, uranium and potassium in tektites. *Geochim. Cosmochim. Acta* **27,** 1089–1096.

Gentner W., Lippolt H. J. and Müller W. H. (1964) Das Kalium–Argon-Alter des Bosumtwi-Kraters in Ghana und die chemische Beschaffenheit seiner Gläser. *Z. Naturforsch.* **19a,** 150–153.

Maclaren M. (1931) Lake Bosumtwi, Ashanti. *Geogr. J.* **78,** 270–76.

O'Keefe J. (editor) (1963) *Tektites*, p. 109. University of Chicago Press.

Rybach L., von Raumer J. and Adams J. A. S. (1966) A gamma spectrometric study of Mont-Blanc granite samples. *Pure Appl. Geophys.* **63,** 172–178.

Schnetzler C. C., Pinson W. H. and Hurley P. M. (1966) Rubidium–strontium age of the Bosumtwi Crater area, Ghana, compared with the age of the Ivory Coast tektites. *Science* **151,** 817–819.

Stöffler D. (1966) Zones of impact metamorphism in the crystalline rocks of the Nördlinger Ries Crater. *Contribs. Mineral. Petrol.* **12,** 15–24.

Taylor H. P. and Epstein S. (1966) Oxygen isotope studies of Ivory Coast tektites and impactite glass from the Bosumtwi Crater, Ghana. *Science* **153,** 173–175.

Wasserburg G. J., MacDonald G. J. F., Hoyle F. and Fowler W. A. (1964) Relative contributions of uranium, thorium and potassium to heat production in the Earth. *Science* **143,** 456–467.

Editors' Comments on Papers 21 and 22

Three reports were contemplated for inclusion here on the subject of isotopic composition of tektites, but space limitations prevented reproduction of the paper by Zahringer (1963). He reviewed the status of isotope measurements on tektites up to 1963. Zahringer concluded, concerning cosmic-ray-produced isotopes, "either that the time of exposure to cosmic rays has been short or that the proto-tektite body was large. This supports the conclusion, drawn from the K/Ar ages, that tektites cannot derive from bodies more distant than the moon."

Although age determinations by the K^{40}–Ar^{40} method for tektites and impact glasses have been widely accepted, there are some problems. For example, Clarke et al. (1966) melted various geologically old materials and found by the K^{40}–Ar^{40} method that the resulting glasses have apparent ages from zero to over one million years. This finding, and the finding that the geologic age of australites (Lovering et al., 1972)* is much younger than the K–Ar age, introduce doubt as to the reliability of K–Ar dating.

In the first paper, Tilton* decisively demonstrates that lead isotopes in tektites are similar to those in terrestrial rocks and that a source material could be argillaceous sediments. Silverman (1951), in a study of oxygen isotopes, demonstrated that tektites have isotope ratios for oxygen within the range of terrestrial sedimentary rocks. These papers have been largely ignored by workers who favored the moon as the source of tektites.

Although Tilton* has shown that the lead in tektites is terrestrial lead, Starik et al. (1961, 1963)*, using similar data, concluded that tektites are extraterrestrial in origin as noted in the second paper, by Wampler, Smith, and Cameron,* who support Tilton's original conclusion. They explored as well the postulate that Ivory Coast tektites are related to the Bosumtwi Crater and that moldavites are related to the Ries Crater. In the Ivory Coast, in the area of tektite occurrence, the rocks are of the same kind and age as those in which Bosumtwi Crater is excavated. Therefore, if a connate source of energy had been available, the Ivory Coast tektites could have formed locally so far as source material is concerned.

To conserve space, the "Experimental" section of the paper by Tilton* and the section on "Analytical procedures" and "Appendix 1—Description of Samples" of the paper by Wampler, Smith, and Cameron* have been omitted.

Dr. G. R. Tilton is now Professor, Department of Geological Sciences, University of California, Santa Barbara, California, 93106.

Dr. J. M. Wampler is now Associate Professor, School of Geophysical Sciences, Georgia Institute of Technology, Atlanta, Georgia, 30332. His co-authors, Dr. D. H. Smith and Dr. A. E. Cameron, are at Oak Ridge National Laboratory, in the Analytical Chemistry Division. Although Dr. Cameron is retired, he still receives his mail at the Laboratory.

References

Clarke, R. S., Jr., Wosinski, J. F., Marvin, R. F., and Friedman, I. (1966). Potassium–argon ages of artificial tektite glass. *Trans. Amer. Geophys. Union,* **47,** 144.

*Lovering, J. F., Mason, B., Williams, G. E., and McColl, D. H. (1972). Stratigraphical evidence for the terrestrial age of australites. *J. Geol. Soc. Australia,* **18,** 409–418.

Silvermann, S. R. (1951). The isotope geology of oxygen. *Geochim. Cosmochim. Acta,* **2,** 26–42.

Starik, I. E., Sobotovich, E. V., Shats, M. M., and Lovtsyus, G. P. (1961). Uranium and lead in tektites. *Meteoritika,* **20,** 204–*207.*

Starik, I. E., Sobotovich, E. V., and Shats, M. M. (1963). The problem of the origin of meteorites and tektites. "Geochemistry" (Engl. translation) pp. 261–270.

Zahringer, J. (1963). Isotopes in tektites. *In* "Tektites" (J. A. O'Keefe, ed.), pp. 137–149. Univ. Chicago Press, Chicago, Illinois.

21

Geochimica et Cosmochimica Acta, 1958, Vol. 14, pp. 323 to 330. Pergamon Press Ltd., London

Isotopic composition of lead from tektites

G. R. TILTON
Geophysical Laboratory, Carnegie Institution of Washington, Washington, D.C.

Abstract—The isotopic composition of lead in three tektites and Libyan Desert glass is compared with that in known terrestrial and extra-terrestrial sources. The lead contained in the glasses is similar to modern terrestrial lead, particularly lead from modern oceanic sediments. The uranium, thorium and lead concentrations were determined for one of the glasses, an australite. Evidence is given which indicates that within the last tens of millions of years differentiation of uranium, thorium and lead occurred in the parent material of the australite. These results are difficult to explain in terms of any extra-terrestrial origin involving fusion of materials from the moon, meteorites or comets, but they are readily explained if tektites are of terrestrial origin involving fusion of argillaceous sediments in some unspecified way.

INTRODUCTION

THE isotopic compositions of modern terrestrial lead isolated from various sources such as young basalt, oceanic clay, and galenas (excluding "anomalous" galenas) are quite similar. The narrow range of isotopic composition (see Table 3 below) indicates that the lead was produced in environments having nearly equal uranium–lead and thorium–lead ratios. Moreover, information obtained from studies of the isotopic composition of galena lead as a function of time indicates that there has been little, if any, chemical separation of uranium and thorium from lead in the parent material of galenas during the past two or three billion years. Since average terrestrial leads are to be associated with closely defined ratios of uranium and thorium to lead, extra-terrestrial bodies which have had different proportions of these elements over periods of time which are of the order of the age of the solar system will yield lead of unusual isotopic composition. An example of this is the lead found in the Nuevo Laredo stone meteorite by PATTERSON (1955).

Since tektites have been considered to have a possible extra-terrestrial origin, an examination of the isotopic composition of their lead appeared profitable. Accordingly, lead was isolated from three tektites and a sample of Libyan Desert glass, analysed isotopically, and found to be similar to modern terrestrial lead in composition. The concentrations of uranium, thorium and lead were then determined for one of the glasses in order to see whether the uranium–lead and thorium–lead ratios had been changed recently, as might be expected if the source material was a terrestrial sediment. These studies constitute a preliminary investigation into the natural glass problem by the lead method, but the results are reported here since they provide information in an area where data have been completely lacking.

EXPERIMENTAL

The glasses used were supplied to the author by I. FRIEDMAN of the U.S. Geological Survey. The following descriptions accompanied the samples:

Australite: Collected near railroad line from Abminga to Finke, South Australia, by W. A. CASSIDY. $d = 2{\cdot}46$; no. 97-2.

Moldavite:	Habri in S. Bohemia. Courtesy of the Natural History Museum of Vienna. $d = 2{\cdot}37$; no. 97-3.
Philippinite:	Bulacan. American Museum of Natural History sample no. 3790. $d = 2{\cdot}45$.
Libyan Desert glass:	Courtesy of L. J. SPENCER. Described by SPENCER (1939).

The surfaces of the glasses had been sandblasted by FRIEDMAN. The samples were washed for several minutes in hot 6 M hydrochloric acid, rinsed in lead-free distilled water, and dried. They were then crushed in a steel mortar which had been used only for grinding basalts and ultra-basic rocks. The powders were decomposed in platinum dishes with hydrofluoric and perchloric acids. Lead was recovered by dithizone extractions, first at a pH of 8·5 in the presence of citrate and then at a pH of 9 in the presence of potassium cyanide. The samples yielded from 15 to 20 μg of lead, which may be compared to blanks of from 0·2 to 0·3 μg as measured by isotope dilution. The lead, uranium and thorium concentrations in the australite were all determined by the method of isotope dilution and are thought to be accurate within ± 1 per cent. These procedures have been described in previous papers (TILTON, ALDRICH and INGHRAM, 1954; TILTON *et al.*, 1955; ALDRICH *et al.*, 1956). Mass spectrometer runs were made by surface ionization using an instrument equipped with an electron multiplier.

In general, the glasses were nearly ideal materials as far as the analytical methods were concerned. They dissolved easily and the major constituent, silica, was removed during the decomposition. This permitted the use of very dilute solutions for subsequent chemical operations, leaving little danger of interferences from other elements during the extractions. The analyses are by no means as difficult as those for basalts or meteorites.

RESULTS

Tables 1 and 2 give the isotopic composition of lead from the four glasses and the uranium, thorium and lead concentrations for the australite. The ratios in Table 1 are believed to have absolute errors of $\pm 0{\cdot}5$ per cent with regard to instrumental uncertainties. This is based on replicate analyses of two standard leads made up from mixtures of gravimetrically standardized solutions of nearly pure Pb^{206} and Pb^{208}. The standards have Pb^{206}/Pb^{208} ratios of approximately 1 and 20 in order to determine possible non-linearities in the mass spectrometer. It has so far been unnecessary to apply either electron multiplier or non-linearity corrections to the lead ratios. A more complete report of this work is planned for future publication. The ratios in Table 1 are based on Pb^{206} rather than on Pb^{204} as has generally been done in the past. The new method is a more accurate way of presenting the data since it bases the results on a high intensity peak of intermediate mass rather than on a low intensity peak of lowest mass. A Pb^{206}/Pb^{207} ratio can nearly always be measured more accurately than a Pb^{207}/Pb^{204} ratio, for instance, yet the Pb^{206}/Pb^{204} and Pb^{206}/Pb^{207} ratios for a lead contain the same essential information as do the Pb^{206}/Pb^{204} and Pb^{207}/Pb^{204} ratios.

The concentration values in Table 2 are believed to have analytical errors of ± 2 per cent for lead and ± 1 per cent for uranium and thorium. The lead result

Table 1. Isotopic composition of lead from some natural glasses

Source	Pb^{206}/Pb^{204}	Pb^{206}/Pb^{207}	Pb^{206}/Pb^{208}
Australite	18·81	1·206	0·486
Moldavite	18·55	1·185	0·483
Philippinite	18·83	1·199	0·482
Libyan Desert glass	19·11	1·213	0·488

Table 2. Concentrations of uranium, thorium and lead in the australite

Element	Amount taken for analysis (g)	Concentration (p.p.m.)
Uranium	0·587	1·74
Thorium	0·587	9·19
Lead	2·346	2·90

is corrected for a blank of 0·25 μg, uranium for a blank of 0·007 μg, and thorium for 0·02 μg. It is possible to estimate the lead concentrations in the moldavite, philippinite, and Libyan Desert glass from the amounts of dithizone which were required for the final lead extractions. An estimate of this type made for the australite agreed with the final isotope dilution value within 10 per cent. The following estimates are probably accurate within ± 20 per cent: moldavite, 6 p.p.m.; philippinite, 5 p.p.m.; Libyan Desert glass, 5 p.p.m.

Discussion

The data obtained on the glasses are in complete harmony with a terrestrial origin. First, the isotopic composition of the lead is similar to that of modern terrestrial lead. The leads in Table 1 are quite similar to each other although small differences probably exist. In particular the Libyan desert glass appears to have more Pb^{206} with respect to Pb^{204} and Pb^{207} than do the other glasses. Replicate determinations would be necessary in order to establish these differences with certainty. With the exception of the australite the samples were completely consumed in the analytical work reported in Tables 1 and 2. For purposes of comparison the isotopic compositions of some modern terrestrial leads are given in Table 3. The similarity of the leads in Table 1 to those in Table 3 is obvious. The leads from the glasses are particularly similar to those found in the Pacific Ocean clay and the manganese nodule. The discussions of this paper will ignore "anomalous" leads, that is, leads which have been associated with an unusually large amount of uranium (and sometimes thorium as well) with respect to lead for periods of time which are short with respect to the age of the earth. Anomalous leads often occur in known uraniferous provinces, for example.

A second similarity to terrestrial materials concerns the uranium–thorium–lead balance observed for the australite. That is, the relative amounts of uranium and thorium with respect to lead found in the glass can be compared with the uranium–lead and thorium–lead ratios calculated for the environments which have produced modern terrestrial lead. This comparison will yield information

Table 3. Isotopic compositions of some modern terrestrial leads

Source of lead	Reference	Pb^{206}/Pb^{204}	Pb^{206}/Pb^{207}	Pb^{206}/Pb^{208}
Snake River basalt	PATTERSON *et al.*, 1955	18·12	1·173	0·476
Red Clay, Pacific Ocean	PATTERSON *et al.*, 1955	18·95	1·202	0·487
Manganese nodule, Pacific Ocean	PATTERSON *et al.*, 1955	18·91	1·205	0·489
Fumarolic magnetite, Valley of Ten Thousand Smokes, Alaska	This paper	18·99	1·221	0·493
Average modern galena	BATE, 1955	18·9	1·19	0·482

concerning possible recent changes in the ratios of thorium to lead and uranium to lead which might have occurred in the source material of the australite. There are two ways in which the uranium–lead and thorium–lead ratios which have governed the evolution of terrestrial lead may be calculated. The first is to plot the isotopic composition of lead from galena deposits as a function of their age. Studies of this type indicate that galena leads have been produced in environments which have a U^{238}/Pb^{204} atom ratio of 9·8 and a Th^{232}/U^{238} ratio of 3·9 at the present time (COLLINS, RUSSELL and FARQUHAR, 1953). As mentioned previously the observed variation of the isotopic composition of the leads is very nearly that which would be calculated for a source in which no differentiation of uranium, thorium and lead by processes other than radioactive decay have taken place.

A second method of calculating the uranium–lead and thorium–lead ratios of the source material which has produced terrestrial leads is to assume that the earth originated 4·6 billion years ago with lead of the same isotopic composition as that found in Canyon Diablo and Henbury troilites by PATTERSON (1956). PATTERSON has discussed the plausibility of this model in some detail and no further treatment will be given here. Using this model together with the isotopic composition of lead for an average modern galena as given in Table 3, the following atom ratios are calculated for the present-day environment which has produced the lead: $U^{238}/Pb^{204} = 9{\cdot}3$ and $Th^{232}/U^{238} = 4{\cdot}1$. These values are in quite good agreement with those calculated from the observed variation of the isotopic composition of lead in galenas with time.

Table 2 gives the concentrations of uranium, thorium and lead observed in the australite. From this data and the isotopic composition of the lead in Table 1, it is found that the U^{238}/Pb^{204} atom ratio of the glass is 38·5 and the Th^{232}/U^{238} atom ratio is 5·4. These values differ substantially from those calculated above.

Since the isotopic composition of the lead in the glass is that of a modern terrestrial lead, it must have been in an environment with "normal" ratios of uranium and thorium to lead until recent times, assuming a terrestrial origin for the present discussion. Within the last 50 to 100 million years at most uranium was enriched by a factor of 4 with respect to lead and thorium was enriched by a factor of 1·4 with respect to uranium.

The similarity of the bulk chemical composition of tektites to that of argillaceous sediments has been emphasized by many writers. The enrichment of thorium with respect to uranium in the australite further emphasizes this resemblance. The thorium would be expected to accumulate relative to uranium as a result of precipitation or adsorption processes taking place during the formation of an argillaceous sediment. Leaching processes would remove uranium preferentially to thorium. Although the arguments of this section apply only to the australite, ADAMS (1956) has made comparisons of the ratios of alpha activity to uranium concentration for a number of tektites and other glasses. ADAMS' data suggest that all tektites have high Th/U ratios similar to those found in oxidized and weathered sediments. Thus the conclusions regarding thorium–uranium differentiation probably apply to all tektites, although the magnitude of the effect can be expected to vary somewhat from one glass to another.

SUESS, HAYDEN and INGHRAM (1951) have determined argon in two philippinites and an australite and set upper limits on the argon contents which correspond to maximal potassium–argon ages of from 10 to 75 million years. They believe the actual ages are probably considerably less than the quoted limits. GERLING and YASHCHENKO (1952) similarly set upper limits on the potassium–argon ages of a philippinite indochinite, and moldavite of from 3 to 12 million years. These are the maximum periods of time elapsed since fusion of the glasses. The uranium–thorium–lead differentiation must have preceded the fusion of the glasses, since melting alone could not be expected to produce the observed separations. For a terrestrial origin the past history of the tektites must include a chemical differentiation of elements from some source material within the last 100 million years followed by a fusion of the differentiated product within the last tens of millions of years at most. Both of these processes may actually have occurred within the last few thousands of years.

The data may be tested against possible extra-terrestrial origins for tektites. The possibilities which have been considered in the literature involve origins from meteorites, the moon or comets. The possibility of a meteoritic origin will be considered first. Simple fusion of stone meteorites is not an acceptable mode of origin. Stone meteorites with chemical compositions which correspond to those of tektites are not known, nor can a tektitic composition be produced from a stone meteorite by removal of volatile constituents. In addition tektites contain about 100 times more uranium and thorium than do stone meteorites. Thus if tektites have a meteoritic origin they must represent some special class of meteorites. The possibility that tektites are fused sediments from another planet has recently been considered by BARNES (1958). The atmospheres of all the other planets in the solar system are grossly different from that of the earth

(Kuiper, 1949). The differences make it seem unlikely that terrestrial sedimentary processes could have been duplicated on another planet.

There is moreover the problem of transporting the sediments from interplanetary space to earth in a compact swarm. Barnes uses comets for this purpose. Some evidence against comets as a source material for tektites is given below. Although it is impossible to disprove rigorously an extra-terrestrial sedimentary origin for tektites, it would seem that such a theory raises as many problems as it answers.

It appears to be difficult to reconcile the lead data with a lunar origin. Meteorites and the earth give lead results which indicate that they have evolved from matter which 4·5 billion years ago possessed lead of the same isotopic composition as that found today in Canyon Diablo and Henbury troilites. In subsequent discussion this will be referred to as the "earth–meteorite lead system." If the moon is also a member of this system, it becomes difficult to explain the uranium–thorium–lead balance for the australite. It will be recalled that the present day concentrations of uranium and thorium with respect to lead are respectively four and six times those which are required to account for the isotopic composition of the lead on the basis of the above model. The uranium–lead and thorium–lead ratios must have increased within the last few millions of years to account for the isotopic composition of the lead. It is unlikely that such a process could have taken place on the moon since it is improbable that igneous or sedimentary processes are presently taking place there. Urey (1956) recently discussed the surface features of the moon and gives evidence for believing that the craters seen on the moon were formed at a time earlier than 3 billion years ago. There are no orogenic belts on the moon such as are known on the earth, although small lunar volcanoes might exist and escape detection. Even if volcanoes are active on the moon, no terrestrial igneous processes are known which produce materials of the chemical composition of tektites so that quite different conditions would be required on the moon as compared to the earth in this case. The moon has an atmosphere of less than from 10^{-5} to 10^{-6} terrestrial atm on the hemisphere facing the sun (Struve, 1944), so that sedimentary processes must be ruled out. Alternatively, the moon may have captured large meteorites within the last tens of millions of years since it is known that the earth has done so—the Canyon Diablo meteorite, for example. The energy of these collisions would undoubtedly have produced some melted silicates such as are found near the large terrestrial meteorite craters, but it is difficult to see how this process could separate thorium from uranium in the resulting glass. The glasses found near terrestrial craters contain diagnostic quantities of nickel which are lacking in tektites (Spencer and Hey, 1933). There is, moreover, the question as to whether a group of glass particles originating from an impact on the moon could arrive at the earth in a compact swarm.

An alternative assumption is that the moon has been made in some way from matter with lead of an isotopic composition which bears no relation to that observed in the earth–meteorite lead system. Then it might be expected to be a remarkable coincidence that the lead found in tektites has an isotopic composition which closely resembles modern terrestrial lead.

Little is known about comets so that it is admittedly difficult to exclude them as a parent material for tektites. Nickel is observed in the spectra of comets along with iron, chromium and sodium when they pass near the sun (WHIPPLE, 1950). Perhaps the lack of nickel in tektites is the best evidence against their originating in comets. It might also be an unexpected coincidence if comets were to contain lead of the same isotopic composition as modern terrestrial lead.

The effect of all the arguments just discussed is to place the lead evidence concerning tektites in favour of a terrestrial origin. The difficulties encountered in an extra-terrestrial origin could be avoided by considering extra-terrestrial bodies with grossly different chemical characteristics from those of the earth–meteorite system but there is no evidence to justify such an approach at present. It is true that many questions remain unanswered once a terrestrial origin is accepted. The most serious of these is a mechanism for melting the parent materials so as to produce the observed geographic distribution of tektites. It should be noted, however, that the geographic distribution is a stumbling block to extra-terrestrial modes of origin as well. It is difficult to specify a mechanism by which tektites can arrive at the earth in a compact swarm from the moon or from interplanetary space (UREY, 1955). The problem is that if the swarm is given sufficient density in order to be gravitationally stable in the sun's gravitational field at the distance of the earth from the sun, then australites, for example, would have to be many orders of magnitude more abundant than they are.

In conclusion a general statement concerning all tektite experiments is perhaps in order. If tektites are of extra-terrestrial origin it is possible that this may be proved unambiguously by demonstrating that they have certain properties which are clearly outside the range of experience with terrestrial materials. If, on the other hand, they are of terrestrial origin, it will be impossible to prove this with certainty since our knowledge of extra-terrestrial materials is limited. The possibility will always remain that the processes which produced tektites could have been duplicated in some extra-terrestrial environment. If they are terrestrially produced objects, further experiments will only serve (in principle) to define additional areas of similarity between terrestrial materials and tektites; the greater the number of areas of similarity established, the more probable a terrestrial origin will become. This would seem to be the trend of most tektite investigations which have been conducted to date.

Acknowledgements—J. A. S. ADAMS of The Rice Institute provided the author with a copy of his unpublished data on the thorium–uranium ratios in sediments. P. H. ABELSON of this Laboratory reviewed the manuscript critically and contributed suggestions which materially improved several aspects of the discussions.

REFERENCES

ADAMS J. A. S. (1956) *Uranium Contents and Alpha Particle Activities of Tektites.* International Geological Congress, Mexico City.

ALDRICH L. T., DAVIS G. L., TILTON G. R. and WETHERILL G. W. (1956) Radioactive ages of minerals from the Brown Derby Mine and the Quartz Creek granite near Gunnison, Colorado. *J. Geophys. Res.* **61,** 215.

BARNES V. E. (1940) North American tektites. *Univ. Tex. Publ.* No. 3945, 477.

BARNES V. E. (1958) *Geochim. et Cosmochim. Acta* **14,** 267.

BATE G. L. (1955) *Variations in the Isotopic Composition of Common Lead and the History of the Crust of the Earth.* Ph.D. Thesis, Columbia University.

COLLINS C. B., RUSSELL R. D. and FARQUHAR R. M. (1953) The maximum age of the elements and the age of the earth's crust. *Canad. J. Phys.* **31,** 402.

GERLING E. K. and YASHCHENKO M. L. (1952) On the age of tektites. *Dokl. Akad. Nauk SSSR* **83,** 901.

KUIPER G. P. (1949) *The Atmospheres of the Earth and Planets* pp. 304–346. University of Chicago Press, Chicago.

PATTERSON C. (1955) The Pb^{207}/Pb^{206} age of some stone meteorites. *Geochim. et Cosmochim. Acta* **7,** 151.

PATTERSON C. (1956) Age of meteorites and the earth. *Geochim. et Cosmochim. Acta* **10,** 230.

PATTERSON C., TILTON G. R. and INGHRAM M. G. (1955) Age of the earth. *Science* **121,** 69.

SPENCER L. J. (1939) Tektites and silica glass. *Miner. Mag.* **25,** 425.

SPENCER L. J. and HEY M. H. (1933) Meteoritic iron and silica glass from the meteorite craters Henbury (central Australia) and Wabar (Arabia). *Miner. Mag.* **23,** 387.

STRUVE O. (1944) An upper limit for the mass of the lunar atmosphere. *Astrophys. J.* **100,** 104.

SUESS H. E., HAYDEN R. L. and INGHRAM M. G. (1951) Age of tektites. *Nature, Lond.* **168,** 432.

TILTON G. R., ALDRICH L. T. and INGHRAM M. G. (1954) Mass spectrometric determination of thorium. *Analyt. Chem.* **26,** 894.

TILTON G. R., PATTERSON C., BROWN H. S., INGHRAM M. G., HAYDEN R. J., HESS D. H. and LARSEN E. S., JR. (1955) Isotopic composition and distribution of lead, uranium and thorium in a Precambrian granite. *Bull. Geol. Soc. Amer.* **66,** 1131.

UREY H. C. (1955) On the origin of tektites. *Proc. Nat. Acad. Sci., Wash.* **41,** 27.

UREY H. C. (1956) The origin of the moon surface features. *Sky & Telesc.* **15,** 2.

WHIPPLE F. L. (1950) A comet model. I: the acceleration of comet Encke. *Astrophys. J.* **111,** 375.

22

Geochimica et Cosmochimica Acta, 1969, Vol. 33, pp. 1045 to 1055. Pergamon Press. Printed in Northern Ireland

Isotopic comparison of lead in tektites with lead in earth materials*

J. M. WAMPLER,† D. H. SMITH and A. E. CAMERON
Oak Ridge National Laboratory, Oak Ridge, Tennessee 37830

(*Received* 11 *March* 1969; *accepted* 31 *March* 1969)

Abstract—Lead samples from moldavites and from Ries glass are isotopically similar, except that the Ries glass tends to be slightly enriched in the radiogenic lead isotopes relative to moldavites. The lead in these materials is similar to the lead found in oceanic sediments and falls on the growth curves typical of most terrestrial rock and ore lead samples. This is true of lead in other tektite groups, with one exception. Lead from Ivory Coast tektites does not fall on the growth trend of modern terrestrial lead. Lead samples from Bosumtwi Crater dark glass and from the gray phyllite (country rock) at Bosumtwi Crater are similarly displaced from the typical growth trend. These features may be explained in terms of lead isotopic development over long periods of time in environments which differ somewhat from the normal in terms of relative U, Th, and Pb abundances, which suggests that the tektites as well as the impact glass were derived from the Precambrian rocks at Bosumtwi Crater. High U/Pb ratios ($U^{238}/Pb^{204} > 20$) were found in all tektites analysed, indicating enrichment of U relative to Pb either at the time of tektite formation or within a time period not exceeding a few hundred million years prior to tektite formation. Lower U/Pb ratios were found in the impact glasses. Lead samples from a bediasite and a Georgia tektite are the same, isotopically, and similar to lead from other tektite groups (except the Ivory Coast tektites).

INTRODUCTION

THE first isotopic analyses of lead in tektites were carried out by TILTON (1958). On the basis of the very close similarity of lead in the tektites to typical modern terrestrial leads, and on the evidence from the abundances of uranium, thorium, and lead in one sample (an australite) which indicated fairly recent changes in the relative abundances of these elements, TILTON concluded that the tektites were probably of terrestrial origin.

STARIK and his co-workers (STARIK *et al.*, 1961, 1963) determined lead isotopic compositions and uranium, thorium, and lead abundances in a number of additional tektites, and suggested that tektites are of extraterrestrial origin. Their data, however, are essentially in agreement with those of TILTON, i.e. lead isotopic compositions typical of modern terrestrial lead (with one exception), and high U/Pb and Th/U ratios. As ZÄHRINGER (1963) has pointed out, interpretation of these results in favor of an extraterrestrial origin is difficult.

Since COHEN (1963) suggested that both moldavites and Ivory Coast tektites may have been derived from terrestrial impact craters (the Nördlingen Ries Crater in Germany and the Lake Bosumtwi Crater in Ghana, respectively), a remarkable body of evidence has developed in support of this hypothesis, particularly with

* Research sponsored by the U.S. Atomic Energy Commission under Contract with the Union Carbide Corporation.

† Permanent address: Georgia Institute of Technology, Atlanta, Georgia 30332.

respect to the relationship between Ivory Coast tektites and materials from Bosumtwi Crater. The studies which have provided this evidence have been reviewed by O'KEEFE (1966) and more recently by SCHNETZLER *et al.* (1967), so not all of the original contributions will be cited here. Some studies particularly pertinent to the present work, however, are the studies of the Rb/Sr geochronology of Ivory Coast tektites and of the impact glass and the local country rocks of the Bosumtwi Crater (SCHNETZLER *et al.*, 1966; LIPPOLT and WASSERBURG, 1966; KOLBE *et al.*, 1967). The material from which the Ivory Coast tektites were formed has the same Rb/Sr age, about $2{\cdot}0 \times 10^9$ yr, as the Bosumtwi Crater materials. In spite of this remarkable evidence for the terrestrial origin of the Ivory Coast tektites, some unresolved difficulties about the mechanism of tektite origin from earth surface materials has prevented acceptance of COHEN's hypothesis by all (O'KEEFE, 1966; WALTER, 1967).

The present study was undertaken in order to compare the lead isotope relationships of terrestrial materials at the Ries Crater and at the Bosumtwi Crater, with those of the associated tektites. In particular, it was anticipated that lead in the ancient rocks at the Bosumtwi Crater might have lead isotope relationships sufficiently different from that of relatively young terrestrial rocks that a unique relationship between the Bosumtwi samples and Ivory Coast tektites might be established, if the tektites are of terrestrial origin.

* * * * * * *

The results of the analytical work are presented in Table 2. The determinations of lead and uranium abundances are considered to be accurate to within 10%, except where lead abundances are indicated as upper limits. Upper limits are given where the total lead determined could reasonably be expected to contain more than 10% lead contamination. Unfortunately this applies to all lead determinations on Ivory Coast tektites, because of the small size of samples and the low lead content. We estimate the blank contributing to these samples to be about one-third of the total lead determined, and thus that the true lead content of these samples averages about one part per million.

The precision of the lead isotope ratios is of considerable importance to the following discussion. Duplicate mass spectrometer runs were carried out for a number of the separated lead samples, as shown in Table 2. On the basis of these duplicate analyses the precision of the mass spectrometry may be estimated to be about 0·5% for the Pb^{206}/Pb^{207} ratio and about 1% for the Pb^{208}/Pb^{207} ratio (95% confidence level). However duplicate determinations of the Pb^{206}/Pb^{204} ratio differ by as much as 3%. Furthermore, since all Pb^{206}/Pb^{207} values for Ivory Coast tektites were the same there are not likely any significant differences in the Pb^{206}/Pb^{204} ratio among these samples, yet a maximum difference of 5% is seen between Portion 1 of GSFC #4 and IVC-7-64. Probably some form of extraneous material has added to the signal at mass 204 in some cases, most likely some combination of lighter elements. High 204 values were generally observed near the beginning of an isotopic analysis, although the material causing the extra signal appeared to "burn off" rather quickly

Table 2. Lead and uranium abundances and lead isotope composition

Sample	Pb(ppm)*	U(ppm)*	Pb/U	U^{238}/Pb^{204}	Pb^{206}/Pb^{204}†	Pb^{206}/Pb^{207}	Pb^{208}/Pb^{207}
Moldavites							
Sample 1, Habři	6·2	2·1	3·0	21	18·8	1·191	2·500
2054-1, Slavče	4·6	2·6	1·8	35	18·4	1·194	2·464
Ries Glass							
RCC-3-61: Portion 1	19	6·4	3·0	21	18·8	1·203	2·496
Portion 2	18	5·2	3·5	18	18·5	1·209	2·488
Uncrushed Stock:							
Sample	28	5·3	5·3	12	18·5	1·190	2·479
Leach solution*	2·6	0·1	—	—	—	1·18	2·45
Ivory Coast tektites							
GSFC #4: Portion 1	—	—	—	—	17·2	1·181	2·397
Portion 2	⩽1·1	0·53	⩽2·1	⩾30	17·4	1·184	2·406
					17·9	1·184	2·421
Portion 3	⩽1·3	1·2	⩽1·1	⩾52	—	—	—
IVC-7-64	⩽1·4	0·57	⩽2·5	⩾25	18·1	1·184	2·392
					18·1	1·183	2·404
Bosumtwi Crater samples							
BCC-8A-64, Dark glass:							
Sample	2·7	0·55	4·9	13	19·0	1·234	2·385
					18·7	1·230	2·388
Leach solution*	⩽0·6	0·1	—	—	—	1·22	2·44
BCC-9-64, Phyllite:							
Portion 1: Sample	2·6	0·68	3·8	16	18·0	1·180	2·305
					17·9	1·174	2·317
Leach solution*	5·5	0·5	11	5·6	20·9	1·327	2·756
Portion 2 (leached)	—	—	—	—	18·0	1·191	2·293
Other tektites							
Indochinite	2·8	1·9	1·5	42	18·3	1·197	2·496
Bediasite, 30775-F	—	—	—	—	18·3	1·20	2·46
Georgia Tektite, EY-WB	—	—	—	—	17·8	1·203	2·457

* Abundances of lead and uranium in leach solutions are given as $\mu g/g$ sample, i.e. as parts per million with respect to the weight of sample before leaching.

† Considerable uncertainty is associated with the Pb^{206}/Pb^{204} ratios reported here, as explained in the text.

and the approach to a constant value for the 206/204 ratio was taken as an indication that the 204 signal was due entirely to lead. This criterion may not always have been valid. The experience from lead isotopic reference samples is not applicable here because the reference samples do not have impurities (e.g. organic material from the ion exchange resin) generally present in the analytical samples of this study. Another possible source of the extra signal at mass 204 is cross contamination with Pb^{204} from spiked samples, but some special blank tests for such contamination were negative.

The exact nature of the problem of Pb^{204} measurement is unresolved. The Pb^{206}/Pb^{204} ratios reported in this work may be low by up to several per cent. In Fig. 1, the Pb^{206}/Pb^{204} ratios are plotted vs. Pb^{206}/Pb^{207}. It is highly unlikely that materials of similar origin having similar Pb^{206}/Pb^{207} ratios should differ in their Pb^{206}/Pb^{204} ratio. Thus the indochinite and Ries glass analyses which fall well below the $\mu_0 = 8{\cdot}8$ growth curve presumably do so because of incorrect Pb^{204} measurements, for other analyses (Starik *et al.*, 1961; this work) indicate these materials belong on this growth curve typical of modern terrestrial lead. The highest Pb^{206}/Pb^{204} ratio

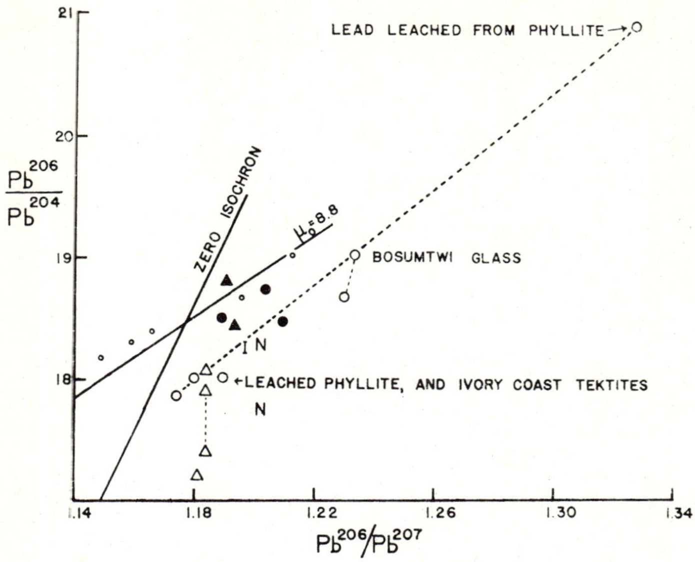

Fig. 1. Relationship of lead isotope ratios determined in this study to a primary growth curve for $\mu_0 = 8 \cdot 8$ and to the primary isochron of zero age for the earth (MURTHY and PATTERSON, 1962). (μ_0 is the present day U^{238}/Pb^{204} ratio.) Symbols not identified on the figure are as follows:

▲ Moldavites
● Ries glass
I Indochinite
N North American tektites
∘ Young conformable lead ores (RUSSELL and FARQUHAR, 1960) and average oceanic lead (CHOW and PATTERSON, 1962a,b).

Dotted lines connect some duplicate analyses. The dashed line is drawn arbitrarily through the analyses of African samples which probably have the most reliable Pb^{206}/Pb^{204} ratios.

obtained for Ivory Coast tektites is 18·1, the value for both analyses of IVC-7-64. The true Pb^{206}/Pb^{204} ratio for these Ivory Coast tektites is probably not less than 18·1. If this value is the true value, the Ivory Coast tektites fall below the $\mu_0 = 8 \cdot 8$ growth curve. Lead samples from Bosumtwi Crater glass and from leached Bosumtwi phyllite also appear to fall below this curve. Because of the unfortunate uncertainty in Pb^{204} measurement we cannot be sure that any of these displacements from the $\mu_0 = 8 \cdot 8$ growth curve are real. The similar apparent displacement of the two North American tektites samples is not likely significant, for the same reason.

There is no problem similar to the Pb^{204} measurement problem for the other, more abundant, lead isotopes. It is necessary, however, to consider the effects of contaminating lead on the isotopic abundances observed for the Ivory Coast tektites. Because of the low lead content and the small size of these samples, contaminating lead probably contributed from 10 to 20% of the lead separated for isotopic analysis. Figure 2 is a plot of Pb^{208}/Pb^{207} vs. Pb^{206}/Pb^{207} on which the average composition of blank lead is indicated. Since the blank lead is not very different from the sample lead, it should have little effect on the observed isotopic compositions. It should be noted

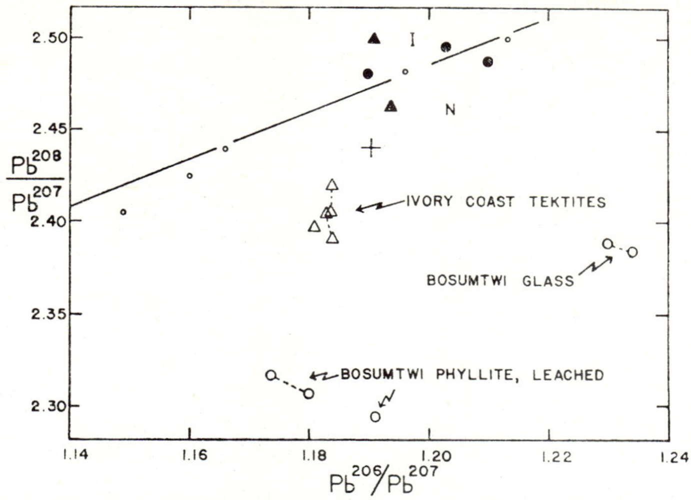

Fig. 2. Pb^{208}/Pb^{207} vs. Pb^{206}/Pb^{207} plot of isotope ratios determined in this study. Symbols are the same as in Fig. 1. The solid line is sketched arbitrarily to indicate the growth trend of typical modern terrestrial lead. A cross (+) indicates the average composition of contaminating lead as determined in the blank runs.

that there is an uncertainty in the blank contribution to the three samples for which lead was separated at Georgia Institute of Technology, since the lead content of these samples was not determined. These samples were large enough, however, that the proportion of contaminating lead was probably small. (We have used a Pb^{208}/Pb^{207} vs. Pb^{206}/Pb^{207} plot in Fig. 2 because such a plot is convenient for showing differences among lead samples which have isotopic compositions in or near the range of typical modern terrestrial lead. Since the rate of production of Pb^{207} has been very small for the past 1-2 billion years, such a plot indicates Pb^{208}, Pb^{206} relationships in a way that is easy to interpret and which is as free as possible of the effects of analytical errors.)

Discussion

The isotopic composition of lead from the Ries glass is similar to that of the moldavites, with one sample of Ries glass showing a slight excess of Pb^{206} with respect to the other materials. Since these samples fall on the growth curves typical of modern terrestrial lead (represented in Fig. 1 and Fig. 2 by some young conformable lead ores and by average lead from oceanic sediments), these data provide no proof that moldavites are derived from Ries crater material, but they do give more evidence that such a correlation is plausible.

The Ivory Coast tektites do not fall on the growth trend typical of modern terrestrial lead. Relative to ordinary lead of similar Pb^{206}/Pb^{207} ratio, the Ivory Coast tektites are depleted in Pb^{208} (Fig. 2). The impact glass and phyllite (after leaching) from the Bosumtwi crater are depleted in Pb^{208} to an even greater extent. Figure 3 shows the considerable amount of lead leached from the phyllite, in relation to the other samples. The large differences in isotopic composition within the phyllite, and the above-mentioned depletion in Pb^{208} with respect to the ordinary

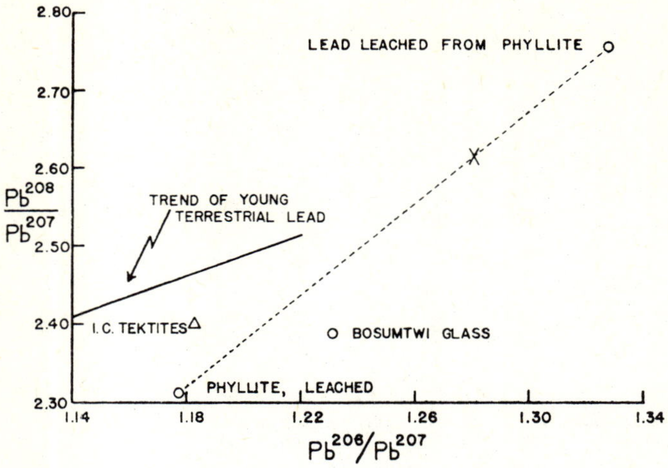

Fig. 3. Relationship of lead leached from Bosumtwi phyllite to other samples. A cross (X) indicates the calculated overall lead isotopic composition of the phyllite.

growth trend, are phenomena typical of situations where lead has developed for long periods of time in crustal rocks, i.e. they are typical of Precambrian terranes. They are not typical of lead from terrestrial rocks of relatively young age. Thus the departure of the Ivory Coast tektites from the typical lead growth curve is not surprising in view of their $2{\cdot}0 \times 10^9$ yr Rb/Sr age. The fact that the departure is in the same sense as that of materials from the Bosumtwi crater is an indication that they were actually formed from material at this crater.

The indication that Ivory Coast tektite lead may fall on a line connecting lead samples from the Bosumtwi crater materials, when plotted as Pb^{206}/Pb^{204} vs. Pb^{206}/Pb^{207} (Fig. 1), would be strong evidence of a common origin for all these materials, if it were established that such a line differs from the typical growth curve of terrestrial lead. Considerably higher precision in measurement of the Pb^{206}/Pb^{204} ratio would be required to determine whether such a distinctive linear relationship exists among these materials. (We might add that even in the event that a distinctive linear relationship should be established for Bosumtwi crater materials and Ivory Coast tektites, the significance of such a relationship would be reduced by a failure of the tektite samples to show a spread along the line. A single point could fall on the line by coincidence.)

It is of some interest to compare the moldavites and Ivory Coast tektites with the possibly related impact glasses in terms of uranium and lead contents. The most striking aspect is the low lead and uranium contents of the African glasses relative to the Ries glass and moldavites. We take this as additional evidence of the suggested genetic relationships between the tektites and the impact glasses.

Finally, the high U/Pb ratios observed in all the tektite samples are of considerable significance. Modern terrestrial lead has developed in an environment with an overall U/Pb ratio represented by a present day U^{238}/Pb^{204} ratio (μ_0) of about 9. All tektites analysed for U and Pb in this work and by Tilton (1958) and by Starik *et al.* (1961) have higher U/Pb ratios, mostly 3–5 times higher, yet all (with one exception) have lead isotopic compositions not very different from typical modern terrestrial lead. As TILTON pointed out, this suggests an increase in the U/Pb ratio within the last 100 million years, if one assumes a rather simple growth of lead isotopes in the tektite parent material. Specifically, one may say the following, assuming that lead in tektite parent material began to "grow" at the same time and with the same (primordial) isotopic composition as lead in the earth and meteorites (MURTHY and PATTERSON, 1962):

1. If there has been but one significant change in the U/Pb ratio, it must have been very recent—otherwise tektite lead would not fall near the primary isochron of zero age.

2. If there has been more than one significant change in U/Pb ratio such a short time limit on the most recent change is not necessary, but the occurrence of all tektite leads near both the primary zero isochron and the typical terrestrial growth curve would require a truly remarkable set of coincidences.

This sort of reasoning is particularly pertinent to the Ivory Coast tektites. Although the Rb/Sr data indicate the parent material is 2 billion years old, a much more recent increase in the U/Pb ratio is essential to explain the U/Pb data. Such a recent change seems reasonable for rocks near the surface of the earth, as a result of ground water leaching or other processes of weathering. The similarity in lead isotopic composition between the tektites and the leached phyllite suggests that the parent material of the tektites may have been an altered equivalent of rock similar to the phyllite sample that we analysed. Such alteration could account for the high U/Pb ratios in the tektites. Somewhat similar reasoning may be applied to the oxygen isotope data of TAYLOR and EPSTEIN (1966).

Another process which might account for the high U/Pb ratios of tektites is selective volatilization of lead during tektite formation (SCHNETZLER *et al.*, 1966; O'KEEFE, 1966; WALTER, 1967). If selective volatilization is the only process which has affected the U/Pb ratio of the Ivory Coast tektite material in the last 2 billion years, then a large proportion of the lead (more than 50%) must have been lost by volatilization. In view of other evidence for a terrestrial origin of these tektites, recent alteration of the parent material seems a preferable explanation.

In conclusion, the data of this study indicate a rather recent increase in the U/Pb ratio of the parent material of Ivory Coast tektites, and indicate in two independent ways a genetic correlation between these tektites and materials from the Bosumtwi crater. Weathered phyllite may have been the parent material. Lead in moldavites and Ries glass is isotopically similar, consistent with the hypothesis of a genetic relationship.

Acknowledgments—Dr. E. C. T. CHAO and Dr. C. C. SCHNETZLER kindly provided the rare Ivory Coast tektites, as well as other samples used in this study. Dr. V. E. BARNES and Mr. T. E. ALLEN contributed the North American tektites. The first author is grateful to the Oak Ridge Institute for Nuclear Studies (now Oak Ridge Associated Universities) and to the Analytical Chemistry Division of ORNL for the opportunity to do this work as a Summer Research Participant.

REFERENCES

CAMERON A. E., SMITH D. H. and WALKER R. L. (1969) Mass spectrometry of nanogram-size samples of lead. *Anal. Chem.* **41**, 525–526.

Chow T. J. and Patterson C. C. (1962a) The occurrence and significance of lead isotopes in pelagic sediments. *Geochim. Cosmochim. Acta* **26,** 263–308.

Chow T. J. and Patterson C. C. (1962b) Correction. *Geochim. Cosmochim. Acta* **26,** 973.

Cohen A. J. (1963) Asteroid—or comet—impact hypothesis of tektite origin: The moldavite strewn-fields. In *Tektites,* (editor J. O'Keefe), Chap. 9, pp. 189–211. University of Chicago Press.

Kolbe P., Pinson W. H., Saul J. M. and Miller E. (1967) Rb–Sr study on country rocks of the Bosumtwi Crater, Ghana. *Geochim. Cosmochim. Acta* **31,** 869–875.

Lippolt H. J. and Wasserburg G. J. (1966) Rubidium–Strontium–Messungen an Glasern von Bosumtwi-Krater und an Elfenbeinküsten-Tektiten. *Z. Naturforsch.* **21a,** 226–231.

Murthy V. R. and Patterson C. C. (1962) Primary isochron of zero age for meteorites and the earth. *J. Geophys. Res.* **67,** 1161–1167.

O'Keefe J. A. (1966) The origin of tektites. *Space Sci. Rev.* **6,** 174–221.

Russell R. D. and Farquhar R. M. (1960) Dating galenas by means of their isotopic constitutions—II. *Geochim. Cosmochim. Acta* **19,** 41–52.

Schnetzler C. C., Pinson W. H. and Hurley P. M. (1966) Rubidium–strontium age of the Bosumtwi crater area, Ghana, compared with the age of the Ivory Coast tektites. *Science* **151,** 817–819.

Schnetzler C. C., Philpotts J. A. and Thomas H. H. (1967) Rare-earth and barium abundances in Ivory Coast tektites and rocks from the Bosumtwi Crater area, Ghana. *Geochim. Cosmochim. Acta* **31,** 1987–1993.

Starik I. E., Sobotovich E. V., Shats M. M. and Lovtsyus G. P. (1961) Uranium and lead in tektites. *Meteoritika* **20,** 204–207.

Starik I. E., Sobotovich E. V. and Shats M. M. (1963) The problem of the origin of meteorites and tektites. *Geochemistry* (in translation) 261–270.

Taylor H. P., Jr. and Epstein S. (1966) Oxygen isotope studies of Ivory Coast tektites and impactite glass from the Bosumtwi Crater, Ghana. *Science* **153,** 173–175.

Tilton G. R. (1958) Isotopic composition of lead from tektites. *Geochim. Cosmochim. Acta* **14,** 323–330.

Walter L. S. (1967) Tektite compositional trends and experimental vapor fractionation of silicates. *Geochim. Cosmochim. Acta* **31,** 2043–2063.

Zähringer J. (1963) Isotopes in tektites. In *Tektites,* (editor J. O'Keefe), Chap. 6, pp. 137–149. University of Chicago Press.

Erratum: The first sentence of the second paragraph on page 172 should read: "The results of the analytical work are results presented in Table 2."

Editors' Comments on Paper 23

Concerning the source material for tektites, Taylor and Epstein* conclude: "The great bulk of soils, sandstones, shales, and other sedimentary and metasedimentary rocks on the earth definitely do not qualify (except in the case of the Ivory Coast tektites)." However, in their 1963 publication it was shown that A-zone soil in the two examples given has the correct oxygen isotope ratio for the formation of tektites. A-zone soil is continuously disturbed by plant roots and burrowing animals and subjected to winnowing of the clay and silt fraction by rain, with enrichment of sand. The sand is mostly quartz and most of the quartz originally was derived from an igneous rock. Therefore, a suitable source material for the formation of tektites exists over wide areas.

Epstein and Taylor (1971, p. 1425–1426) effectively eliminate a lunar source for tektites on the basis of oxygen isotopes in lunar materials so far examined. They state: "Inasmuch as portions of rock 12013 have SiO_2 contents in the same range as some tektites, it is natural to entertain the possibility that analogous material on the moon might represent the parent materials of tektites. This has been suggested, for example, by O'Keefe (1970). However, the δO^{18} values of 12013 are totally outside the range of values for tektites so this hypothesis must be rejected. Thus, even though high-SiO_2 rocks superficially resembling some tektites in chemical composition are now known to exist on the moon, they cannot be the parent materials of tektites."

Other papers concerning oxygen isotopes in tektites and postulated source materials were published by Taylor and Epstein in 1966 and 1970.

References

Epstein, S., and Taylor, H. P., Jr. (1971). O^{18}/O^{16}, Si^{30}/Si^{28}, D/H, and C^{13}/C^{12} ratios in lunar samples. Proc. 2nd Lunar Sci. Conf., Vol. 2, pp. 1421–1441. M.I.T., Cambridge, Massachusetts.

O'Keefe, J. A. (1970). Tektite glass in Apollo 12 sample. *Science,* **168,** 1209–1210.
Taylor, H. P., Jr., and Epstein, S. (1963). Comparison of oxygen isotope analyses of tektites, soils, and impactite glasses. *In* "Isotopic and Cosmic Chemistry," pp. 181–199. North-Holland Pub. Co., Amsterdam.
Taylor, H. P., Jr., and Epstein, S. (1966). Oxygen isotope studies of Ivory Coast tektites and impactite glass from the Bosumtwi crater, Ghana. *Science,* **153**, 173–175.
Taylor, H. P., Jr., and Epstein, S. (1970). Oxygen and silicon isotope ratios of lunar rock 12013. *Earth Planet. Sci. Lett.,* **9,** 208–210.

JOURNAL OF GEOPHYSICAL RESEARCH VOL. 74, NO. 27, DECEMBER 15, 1969

Correlations between O^{18}/O^{16} Ratios and Chemical Compositions of Tektites[1]

HUGH P. TAYLOR, JR., AND SAMUEL EPSTEIN

Division of Geological Sciences, California Institute of Technology Pasadena, California 91109

The limits of oxygen isotope variation in tektites have been determined by analyzing 33 tektites that cover the widest possible spectrum of chemical compositions. δO^{18} values were obtained for 7 australites (8.9 to 11.0), 4 javaites (9.4 to 10.0), 3 indochinites (9.3 to 10.6), 5 philippinites (9.8 to 10.6), 8 moldavites (10.7 to 11.8), and 6 Ivory Coast tektites (13.0 to 13.6), as well as 2 samples of Darwin glass (14.4 to 14.9). Various tektite groupings based on chemical composition and geographic occurrence all show a systematic increase in O^{18} with decreasing SiO_2 content, similar to that previously found for the bediasites. These systematic correlations must arise either by (1) vapor fractionation of tektite material during impact melting or (2) mixing of a SiO_2-rich igneous component and a low-SiO_2 component formed at a much lower temperature; perhaps the mixing occurred during weathering at the earth's surface or during hydrothermal alteration of silicic igneous rocks either on the earth or the moon.

INTRODUCTION

In three previous papers [*Taylor and Epstein*, 1962*b*; 1964; 1966] we showed that the O^{18}/O^{16} ratios of tektites from the Australasian, North American, and Czechoslovakian strewn fields are all very similar (δO^{18} = 8.9 to 11.3‰, relative to SMOW, standard mean ocean water), whereas the Ivory Coast tektites are distinctly higher in O^{18}/O^{16}. In addition, samples covering essentially the entire known range of chemical variation in the Texas tektites (bediasites) were analyzed, and a simple inverse correlation between SiO_2 content and O^{18}/O^{16} ratios was found.

It has recently become possible to obtain tektites displaying a wide range in chemical composition from the strewn fields in Australia, Java, the Philippine Islands, Indochina, Czechoslovakia, and the Ivory Coast. These new tektite samples were analyzed for O^{18}/O^{16} and the data are given in Table 1 (see the references noted above for details of the analytical procedures). Also included in Table 1 are the first oxygen isotope analyses of Darwin glass from Tasmania (see *Chapman et al.* [1967] for a description of this material) and of impactites from Meteor Crater, Arizona.

DISCUSSION OF THE ISOTOPE DATA

Relation to chemical groupings. The δO^{18} values of tektites are plotted as a function of index of refraction in Figure 1. Also plotted are a few previous analyses, including the bediasite data referred to above. The various tektite groups are delineated in Figure 1 principally according to geographic occurrence, but also with respect to the chemical groupings discovered by *Chapman and Scheiber* [1969].

The chemical groupings of Chapman and Scheiber are very nicely substantiated by the oxygen isotope data. Their HMg group, HCa australites, HNa/K australites, HCa philippinites, and HCu-HB indochinites each form a distinct grouping on a δO^{18}–refractive index diagram like Figure 1. Also, the two geographically separated sets of moldavites, from Bohemia and Moravia, plot in slightly different positions in Figure 1.

Note that *all* the tektite groups (except the Ivory Coast tektites) show a consistent positive correlation between index of refraction and δO^{18}. In the light of the well-established linear correlation between SiO_2 content and refractive index of tektites [*Barnes*, 1940], this implies that for each tektite group the higher SiO_2

[1] Publications of the Division of Geological Sciences, California Institute of Technology, Contribution 1641.

TABLE 1. Oxygen Isotope Ratios of Tektites, Darwin Glass, and Impactites from Meteor Crater, Arizona

Sample Number and Locality	δO^{18},‰*		Index of Refraction	Type
Tektites				
1. AN-288 Snake Creek bore, Australia	8.93 ± 0.17†	(4)‡	1.492	HCa
2. AN-148 Bloods Creek bore, Australia	9.14	(1)	1.501	HCa
3. T-224 Site A, Thailand	9.33	(1)	1.487	HCu-HB
4. J74 Sangiran, Java	9.41 ± 0.10	(3)	1.496	HMg
5. J35 Sangiran, Java	9.43 ± 0.03	(2)	1.513	HMg
6. AN-61 Lake Margareta, Australia	9.60	(1)	1.520	HMg
7. P192 Pasuquin, Luzon, Philippine Is.	9.81 ± 0.04	(2)	1.525	HMg
8. J15 Sangiran, Java	9.92	(1)	1.529	HMg
9. J2 Sangiran, Java	9.96 ± 0.14	(3)	1.539	HMg
10. AN-157 Bloods Creek bore, Australia	10.00 ± 0.04	(2)	1.521	HCa
11. P290 Babalog, Luzon, Philippine Is.	10.04 ± 0.09	(2)	1.506	HCa
12. Dalat, South Vietman DTM composite	10.27 ± 0.05	(6)		
13. P195 Pangranayan, Philippine Is.	10.30 ± 0.07	(2)	1.519	LCa/Al
14. P363 Santa Mesa, Luzon, Philippine Is.	10.40	(1)	1.525	HCa
15. P197 Ortigas site, Philippine Is.	10.60 ± 0.08	(3)	1.534	HCa
16. T-220Dk Ubol, Thailand	10.64 ± 0.02	(2)	1.513	HCu-HB
17. T5322 Slavetice, Moravia	10.68 ± 0.07	(4)	1.489	HRb/Sr
18. AN-318 S. of Lake Wilson, Australia	10.69 ± 0.02	(2)	1.531	HNa/K
19. AN-87 Williams Creek, Australia	10.76 ± 0.14	(3)	1.542	HNa/K
20. T5324 Dukovany, Moravia	10.95 ± 0.04	(3)	1.492	HRb/Sr
21. AN-245 Raspberry bore, Australia	11.03 ± 0.07	(3)	1.536	HNa/K
22. T5314 Slavce, Bohemia	11.05 ± 0.18	(4)	1.488	HRb/Sr
23. T4574 Radomilice, Bohemia	11.16 ± 0.06	(3)		HSi
24. T5319 Nechov, Bohemia	11.25 ± 0.05	(3)	1.489	HRb/Sr
25. Lhenice, Bohemia [*Taylor and Epstein*, 1966]	11.38	(1)	1.492	
26. T5317 Kroclov, Bohemia	11.70 ± 0.03	(3)	1.494	LRb/Sr
27. T5296d Lhenice, Bohemia	11.81 ± 0.09	(3)	1.495	LRb/Sr
28. T5658 Ivory Coast (2.93% MgO)	13.02	(1)		
29. T5702 Ivory Coast (2.86% MgO)	13.16	(1)		
30. T5653 Ivory Coast (4.26% MgO)	13.22	(1)		
31. T5649 Ivory Coast (3.85% MgO)	13.24	(1)		
32. T5686 Ivory Coast (4.20% MgO)	13.43	(1)		
33. T5706 Ivory Coast (3.56% MgO)	13.55	(1)		
Darwin Glass				
34. DG 3 Ten Mile Hill, Mt. Darwin, Tasmania	14.36	(1)	1.485	Dark
35. DG 24 Ten Mile Hill, Mt. Darwin, Tasmania	14.93	(1)	1.476	Light
Shocked Coconino Sandstone, Meteor Crater				
36. SS-2, frothy lechatelierite, shaft 2, 105-ft depth	12.57 ± 0.03	(2)		H
37. SS-3, bottom of crater, main shaft	12.57 ± 0.04	(2)		H
38. SS-4, original Coconino sandstone	12.84 ± 0.05	(2)		U
39. SS-6, south talus slope, bedding retained	13.02 ± 0.02	(2)		L
40. SS-1, moderately shocked, south talus slope	13.81	(1)		M

* Oxygen isotope analyses are relative to standard mean ocean water (SMOW) in per mille.
† Analytical error shown is average deviation from the mean.
‡ Numbers in parentheses indicate number of separate determinations.

Notes.

Samples 1 through 16, 18, 19, 21, 34, and 35 were obtained from Dean R. Chapman, who also supplied the indices of refraction and the chemical classifications, as follows: HCa, high calcium australites or philippinites; HMg, high magnesium group; HCu-HB, high copper, high boron indochinites; LCa/Al, philippinite with low Ca : Al ratio; HNa/K, australites with high Na : K ratios.

Samples 17, 20, 22 through 27, and 28 through 33 were obtained from W. H. Pinson, Jr., and the idices of refraction and chemical classifications are taken from *Philpotts and Pinson* [1966] and *Schnetzler and Pinson* [1964]: HRb/Sr, moldavites with high Rb : Sr ratios; LRb/Sr, moldavites with low Rb : Sr ratios; HSi, moldavite with highest SiO_2 content of any known tektite (84.5 wt%).

Samples 36 to 40 were obtained from Eugene M. Shoemaker and Susan Kieffer. H, highly shocked; M, moderately shocked; L, lightly shocked; U, unshocked parent material.

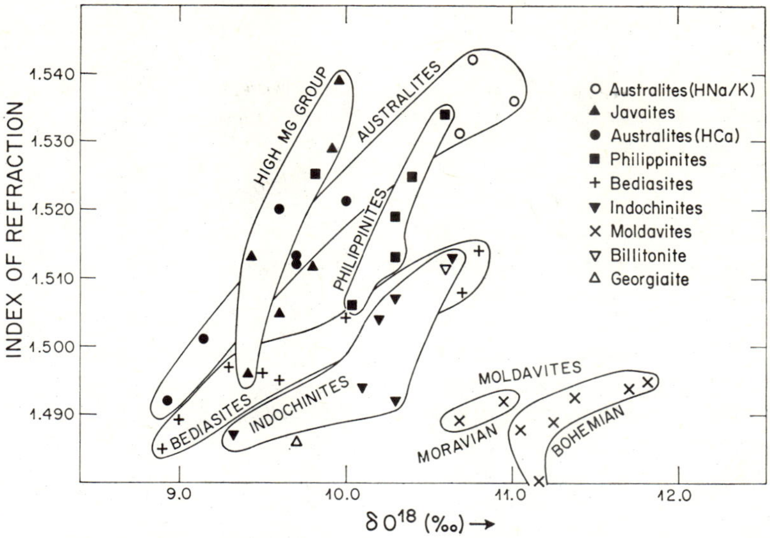

Fig. 1. Plot of δO^{18} versus index of refraction for the tektites analyzed in the present study. Also included are data on 8 bediasites, 2 javaites, 2 australites, 2 philippinites, 3 indochinites, a georgiaite, and a billitonite from *Taylor and Epstein* [1964].

contents are associated with lower O^{18}/O^{16} ratios. Besides these systematic variations in δO^{18} within each group, the new data indicate that the total range in δO^{18} in every one of the Australasian tektite groups is within the range shown by the bediasites and the australites. It is truly remarkable that the O^{18}/O^{16} variation exhibited by the 700,000-yr-old australites is exactly the same as that shown by the 33-m.y.-old bediasites.

The HMg tektite group shows very little variation in O^{18}/O^{16} over a wide range of SiO_2 content, changing only from 9.41 to 9.96, whereas the SiO_2 content changes by almost 15 wt %. The HCa philippinites show a similarly small variation in δO^{18}, but all other tektite groupings show a much more rapid increase in O^{18} with decreasing SiO_2, notably the bediasites, moldavites, and indochinites. Inasmuch as essentially the entire range of SiO_2 contents in the various tektite strewn fields are included in Figure 1, we believe that the correlations shown are a fundamental feature of tektite populations, and they are unlikely to be seriously modified by future work. For the same reason, it is considered unlikely that any other Australasian, North American, or Czechoslovakian tektites will be found to have O^{18}/O^{16} ratios appreciably different from those plotted in Figure 1. In other words, this figure probably indicates the *limits* of δO^{18} in the three major tektite fields.

Unforunately, the newly discovered microtektites [*Glass*, 1967] are too small to be readily analyzed for O^{18}/O^{16}, and, because their SiO_2 contents range to lower values than those of land tektites do, the actual range of δO^{18} might be larger than that shown in Figure 1. We may note, however, that the 'bottle-green' microtektites [*Cassidy et al.*, 1969] are simply an extension of the HMg group of *Chapman and Scheiber* [1969]; even if the HMg trend shown in Figure 1 is extrapolated to SiO_2 contents as low as 48 wt % (the lowest value known for a microtektite), the range of δO^{18} values of the HMg-'bottle-green' group would be within the 8.9 to 11.0‰ range shown by the other Australasian tektites. When sufficient microtektite material is available for analysis, it would be highly desirable to attempt O^{18} analyses on composite samples or perhaps even on unusually large individual specimens.

The new analyses of Ivory Coast tektites are very similar to one another, and they lie within the range of O^{18}/O^{16} previously determined for Ivory Coast samples [*Taylor and Epstein*, 1966]. These new samples cover the entire spectrum of chemical variations found in a suite of Ivory Coast tektites analyzed by W. H. Pinson. Their chemical variations are much smaller than in the tektites of the major strewn fields listed in Table 1, however, so the small range (only 0.5‰) in O^{18}/O^{16} is not unexpected. The SiO_2 contents of the Ivory Coast specimens, for example, vary from only 66.8 to 68.6 wt % [*Pinson and Kolbe,* 1969]. Note that the two Ivory Coast tektites that are lowest in MgO also have the lowest δO^{18} values. Although this trend is not completely clear, it is similar to the rather definite trends observed in the three major tektite groups.

Comparison with terrestrial impactite glass. In Figure 2, the tektite data are compared with available oxygen isotope analyses of various forms of terrestrial impactite glass and other glass of unknown origin (Libyan glass and Darwin glass). The impactite glasses cover a wide spectrum of SiO_2 contents and O^{18}/O^{16} ratios, just as would be expected as a result of random fusion of the earth's surface materials by meteorite impact. The Ivory Coast tektites are distinctly different in O^{18}/O^{16} from the other tektites, and it has been suggested that they are terrestrial impact glasses ejected from Bosumtwi crater [*Cohen,* 1961; *Schnetzler et al.,* 1966; *Taylor and Epstein,* 1966].

The oxygen isotope ratios of Darwin glass are typical of values obtained for terrestrial sandstones, shales, or most sedimentary and metasedimentary rocks. In terms of their O^{18}/O^{16} ratios the Darwin glass samples have nothing in common with the nearby Australian tektites. In fact, on a δO^{18} refractive index diagram they exhibit a trend opposite to that shown by the australites (Figure 2).

Isotopic analyses of impact lechatelierite and shocked sandstone from Meteor Crater, Arizona, are practically identical with the parent Coconino sandstone. This result, added to the data presented by *Taylor and Epstein* [1964], indicates that impact melting at moderate temperatures produces little change in the O^{18}/O^{16} ratios of the fused materials.

Comparison with igneous rocks. In Figure 3 we show the δO^{18} variations in tektites as a function of chemical composition, employing the quantity $\frac{1}{3}SiO_2 + K_2O - CaO - MgO - FeO$ (Larsen 'differentiation index'). Data on terrestrial igneous rocks from Figure 20 of *Taylor* [1968] are also plotted in Figure 3. The trends shown by the various tektite groups are completely different from the trends shown

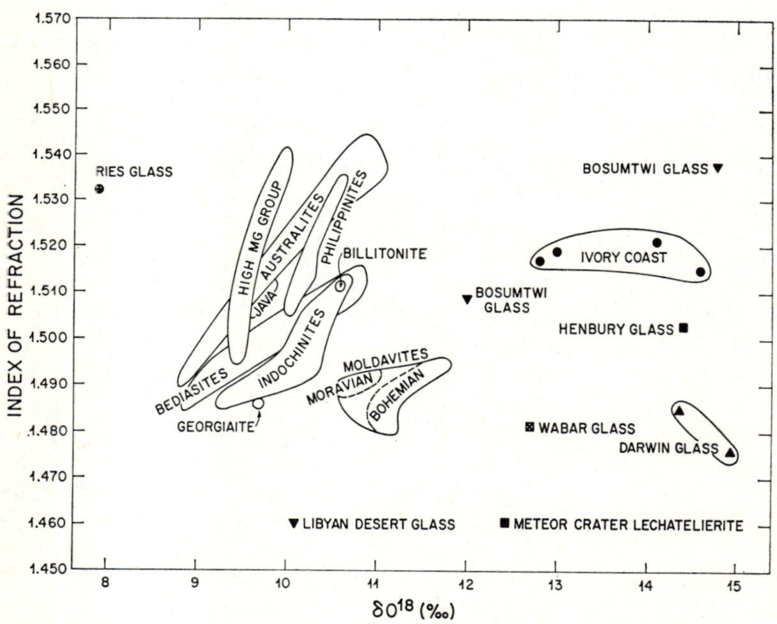

Fig. 2. Plot of δO^{18} versus index of refraction for Ivory Coast tektites and various samples of terrestrial impactite glass [*Taylor and Epstein,* 1962*b*; 1964; 1966]. These data are compared with the tektite fields delineated in Figure 1 and with the analyses of Darwin glass and Meteor Crater lechatelierite.

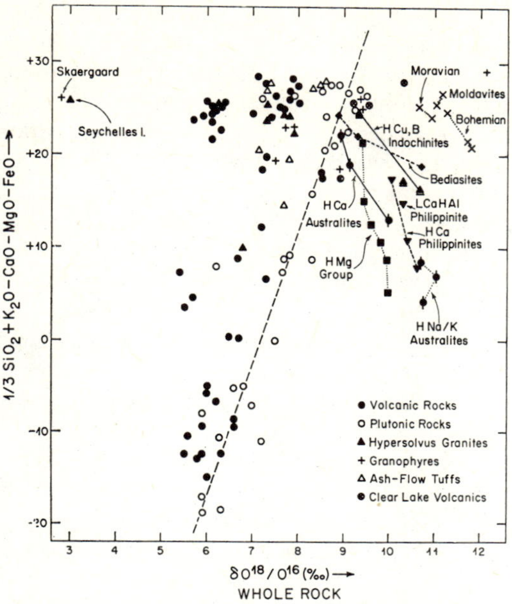

Fig. 3. Plot of δO^{18} versus Larsen 'differentiation index' for tektites and igneous rocks. The data on terrestrial igneous rocks are from Figure 20 of *Taylor* [1968]. The chemical data on the tektites were kindly supplied by D. R. Chapman.

by most of the igneous rocks. However, with increasing Larsen index, the tektite trends and igneous rock trends approach one another.

Isotopically, the SiO_2-rich tektites are practically indistinguishable from SiO_2-rich igneous rocks such as granites, granodiorites, granophyres, and many rhyolitic ash-flow tuffs. Obsidians and a rhyodacite from the Clear Lake, California, volcanic field also plot in the same range of δO^{18}, but most rhyolite obsidians have distinctly lower O^{18}/O^{16} ratios.

During igneous differentiation on the earth, O^{18}/O^{16} ratios either remain essentially constant, as in basalt-andesite-rhyolite associations, or they increase by only 2 to 3‰, as in gabbro-tonalite-granodiorite provinces like the Southern California batholith (dashed line of Figure 3). We therefore conclude that, if the O^{18}/O^{16} ratios and chemical compositions of tektites reflect the nature of their parent materials before impact melting, these parent materials cannot be a suite of igneous rocks related to one another by simple magmatic differentiation. All known meteorites have δO^{18} values lower than 6.5, except types 1 and 2 carbonaceous chondrites and ureilites [*Reuter et al.*, 1965; *Taylor et al.*, 1965], so that we can extend the above conclusion to include the igneous meteorites as well as terrestrial igneous rocks.

Nevertheless, the coincidence of δO^{18} in the high-SiO_2 tektites and igneous rocks strongly suggests that a significant portion of the tektite parent materials are of igneous derivation, in particular the high-SiO_2 part. Igneous quartz is thus a very likely constituent of tektites, as there is abundant evidence (lechatelierite particles) that quartz grains were present in tektite parent materials.

Comparison with altered volcanic rocks, soils, and sediments. In Figure 4 the δO^{18} values of tektites are compared with those of terrestrial soils, shales, phyllites and pelitic schists (metamorphosed shales), and various SiO_2-rich volcanic rocks. Observe that, except for a few of the altered volcanic flows and tuffs, the entire range of samples from the three major tektite fields is fairly well delineated from most of the terrestrial rocks plotted in Figure 4. It is very common for volcanic rocks, particularly ash-flow tuffs, to have undergone a low-temperature alteration process. This alteration occurs during and after cooling of the volcanic units as a result of interactions with late-stage gasses and aqueous solutions that pass through the rocks and exchange oxygen with the primary minerals and glass. The effects of this alteration process are (1) hydrated volcanic glass, (2) devitrified volcanic glass, and (3) deuteric or hydrothermal alteration of primary minerals, producing chlorite, epidote, clay minerals, etc. Because oxygen isotope fractionations between minerals and H_2O increase sharply with decreasing temperature, both surface weathering and low-temperature hydrothermal alteration usually produce an O^{18} enrichment in the affected rocks.

Except in localities where the meteoric waters are very low in O^{18} (only at high latitudes or high elevations [*Craig*, 1961]), the weathering of an igneous rock always produces an increase in O^{18}/O^{16} ratio in the rock. Weathered igneous rocks or mixtures of soils and fresh igneous rocks therefore qualify as parent materials of tektites on the basis of their O^{18}/O^{16} ratios, as pointed out previous by *Taylor and Epstein* [1964]. However, again excepting situations in which low-O^{18} meteoric waters are involved, hydrothermally altered igneous rocks are equally suitable tektite parent materials.

Mixing Curves

Least-squares trend lines. The above discussion indicates that tektites could have formed from a parent material composed of a mixture of (1) fresh SiO_2-rich igneous rock and (2) a low-temperature alteration product of such rocks. Thus it is interesting to inquire more fully whether the oxygen isotope trend lines shown in Figures 1 and 3 could be mixing lines involving an SiO_2-rich component and an SiO_2-poor component.

In Figure 5, least-squares lines have been drawn through the data points for each tektite group. Included is a least-squares line through the bediasite data of *Taylor and Epstein* [1964]. Note that one peculiar moldavite has been discarded in deriving the least-squares line for the Bohemian moldavites; this sample has the highest SiO_2 content of any known tektite [*Schnetzler and Pinson,* 1964], and it clearly departs from the rather consistent trend exhibited by the other Bohemian tektites.

Not all of the tektites plotted in Figure 5 have been chemically analyzed, but, in view of the systematic trends shown, it is useful to plot those with known SiO_2 content on a δO^{18}-SiO_2 diagram. New least-squares lines are drawn through these more restricted sets of data points in Figure 6, and the lines are extrapolated to higher and lower SiO_2 contents. There is a rough tendency for the least-squares lines to converge in the high-SiO_2 part of the diagram, and they markedly diverge at low-SiO_2 contents. These relationships are exactly what would be predicted if the lines represented mixing curves between (1) a high-SiO_2 igneous component having a relatively uniform δO^{18} of about 8.0 to 10.0‰ and (2) a low-SiO_2 low-temperature component having a highly variable δO^{18} of about 11 to more than 20‰.

Quartz is known to be very resistant to exchange and recrystallization in the presence of low-temperature aqueous fluids [*Taylor,* 1968; *Clayton et al.,* 1968], whereas feldspars and

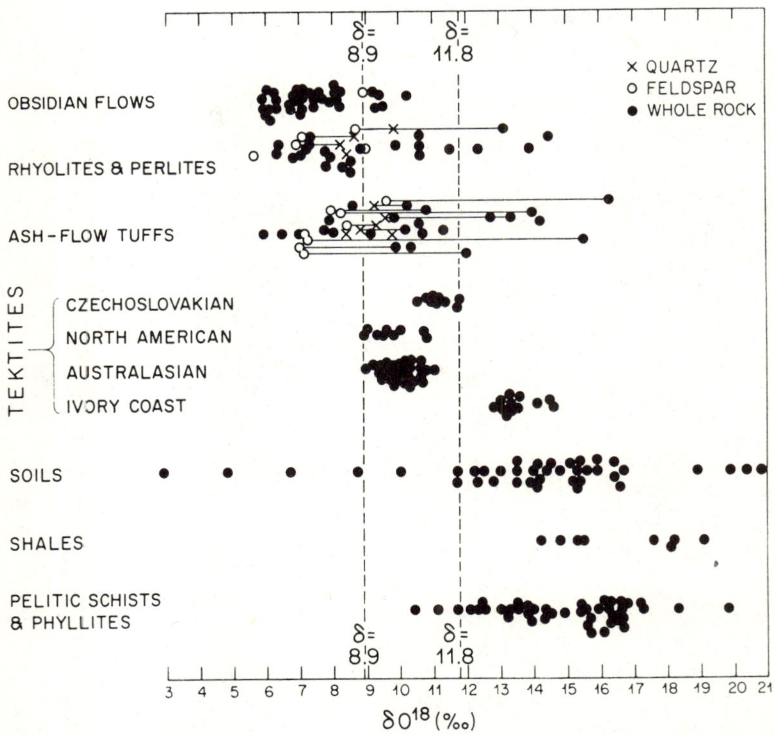

Fig. 4. Comparison of δO^{18} data on tektites with data on silicic volcanic rocks [*Taylor,* 1968], soils [*Taylor and Epstein,* 1964; unpublished data by J. R. Lawrence], and sedimentary and metasedimentary rocks [*Taylor and Epstein,* 1964; *Garlick and Epstein,* 1967; *Savin and Epstein,* 1969]. Practically all of the volcanic rocks having δO^{18} values higher than 9.5 have undergone O^{18} enrichment during low-temperature hydrous alteration.

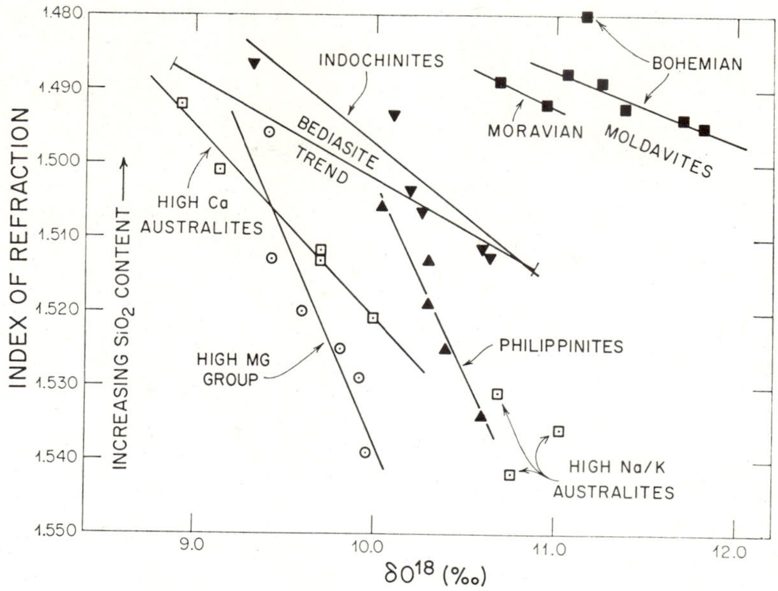

Fig. 5. Plot of δO^{18} versus index of refraction for tektites, showing least-squares lines drawn through the data points obtained in the present study. Also included is the least-squares line through the bediasite data of *Taylor and Epstein* [1964].

mafic minerals are very susceptible to oxygen isotope exchange during hydrothermal alteration or weathering in forming chlorite, kaolinite, epidote, and other hydrous alteration minerals. The effects of such processes on granitic or rhyolitic igneous rocks would be to produce variations on a δO^{18}-SiO_2 diagram that are similar to those shown in Figure 6. It should be noted that, if the trend lines of Figure 6 are extrapolated all the way to 100% SiO_2 (pure quartz), δO^{18} values of 7 to 9‰ are obtained. This is a little lower than the value for typical quartz in plutonic granitic rocks but is almost identical with the quartz in most terrestrial rhyolites [*Taylor and Epstein*, 1962*a*; *Taylor*, 1968].

Possible mineralogy of tektite parent materials. If we accept the possibility that the least-squares lines of Figures 5 and 6 are mixing curves of the type outlined above, it is instructive to consider possible mineralogical compositions of tektite parent materials. Let us make the simplifying assumptions that the only 'igneous' mineral present is quartz and that the other minerals are all typical low-temperature alteration products. We also ignore any chemical variations arising from vapor fractionation of alkalies, SiO_2, etc., that might occur during the formation of tektite glass. If we do this, and if we remove all H_2O because of the anhydrous nature of tektite glass, then a kind of 'norm' can be calculated for any tektite chemical analysis; some typical examples are shown in Table 2, along with tabulations for a few microtektites that are very unusual in chemical composition.

Table 2 is useful primarily to test whether the kinds of mixing curves postulated on the basis of the oxygen isotope data are at all feasible. For example, if quartz is assigned a δO^{18} of 8.0, the δO^{18} of the rest of the assemblage can be calculated by material balance, utilizing appropriate O^{18}/O^{16} ratios for the various tektites. When this is done, we obtain the following δ values for the nonquartz portion of the various tektites in Table 2: moldavite, 15.4; bediasite, 11.9; australite, 11.3; J86 HMg tektite, 11.4‰. Except for the moldavite, these values are considerably lower than those of weathering products formed at the earth's surface [*Savin and Epstein,* 1969]. This means that either (1) primary igneous minerals other than quartz are present in appreciable quantities in the tektite parent materials along with the weathering products or (2) the alteration process occurred at higher temperatures than that of the earth's surface.

In particular, observe that chemical variations in the HMg 'bottle-green' tektite group can be largely explained by adding chlorite and epidote and substracting muscovite (or K feldspar) from the more 'normal' tektites. This is in the right direction to explain the steeper slope of the HMg least-squares lines in Figures 5 and 6 because in a given mineral assemblage chlorite and epidote are invariably lower in O^{18} than coexisting muscovite or K feldspar [*Garlick and Epstein,* 1967]. Similarly, this might also explain the steeper slope of the least-squares line for the HCa philippinites shown in Figures 5 and 6.

It is worth pointing out that many, if not most, granophyres that have apparently formed as late-stage differentiates of large layered gabbro plutons or diabase sills have δO^{18} values similar to tektites. As has been shown by *Taylor* [1967; 1968], these granophyres typically have undergone a deuteric (hydrothermal) alteration that has enriched the whole rock in O^{18} while leaving the quartz nearly unaffected. Granophyres from the Precambrian Muskox and Duluth complexes in particular seem to have all the attributes of the postulated tektite parent material, at least in terms of the oxygen isotope relationships. They contain, however, exchanged alkali feldspar instead of muscovite, as do the examples in Table 2, and of course their quartz contents are lower than the average bediasite, moldavite, and australite shown in Table 2.

Vapor Fractionation in Tektites

Walter [1967] has presented evidence that much of the chemical variation in tektites can be explained by fractional vaporization of the tektite liquid at the time of impact fusion. A single experiment by *Walter and Clayton* [1967] indicated that fusion of a sample containing

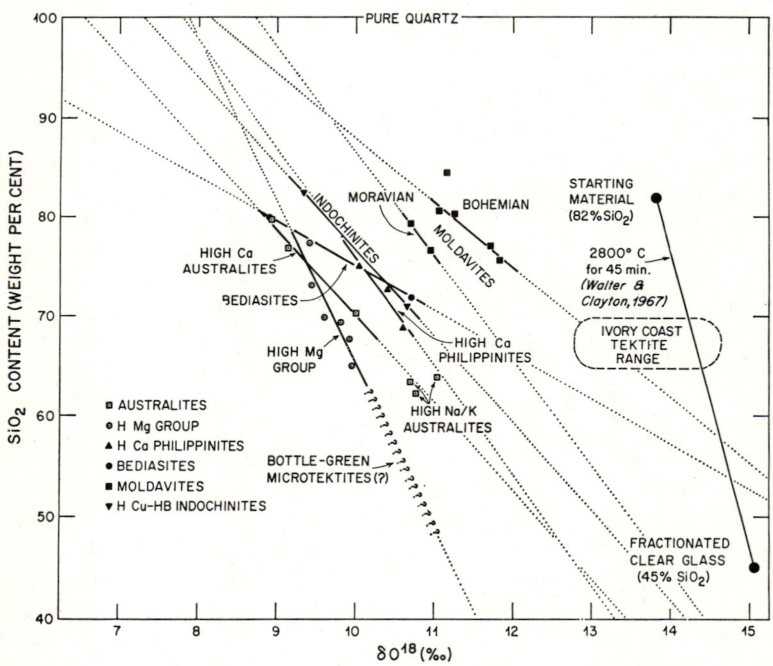

Fig. 6. Plot of δO^{18} versus SiO_2 content, showing least-squares lines drawn through data points for which chemical analyses are available [*Chapman and Scheiber*, 1969; *Schnetzler and Pinson*, 1964; *Philpotts and Pinson*, 1966]. The least-squares trend lines are very similar to those of Figure 5, which are based on a larger number of data points. Note that the bediasite line is drawn through data points for only two end member samples, BM-1 and B-90 [see *Taylor and Epstein*, 1964; *Chao*, 1963]. The lines are extrapolated toward higher and lower SiO_2 contents, illustrating the possibility that they represent mixing lines involving a high-SiO_2, low-O^{18} component and a low-SiO_2, high-O^{18} component. The lines are, however, qualitatively similar in slope to a vapor fractionation curve obtained by *Walter and Clayton* [1967] during 2800°C fusion of a high-SiO_2 glass.

TABLE 2. Calculated Mineralogical Compositions of Tektites (Weight per cent)

Mineral	Tektites				Microtektites		
	Average* Bediasite	Average† Moldavite	Average† Australite	HMg Tektite‡ J86	#63§	#64§	#36§
Quartz	56	57	49	41	10	25	29
Chlorite‖	2	1	3	18	59	45	10
Epidote¶		6	11	10	19	12	19
Muscovite**	17	24	18	12	1	3	5
Albite	13	4	11	7	4	5	9
Magnetite††	4	3	5	9	5	8	4
Kaolinite	6	3	1	1			22
Sphene	2	2	2	2	2	2	2

* Table 5, *Chao* [1963].
† Table 4, *Schnetzler and Pinson* [1963].
‡ Table 6, *Cassidy et al.* [1969].
§ Table 4, *Cassidy et al.* [1969], two 'bottle-green' samples (#63 and #64), and one high Al_2O_3 sample (#36).
‖ Chlorite varies from $Mg_5Al(OH)_8AlSi_3O_{10}$ in high Al samples to $Mg_{5.75}Al_{0.5}Si_{3.75}O_{10}(OH)_8$
¶ Epidote is $Ca_2Al_3(OH)(SiO_4)_3$.
** Muscovite is assumed to be a phengitic sericite, $KMg_{0.5}Al_2Si_{3.5}O_{10}(OH)_2$.
†† Magnetite calculated as FeO; note that the Fe could equally well have been assigned to the silicates, and if calculated as $FeSiO_3$, for example, the quartz contents of tektites are reduced to 53, 54, 45, 33, 6, 18, and 26, respectively.

Notes.

Calculations made as follows: all Na goes into albite; all Fe into magnetite; all K into muscovite; all Ti into sphene; all Ca left over from the sphene goes into epidote; all Mg left over from the muscovite goes into chlorite; any Al left over from albite, muscovite, epidote, and chlorite goes into kaolinite; the remaining Si goes into quartz.

82% SiO_2 at 2800°C for 45 min produced a lowering of SiO_2 content to 45% and an enrichment in O^{18} of about 1.2‰ (see Figure 6). This change is qualitatively in the same direction as we observe in the various tektite groups, and the rate of change of O^{18}/O^{16} with SiO_2 content is quantitatively very similar to that observed for the HMg group and for the philippinites in the present work. The oxygen isotope variations are thus compatible with the vapor fractionation model of *Walter* [1967], as it is possible that variations in temperature or rate of heating might give the more rapid rate of change of δO^{18} with SiO_2 content demanded by the other tektite groups, particularly the bediasites and moldavites. More experiments along these lines would be valuable.

If the mechanism of fractional vaporization is the correct explanation of tektite formation, the tektite parent materials for all three major strewn fields must have an SiO_2 content of about 82 wt % [*Walter*, 1967] and an O^{18}/O^{16} ratio of about 8.9 for the bediasites and Australasian tektites and about 10.7 for the moldavites. There are no even remotely widespread materials on the earth that have the above characteristics and this might be construed as evidence for a lunar origin.

It is very difficult, however, to explain the chemical data on the HMg 'bottle-green' tektites by any process of vapor fractionation [*Cassidy et al.*, 1969, p. 1021], even though this is one of the groups quantitatively most similar to the experimental O^{18}/O^{16} variations observed during vapor fractionation. In addition, *Chapman and Scheiber* [1969] have observed several other chemical trends among the Australasian tektite populations that are very difficult to ascribe to fractional vaporization of tektite material. Certainly, in view of the consistency in the patterns shown by all the tektite groups in Figure 6, whatever is the explanation of the trends it must surely apply to all. Hence, if vapor fractionation is deemed impossible as an explanation of even one of the groups in Figure 6, it must be considered doubtful for all.

Loss of H_2O during relatively low temperature fusion (1700°C) of hydrous silicates also produces an enrichment in O^{18} [*Taylor and Epstein*, 1964]. Also, even though vapor fractionation may not be the explanation of the entire chemical variations in tektites, it very likely has occurred to a limited extent, and we can therefore expect that the original parent material of tektites is probably lower in O^{18} than the tektites themselves. The magnitude of this effect is unlikely to be more than about 1‰, but it is in a direction that increases the disparity between tektites and terrestrial soils or sedimentary rocks and reduces the isotopic differences between tektites and igneous rocks.

Summary and Conclusions

In summarizing the present study we must report that the oxygen isotope data still provide no definite answer to the elusive question of a lunar versus terrestrial origin for tektites. They do, however, place certain restraints on tektite parent materials, as follows:

1. It is practically inconceivable that the correlations observed in δO^{18} and SiO_2 content of tektites could be due to any process of magmatic differentiation involving simple crystal-melt equilibria.

2. At least a portion of the tektite material is almost certainly of igneous origin (the high-SiO_2 part); in particular, igneous quartz is probably an abundant constituent of the parent materials.

3. The consistent patterns showing an increase in O^{18}/O^{16} ratio with decreasing SiO_2 content in several tektite groups must be due to (*a*) vapor fractionation of the liquid during impact fusion or (*b*) mixing between a high-SiO_2 igneous component and a low-SiO_2 low-temperature component formed during weathering at the earth's surface or during hydrothermal alteration at moderate temperatures.

4. Weathered or hydrothermally altered quartz-bearing terrestrial igneous rocks are therefore possible parent materials of tektites. This includes granites, granodiorites, granophyres, and silicic volcanic rocks, as well as possibly some sediments directly derived from igneous rocks, such as certain arkoses and graywackes. The great bulk of soils, sandstones, shales, and other sedimentary and metasedimentary rocks on the earth definitely do not qualify (except in the case of the Ivory Coast tektites).

5. The bediasites and Australasian tektites must have formed from practically identical parent materials. The similarities are even more astonishing if the HNa/K australites were formed during a separate, older event than the other Australasian tektites, as is suggested by fission-track dating [*Fleischer et al.*, 1969]; these peculiar, low-SiO_2 tektites plot on the same trend as do the other australites (Figure 1). The parent materials of moldavites were apparently about 1‰ richer in O^{18} than the other tektite parent materials, but this still represents great uniformity considering the known variation of O^{18} in earth surface materials and meteorites. The tektite parent materials are *highly selected* in some fashion; they are far from being random samples of the SiO_2-rich rocks of the earth's surface. This is the only aspect of the oxygen isotpe data that could prevent us from wholeheartedly accepting a terrestrial origin of tektites.

6. If tektites originate on the moon, and if the observed chemical and isotopic variations are a property of their parent materials, a low-temperature hydrothermal alteration event of some type must have affected the SiO_2-rich igneous rocks that are presumed to exist on the lunar surface. This conceivably could have happened during the degassing that probably would have accompanied volcanism, or perhaps during an interval when a transient atmosphere was present on the moon. We should note that the only extraterrestrial objects similar in O^{18}/O^{16} ratio to tektites are the types 1 and 2 carbonaceous chondrites, which are the only meteorites that have clearly suffered a low-temperature hydrous alteration.

7. Regardless whether tektites are from the moon or the earth, there must be a sorting mechanism in which the high-SiO_2 portions are at least partially separated from the rest of the tektite parent material, in order to produce the spectrum of compositions observed in tektites. Such sorting processes obviously exist on the earth, and if the moon had a transient atmosphere they might also have existed there. It is perhaps also possible that meteorite bombardment, slumping, and downslope movement may produce sorting effects on the moon. However, the necessity for such a sorting process conceivably could be overcome if (*a*) the chemical variations in tektite materials are mainly due to quartz content and (*b*) a rather uniform low-temperature alteration affects the rocks. In such a case, the rocks would all be

enriched in O^{18}, but the high-SiO_2 rocks would be affected least because of the more abundant quartz, which is relatively inert in comparison with the coexisting minerals likely to be found in igneous rocks. In such an example the chemical variations could be a result of any plausible process, and the O^{18}/O^{16} ratios would be merely superimposed later.

Acknowledgments. We are very grateful to Dean R. Chapman and William H. Pinson, Jr., for contributing the samples analyzed in this study. In particular, Dean Chapman suggested that we do further isotopic work on tektites, and he provided carefully selected samples chosen specifically to cover the entire spectrum of chemical compositions and types that he has delineated among the thousands of Australasian tetkites examined by him. We also wish to thank Eugene M. Shoemaker and Susan Kieffer for supplying the Meteor Crater samples, and we are grateful for stimulating discussions with E. C. T. Chao and L. S. Walter.

This work was supported by NASA grant NGR 05-002-085 and Atomic Energy Commission contract AT (04-3)-427, CALT-427-41.

References

Barnes, V. E., North American tektites, *Texas Univ. Publ. 3945,* 477-582, 1940.

Cassidy, W. A., B. Glass, and B. C. Heezen, Physical and chemical properties of Australasian microtektites, *J. Geophys. Res., 74*(4), 1008–1025, 1969.

Chao, E. C. T., The petrographic and chemical characteristics of tektites, in *Tektites,* edited by J. A. O'Keefe, pp. 51–94, University of Chicago Press, 1963.

Chapman, D. R., K. Keil, and C. Annell, Comparison of Macedon and Darwin glass, *Geochim. Cosmochim. Acta, 31,* 1595–1603, 1967.

Chapman, D. R., and L. C. Scheiber, Chemical investigation of Australasian tektites, *J. Geophys. Res., 74*(27), 1969.

Clayton, R. N., L. P. J. Muffler, and D. E. White, Oxygen isotope study of calcite and silicates, Salton Sea geothermal field, *Am. J. Sci., 266,* 968–979, 1968.

Cohen, A. J., A semi-quantitative hypothesis of tektite origin by asteroid impact (abstract), *J. Geophys. Res., 66,* 2521, 1961.

Craig, H., Isotopic variations in meteoric waters, *Science, 133,* 1702–1703, 1961.

Fleischer, R. L., P. B. Price, J. R. M. Viertl, and R. T. Woods, Ages of far eastern tektites, *Intern. Tektite Symp., 3rd,* Corning, N. Y., 1969.

Garlick, G. D., and S. Epstein, Oxygen isotope ratios of coexisting minerals in regionally metamorphosed rocks, *Geochim. Cosmochim. Acta, 31,* 181–214, 1967.

Glass, B., Microtektites in deep-sea sediments, *Nature, 214,* 372–374, 1967.

Philpotts, J. A., and W. H. Pinson, Jr., New data on the chemical composition and origin of moldavites, *Geochim. Cosmochim. Acta, 30,* 253–266, 1966.

Pinson, W. H., Jr., and P. Kolbe, Chemistry of Ivory Coast tektites, *Intern. Tektite Symp., 3rd,* Corning, N. Y., 1969.

Reuter, J. H., S. Epstein, and H. P. Taylor, Jr., O^{18}/O^{16} ratios of some chondritic meteorites and terrestrial ultramafic rocks, *Geochim. Cosmochim. Acta, 29,* 481–488, 1965.

Savin, S. M., and S. Epstein, Oxygen and hydrogen isotope ratios in sedimentary rocks and minerals, *Geochim. Cosmochim. Acta,* in press, 1969.

Schnetzler, C. C., P. M. Hurley, and W. H. Pinson, Jr., Rubidium-strontium age of the Bosumtwi crater area, Ghana, compared with the age of the Ivory Coast tektites, *Science, 151,* 817–819, 1966.

Schnetzler, C. C., and W. H. Pinson, Jr., The chemical composition of tektites, in *Tektites,* edited by J. A. O'Keefe, pp. 95–129, University of Chicago Press, 1963.

Schnetzler, C. C., and W. H. Pinson, Jr., A report on some recent major element analyses of tektites, *Geochim. Cosmochim. Acta, 28,* 793–806, 1964.

Taylor, H. P., Jr., Origin of red-rock granophyres (abstract), *Trans. Am. Geophys. Union, 48,* 245–246, 1967.

Taylor, H. P., Jr., The oxygen isotope geochemistry of igneous rocks, *Contrib. Mineral. Petrol., 19,* 1–71, 1968.

Taylor, H. P., Jr., M. B. Duke, L. T. Silver, and S. Epstein, Oxygen isotope studies of minerals in stony meteorites, *Geochim. Cosmochim. Acta, 29,* 489–512, 1965.

Taylor, H. P., Jr., and S. Epstein, Relationship between O^{18}/O^{16} ratios in coexisting minerals of igneous and metamorphic rocks, *Bull. Geol. Soc. Am., 73,* 461–480, 1962*a*.

Taylor, H. P., Jr., and S. Epstein, Oxygen isotope studies on the origin of tektites, *J. Geophys. Res., 67,* 4485–4490, 1962*b*.

Taylor, H. P., Jr., and S. Epstein, Comparison of oxygen isotope analyses of tektites, soils and impactite glasses, in *Isotopic and Cosmic Chemistry,* edited by H. Craig, S. L. Miller, and G. J. Wasserburg, pp. 181–199, North-Holland Publishing Co., Amsterdam, 1964.

Taylor, H. P., Jr., and S. Epstein, Oxygen isotope studies of Ivory Coast tektites and impactite glasses from the Bosumtwi crater, Ghana, *Science, 153,* 173–175, 1966.

Walter, L. S., Tektite compositional trends and experimental vapor fractionation of silicates, *Geochim. Cosmochim. Acta, 31,* 2043–2063, 1967.

Walter, L. S., and R. N. Clayton, Oxygen isotopes: Experimental vapor fractionation and variations in tektites, *Science, 156,* 1357–1358, 1967.

(Received June 27, 1969.)

Ages of Tektites

III

Editors' Comments on Papers 24, 25, and 26

These three papers have been chosen to illustrate the relatively new technique of fission-track dating and to point out some inconsistencies which have been found. Because of the discrepancy between the geologic age on earth of australites and the potassium-40–argon-40 age of australites since their last melting, other methods of age determination are needed. However, the fission-track method of dating tektites, introduced by Fleischer and Price (1964), leaves something to be desired because of the uncertainty introduced by fission-track fading through annealing. Because of this annealing effect only the oldest ages found are usable, and rather than corroborating the potassium–argon method, fission-track dating depends on that method for the confirmation of the age obtained.

Some of the uncertainty of fission-track dating has been removed by the discovery that fission tracks become smaller as they fade. Fossil fission tracks can be compared with induced fission tracks and an age correction applied.

The first paper, by Fleischer, Price, and Walker,* illustrates the risk in the use of the fission-track method of age determination for tektites. The authors confirmed (Fig. 2)* previous age determinations of about 34 million years for bediasites, Libyan Desert glass, and Clearwater Lakes, Quebec impact glass (Fleischer and Price, 1964), and concluded that all three glasses were formed by the same event, and that the great distances between these occurrences is compatible with a lunar origin of tektites. However, Wanless et al. (1965) found K–Ar ages of 285 and 300 million years for the Clearwater Lakes glass. Later, Fleischer et al. (1969) concluded that fission-track fading in the Clearwater Lakes glass had caused erroneously low values.

In the second paper Gentner, Storzer, and Wagner* report that Libyan Desert glass is younger than the bediasites. On this basis three events were involved, instead of one, to produce bediasites, Libyan Desert glass, and the Clearwater Lakes glass. They also found that the age of australites, Darwin glass, and Muong Nong-type indochinites agree, a conclusion reached independently by Fleischer et al. (1969), as well as by earlier workers who used the K^{40}–Ar^{40} method of age determination.

Using the fission-track method Gentner et al. (1970) found ages of 0.71 and 1.09 million years, and Durrani and Khan (1971) 0.68 and 0.86 million years for Australasian and Ivory Coast microtektites, respectively.

In the third paper of this group Fleischer, Price, and Woods,* using the fission-track method, announced a second group of ablated australites four million years old corresponding to the high-sodium australites of Chapman and Scheiber (1969). Caution should be exercised in accepting this finding until K^{40}–Ar^{40} age determinations have been made on the same samples, and preferably as well on other samples of high-sodium australites, using both methods of dating. If this second group of australites is validated beyond reasonable doubt, another element to consider will have been added to the already confusing array of data on australites. Why do these specimens show the same amount of etching as the younger australites and why, with the whole world as a target, did these tektites come to rest in an area of similar tektites?

Relating impact craters on earth to tektite strewn-fields has been considered by most workers as favoring a terrestrial origin for tektites; however, in another paper reporting fission-track work, Fleischer et al. (1969) suggest that tektites and the objects which made the craters may have been related, coming to earth at about the same time. As part of their argument they pair four strewn-fields to four impacts—moldavites with the Ries, Ivory Coast tektites with Bosumtwi, australites with Darwin glass (probably an impact glass), and bediasites with Libyan Desert glass. However, as noted previously it is likely that bediasites and Libyan Desert glass are of different ages; furthermore, Libyan Desert glass is a Muong Nong-type tektite and not an impact glass of the type directly associated with impact craters. It represents instead an individual tektite strewn-field.

Dr. Wolfgang Gentner, Professor and Director, Max-Planck-Institut für Kernphysik, Heidelberg, Germany is Vice President, Max-Planck-Gesellschaft and President of the Council, CERN, Geneva. He also holds numerous other appointments and has received various awards and honors. His co-authors are associated with the Max-Planck-Institut für Kernphysik at Heidelberg.

Dr. P. B. Price is now Professor, Department of Physics, University of California, Berkeley, California 94720. Dr. R. M. Walker is now Professor, Department of Physics, Washington University, St. Louis, Missouri 63130. Dr. R. T. Woods is now at the State University of New York, Albany, New York 12203.

References

Chapman, D. R., and Scheiber, L. C. (1969). Chemical investigation of Australasian tektites. *J. Geophys. Res.*, **74,** 6737–6776.

Durrani, S A., and Khan, H. A. (1971). Ivory Coast microtektites: fission track age and geomagnetic reversals. *Nature,* **232,** 320–323.

Fleischer, R. L., and Price, P. B. (1964). Fission track evidence for the simultaneous origin of tektites and other natural glasses. *Geochim. Cosmochim. Acta,* **28,** 755–760.

Fleischer, R. L. Viertl, J. R. M., and Price, P. B. (1969). Age of the Manicouagan and Clearwater Lakes craters. *Geochim. Cosmochim. Acta,* **33,** 523–527.

Gentner, W., Glass, B. P., Storzer, D., and Wagner, G. A. (1970). Fission track ages and ages of deposition of deep-sea microtektites. *Science,* **168,** 359–361.

Wanless, R. K., Stevens, R. D., Lachance, G. R., and Rimsaite, R. Y. H. (1965). Age determinations and geological studies—Part I, Isotopic Ages. Rept. No. 5, Geol. Surv. Canada, Paper 64–17, pp. 105–106.

24

Geochimica et Cosmochimica Acta 1965, Vol. 29, pp. 161 to 166. Pergamon Press Ltd. Printed in Northern Ireland

On the simultaneous origin of tektites and other natural glasses

R. L. Fleischer, P. B. Price and R. M. Walker
General Electric Research Laboratory, Schenectady, New York

(*Received* 15 *October* 1964; *in revised form* 11 *November* 1964)

Abstract—Further fission-track dating experiments have been done to examine the associations between tektite falls and various impact glasses. The previous ages found for Libyan desert glass by fission track dating, and for Ries Kessel glass and the Martha's Vineyard tektite by K–Ar dating, are confirmed. Although some low ages are found for samples from the Ivory Coast and the Ashanti Crater, the results support the value of 1·3 million years found by Gentner, Lippolt, Möller and Zähringer. Likely maximum ages are given for several impact glasses which are apparently not associated with tektite falls.

Introduction

In establishing the chronology of tektite formation, geochronological tools have given two primary types of information: They have shown that the geographical groups of tektites are also chronological groups, samples from a given continental area having the same formation age (Gentner and Zähringer, 1960; Zähringer, 1963; Fleischer and Price, 1964a, b). Secondly, it has been found that various other natural glasses, generally regarded as impact glasses, have ages that are indistinguishable from those of the various tektite groups (Gentner, Lippolt and Schaeffer, 1963; Fleischer and Price 1964a, b; Gentner, Lippolt and Müller, 1964). Hence the tektites fell in groups at nearly the same times as much more massive objects created impact glasses.

In the present work we report further ages obtained by the fission track method (Price and Walker, 1963). Results will be presented for North American and Ivory Coast tektites and various impact glasses. Some of the measurements substantiate existing K–Ar results, and others are determinations on previously undated materials.

Sometimes we find that several samples from the same tektite fall give a spread of ages by fission track dating. Such results are to be expected for tektites which have been heated briefly above 550°C (Fleischer and Price, 1964a), a temperature which would easily be attained by surface material during a brush or forest fire. When scattered fission track ages are found, we will be able to decide a "true" age in either of two situations: (1) If the highest fission track age is found for several of the samples, as was the case for the australites (Fleischer and Price, 1964b), or (2) if the highest fission track age agrees with the K–Ar result for the same group of samples. When neither of these situations is the case, we have no basis for giving a firm answer.

Method

In fission track dating, spontaneous fission tracks are displayed by chemical etching and counted in the natural material; the uranium content is found subsequently by a count of tracks induced by neutron irradiation. Although the techniques used for fission track dating of glass have been described in considerable detail by FLEISCHER and PRICE (1964a), a few further comments should be made about the procedure followed: Fig. 1 shows an example of fission tracks in an irradiated impact glass. Etching may be carried out either on a clean fracture surface or on an optical-quality, mechanically polished surface. With a reasonably large spontaneous fission track count ($> 1000/cm^2$) the fracture surface method is generally the easier one. On the other hand, for counting densities of less than a few hundred per cm^2, a level, polished surface allows a large area to be scanned without constant refocusing of the microscope. If the uranium distribution is not uniform, it is vital

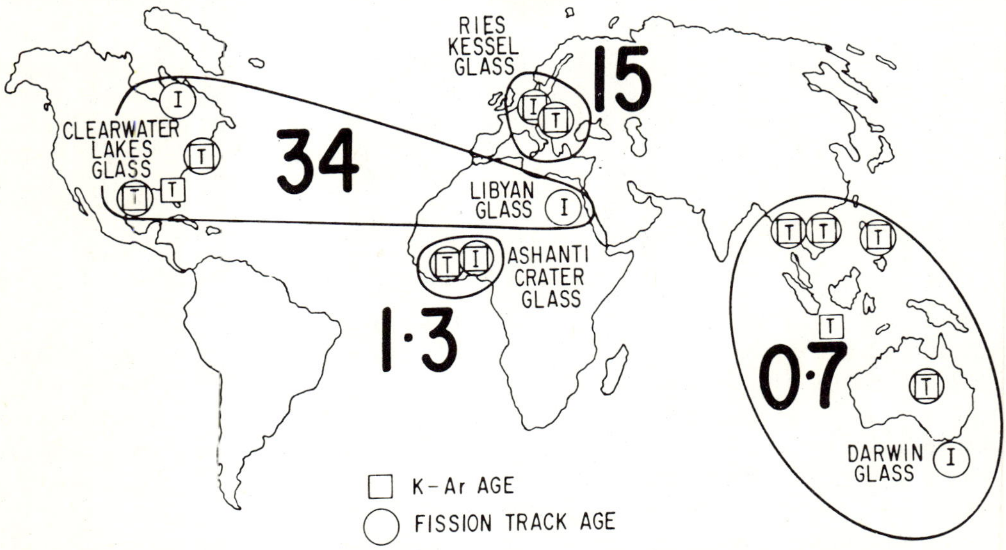

Fig. 2. Distribution of tektite groups and associated impact glasses. The numbers are the ages in millions of years of the glasses indicated within the encircled areas. In all cases where both K–Ar and fission track dating have been done (except for the Georgia tektites), the fission track dating confirms the K–Ar results. In no case is there a known contradiction.

to count induced tracks on the same area of material following irradiation as was examined previously for spontaneous tracks. In order to approximate this condition the glass sample is etched in such a way as to remove a layer of only 10–12 μ (the range of fission fragments in most glasses); in this procedure very nearly the same surface is available for inspection. If small particles of glass are to be examined, they may be imbedded in plastic, sectioned, polished, and thereafter treated as any bulk sample.

The fission decay constant used for the age determination was $6{\cdot}85 \times 10^{-17} yr^{-1}$ (FLEISCHER and PRICE, 1964c). Neutron doses were measured either by Ba^{140} counts done by a commercial firm or by fission-track etch pit counts in a calibrated glass (FLEISCHER, PRICE and WALKER, 1964).

RESULTS

Table 1 presents results on some tektites and chronologically related impact glasses; Table 2 gives results on other impact glasses. The Ries Kessel, Martha's Vineyard and Libyan glass samples are all in good agreement with previous results

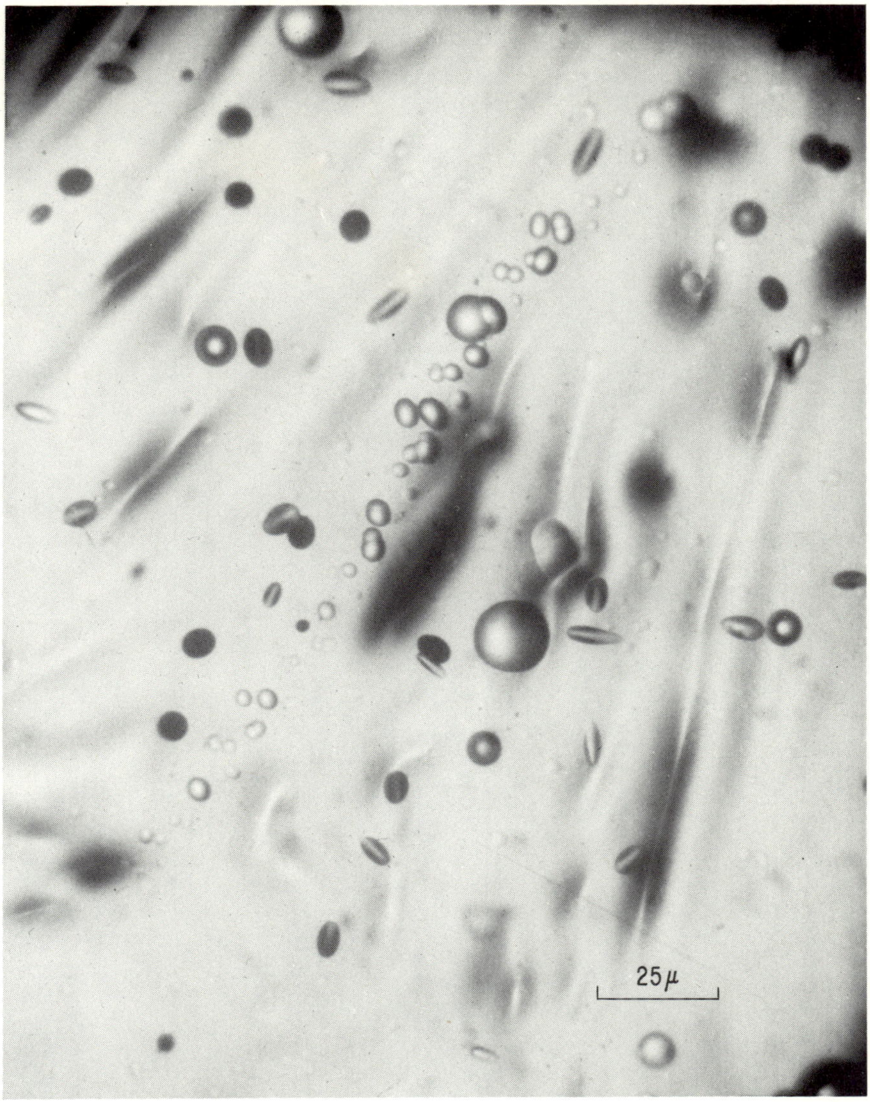

Fig. 1. Fission track etch pits in a neutron-irradiated sample of Aouelloul glass. The string of light-bottomed pits is composed of bubbles; the dark circles and light-centered ellipses are fission track pits. Dose: 5×10^{15} neutrons/cm^2.

Table 1. Ages of some tektites and chronologically related impact glasses

Type of sample	From		Loaned by	$\rho_s \left[\frac{\text{tracks}}{\text{cm}^2}\right]$	$\rho_i/\phi \left[\frac{\text{tracks}}{\text{neutron}} \times 10^{11}\right]$	Fission track age (m.y.)	K–Ar age (m.y.)
Tektite	Ivory Coast	(IVC-13-64)	J. A. O'Keefe	77 (±9%)	0·67	0·71 (±0·07)	1·3 (c)
		(Paris Mus. 2202)	M. Pellas	113 (±9%)	0·72	0·96 (±0·18)	
		(372)	A. J. Cohen	124 (±8%)	0·69	1·10 (±0·16)	
		(IVC-3-64)	E. C. T. Chao	103 (±8%)	0·46	1·41 (±0·13)	
Impact glass	Ashanti Crater	(IA)	W. Gentner	42 (±13%)	1·6	0·16 (±0·05) (d)	1·3 ± 0·3 (f)
		(IB)	W. Gentner	49 (±16%)	0·61	0·49 (±0·09)	
		(IC)	W. Gentner	27 (±30%)	0·33	0·50 (±0·16)	
		(II)	W. Gentner	37 (±33%)	0·70	0·32 (±0·14) (d)	
		(III)	W. Gentner	36 (±50%)	0·72	0·30 (±0·19) (d)	
		(Bcc8a-64)	E. C. T. Chao	18·5 (±28%)	0·079	1·48 (±0·47)	
Impact glass	Ries Kessel		H. Faul	12,200 (±8%)	4·9	15·2 (±1·6)	14·8 ± 0·7 (g)
Tektite	Martha's Vineyard (USNM2082)		R. S. Clark Jr.	11,750 (±5%)	2·0	35·5 (±8·3)	33·7 ± 2·2 (e)
Tektite	Georgia	(USNM1396)	R. S. Clark Jr.	<1 (c)	2·0	<0·003	33·9 ± 1·7 (e)
		(USNM2136)	R. S. Clark Jr.	235 (±18%)	(a)	0·57 (±0·10)	
		(DGA-1)	E. A. King Jr.	250 (±20%)	2·9	0·60 (±0·13)	
		(E.Y.-W.B. (b))	T. Allen	1610 (±10%)	(a)	3·8 (±0·4)	
		(DGA-2)	E. A. King Jr.	6060 (±5%)	2·4	15·2 (±1·5)	
		(RSC-TSB)(h)	T. Allen	3900 (±6%)	1·5	16·1 (±1·4)	
Impact glass	Libyan Desert		Paris Museum	8690 (±6%)	1·4	37·5 (±2·4)	[33·8 ± 2·5] (i)

(a) Assumed to be the same as DGA-1. *Note added in proof:* This assumption is not a valid one, as the result on Georgiaite RSC-TSB has shown since this article was submitted.
(b) Gibbons School, Dodge County.
(c) If no track is found in a surveyed area of $\frac{1}{2}$cm², $\rho_s < 1/\text{cm}^2$ is assumed.
(d) Discrepancy between Ba^{140} and glass slide doses added an error of 12 per cent above the statistical standard deviation.
(e) Zähringer (1963).
(f) Gentner *et al.* (1964).
(g) Gentner *et al.* (1963).
(h) Rock Branch.
(i) Fleischer and Price (1964) fission track result.

(GENTNER, LIPPOLT and SCHAEFFER, 1963; ZÄHRINGER, 1963; FLEISCHER and PRICE, 1964a). The Georgia tektites, on the other hand, show various ages which all are less than the 34 m.y. found by ZÄHRINGER (1963). The *scattered* and *low* results suggest heating and support the suggestion of KING (1964) that Georgia tektites have been stratigraphically disturbed since their fall. The existence of fission tracks in these samples does, however, prove that these are natural rather than man-made glasses. The appearance of heating of tektites is by no means surprising since it is known that at some time in the past they were on the Earth's surface.

Ivory Coast tektites and Ashanti Crater glasses

The ages for Ivory Coast tektites and Ashanti Crater (Lake Bosumtwi) glass are of special interest but, unfortunately, have a considerable spread of values, probably for the reason which was discussed above. GENTNER, LIPPOLT and MÜLLER (1964) found an age of 1·3 ± 0·3 m.y. for the Ashanti Crater glass—in agreement with the age found by ZÄHRINGER (1963) for the Ivory Coast tektites. The fission track data in each case give a series of ages ranging from 0·7 to 1·4 (±0·1) m.y. for the tektites and from 0·2 to 1·5 (±0·5) m.y. for the Ashanti Crater glass. Again, as was found for the Georgiaites and earlier (FLEISCHER and PRICE, 1964b) for australites, heating is the likely cause. Since in each case the highest age is equal to the K–Ar age, these results must be regarded as supporting their validity. The Ashanti Crater result, however, is regarded as deserving of further work since the highest age (1·5 ± 0·5 m.y.) has a large margin of error, because of the exceptional non-uniformity and small volume fraction of glass in the sample used here.

Ages of other impact glasses

As Table 2 shows, we have dated three samples of Aouelloul glass and placed likely upper limits on the ages of Cañon Diablo, Henbury, and Wabar glasses.

Table 2. Ages of impact glasses that are apparently unrelated to known tektites

From	Loaned by	$\rho_s\left[\frac{\text{tracks}}{\text{cm}^2}\right]$	$\rho_i/\phi\left[\frac{\text{tracks}}{\text{neutron}}\cdot 10^{11}\right]$	Age (m.y.)
Aouelloul	A. J. Cohen	123 (±14%)	1·6	0·46 ± 0·10
Aouelloul	D. R. Chapman from W. A. Cassidy	85 (±14%)	1·4	0·38 ± 0·08
Aouelloul	D. R. Chapman from E. T. C. Chao	48 (±14%)	1·8	0·16 ± 0·06 (a)
Cañon Diablo	A. J. Cohen	<12 (b)	0·7	<0·1
Wabar	A. J. Cohen	<1·5 (b)	0·3	<0·03
Wabar (BM1148) of 1932	British Museum	<1·2 (b)	0·3	<0·03
Henbury (BM1546)	British Museum	<3 (b)	0·4	<0·04

(a) Discrepancy between Ba^{140} and glass slide dose measurement added an error of 16 per cent above the statistical standard deviation.

(b) If no track is found in an area of $\frac{1}{2}$cm^2, $\rho_s < 1$ is assumed.

These limits are all consistent with the state of preservation of the craters; but again a word of caution should be added: Since the craters are surface features, the samples could have been exposed to the annealing effect of surface fires.

Discussion

With the completion of this work, samples from all of the four large tektite occurrences have been dated by fission track dating. In all cases except for the Georgiaites, the earlier K–Ar results have been confirmed. For the Georgiaites there is, however, no contradiction, but as discussed in the introduction merely a lack of an answer. For each of the major fields there is at least one impact glass whose age is identical within the experimental uncertainties involved. Figure 2 shows the groupings of tektite and impactite formation ages.

Geological vs. geochronological ages

Occasionally geological ages based on stratigraphy have been given lower values than have been found by the fission track and K–Ar methods. One suggested explanation (Baker, 1960) is that tektites might somehow have been created long before they arrived on Earth, the geochronological ages then being the formation age and the geological age being the time of fall. However, the age of the flange material of australites, which was formed during their descent to Earth, has been shown (Fleischer and Price, 1964b) to agree with that found for whole samples, so that this suggestion must be incorrect. The alternate view, which we take here, is that because two different methods of age determination give identical results on tektites, differences between these ages and geological ages based on stratigraphy must be resolved by some adjustment in the geological reasoning. Since both K–Ar and fission track dating are performed on the samples themselves while geological dating depends on the *environment* of the samples, it is the environment which must somehow be irrelevant to the time of arrival of tektites.

Relation of dating of impactites to theories of tektite origin

The two major current theories of tektite origin are the lunar (Chapman and Larson, 1963) and terrestrial (Barnes, 1961; Cohen, 1962) theories. In each case the existence of an impact glass of the same age as a tektite fall is a natural consequence of the theory. In the terrestrial theory a body from outer space creates a crater on Earth, thereby throwing out tektites and giving *one* impact glass per tektite fall. In the lunar hypothesis an impact on the Moon may throw out tektites and larger fragments, each of which could give rise to an impact glass. The experimental observation that there are *two* impact glasses of the same ages as the North American tektites is thus compatible with the lunar hypothesis.* However, the terrestrial origin is by no means disproved since it is also possible that a second unrelated body created an impact crater at a time within the experimental uncertainty with which the ages are known. A second consideration is the distance between tektite and impactite locations (2000 miles from Clearwater Lakes to

* Although not associated with a known impact crater, Libyan glass has been described as an impact glass (Cohen, 1963). Cohen's speculation that an impact crater could have been obscured by drifting desert sands (and other erosional processes) is fully consistent with the high age of the glass.

Texas and 7500 miles from Egypt to Texas). Such a spacing is compatible with a lunar origin but poses a problem for the terrestrial origin theory to explain how tektites could be hurled through an atmosphere over such distances (O'KEEFE, 1963; CHAPMAN, 1964). This reasoning merely adds to the problem presented by the great size of the Far Eastern strewn field, as described by O'KEEFE (1963).

CONCLUSIONS

1. The existence of a total of four tektite and impactite occurrences is confirmed.

2. The agreement previously noted by GENTNER, LIPPOLT and MÜLLER (1964) between the age of the Ivory Coast tektites and the Ashanti Crater glass is supported by fission track dating.

3. The existence of two impactite glasses, each far from the chronologically associated Texas tektites, is evidence which is difficult to explain by the theory of terrestrial origin of tektites.

Acknowledgement—We are indebted to those who so graciously supplied the samples used in this study: T. ALLEN, E. C. T. CHAO, D. CHAPMAN, R. S. CLARK JR., A. J. COHEN, H. FAUL, W. GENTNER, E. A. KING JR., J. A. O'KEEFE, M. PELLAS and the British Museum.

This research was supported in part by the Air Force Cambridge Research Center.

REFERENCES

BAKER G. (1960) Origin of tektites. *Nature, Lond.* **185,** 291.

BARNES V. E. (1961) Tektites. *Sci. Amer.* **205,** 58.

CHAPMAN D. R. and LARSON H. K. (1963) On the lunar origin of tektites. *J. Geophys. Res.* **68,** 4305.

CHAPMAN D. R. (1964) Amer. Geophys. Union, April 23, unpublished.

COHEN A. J. (1962) Asteroid impact hypothesis of tektite origin. *Proc. 3rd. Int. Space Sci. Symp., Washington.* North Holland, Amsterdam.

COHEN A. J. (1963) Fossil glasses produced by impact of meteorites, asteroids and possibly comets with the planet earth. *Advanc. Glass Tech.* Part 2, p. 360.

FLEISCHER R. L. and PRICE P. B. (1964a) Glass dating by fission fragment tracks. *J. Geophys. Res.* **69,** 331.

FLEISCHER R. L. and PRICE P. B. (1964b) Fission track evidence for the simultaneous origin of tektites and other natural glasses. *Geochim. et Cosmochim. Acta* **28,** 755.

FLEISCHER R. L. and PRICE P. B. (1964c) Decay constant for spontaneous fission of U^{238}. *Phys. Rev.* **133B,** 63.

FLEISCHER R. L., PRICE P. B. and WALKER R. M. (1964) Neutron flux measurements by fission tracks in solids (in press).

GENTNER W. and ZÄHRINGER J. (1960) Das Kalium–Argon-Alter von Tektiten. *Z. Naturf.* **15a,** 93–99.

GENTNER W., LIPPOLT H. J. and SCHAEFFER O. A. (1963) Argonbestimmungen am Kaliummineralien—XI: Die Kalium–Argon-Alter der Gläser des Nördlinger Rieses und der Böhmisch-mahrischen Tektite. *Geochim. et Cosmochim. Acta* **27,** 191.

GENTNER W., LIPPOLT H. J. and MÜLLER O. (1964) Kalium–Argon-Alter des Bosumtwi-Kraters in Ghana und die Chemische Beschaffenheit Seiner Gläser. *Z. Naturf.* **19a,** 150.

KING JR. E. A. (1964) New data on Georgia tektites. *Geochim. et Cosmochim. Acta* **28,** 915.

O'KEEFE J. A. (1963) The origin of tektites. *Tektites*, p. 167. University of Chicago, Chicago.

PRICE P. B. and WALKER R. M. (1963) Fossil tracks of charged particles in mica and the age of minerals. *J. Geophys. Res.* **68,** 4847.

ZÄHRINGER J. (1963) K–Ar measurements of tektites. *Radioactive Dating*, p. 289. Int. Atomic Energy Agency, Vienna.

25

Geochimica et Cosmochimica Acta, 1969, Vol. 33, pp. 1075 to 1081. Pergamon Press. Printed in Northern Ireland

New fission track ages of tektites and related glasses

W. GENTNER, D. STORZER and G. A. WAGNER
Max-Planck-Institut für Kernphysik, Heidelberg, Germany

(*Received* 25 *February* 1969; *accepted in revised form* 8 *April* 1969)

Abstract—Fission track dating is applied to a large number of australites, Muong Nong type tektites, Darwin glasses and Libyan Desert glasses. It is found that these glasses often have lowered fission track ages due to annealing effects. The same specimens have also smaller fission track etch pits.

According to the diminishing percentage of etched fossil fission track diameters, lowered fission track ages are corrected. A mean age of 0·7 m.y. is found for eleven australites, six Darwin glasses and seven Muong Nong type tektites (from Laos and Thailand), indicating a related genesis of these glasses. Uncorrected fission track ages of five Libyan Desert glasses $\left(26{\cdot}6\ {}^{+\,6\cdot3}_{-\,1\cdot3}\ \text{m.y.}\right)$ and two bediasites $\left(33{\cdot}2\ {}^{+\,8\cdot3}_{-\,1\cdot2}\ \text{m.y.}\right)$ are found to be different. With regard to these results an interrelationship between these two groups of glasses seems to be improbable.

INTRODUCTION

DISCUSSION on the origin of tektites has been essentially influenced by the application of radiometric age determination methods to tektites and related glasses.

According to the K/Ar-ages GENTNER and ZÄHRINGER (1960) and ZÄHRINGER (1963) were able to classify the tektites in four independent groups. The North American tektites are the oldest ones, having ages of about 34 m.y. The moldavites have a mean age of 15 m.y. Ivory Coast tektites have an age of 1·3 m.y., while the youngest group is formed by the tektites from the Indochina–Australian area with an age of 0·7 m.y. In addition, K/Ar-dating has been extended to related impact glasses. Concordant K/Ar-ages have been found for moldavites and glasses from the Ries–Kessel by GENTNER *et al.* (1963) and for the Ivory Coast tektites and the glasses from the Bosumtwi-crater by GENTNER *et al.* (1964). These findings suggest a terrestrial origin at least for these two groups of tektites.

Since its introduction by PRICE and WALKER (1963) the fission track dating method has proved to be suitable for glasses. Applied to tektites and related glasses, the method yielded ages quite similar to the K/Ar-ages (FLEISCHER and PRICE, 1964a,b; FLEISCHER *et al.*, 1965; GENTNER *et al.*, 1967).

In this paper further fission track ages of tektites and related glasses are presented.

EXPERIMENTAL PROCEDURE

The fission track method itself is described elsewhere (e.g. PRICE and WALKER 1963). Special procedures were the same as described by WAGNER (1966). The highly polished sections of the glasses were etched in 48% hydrofluoric acid at room temperature for different times (between 10 sec and 2 min, depending on the SiO_2-content of the various glasses). For the spontaneous fission decay constant was used $\lambda_f = (8{\cdot}42 \pm 0{\cdot}1) \times 10^{-17}\ \text{yr}^{-1}$ (SPADAVECCHIA and HAHN 1967).

The values for the decay constant of spontaneous ^{238}U-fission, given by many authors, are broadly ranged. For our age calculations the decay constant of SPADAVECCHIA and HAHN (1967) was chosen, because this value is an independent 4π-measurement without any correction for angle distribution. Further, thermally unaffected fission track ages of glasses, calculated with this constant, agree very well with K/Ar-ages. In addition, KAUFHOLD and HERR (1967) concluded from their fission track dating on Early Victorian uranium-glasses that $\lambda_f = (8{\cdot}2 \pm 0{\cdot}8) \times 10^{-17}\ \mathrm{yr}^{-1}$ (PARKER and KURODA, 1956) is more probable than $\lambda_f = (6{\cdot}9 \pm 0{\cdot}2) \times 10^{-1}\ \mathrm{yr}^{-1}$ (FLEISCHER and PRICE, 1964c).

Table 1. Fission track ages of australites, Darwin glasses and Muong Nong type tektites

No. of sample	Measured fission track age in m.y.*	Corrected fission track age in m.y.†	$\left[\frac{\text{Etched tracks}}{10^{11}\text{ neutrons}}\right]$ (p_i/n)‡
austr. 660-1	$0{\cdot}66^{\delta} \pm 0{\cdot}02$	—	3·6
austr. 660-2	$0{\cdot}71^{\delta} \pm 0{\cdot}03$	—	3·6
austr. 660-3	$0{\cdot}24^{\delta} \pm 0{\cdot}02$	$0{\cdot}71^{\delta}$	3·3
austr. 660-4	$0{\cdot}54^{\delta} \pm 0{\cdot}02$	$0{\cdot}71^{\delta}$	3·4
austr. 660-5	$0{\cdot}33^{\delta} \pm 0{\cdot}04$	$0{\cdot}66^{\delta}$	4·0
austr. 660-6 core	$0{\cdot}34^{\delta} \pm 0{\cdot}03$	$0{\cdot}70^{\delta}$	3·5
flange	$0{\cdot}13^{\delta} \pm 0{\cdot}02$	$0{\cdot}74^{\delta}$	
austr. 660-7 core I	$0{\cdot}40^{\delta} \pm 0{\cdot}05$	$0{\cdot}74^{\delta}$	
core II	$0{\cdot}63^{\delta} \pm 0{\cdot}04$	$0{\cdot}72^{\delta}$	2·7
flange	$0{\cdot}48^{\delta} \pm 0{\cdot}06$	$0{\cdot}71^{\delta}$	
austr. 660-8 core	$0{\cdot}31^{\delta} \pm 0{\cdot}04$	$0{\cdot}71^{\delta}$	3·8
flange	$0{\cdot}33^{\delta} \pm 0{\cdot}03$	$0{\cdot}68^{\delta}$	
austr. 660-9 core	$0{\cdot}71^{\delta} \pm 0{\cdot}04$	—	3·4
flange	$0{\cdot}70^{\delta} \pm 0{\cdot}04$	—	
austr. 133	$0{\cdot}68^{\beta} \pm 0{\cdot}1$	—	3·4
austr. 134	$0{\cdot}63^{\beta} \pm 0{\cdot}1$	0·73	2·6
austr. 131	$0{\cdot}11 \pm 0{\cdot}02$	0·30	4·2
Darwin gl. 646-1	$0{\cdot}73 \pm 0{\cdot}04$	—	3·4
Darwin gl. 646-2	$0{\cdot}73 \pm 0{\cdot}04$	—	3·1
Darwin gl. 646-3	$0{\cdot}72 \pm 0{\cdot}04$	—	2·9
Darwin gl. 646-4	$0{\cdot}69 \pm 0{\cdot}04$	—	3·8
Darwin gl. 646-5	$0{\cdot}74 \pm 0{\cdot}04$	—	3·3
Darwin gl. 646-6	$0{\cdot}71 \pm 0{\cdot}04$	—	3·5
Muong Nong 572-1	$0{\cdot}68 \pm 0{\cdot}04$	—	3·1
Muong Nong 572-2	$0{\cdot}72 \pm 0{\cdot}04$	—	2·9
Muong Nong 597-1	$0{\cdot}61^{\sigma} \pm 0{\cdot}04$	—	3·1
Muong Nong 597-2	$0{\cdot}67 \pm 0{\cdot}04$	—	3·4
Muong Nong 597-3	$0{\cdot}65 \pm 0{\cdot}04$	—	3·0
Muong Nong 58	$0{\cdot}64^{\sigma} \pm 0{\cdot}04$	—	3·2
Kemaraj	$0{\cdot}67 \pm 0{\cdot}04$	—	3·5

* Statistical error of counting fission tracks.
† The limit of error for corrected ages is about 20% and is composed of the statistical counting error and the error of the correction.
‡ Limit of error ±5%.
$^{\delta}$ Fission track ages from STORZER and WAGNER (1969).
$^{\beta}$ Fission track ages from WAGNER (1966), λ_f adjusted.
$^{\sigma}$ Samples slightly thermally affected.

Table 2. Fission track ages of Libyan Desert glasses

Sample No.	Measured fission track age in m.y.*	Corrected fission track age in m.y.†	$\left[\frac{\text{Etched tracks}}{10^{11}\text{ neutrons}}\right]$ (p_i/n)‡
699-1	$26{\cdot}5 \pm 1{\cdot}3$	28·7	2·1
699-2	$26{\cdot}9 \pm 1{\cdot}3$	29·3	2·3
699-3	$26{\cdot}2 \pm 1{\cdot}3$	28·3	1·9
698	$24{\cdot}9 \pm 1{\cdot}3$	28·0	1·4
593§	$28{\cdot}4 \pm 1{\cdot}3$	—	1·6

§ The same sample was also measured by WAGNER (1966).

Tables 1 and 2 give values for the ratios of density of etched induced fission tracks (p_i) to the dose of thermal neutrons (n). This ratio is proportional to the uranium content, and can be calibrated using samples of known uranium content. The measured p_i/n ratios show the variation of uranium contents from sample to sample with similar chemical composition.

Results and Discussion

Australites

Fission track ages of twelve australites from different localities were measured (samples 660 from the Nullabor Plains, S. Austr., 131 from Florieton, S. Austr.,

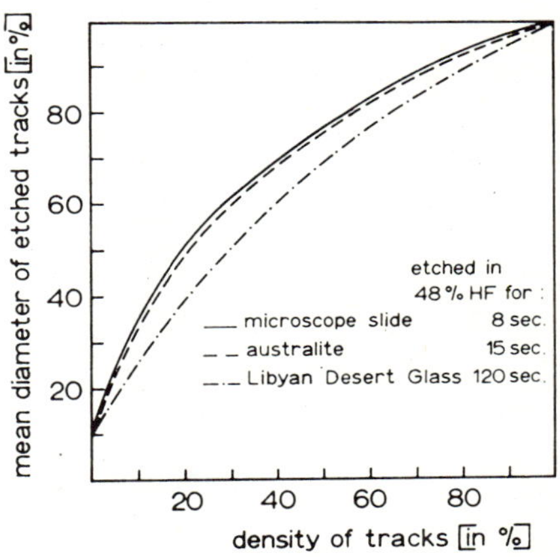

Fig. 1. The diagram shows the relationship between the reduction percentage of mean diameters of etched fission tracks and the degree of reduction of track density. Each mean diameter of etched tracks was obtained by measuring about 200 etch pit diameters. In case of oval pits, the longer diameter was used. These curves were simulated in laboratory annealing experiments for various glasses.

133 from Pine Creek, S. Austr., 134 from Charlotte Waters, C. Austr.). The results are compiled in Table 1. All samples consisted of dense dark-brownish glass. Most of the australites were boat- or lens-shaped. Four specimens (660-6, 660-7, 660-8, 660-9) had aerodynamic flanges. In these cases both, the cores and the flanges were dated. The resulting fission track ages range from 0·11 m.y. up to 0·71 m.y. Fleischer and Price (1964b) dated eight australites and got similar results (0·09–0·65 m.y., adjusted to $\lambda_f = 8{\cdot}42 \times 10^{-17}\,\mathrm{yr}^{-1}$). These ages are interpreted as thermally lowered. In samples with lowered fission track ages etched fossil fission tracks have different appearance compared to the induced ones (Fleischer and Price, 1964b). It was shown in laboratory experiments that at increased temperatures the diameters of fission track etch pits decrease in a characteristic manner related to the degree of reduction of track density (Storzer and Wagner, 1969). In Fig. 1 the percentage of mean diameter reduction is plotted vs. the reduction degree of track density for various glasses. Thus, measuring the mean diameter reduction of fossil tracks it is possible to correct the thermally lowered fission track ages by means of the calibration curve. Table 1 also contains the corrected fission track ages. They agree with

the fission track ages of the thermally unaffected australites (samples 660-1, 660-2, 660-9, 133). Hence, the fission track method gives a mean age of 0·7 m.y. for australites. This value is in good agreement with K/Ar-ages of australites as measured by ZÄHRINGER (1963).

As an exception, the corrected age of sample 131 is 0·30 m.y. only. In the simplest model, the corrected fission track age of a glass is either its solidification age or it dates the last strong annealing combined with complete disappearance of tracks. One might then conclude that australite 131 underwent at least two thermal events, a recent one and a fossil one. The K/Ar-age of the same sample is 0·73 m.y. (ZÄHRINGER, 1963). As deduced from laboratory annealing experiments, age reductions like the observed ones may either be due to natural or artificial thermal events. This problem is discussed in more detail by STORZER and WAGNER (1969).

Fission track dating on the button-shaped tektites gave concordance between cores and flanges for two australites (660-8 and 660-9). In two other cases (660-6 and 660-7), ages for the flanges are lower than for the cores. However, the corrected ages for the cores and flanges are the same. FLEISCHER and PRICE (1964b) found also same ages for cores and flanges of australites.

Darwin glasses

Pieces of irregularly shaped, lightly greenish Darwin glass (from Ten Mile Hill, Tasmania) were dated. Since Darwin glass is rich in bubbles, only the six most compact samples were selected.

The fission track ages are given in Table 1. In average they are 0·72 m.y. A lower fission track age was found by FLEISCHER and PRICE (1964b) for one sample of Darwin glass (0·53 m.y., λ_f adjusted). Size studies of etched fossil and induced fission tracks showed that no strong thermal effects occurred during the history of these glasses.

It is interesting that this age corresponds to the mean fission track age of australites and indicates a possible genetic connection between Darwin glass and australites. Due to chemical differences between these two kinds of glasses, TAYLOR and SOLOMON (1964) regarded them as unrelated. On the other hand agreement of their fission track ages is suggestive of a relationship.

Muong Nong type tektites

Seven specimens of Muong Nong type tektites were dated by the fission track method. Samples 58, 572, 597 come from the locality Muong Nong in Laos and the sample Kemaraj is from Kemaraj site (Thailand). The disks for dating were cut from large, compact glass lumps.

The measured ages range between 0·61 m.y. and 0·72 m.y. (Table 1). In two cases (597-1 resp. 58) diminished diameters of etched fossil fission tracks were observed, indicating a small thermal affection. Unfortunately, an age correction curve as for australites is not yet available.

The fission track ages seem to support that Muong Nong type tektites belong genetically to the Far-East strewn-field.

Libyan Desert glasses

Fission track age determination was applied to five specimens of Libyan Desert glass. The glasses were derived from different sources. From the milky, bubble-rich sample 698 the clearest layer was selected, since small bubbles (similar in size to etched fission tracks) may cause counting errors. The other four glasses were compact and poor in bubbles.

The measured fission track ages range between 24·9 m.y. and 28·4. m.y. Similar fission track ages for two Libyan Desert glasses were published by FLEISCHER and PRICE (1964a) and FLEISCHER *et al.* (1965) (after λ_f-adjustment).

The distribution histograms of etched fossil fission track diameters reveal different degrees of annealing. Sample 593 shows identical size distributions

Table 3. Temperatures and times of annealing which are necessary to reduce the track density to 90% resp. 80% in Libyan Desert glasses extrapolated from annealing experiments on sample 699–2

Time (yr)	10^3	10^4	10^5	10^6	10^7	10^8	Reduction percentage of track density to
	95	75	56	40	25	12	90%
Temp. (°C)	121	101	81	64	45	35	80%

for both the fossil and the induced tracks. The smallest fossil fission tracks were observed in sample 698. Although the age correction curve for thermally lowered fission track ages is exactly valid only for recently annealed samples, it can be shown that its application to a slight continuous annealing during its history, gives a fairly good approximation. The corrected ages are given in Table 2 with a mean age of 28·5 m.y. Track density reductions of 10%, as indicated by the diminishing percentage of etched fossil fission track diameters

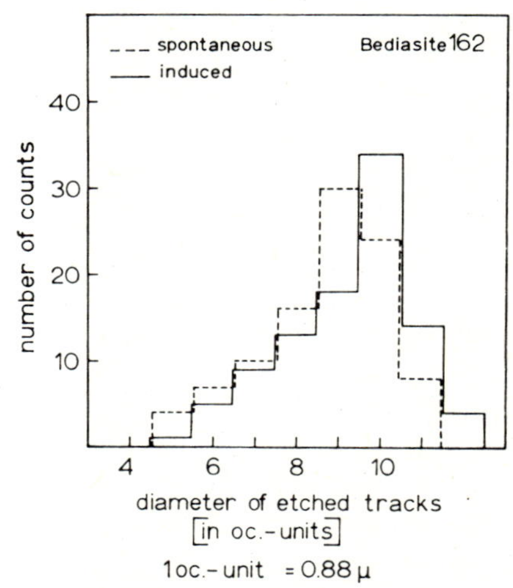

Fig. 2. Diameter size distribution histogram for etched fossil and induced fission tracks in bediasite 162. The etched fossil fission tracks show smaller diameters in contrast to the induced ones, revealing thermal effects.

agree with reasonable temperatures if one extrapolates experimental data to geological times (Table 3). From all experimentally annealed glasses Libyan Desert glass shows the highest track stability.

FLEISCHER *et al.* (1965) suggested a genetic relationship between Libyan Desert glasses and North American tektites from their concordant fission track ages. The present results do not support this hypothesis. Two bediasites already dated by WAGNER (1966) (sample 161: 34·2 m.y. resp. sample 162: 32·2 m.y.) were examined for the size distributions of the etched fossil fission tracks. It was observed that sample 161 was less and sample 162 (Fig. 2) was more thermally affected. Unfortunately, a correction curve is not yet available.

There is a systematic difference at least for the measured and it seems to be also for the corrected fission track ages between these two groups of glasses. Annealing experiments and further fission track age determinations of North American tektites are in preparation.

CONCLUSIONS

(a) For australites, Darwin glasses and Muong Nong type tektites, concordant fission track ages were found within narrow limits of error (0·7 $\pm$ 0·04 m.y.). This indicates a common origin of these glasses.

(b) Fission track ages of Libyan Desert glasses $\left(26{\cdot}6 \begin{array}{c} +\ 6{\cdot}3 \\ -\ 1{\cdot}3 \end{array} \text{m.y.}\right)$ are lower than those of the North American tektites $\left(33{\cdot}2 \begin{array}{c} +\ 8{\cdot}3 \\ -\ 1{\cdot}2 \end{array} \text{m.y.}\right)$. The nonsymmetrical errors of the measured fission track ages are caused by proved annealing effects. A genetic interrelationship between these two groups of glasses seems to be improbable.

(c) Fission track ages of the investigated australites are often lowered by thermal events.

(d) A continous slight annealing effect was observed in many tektites and related glasses. The intensity of this effect depends on the age, the temperature during the past and the chemical composition of a sample. In australites and Muong Nong glasses this effect is in general small. Darwin glasses revealed no thermal affection. For Libyan Desert glasses the effect was found in four out of five samples (age reduction about 10%). North American tektites also show this effect in similar intensity. One may conclude that the continuous annealing effect is more frequent and intensive with increasing age of a sample.

(e) Since fission track ages are in general too low, it is not accurate to calibrate the decay constant λ_f by requiring concordance between fission track ages and K/Ar- or Rb/Sr-ages without considering thermal effects, as it was done by FLEISCHER and PRICE (1964c). A decay constant deduced in this way must automatically be too low.

Acknowledgement—The authors are pleased to thank Prof. ZÄHRINGER and Dr. KIRSTEN for stimulating and helpful discussions. The technical assistance of Mr. HAIDMANN is appreciated.

Specimens of the clear variety of the Libyan Desert glass were kindly given to us through the courtesy of Dr. G. F. CLARINGBULL, British Museum London; those of the "layered" variety were donated by the Geological Museum of Cairo.

References

Fleischer R. L. and Price P. B. (1964a) Glass dating by fission fragment tracks. *J. Geophys. Res.* **69,** 331–333.

Fleischer R. L. and Price P. B. (1964b) Fission track evidence for the simultaneous origin of tektites and other natural glasses. *Geochim. Cosmochim. Acta* **28,** 755–760.

Fleischer R. L. and Price P. B. (1964c) Decay-constant for spontaneous fission of ^{238}U. *Phys. Rev.* **133,** B63–B64.

Fleischer R. L., Price P. B. and Walker R. M. (1965) On the simultaneous origin of tektites and other natural glasses. *Geochim. Cosmochim. Acta* **29,** 161–166.

Gentner W. and Zähringer J. (1960) Das Kalium–Argon-Alter von Tektiten. *Z. Naturforsch.* **15a,** 93–99.

Gentner W., Lippolt H. J. and Schaeffer O. A. (1963) Die Kalium–Argon-Alter der Gläser des Nördlinger Rieses und der böhmisch-mährischen Tektite. *Geochim. Cosmochim. Acta* **27,** 191–200.

Gentner W., Lippolt H. J. and Müller O. (1964) K–Ar-Alter des Bosumtwi-Kraters in Ghana und die chemische Beschaffenheit seiner Gläser. *Z. Naturforsch.* **19a,** 150–153.

Gentner W., Kleinmann B. and Wagner G. A. (1967) New K–Ar- and fission track ages of impact glasses and tektites. *Earth Planet. Sci. Lett.* **2,** 83–86.

Kaufhold J. and Herr W. (1967) Influence of experimental factors on dating by the fission track method. In *Radioactive Dating*, pp. 403–413. IAEA, Wien.

Price P. B. and Walker R. M. (1963) Fossil tracks of charged particles in mica and the age of minerals. *J. Geophys. Res.* **68,** 4847–4862.

Spadavecchia A. and Hahn B. (1967) Die Rotationskammer und einige Anwendungen. *Helv. Phys. Acta* **40,** 1063–1079.

Storzer D. and Wagner G. A. (1969) Correction of thermally lowered fission track ages of tektites. *Earth Planet. Sci. Lett.*, **5,** 463–468.

Taylor S. R. and Solomon M. (1964) The geochemistry of Darwin glass. *Geochim. Cosmochim. Acta* **28,** 471–494.

Wagner G. A. (1966) Altersbestimmungen an Tektiten und anderen natürlichen Gläsern mittels Spuren der spontanen Spaltung des ^{238}Urans. *Z. Naturforsch.* **21a,** 733–745.

Zähringer J. (1963) K–Ar *Measurements of Tektites*, pp. 289–303. IAEA, Wien.

26

EARTH AND PLANETARY SCIENCE LETTERS 7 (1969) 51–52. NORTH-HOLLAND PUBLISHING COMP., AMSTERDAM

A SECOND TEKTITE FALL IN AUSTRALIA

R.L.FLEISCHER, P.B.PRICE and R.T.WOODS

General Electric Research and Development Center, Schenectady, New York, USA

Received 18 August 1969

A chemical subgroup of tektites found in Australia (Chapman's "high sodium" tektites) have fission track ages of approximately 4 m.y. and hence are part of a distinct tektite fall, one that preceded the formation of the well known Australasian strewn field 0.7 m.y. ago.

One of the striking facts about tektites is their distribution over four rather well-defined geographical areas, each one of which includes samples of a single, discrete age group – 0.7 m.y. in the Far East, ~ 1 m.y. in the Ivory Coast, 15 m.y. in Czechoslovakia, and 34 m.y. in the United States [1–5]. These results have been taken as evidence of four distinct impact events, separated both in space and time. Here we point out that a chemical subgroup of tektites recently found in Australia must be the result of a fifth impact event, which in fact preceded the one which created the vast Far Eastern strewn field 0.7 m.y. ago.

Chapman and Scheiber [6] have identified a chemical subgroup ("high sodium tektites") of nine glass samples from Australia as being tektites on the basis of three characteristics: (1) They are glass samples having a single, unique composition found over a rather wide area. (2) They have the common shapes and sizes of tektites. (3) They contain lechatelierite inclusions.

Chapman supplied us with samples of three of the nine known tektites of the group for fission track dating [2,7,8], the results of which are given in table 1. The measured ages range from ~3 to ~4 m.y. and are unambiguously greater than those of tektites normally found in Australia (0.7 m.y.).

This age discordancy being unique among tektite fields, we must ask whether there is any means by which a spuriously high apparent age could have been found. Although for terrestrial samples spuriously high ages are not possible [7], for extra-terrestrial materials there are several potential sources of excess tracks as a result of the cosmic radiation [9]. Most of the natural tracks we found might in principle have resulted from fission of heavy elements, induced by cosmic ray protons and heavier nuclei. We have shown that such irradiation produces identifiable V-shaped events in tektites [10]. We searched AN-245 for such events and found none. Alternatively, the tracks might be from spallation recoil nuclei. Again we have previously identified such tracks [11]; they are too short to be confused with fission tracks.

Finally the tracks could be due to heavy cosmic ray nuclei, many of which might produce tracks of the same length as do fission events. If the tracks were of this origin, the track distribution through the

Table 1
Age data on high sodium tektites from Australia.

Sample *	Uranium (ppm)	ρ_s (cm^{-2})	Age (m.y.)
AN 87	0.7	412	4.25 (±0.43)
(Williams Creek)	0.9	370	4.65 (±0.50)
AN 245 (Raspberry Bore)	1.1	297	3.10 (±0.30)
AN 370 (S. of Lake Wilson)	1.3	346	3.54 (±0.35)

* Supplied by D.R.Chapman.

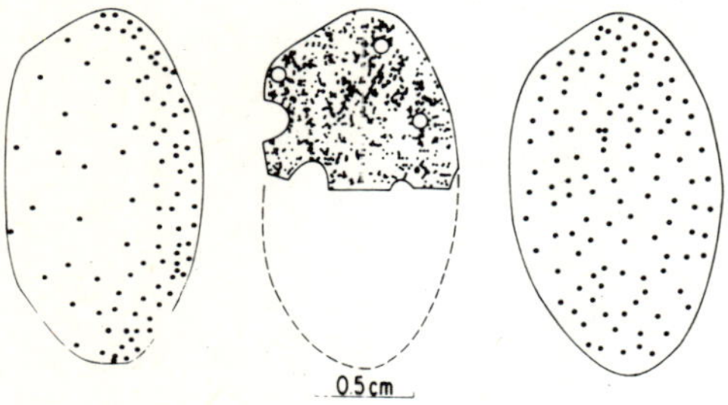

Fig. 1. The track distribution in sample AN-245 (center) compared with those expected from cosmic ray tracks (left) and spontaneous fission tracks (right). The results agree with those expected for spontaneous fission.

tektites would be non-uniform [10,11], due to the nature of the cosmic ray energy spectrum (greater abundance and more rapid slowing down at lower energies) and due to the short interaction mean free path of heavy nuclei (typically ~8 cm). Fig. 1 shows the spatial distribution of tracks to be expected in a tektite "core" if tracks originated from cosmic ray particles, if from spontaneous fission after arrival on earth, and as observed in sample AN-245.

A core is the shape remaining after one side of a sample is first ablated and then rapidly cooled during atmospheric entry. The heated and cooled layer (which was adjacent to the left side as viewed) has spalled off by the high internal stresses. The right hand side is the original pre-atmospheric surface. The gradient expected in the cosmic ray track density arises here primarily from the factor $d\epsilon/dx$ in eq. (6) of ref. [9] which varies by a factor of more than 4 between the surface and a position 1 cm in the interior. If in addition the cosmic ray spectrum for the energies involved (0 to 190 MeV/nucleon) is a simple extrapolation of the sparce data just above this energy range [12], the gradient in track density would be further enhanced. The observations shown in fig. 1 are clearly incompatible with the cosmic ray hypothesis. We concluded that, as assumed in calculating the ages appearing in table 1, the tracks are from spontaneous fission. It follows that the high-sodium group represents a distinct separate fall whose age is at least 4 m.y.

To many observers it may appear odd that a newly identified fall should exist in the very region where a previously known fall had occurred. Presumably, however, this second fall in Australia was identified solely because of the extensive search for more tektites and examination of them in an area already known to contain many samples. It would appear likely that other areas of the earth contain tektites from other, as yet unidentified, falls.

Acknowledgement

We are pleased to give thanks to E.Stella for experimental assistance, to D.R.Chapman for helpful comments and for supplying the three high sodium tektites, and to NASA for support under contract NAS 9-7898.

References

[1] J.Zahringer, Radioactive Dating (Int. Atomic Energy Agency, Vienna, 1963) p. 289.
[2] R.L.Fleischer and P.B.Price, J. Geophys. Res. 69 (1964) 331.
[3] R.L.Fleischer and P.B.Price, Geochim. Cosmochim. 28 (1964) 755.
[4] R.L.Fleischer, P.B.Price and R.M.Walker, Geochim. Cosmochim. 29 (1965) 161.
[5] W.Gentner, B.Kleinmann and G.A.Wagner, Earth Planet. Sci. Letters 2 (1967) 83.
[6] D.R.Chapman and L.C.Scheiber, "Australasian Tektite Chemistry and Geographic Pattern", 3rd Int. Tektite Conf., Corning, N.Y., April 16-18, 1969, to be published.
[7] P.B.Price and R.M.Walker, J. Geophys. Res. 68 (1963) 4847.
[8] R.L.Fleischer and P.B.Price, Geochim. Cosmochim. Acta 28 (1964) 1705.
[9] R.L.Fleischer, P.B.Price, R.M.Walker and M.Maurette, J. Geophys. Res. 72 (1967) 331.
[10] R.L.Fleischer, C.W.Naeser, P.B.Price, R.M.Walker and M.Maurette, J. Geophys. Res. 70 (1965) 1491.
[11] R.L.Fleischer, P.B.Price, R.M.Walker, M.Maurette and G.Morgan, J. Geophys. Res. 72 (1967) 355.
[12] G.M.Comstock, C.Y.Fan and J.A.Simpson, Astrophys. J. 155 (1969) 609.

Editors' Comments on Papers 27 and 28

Two papers have been chosen to illustrate recent geological-age dating of tektites. Fossil evidence, cited in the first paper, gives the approximate time one indochinite had been in the sea, although the actual age of the specimen is not known. The second paper, based on stratigraphical and radiocarbon evidence, gives the age of infall of australites.

Janoschek (1934) may have been the first to date a tektite strewn-field paleontologically when he identified the Helvetian (Middle Miocene) age of moldavites. However, Dvorsky noted before the turn of the century that a silicified fresh-water limestone containing snail shells overlies quartz gravel containing moldavites. Bouška (1964) discounts the Miocene age of moldavites. He believes they are Pleistocene or latest Pliocene. In Java, von Koenigswald (1935) found tektites associated with vertebrate fossils in alluvial gravel of Middle Pleistocene age (Trinil beds). Later, microfossils were used in dating the magnetic reversal with which microtektites are associated (Glass and Heezen, 1967).

The encrusted specimen from the South China Sea, described in the first paper, was given to us by Professor Edmond Saurin when we visited him in Saigon in 1960 (Saurin and Millies-Lacroix, 1961). The encrusting fossils were in growth position on a surface that had been etched previously, and on which very little etching had taken place since. In view of the fact that the fossils on the encrusted tektite grew about one million years ago on an etched surface, it is likely that the tektite entered the sea after it had been etched on land. Either it was submerged by a change of sea level or was carried into the sea by a stream. Therefore, the fossils on the tektite date essentially the length of time the specimen was in the sea, but do not date the time of its formation.

The precise geological-age dating of tektites is usually difficult. For example, the geological age of bediasites, predicted to be coming from Eocene rocks because of their distribution, was not finally verified until King (1968) found one by excavation in gravely sand in the Wellborn Formation of the Eocene Jackson Group. Although this specimen is definitely at least as old as the bed in which it was found it is not in its primary site of deposition, as shown by its pebble-like character and its association with gravel that has been transported by currents.

The unworn australites described by Lovering, Mason, Williams, and McColl,* as shown in the second paper, are in their primary site of deposition and their infall took place in the late Quaternary, confirming the findings of others who have worked with the geological occurrence of australites.

In a previous attempt to test the geological age of australites on earth, McDougall and Lovering (1969) determined K^{40}–Ar^{40} ages for both bodies and flanges of the same australites. Surprisingly, instead of being younger, as they would be if their aerodynamic remelting took place upon arrival during the late Quaternary, the flanges were found to have greater ages than the 860,000-year age found for the bodies of australites. Obviously, the flanges could not have formed prior to entry into the atmosphere; therefore the K^{40}–Ar^{40} age dating of australites does not give the time of their arrival on earth. Although the authors do not comment on the differences in australite age that they found by the two methods, in this paper they have built an incontrovertible case for the geologically young age of australite arrival on earth.

These two papers give evidence that two strewn-fields of widely different ages exist in the area of Australasian tektite occurrence. If the four-million-year-old group of australites announced by Fleischer et al. (1969)* is confirmed, a third strewn-field might include these older australites, as well as the javanites, billitonites, Borneo tektites, and some of the Malay Peninsula tektites. Of these three possible strewn-fields, the presently known microtektites come nearest to matching in age the tektites of the indochinite–philippinite strewn-field. If three strewn-fields of different ages exist in this area, the finding of three levels of microtektites in sea-floor cores would be confirming evidence.

References

Bouška, V. (1964). Geology and stratigraphy of moldavite occurrences. *Geochim. Cosmochim. Acta,* **28,** 921–930.

*Fleischer, R. L., Price, P. B., and Woods, R. T. (1969). A second tektite fall in Australia. *Earth Planet. Sci. Lett.,* **7,** 51–52.

Glass, B. P., and Heezen, B. C. (1967). Tektites and geomagnetic reversals. *Scient. Amer.* **217,** 32–38.

Janoschek. R. (1934). Das Alter der Moldavitschotter in Mähren. *Anz. Akad. Wiss. Wien, Math.-naturw.* **71** (17), 195–197.

King, E. A., Jr. (1968). Stratigraphic occurrence of bediasites (abstract). Geol. Soc. Amer., Program 1968 Ann. Meeting, pp. 160–161.

Koenigswald, G. H. R. von (1935). Vorläufige Mitteilung über das Vorkommen von Tectiten auf Java. *Proc. Koninkl Akad. Wet. Amsterdam,* **38,** 287–289.

McDougall, I., and Lovering, J. F. (1969). Apparent K–Ar dates on cores and excess Ar in flanges of australites. *Geochim. Cosmochim. Acta,* **33,** 1057–1070.

Saurin, E., and Millies-Lacroix, A. (1961). Tectites par 1270 m de fond au large du Vietnam. C. R. *Geol. Soc. France,* 128–129.

27

VIRGIL E. BARNES *Bureau of Economic Geology, The University of Texas at Austin, Austin, Texas 78712*

Age of Asian Tektites

ABSTRACT

Nannofossils in a calcareous crust firmly attached to a tektite dredged from the South China Sea, indicate a minimum age of about 1 m.y. for indochinite tektites on the basis of a 3-m.y. duration for the Quaternary Period.

INTRODUCTION

Four tektites were recovered February 29, 1960, from the South China Sea by the Scripps Institute oceanographic vessel *Stranger* during a 10-min period of dredging at depths of 1264 to 1274 m (Saurin and Millies-Lacroix, 1961). The nearest land is Triton Island of the Paracels group about 70 km to the north; Vietnam is about 270 km to the west. The specimen discussed weighs 14.25 gm and is a 2.5 cm segment of a flattened tear drop 3.2 cm wide. It has a refractive index of 1.5112 ± 0.0003 and a specific gravity of 2.447 ± 0.001.

PALEONTOLOGIC AGE

A firm crust of biogenic calcium carbonate on the convex side of the specimen indicates that this side was up on the sea bottom. Since this crust was deposited after arrival of the tektite at the sea bottom and before its disturbance by dredging, the time of formation of the crust, if it could be dated, would give a minimal age for the tektite. The calcium carbonate crust, therefore, was removed for paleontologic study by Lee A. Smith, Esso Production Research Company, Houston, Texas, and his findings were reported at the Third International Tektite Symposium held in Corning, New York (Smith and Barnes, 1969). Smith's data follow.[1]

All of the fossils in the following list were in the calcium carbonate removed from the largest pit on the concave side at the smaller end of the tektite. The remaining material removed from the entire exterior contains a sparse flora, mainly of the smaller forms listed.

Coccolithus pelagicus (Wallich), 1877
Coccolithus sp. cf. *C. minutulus* Gartner, 1967
Reticulofenestra? sp.
Emiliania? sp.
Gephyrocapsa caribbeanica Boudreaux and Hay, 1967
Gephyrocapsa oceanica Kamptner, 1943
Cyclococcolithus leptoporus (Murray and Blackman), 1898
Cyclococcolithus sp. cf. *C. multiradiatus* Kamptner, 1963
Oolithotus antillarum (Cohen), 1964
Rhabdosphaera claviger Murray and Blackman, 1898
Discolithina sp.
Helicopontosphaera kamptneri Hay and Mohler, 1967
Helicopontosphaera sp. cf. *H. intermedia* (Martini), 1965
Scyphosphaera apsteini Lohmann, 1902

The dominant forms in this assemblage are those assigned to the genus *Gephyrocapsa* Kamptner. According to McIntyre and Bé (1967) the dominance of *Gephyrocapsa* suggests a mid-latitude position for the site of deposition; that is, at the temperatures now found between 35° and 50° north latitude in the Atlantic Ocean. The diversity of species in the assemblage, however, suggests a position near the southern end of the latitude range; that is, a temperature range similar to that now found at approximately 35° north latitude in the Atlantic. As the tektite was actually collected near 15° north latitude, the indications are that the oceanic temperatures at the time the crust formed on the tektite were significantly cooler than today; that it is presumably not of interstadial origin.

The sample removed from the tektite contains both *Gephyrocapsa caribbeanica* Boudreaux and Hay, and *G. oceanica* Kemptner. It is believed, therefore, that the indochinite in question received its crust of calcium carbonate on the sea floor between 1.0 and 1.3 m.y. ago. The relative coolness indicated by the calcareous nannofossil assemblage suggests an "Illinoian" rather than a "Yarmouth" age for the assem-

[1] Three paragraphs quoted from Smith.

Geological Society of America Bulletin, v. 82, p. 1995-1996, July 1971

blage, and indicates an age for the calcium carbonate crust on the order of 1 m.y. As the age of the *Gephyrocapsa Caribbeanica G. oceanica* overlap is not precisely known, the calcium carbonate crust could be somewhat younger.

DISCUSSION

The tektite dredged from the South China Sea had been on earth long enough for an etch pattern to be fully developed before the calcium carbonate crust was deposited on it; therefore, the tektite on the basis of the nannofossils in the crust, is considerably older than the 0.7-m.y. age for indochinites and australites found by Zahringer and Gentner (1963) using the potassium-argon method. This method which dates the last time tektites were melted, generally has been assumed to be the age of tektites on earth. However, Lovering and others (1969) and Gill (1970), using the carbon-14 method of dating associated materials, found that Australian tektites could not have been on earth longer than about 16,000 yrs, thus introducing serious doubt of the validity of the potassium-argon method in determining the length of time tektites have been on earth.

The paleontologic age of 1 m.y. for the calcium carbonate crust on the South China Sea tektite is based on a 3-m.y. length for the Quaternary Period. Even on the basis of 1 m.y. for the Quaternary, the indochinites would have been on earth much longer than the 16,000 yrs estimated for australites. The evidence furnished by the nannofossils, and the occurrence of indochinites in marine sedimentary intercalations between late Pleistocene basalt eruptions in the southern part of the Leichow Peninsula (Lee, 1963) show that at least two tektite-forming events occurred in Australasia, instead of one as indicated by the potassium-argon method.

ACKNOWLEDGMENTS

I thank Edmund Saurin and Adrien Millies-Lacroix for their generosity in furnishing the specimen, Lee A. Smith and Esso Production Research Company for the paleontologic study, Dean R. Chapman and associates, Ames Research Center, for the specific gravity determination, Preston E. Cloud for critical review, and the National Science Foundation for support of my tektite research through grants G-10236, GP-408, and GA-914.

REFERENCES CITED

Gill, E. D., 1970, Age of australite fall: Jour. Geophys. Research, v. 75, p. 996-1002.

Lee, Da-Ming, 1963, A preliminary survey and study of the tektites—lei-gong-mo—from Leichow Peninsula and Hainan Island, China: Scientia Geologica Sinica, February, no. 1, p. 42-49 (for translation of excerpts from this paper by C. J. Peng and commentary *by* V. E. Barnes *see* Progress of tektite studies in China: Am. Geophys. Union Trans., v. 50, p. 704-709, 1969).

Lovering, J. F., Mason, Brian, Williams, G. E., and McColl, H. D., 1969, Stratigraphic evidence for the terrestrial age of australites [abs.]: 3rd Internat. Tektite Symposium, Corning, New York, p. 10.

McIntyre, A., and Bé, A.W.H., 1967, Modern Coccolithophoridae of the Atlantic Ocean—I. Placoliths and Cyrtoliths: Deep-Sea Research, v. 14, p. 561-597.

Saurin, E., and Millies-Lacroix, A., 1961, Tectites par 1270 m de fond au large du Vietnam: Geol. Soc. France, sess. no. 5, p. 128-129.

Smith, L. A., and Barnes, V. E., 1969, Age of South China Sea tektites [abs.]: 3rd Internat. Tektite Symposium, Corning, New York, p. 34.

Zahringer, J., and Gentner, W., 1963, Radiogenic and atmospheric argon content of tektites: Nature, v. 199, p. 583.

Manuscript Received by The Society January 25, 1971

28

STRATIGRAPHICAL EVIDENCE FOR THE TERRESTRIAL AGE OF AUSTRALITES

By J. F. LOVERING, B. MASON, G. E. WILLIAMS & D. H. McCOLL

(With 2 Plates, 3 Text-Figures and 2 Tables)

(*Received 27 September 1971: read by title at Perth, 4 October 1971*)

ABSTRACT

Australites in an excellent state of preservation are common (up to 1 specimen per 300 m^2) in lag gravels flooring corridors between seif dunes in the Motpena and Myrtle Springs areas of the Lake Torrens plain, South Australia. A study of the Quaternary stratigraphy of the region indicates that late Wisconsinan relict dunes (Lake Torrens Formation) are the most likely source of the australites. Radiocarbon dating indicates that the Lake Torrens Formation was deposited between about 24,000 and 16,000 years B.P. As the excellent preservation of most of the australites indicates that they have undergone negligible transport since their infall, it is concluded that the australites fell into the dune field sometime between about 24,000 and 16,000 years B.P.

INTRODUCTION

The time of infall of australites (Australian tektites) is clouded with uncertainty. On the one hand, K-Ar dating of australite glass places australites at 700,000 to 860,000 years B.P. (Zähringer, 1963; McDougall & Lovering, 1969). In apparent support of such an age, 'microtektites' in Indian Ocean cores have been linked with australites and correlated with a reversal in the Earth's magnetic field 700,000 years ago (Glass, 1967; Glass & Heezen, 1967). The Indian Ocean microtektites have a fission track age of 710,000 years (Gentner *et al.*, 1970), which is in good agreement with the palaeomagnetic data. Fission track ages for australites are more variable, grouping at around 800,000 years B.P. (Fleischer & Price, 1964) and 4 m.y. B.P. (Fleischer, Price & Woods, 1969). In the absence of cosmogenic radionuclides from the australite glass, the K-Ar and fission-track ages are commonly accepted as approximating the time of australite infall. Fleischer, Price & Woods (1969) have in fact suggested two distinct infalls, the first at 4 m.y. B.P. and the second at about 700,000 years B.P.

In contrast, stratigraphical evidence from widely spaced strewnfields on the Australian mainland suggests that the australites fell during the late Pleistocene or Holocene (Baker, 1962; Gill, 1965, 1970; Corbett, 1967; Polach, Lovering & Bowler, 1970). Following detailed stratigraphic study, Gill (1965, 1970) placed the age of infall of australites at Port Campbell, Victoria, at between last glacial (Wisconsinan) and mid-Holocene. Moreover, the distribution pattern of australites in southern central Australia (McColl & Williams, 1970), where most dated specimens were obtained, strongly points to only one period of infall. Hence, geological evidence from the Australian mainland is at variance, both as to infall frequency and age, with K-Ar and fission-track dating.

The present paper provides stratigraphical and radiocarbon evidence for the age of infall of australites in arid South Australia. In recent years large numbers of well-preserved, delicately shaped australites (Plate 27) have been collected from several areas on the western, basinward side of the Lake Torrens piedmont plain (Chalmers, 1964; Corbett, 1967). The majority of specimens appear to have undergone negligible transport since their infall. Geological evidence and radiocarbon dating point to an age of infall some time between about 24,000 and 16,000 years B.P.

THE LAKE TORRENS STREWNFIELD

The Lake Torrens piedmont plain slopes westwards from the western flank of the Flinders Ranges to the Lake Torrens salt pan (Fig. 1). The plain has an arid climate, mean annual rainfall ranging from 240 mm at Port Augusta to 170 mm at Myrtle Springs. Vegetative cover is sparse, especially on the drier, northern part of the plain.

Journal of the Geological Society of Australia, Vol. 18, Pt. 4, pp. 409-418, Pls. 27-28, June, 1972.

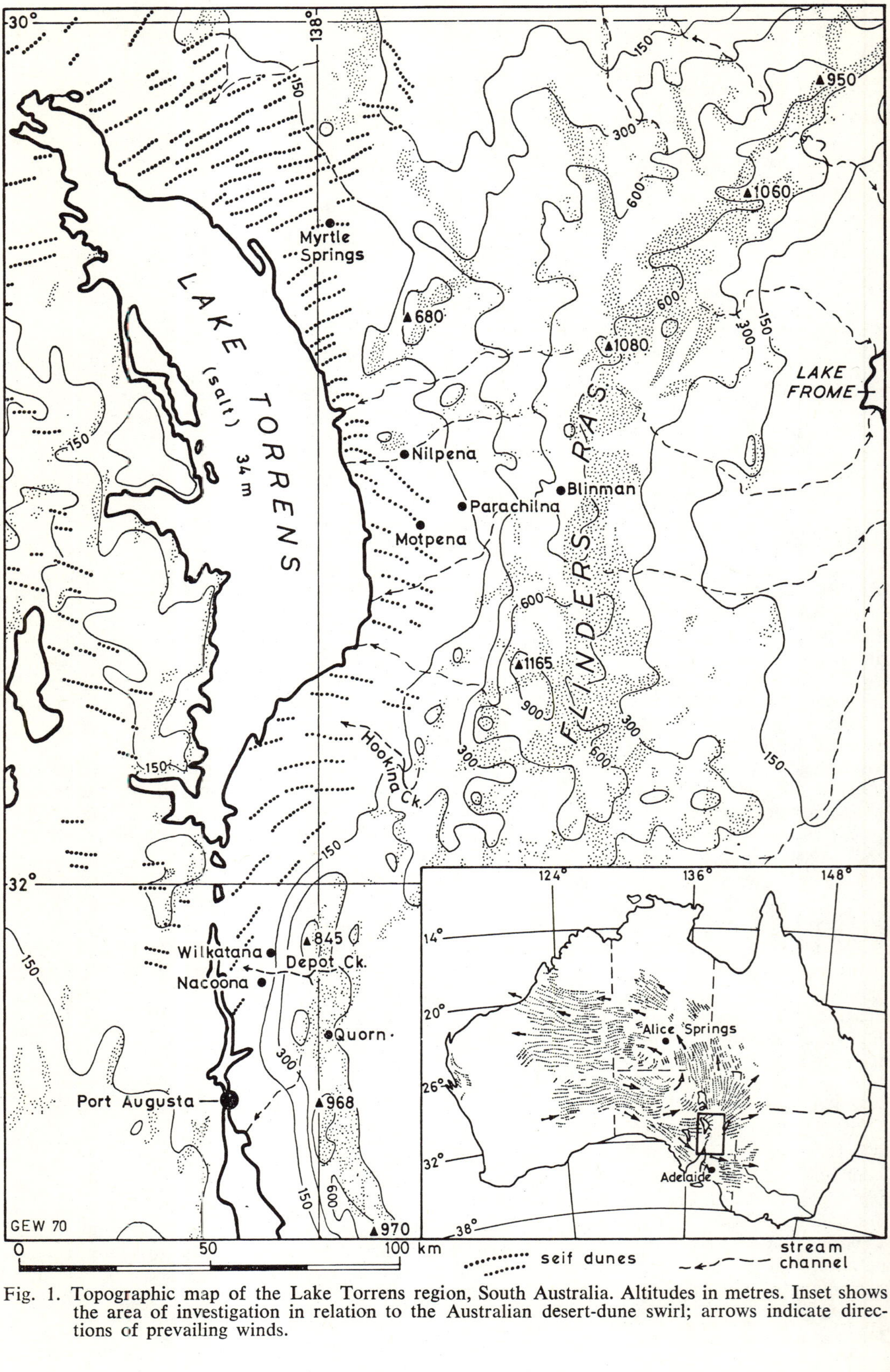

Fig. 1. Topographic map of the Lake Torrens region, South Australia. Altitudes in metres. Inset shows the area of investigation in relation to the Australian desert-dune swirl; arrows indicate directions of prevailing winds.

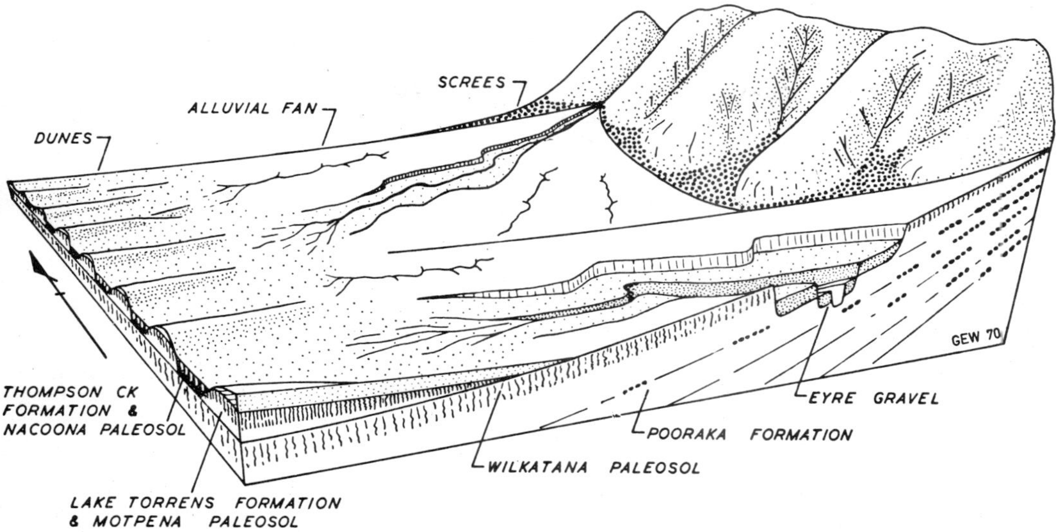

Fig. 2. Block diagram of the late Quaternary piedmont deposits and paleosols on the Lake Torrens plain, South Australia.

TABLE I

Suggested late Quaternary chronology, Lake Torrens plain, South Australia

Years B.P.	Event
0	
	Fan dissection. Local alluviation (Thompson Creek Fm. Member 2, in part) and dune building at base of fans.
ca 1800	
	Alluviation (Eyre Gravel Member 2; Thompson Creek Fm. Member 2, in part?)
ca 3500	
	Fan dissection. Soil formation (Nacoona paleosol).
ca 5000	
	Alluviation (Eyre Gravel Member 1; Thompson Creek Fm. Member 1). Local dune building.
ca 12,000	
	Soil formation (Motpena paleosol).
ca 16,000	
	Deposition of seif dunes at base of fans (Lake Torrens Fm.). Fan dissection.
ca 24,000	
	Soil formation (Wilkatana paleosol).
ca 30,000	
	Fan building (Pooraka Fm.). Talus breccias. Local dune building.
> 38,000	

Lake Quaternary stratigraphy

The late Quaternary stratigraphy of the Lake Torrens plain is illustrated schematically in Figure 2 and summarized in Table I. Full details of the stratigraphy and radiocarbon dating of the sediments and paleosols are given by Williams & Polach (1971) and Williams (*in press*). Relevant radiocarbon dates are listed in Table II. Briefly, the late Quaternary history is as follows.

1. Alluvial-fan building, with deposition of red pebbly-mudstones, gravels, sands and clays (*Pooraka Formation*, maximum exposed thickness 15 m). Away from the mountains the fan deposits pass into aeolian sands, as at Port Augusta and Myrtle Springs.
   Three radiocarbon dates for carbonized detrital wood are available for the Pooraka Formation: $33{,}270\ {}^{+2130}_{-1680}$ years B.P. (ANU-300) for material from 8-9 m below the top of the formation at Hookina Creek; > 37,800 years B.P. (N-840) for material from about 15 m below the top of the formation near the apex of the Wilkatana fan; and 34,600 ± 2700 years B.P. (N-633) for material from 3 m below the top of the formation at Pooraka, Adelaide (Williams, 1969). A mid-Wisconsinan age is thus indicated for the Pooraka Formation. Fan building probably continued until about 30,000 years B.P.

Journal of the Geological Society of Australia, Vol. 18, Pt. 4, pp. 409-418, Pls. 27-28, June, 1972.

TABLE II

Details of radiocarbon dates for late Quaternary sediments and paleosols, Lake Torrens plain, South Australia

Formation or paleosol	Locality	Sample type	C^{14} age	Reference	Code No.
Thompson Creek Fm. (Member 2)	Nacoona	Charcoal	1430 ± 80	ANU-IV	ANU-220
Eyre Gravel (Member 2)	Depot Creek	Charcoal	1880 ± 85	ANU-IV	ANU-217
Eyre Gravel (Member 2)	Wilkatana	Charcoal	1790 ± 400	ANU-IV	ANU-215
Eyre Gravel (Member 2)	Wilkatana	Charcoal	2900 ± 200	ANU-IV	ANU-213
Eyre Gravel (Member 2)	Wilkatana	Charcoal	3430 ± 80	ANU-IV	ANU-216
Nacoona paleosol	Nacoona	Whole-soil carbonate	6490 ± 80	ANU-IV,VI	ANU-224
Nacoona paleosol	Nacoona	Whole-soil carbonate	8450 ± 140	ANU-VI	ANU-224/2
Nacoona paleosol	Wilkatana	Whole-soil carbonate	7950 ± 110	ANU-IV	ANU-223*
Nacoona paleosol	Wilkatana	Carbonate nodules	8110 ± 130	ANU-VI	ANU-223/2
Thompson Creek Fm. (Member 1)	Nacoona	Charcoal	6000 ± 100	ANU-IV	ANU-219
Thompson Creek Fm. (Member 1)	Motpena	Charcoal	6000 ± 100	ANU-IV	ANU-265
Eyre Gravel (Member 1)	Depot Creek	Charcoal	5130 ± 100	ANU-IV	ANU-214
Motpena paleosol	Motpena	Carbonate nodules	$12{,}050 \pm 160$	ANU-IV	ANU-281*
Motpena paleosol	Nacoona	Carbonate nodules	$16{,}160 \pm 250$	ANU-IV,VI	ANU-280
Motpena paleosol	Myrtle Springs	Carbonate nodules	$14{,}840 \pm 250$	ANU-IV	ANU-100*
Motpena paleosol	Nilpena	Carbonate nodules	$12{,}540 \pm 150$	ANU-IV	ANU-132*
Wilkatana paleosol	Depot Creek	Carbonate nodules	$30{,}880 \pm 1200$	ANU-VI	ANU-295
Wilkatana paleosol	Depot Creek	Carbonate nodules	$35{,}000^{+4090}_{-2860}$	ANU-VI	ANU-296
Wilkatana paleosol	Depot Creek	Carbonate nodules	$35{,}100^{+2300}_{-1700}$	ANU-VI	ANU-226/2
Wilkatana paleosol	Nacoona	Carbonate nodules	$28{,}780 \pm 1000$	ANU-VI	ANU-227/2
Wilkatana paleosol	Motpena	Carbonate nodules	5820 ± 100	ANU-IV,VI	ANU-282
Wilkatana paleosol	Port Augusta	Carbonate nodules	$32{,}180^{+2400}_{-1840}$	ANU-VI	ANU-297
Wilkatana paleosol	Port Augusta	Carbonate nodules	$22{,}160 \pm 500$	ANU-VI	ANU-298
Wilkatana paleosol	Port Augusta	Carbonate nodules	$27{,}600 \pm 1000$	ANU-VI	ANU-299
Wilkatana paleosol	Myrtle Springs	Carbonate nodules	$20{,}310 \pm 360$	ANU-IV	ANU-127*
Wilkatana paleosol	Depot Creek	Whole-soil carbonate	$31{,}360 \pm 1000$	ANU-IV	ANU-226*
Wilkatana paleosol	Nacoona	Whole-soil carbonate	$22{,}290 \pm 450$	ANU-IV,VI	ANU-227
Wilkatana paleosol	Wilkatana	Whole-soil carbonate	$25{,}600 \pm 680$	ANU-IV	ANU-225*
Wilkatana paleosol	Motpena	Whole-soil carbonate	$22{,}900 \pm 500$	ANU-IV	ANU-264*
Pooraka Formation	Hookina Creek	Carbonized wood	$33{,}270^{+2130}_{-1680}$	ANU-VI	ANU-300
Pooraka Formation	Wilkatana	Carbonized wood	$>37{,}800$		N-840

Note: (1) All samples collected by G. E. Williams in 1967-1970 with the exception of ANU-100, 127 and 132, collected by B. Mason in 1967.

(2) All dates reported relative to Libby half-life (5570 ± 30 years).

(3) Radiocarbon ages for carbonate reported in ANU Date List IV (Polach *et al.*, 1970) were corrected for isotopic fractionation using assumed δC^{13} values. Recalculated ages using measured δC^{13} values, together with details of additional dates, to be reported in ANU Date List VI (in preparation). Asterisk (*) denotes radiocarbon age unchanged from that reported in ANU-IV.

2. Formation of a calcareous red-brown earth (*Wilkatana paleosol*),† containing abundant white nodules and cylindroids of micritic calcite, within the Pooraka Formation and equivalent aeolian deposits.

Thirteen radiocarbon dates have been determined for carbonate-nodule and whole-soil carbonate samples collected from the Wilkatana paleosol in a variety of geomorphological and stratigraphical settings. Dates for nodules, with one exception (ANU-282), range from about 35,000 to 20,000 years B.P., averaging 27,450 years B.P. The mean date is thus consistent with the stratigraphy, and suggests a Wisconsinan age (*ca* 30,000 to 24,000 years B.P.) for soil formation.

† The term 'paleosol' is used here informally for fossil Cca horizons.

3. Deposition of seif dunes of fine to medium quartz sand (*Lake Torrens Formation*, 0-5 m thick) above the Wilkatana paleosol at the base of the fans.
   No datable organic material was found in the Lake Torrens Formation. Elsewhere in southeastern Australia, however, the last period of instability of the inland dune system has been dated at 20,000 to 16,000 years B.P. (Polach *et al.*, 1970, p. 2). These age limits probably approximate those for the deposition of the Lake Torrens Formation.
4. Development of a calcareous red-brown earth (*Motpena paleosol*), containing white nodules, cylindroids, tubules and plates of calcite-cemented sand, within the Lake Torrens Formation.
   Four dates for carbonate nodules from the Motpena paleosol range from about 16,000 to 12,000 years B.P., averaging 13,900 years B.P. These dates accord with the likely time limits for pedogenesis.
5. During the Holocene, deposition of the conglomeratic *Eyre Gravel* on the fans, and the *Thompson Creek Formation* (mainly sand and clay) overlying the Wilkatana and Motpena paleosols within interdune corridors at the base of the fans. Moderate aeolian activity and fan dissection also occurred at times during the Holocene.
6. Development of an immature desert soil (*Nacoona paleosol*) within the Thompson Creek Formation (Member 1). At Nacoona the paleosol is developed within sediments dated by charcoal at 6000 ± 100 years B.P. (ANU-219), but is not developed within Member 2 of the Thompson Creek Formation, charcoal from which has been dated at 1430 ± 80 years B.P. (ANU-220). Pedogenesis thus occurred sometime between 6000 and about 1500 years B.P. The Nacoona paleosol contains small patches of authigenic earthy carbonate (calcite and dolomite together).

It is evident from Figure 2 that much of the western, basinward side of the Lake Torrens plain—that part of the plain locally yielding many australites—was aggraded almost continuously during the late Quaternary. As the western side of the plain is underlain by about 200 m of piedmont and lacustrine deposits dating back to the Tertiary (Johns, 1968), aggradation probably has prevailed during most of Quaternary time.

Occurrence of australites

Australites have been found at numerous localities in the vicinity of Lake Torrens. Only at Motpena and Myrtle Springs, however, have many well-preserved specimens, such as those shown in Plate 27, been collected. Regionally, these two localities form part of a line of australite strewnfields which extends northwestwards from Victoria into central Australia (McColl & Williams, 1970).

The stratigraphy and occurrence of australites at Motpena and Myrtle Springs are shown in Figure 3. The great majority of australites occur as lag, together with small nodules of soil carbonate and water-worn pebbles, on deflated interdune corridors (Plate 28A). In the larger corridors (up to 200 m wide) australite density ranges from practically nil to about one specimen per 300 m^2.

In deeply deflated corridors australites may lie directly on the truncated Wilkatana paleosol. At Motpena, however, most australites occur as lag on the lower slopes of the relict seif dunes (Lake Torrens Formation), whose deflated surfaces exhibit calcareous nodules and truncated vertical tubules of the Motpena paleosol. One australite was found partly embedded in this paleosol (Plate 28B, C). The specimen shown is incomplete, indicating fracturing and some transport prior to its burial.

Some interdune corridors at Motpena are floored by moderately consolidated clayey sand up to 40 cm thick. Charcoal obtained from this horizon has been dated at 6000 ± 100 years B.P. (ANU-265). The sand is interpreted as mid-Holocene alluvial fill derived in part from the ranges and probably also by rainwash from the adjacent relict ridges of the Lake Torrens Formation; stratigraphically it correlates with Member 1 of the Thompson Creek Formation. Mr N. Bartlett (*pers. comm.*, 1968) has found several australites partly embedded in this sand. Both the Lake Torrens Formation and the mid-Holocene sand are underlain along a sharp contact by the Wilkatana paleosol.

The writers have found no australite lying at the top of the relict or modern dune ridges at Motpena. However, several large australites have been collected from the modern dunes by Mr N. Bartlett (*pers. comm.*, 1968).

The geometry of the deposits at Myrtle Springs is slightly different from that at Motpena. In the Pine Dam area of Myrtle Springs Station most interdune corridors are floored by the Wilkatana paleosol. Moreover, the Motpena paleosol commonly is capped by 1 m or

Journal of the Geological Society of Australia, Vol. 18, Pt. 4, pp. 409-418, Pls. 27-28, June, 1972.

so of moderately consolidated, red, clayey sand (the 'red silt' of Corbett, 1967). The presence in the sand of aboriginal implements referable to the Pirrian culture (Corbett, 1967) indicates a mid-Holocene age (*ca* 5000 years B.P.) of deposition. The sand thus correlates with or is slightly younger than the alluvium locally flooring interdune corridors at Motpena.

Australites at Myrtle Springs are commonly found as lag on the mid-Holocene sand or the truncated Motpena paleosol at the margins of interdune corridors. In addition, Corbett (1967, p. 565) found two australites partly embedded in the mid-Holocene sand, and R. O. Chalmers (*pers. comm.*, 1971) found a well-preserved dumbbell australite apparently embedded in the Motpena paleosol at Pine Dam. Few australites rest on the Wilkatana paleosol exposed in interdune corridors, and none was found on the modern dunes.

State of preservation of australites

The australites from the Lake Torrens strewnfields are noteworthy for their generally excellent state of preservation (Plate 27; see

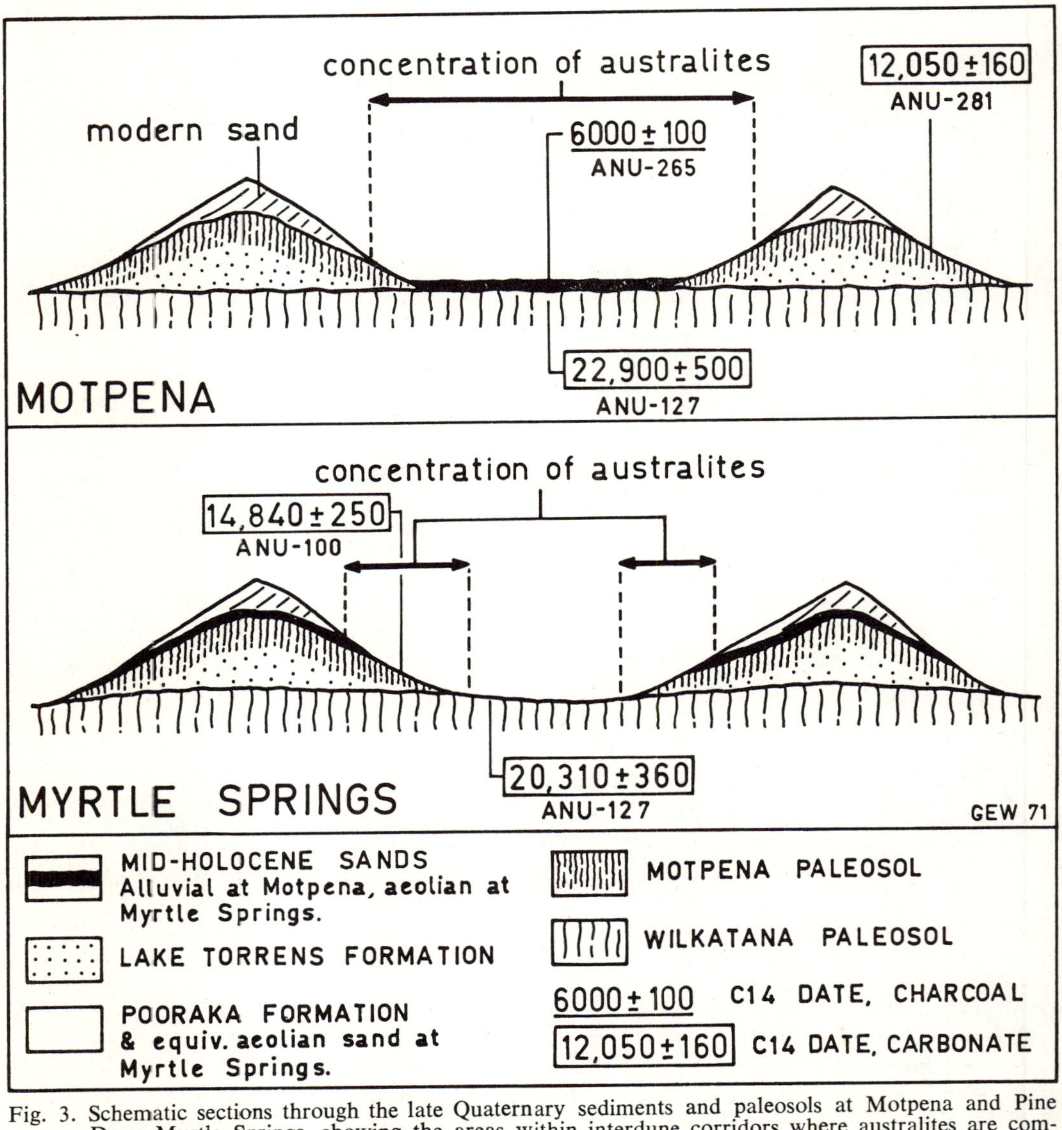

Fig. 3. Schematic sections through the late Quaternary sediments and paleosols at Motpena and Pine Dam, Myrtle Springs, showing the areas within interdune corridors where australites are commonly found.

also Corbett, 1967, Figs 4 & 5). Some specimens, however, show evidence of weathering and abrasion. Several of those collected had been faceted by sand-blasting. More commonly, australites exhibit solution etching along schlieren and strain fractures. Many of such etched pits and grooves are filled with indurated, micritic white calcite (X-ray diffraction identification) which in some cases encloses rounded grains of quartz jammed in the grooves. A crust of white calcite adheres to the surface of several australites collected. Similar calcite crusts also occur on pebbles in the lowest metre of the Lake Torrens Formation. Such pebbles are water-worn and clearly had undergone prolonged transport prior to their burial, contrasting with the unabraded state of most australites.

Deep grooves in several australites were filed with a vitreous, white mineral identified by X-ray diffraction as a zeolite related to sodium harmotome. It is noteworthy that zeolite (analcime) has been identified in strongly alkaline soils in arid California (Baldar & Whittig, 1968), and that phillipsite and erionite are the dominant alteration products of silicic glass in saline, alkaline lakes (Hay, 1966; Mariner & Surdam, 1970). The zeolite in the grooved australites is thus regarded as a weathering product resulting from the reaction of alkaline soil-waters with tektite glass.

The excellent preservation of most of the australites precludes long exposure or extended weathering for the majority of specimens. The excellent preservation of delicate flanges on many specimens (Plate 27) also indicates little or no transport prior to their burial.

Source horizon of australites

The following points have bearing on the ultimate source horizon of the australites.

1. Exposure of the Wilkatana paleosol is not necessary for an area or interdune corridor to yield australites. Further, no australite has been found on the surface of the alluvial fans where the Wilkatana paleosol is widely exposed.
2. At Motpena, some australites have been found in corridors floored entirely by the Motpena paleosol developed within the Lake Torrens Formation. Several large australites have been found on the slopes of this paleosol, where wind and water action could not move them from lower levels.
3. Some australites are partly encrusted with indurated, micritic calcite; pebbles derived from the base of the Lake Torrens Formation are similarly encrusted with calcite. Other australites exhibit pits and grooves containing rounded quartz grains cemented with calcite.
4. One australite has been found partly embedded in the Motpena paleosol at Motpena, and another in the Motpena paleosol at Pine Dam, Myrtle Springs.
5. Several australites have been found partly embedded in mid-Holocene alluvial and aeolian deposits at Motpena and Myrtle Springs.

It is evident that some of the australites were derived from a calcareous, sandy soil in which specimens were corroded and encrusted with calcite. Australite distribution on the plain strongly suggests that this soil is the Motpena paleosol. Moreover, the carbonate crusts enclosing rounded grains of quartz within fissures in some australites compare texturally and mineralogically with the nodules of calcite-cemented sand in the Motpena paleosol.

The overall evidence points to the relict dune sands of the Lake Torrens Formation as the ultimate source horizon of the australites in the Motpena and Myrtle Springs strewnfields. It is envisaged that after the Motpena soil-forming episode, erosion of these dune ridges flushed out australites onto the interdune corridors where they were incorporated in mid-Holocene alluvial and aeolian deposits. The isolated occurrence of australites on the modern dunes may be attributed to transport by aborigines; rare artifacts of flaked australites confirm their utilization by natives.

TIME OF AUSTRALITE INFALL

The question remains as to whether the presence of well-preserved australites within aeolian sands of late Wisconsinan age (about 24,000 to 16,000 years B.P.) indicates a comparable late Wisconsinan age for the australite infall. The following points favour such a conclusion.

First, the nearest rocks or erosion surfaces older than the late Pleistocene crop out from 15 to 25 km mountainwards from the Motpena and Myrtle Springs strewnfields. It is inconceivable that the many delicately shaped, fragile australites could have been transported at least 15 to 25 km by vigorous piedmont streamfloods without abrasion. The distribution of australite occurrences (McColl & Williams,

1970) further argues against transport from the Flinders Ranges catchment, as comparable late Quaternary sediments south of Motpena and on the eastern piedmont have yielded few australites. Indeed, the regional distribution pattern of strewnfields in southern central Australia is quite inconsistent with the wide dispersal of australites by terrestrial processes. The australites in the Motpena and Myrtle Springs strewnfields must lie near their place of infall.

Second, the Santos Motpena stratigraphical bore intersected nearly 200 m of Quaternary deposits beneath the Motpena strewnfield (Johns, 1968). If the australites had fallen onto the plain in the early Pleistocene (about 700,000 years ago), they would by now be deeply buried within the Quaternary deposits. The occurrence of australites at the surface indicates a terrestrial age no older than the mid-Wisconsinan Pooraka Formation.

Third, unconsolidated dune sands would have provided a soft landing for the australites.

Fourth, the chances of burial and consequent preservation of the australites would be excellent in such an aggrading environment. With an active dune system, large foreign particles such as australites would quickly gravitate to the lower slopes of the dunes by rolling down steep slopes or by deflation of the surounding sand. (This has been confirmed by experiments in which australites and pebbles were placed near the crests of modern dunes at times of strong winds). Burial by shifting dunes would then occur. The continued aggradation of much of the western part of the Lake Torrens plain during the late Quaternary thus accounts for the high proportion of well-preserved australites found in the Motpena and Myrtle Springs areas.

The state of preservation, the local and regional distribution, and the stratigraphic occurrence of australites on the Lake Torrens plain together argue strongly in favour of a late Wisconsinan age of infall. It is concluded that the australites fell into a system of active seif dunes some time between about 24,000 and 16,000 years B.P. During dune stabilization and formation of the Motpena paleosol approximately 16,000 to 12,000 years B.P. some of the australites were corroded and encrusted with calcite. Subsequently, australites were eroded from the relict dune ridges and incorporated in alluvial and aeolian deposits of mid-Holocene age flooring and bordering interdune corridors. During the late Holocene, deflation of the corridors has been exposing the australites. Land use has probably accelerated this process in modern times.

The time limits given here for the infall of australites near Lake Torrens are in general agreement with the suggested late Quaternary age of infall at Port Campbell, Victoria (Gill, 1965, 1970), which lies 1000 km southwest of Lake Torrens on the same narrow belt of australite strewnfields (McColl & Williams, 1970). Sediments of late Quaternary age intersected by this line of australite occurrences are thus regarded as potentially australite-bearing and should provide fruitful areas for further research.

ACKNOWLEDGMENTS

Financial assistance was received from the National Geographic Society and the Smithsonian Research Foundation (B.M.), and a Queen Elizabeth II Research Fellowship (G.E.W.). G.E.W. and D.H.McC. thank Mr and Mrs Dick Craigie, formerly of Motpena Station, for assistance and hospitality.

REFERENCES

BAKER, G., 1962: The present state of knowledge of the 'age-on-earth' and the 'age-of-formation' of australites. *Ga Miner. Newsl., 15*, pp. 62-83.

BALDAR, N. A., & WHITTIG, L. D., 1968: Occurrence and synthesis of soil zeolites. *Proc. Soil Sci. Soc. Am., 32*, pp. 235-238.

CHALMERS, R. O., 1964: Searching for meteorites. *Aust. Mus. Mag., 14*, pp. 319-321, 356-359.

CORBETT, D. W. P., 1967: Australites from Myrtle Springs Station, South Australia. *Rec. S. Aust. Mus., 15*, pp. 561-574.

FLEISCHER, R. L., & PRICE, P. B., 1964: Fission track evidence for the simultaneous origin of tektites and other natural glasses. *Geochim. cosmochim. Acta*, 28, pp. 755-760.

———, ———, & WOODS, R. T., 1969: A second tektite fall in Australia. *Earth plan. Sci. Lett., 7*, pp. 51-52.

GENTNER, W., GLASS, B. P., STORZER, D., & WAGNER, G. A., 1970: Fission track ages and ages of deposition of deep-sea microtektites. *Science, 168*, pp. 359-361.

GILL, E. D., 1965: Quaternary geology, radiocarbon datings, and the age of australites. *Spec. Pap. geol. Soc. Am., 84*, pp. 415-432.

———, 1970: Age of australite fall. *J. geophys. Res., 75*, pp. 996-1002.

GLASS, B., 1967: Microtektites in deep-sea sediments. *Nature, Lond., 214*, pp. 372-374.

———, & HEEZEN, B. C., 1967: Tektites and geomagnetic reversals. *Nature, Lond., 214*, p. 372.

HAY, R. L., 1966: Zeolites and zeolitic reactions in sedimentary rocks. *Spec. Pap. geol. Soc. Am., 85.*

JOHNS, R. K., 1968: Investigation of Lakes Torrens and Gairdner. *Rep. Invest. geol. Surv. S. Aust., 31.*

MARINER, R. H., & SURDAM, R. C., 1970: Alkalinity and formation of zeolites in saline alkaline lakes. *Science, 170*, pp. 977-980.

MCCOLL, D. H., & WILLIAMS, G. E., 1970: Australite distribution pattern in southern central Australia. *Nature, Lond., 226*, pp. 154-155.

MCDOUGALL, I., & LOVERING, J. F., 1969: Apparent K-Ar dates on cores and excess Ar in flanges of australites. *Geochim. cosmochim. Acta, 33*, pp. 1057-1070.

POLACH, H. A., LOVERING, J. F., & BOWLER, J. M., 1970: ANU radiocarbon date list IV. *Radiocarbon, 12*, pp. 1-18.

WILLIAMS, G. E., 1969: Glacial age of piedmont alluvial deposits in the Adelaide area, South Australia. *Aust. J. Sci., 32*, p. 257.

———, (*in press*): Late Quaternary piedmont sedimentation, soil formation and paleoclimates in arid South Australia. *Z. Geomorph.*

———, & POLACH, H. A. (1971): Radiocarbon dating of arid-zone calcareous paleosols. *Bull. geol. Soc. Am., 82*, pp. 30-69.

ZÄHRINGER, J., 1963: K-Ar measurements of tektites; *in* Radioactive dating, I.A.E.A. Symposium Proc., Athens, 1962, pp. 289-305.

J. F. Lovering,
Department of Geology,
University of Melbourne,
Parkville,
Victoria 3052.

B. Mason,
Division of Meteorites,
Smithsonian Institution,
Washington, D.C.

G. E. Williams,
Geosurveys of Australia Pty Ltd,
57 Todville Street,
Woodville West,
South Australia 5011.

D. H. McColl,
McPhar Geophysics Pty Ltd,
50 Mary Street,
Unley,
South Australia 5061.

Journal of the Geological Society of Australia, Vol. 18, Pt. 4, pp. 409-418, Pls. 27-28, June, 1972.

EXPLANATION OF PLATES

PLATE 27

Delicate flange features preserved on a range of australite types collected in interdune corridors in the Pine Dam area, Myrtle Springs. Numbers refer to specimens in the Tektite Collection, School of Geology, University of Melbourne.

PLATE 28

A. Interdune corridor at Motpena floored by the Motpena paleosol. The stake marks the position of the australite found embedded in the Motpena paleosol (see Plate 28B, C). View looking southeast, with Wilpena Pound on skyline. (*G.E.W. photo*).

B. Australite embedded in the truncated Motpena paleosol at Motpena. Soil structure is visible on the near-horizontal surface in the foreground. Scale in millimetres. (*G.E.W. photo*).

C. Same australite as shown in Plate 28B, with the surrounding soil removed. The australite is part of a large flanged dumbbell, in two fragments, and is encrusted with calcite. Scale in millimetres. (*G.E.W. photo*).

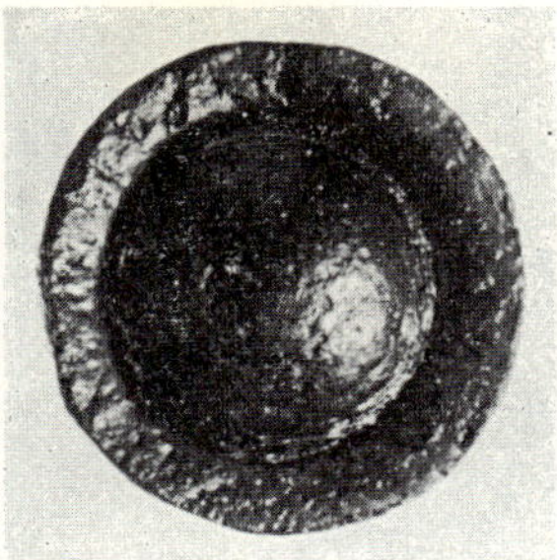

T69

T103

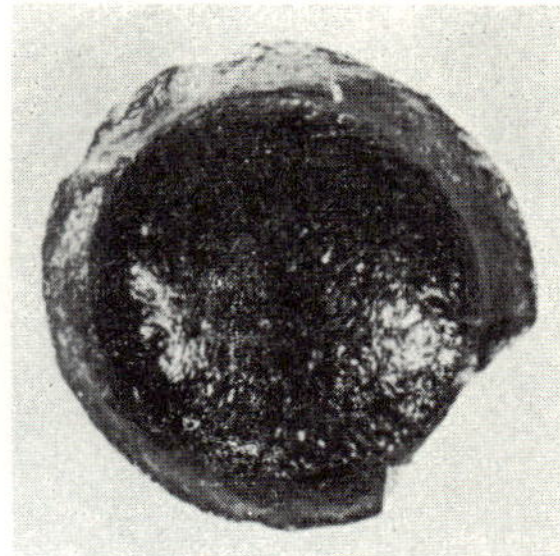

T70

T98

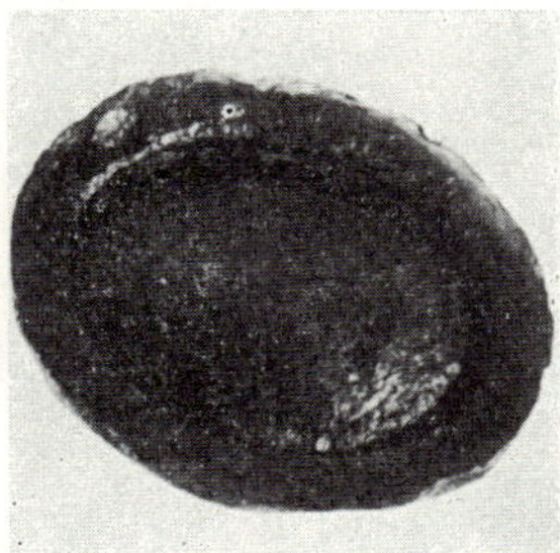

T97

T115

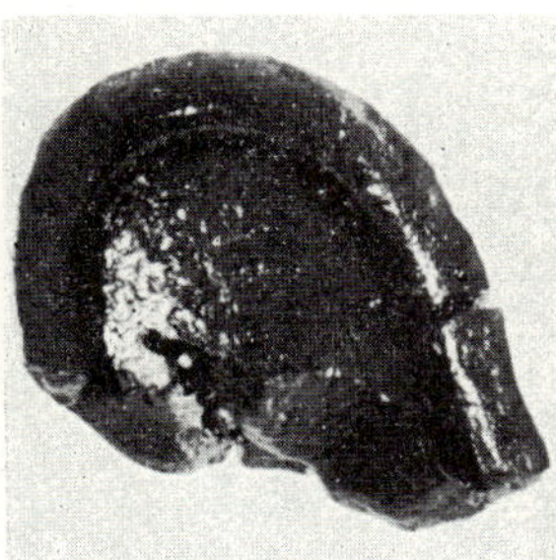

T72

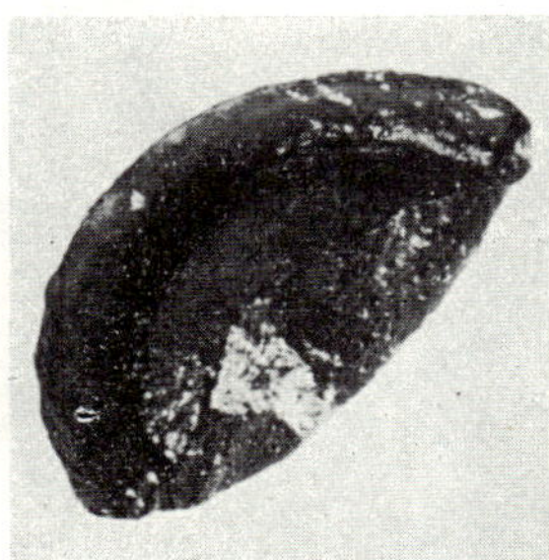

T101

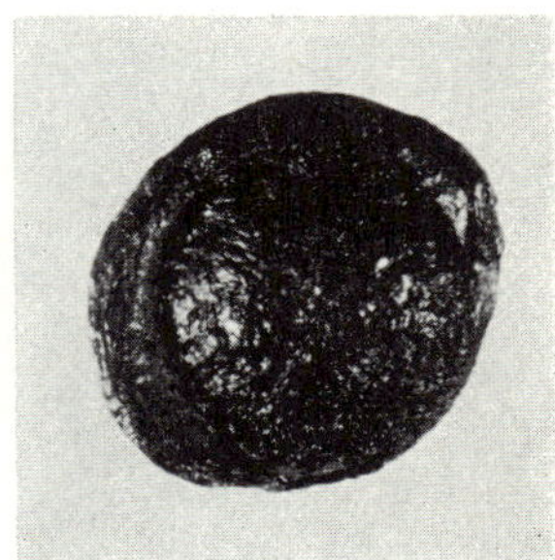

T84

5mm

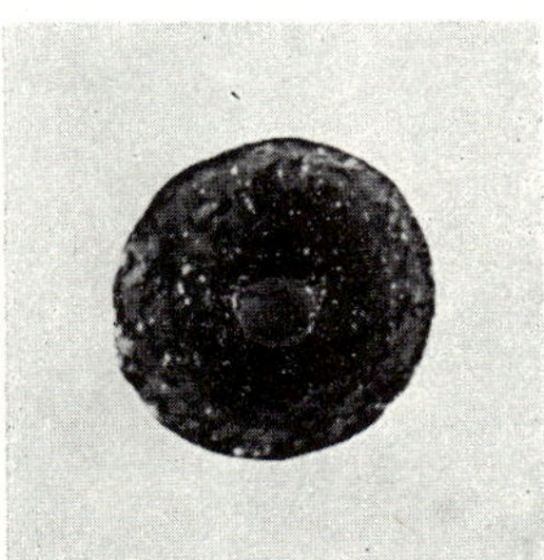

T108

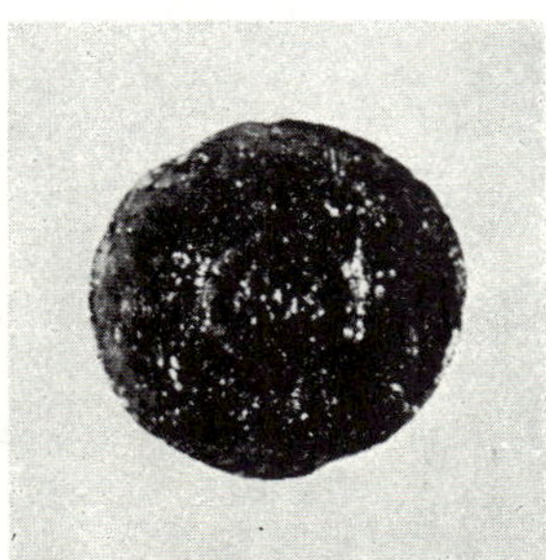

T91

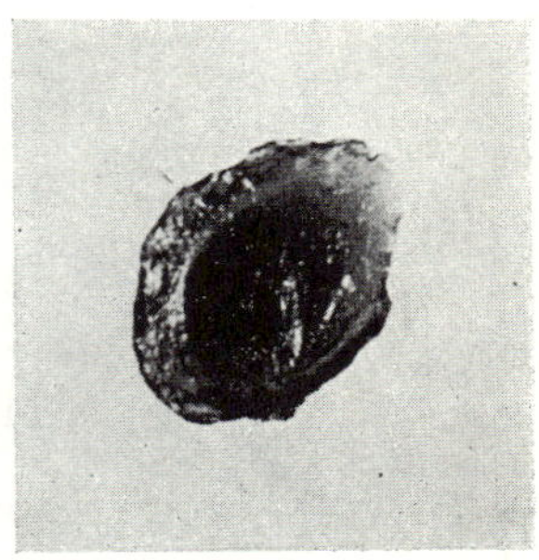

T90

5mm

Journal of the Geological Society of Australia, Vol. 18, Pt. 4, Pls. 27-28, June, 1972.

A

B

C

Sculpture of Tektites

IV

Editors' Comments on Paper 29

Dr. George Baker has been a prodigious worker on tektites for forty years, and his classical Memoir No. 23 of the National Museum of Victoria covered all facets of the tektite problem known when it was published in 1959. Dr. Baker has been especially intrigued by the shapes of australites and how such shapes could have formed, as can be seen by examination of the list of his papers in the bibliography of the Memoir and the references following this commentary (Baker, 1961a, b, c; 1962; 1963a, b; 1964; 1965; 1966a, b, c; 1967; 1968; 1969).

The Memoir, and the paper selected for reproduction in this section, have been guiding influences in ablation experiments carried out at Ames Research Center, Moffett Field, California (Chapman, 1960, 1964), and in shaping thought on the origin of tektites in general. The peculiar button shape common among australites had been subject to various interpretations until Baker's careful analysis correctly portrayed the changes which took place during the passage of an australite through the atmosphere.

Dr. Baker formerly was associated with the Commonwealth Scientific and Industrial Research Organization of Australia. He is now retired, and his present address is 146 Wimbledon Avenue, Mt. Eliza, Victoria 3930, Australia.

References

Baker, G. (1961a). A complete oval australite. *Proc. Roy. Soc. Victoria,* **74** (1), 47–54.

Baker, G. (1961b). A perfectly developed hollow australite. *Amer. J. Sci.,* **259**, 791–800.

Baker, G. (1961c). Australite von Wingellina, West-Australien. *Chem. Erde,* **21,** 118–130.

Baker, G. (1962). The largest known australite and three smaller specimens from Warralakin, Western Australia. *J. Roy. Soc. Western Australia,* **45** (1), 12–17.

Baker, G. (1963a). Disc-, plate-, and bowl-shaped australites. *Meteoritics,* **2** (1), 36–49.

Baker, G. (1963b). Form and sculpture of tektites. *In* "Tektites" (J. A. O'Keefe, ed.), pp. 1–24. Univ. Chicago Press, Chicago, Illinois.
Baker, G. (1964). A thin, flanged, boat-shaped australite from Port Campbell, Victoria, Australia. *Meteoritics,* **2** (2), 141–147.
Baker, G. (1965). Dumbbell shaped australite from Port Campbell, Victoria. *Meteoritics,* **2** (4), 325–335.
Baker, G. (1966a). External form and structure of some hollow australites. *Geochim. Cosmochim. Acta,* **30,** 607–615.
Baker, G. (1966b). Hollow australite button with flange, Hordern Vale, Otway Peninsula, Western Victoria. *Meteoritics,* **3** (1), 35–53.
Baker, G. (1966c). The largest known dumbbell-shaped australite. *J. Roy. Soc. Western Australia,* **49,** 59–63.
Baker, G. (1967). A second large dumbbell-shaped australite, Ongerup, Western Australia, with notes on two other large australites. *J. Roy. Soc. Western Australia,* **50,** 113–120.
Baker, G. (1968). Australites from NNE of Morgan, South Australia. *Mem. Nat. Mus. Victoria,* No. 28, pp. 39–46.
Baker, G. (1969). Five large australites from Victoria, Australia, and their relationships to other large forms. *Mem. Nat. Mus. Victoria,* No. 29, pp. 53–64.
Chapman, D. R. (1960). Recent re-entry research and the cosmic origin of tektites. *Nature,* **188,** 353–355.
Chapman, D. R. (1964). On the unity and origin of the Australasian tektites. *Geochim. Cosmochim. Acta,* **28,** 841–880.

29

[American Journal of Science, Vol. 256, June 1958, P. 369-383]

American Journal of Science

JUNE 1958

THE ROLE OF AERODYNAMICAL PHENOMENA IN SHAPING AND SCULPTURING AUSTRALIAN TEKTITES

GEORGE BAKER

ABSTRACT. Primary australite shapes are considered to have been generated in an extraterrestrial environment, but secondary shapes and front surface sculpture arose as a consequence of the development and action of certain aerodynamical phenomena during ultra-supersonic, non-rotary flight earthwards through the atmosphere. Shock waves, boundary layer laminar flow and turbulence in the atmospheric medium traversed by australites at such high speeds, resulted in high temperatures, skin friction and form drag to which the front surfaces and equatorial regions of australites were subjected for a very short period of time. The primary shapes, of which spheres and spheroids constituted the greatest proportion, were cold on initial entry into the earth's atmosphere, and each member of each shape group already possessed pre-formed complex internal flow patterns and bubble-pitted, flow-swirled surface features. Conflict with the atmoshere during ultra-supersonic speeds of earthward passage resulted in considerable diminution in bulk of each australite. This arose from removal of glass in front surface and equatorial regions under the influence of ablation and thin-film fusion-stripping, accompanied by the building up of circumferential flanges with plane spiral coiling on the peripheral edges of the otherwise unaffected back surfaces, and the production of small troughs and ridges as newly formed sculptural elements upon the front surfaces. The production of the relatively rare flat, thin varieties of australites, hitherto regarded as difficult of explanation, is readily envisaged under the terms of aerodynamical control at high speeds of flight.

INTRODUCTION

Many diverse and unsuccessful attempts have been made to explain the mode of origin of the shapes and sculpture patterns of australites, since the days when the Rev. W. B. Clarke (1855) wrote that australites appeared "as if cast in a mould", and Walcott (1898) recorded that some people believed australites were originally plastic materials that had been pressed by a saucer-shaped mould in the ground.

Suggestions advanced as explanatory of the shapes and sculptural elements, are inevitably bound up with some particularly favored, but unproven source, advocated for the glassy material composing australites by adherents to the different hypotheses. No one theory has yet succeeded in formulating an adequate explanation for the origin of australite glass and the processes responsible for fashioning it into the shape types found throughout Southern Australia and Tasmania.

Theories put forward during the latter half of the nineteenth and earlier part of the twentieth centuries, have called upon artificial, natural terrestrial and extraterrestrial sources, and still the problem remains incompletely solved. Most earlier theories have been discarded as untenable. Among those receiving favorable consideration, theories relating to an extraterrestrial origin have most to offer in attempting to unravel the mystery of australite formation. An extraterrestrial source for australites is the most adequate for providing explanations of their original shapes, sculpture patterns, internal structures and other features, such as distribution, etc.

A theory that ultimate secondary shapes were produced by the action of aerodynamical phenomena during high speeds of flight is herewith advanced in the belief that it is the most logical and most acceptable, and that it lends strong support to the hypothesis of an extraterrestrial origin for Australian tektite glass. This discussion of the effects of aerodynamical phenomena on australites must of necessity be qualitative, since the aerodynamicists have not yet formulated the answers to all the problems involved.

If australites represent forms of extraterrestrial silicate glass that were already solid on reaching the earth's atmosphere at cosmic velocities the question arises whether they traversed the atmosphere with or without a spinning trajectory.

Up to now, most theories relating to the mode of generation of the shapes of australites (e.g. Walcott, 1898; Fenner, 1934) have favored rotation normal to the direction of propagation through the atmosphere.

The new theory presented herein, does not require rotary motion during atmospheric flight. The evidence from studies of the shapes, structures, sculpture, internal and external flow patterns ("schlieren", etc.) of some 2,000 well-preserved australites (Baker, 1944; 1955a; 1955b; 1956), completely militates against theories invoking rotation through the earth's atmosphere. The greater proportion (approximately 80 percent) of australites were originally primary spheres or oblate spheroids close to spheres, (Baker, 1955; 1956) formed by rapid fusion followed by rapid cooling. Spheres require no rotation for their formation, and there is no reason to suppose that australite spheres rotated on entering or traversing the atmosphere. The remaining 20 percent of australite shape types were obviously originally forms of revolution, but the important point is that they were produced as such at their extraterrestrial birthplace. As pre-formed shapes, they also became modified without rotation playing any part during the atmospheric phase of earthward flight.

Opinions have sharply differed concerning whether tektites could become completely liquid during atmospheric transit. Grant (1909) calculated that if only 1 percent of the energy possessed by meteors entering the atmosphere at 40 miles/sec. was converted into heat by air friction, and this was retained by the body, it would be sufficient to raise the substance of an australite to its melting temperature (1,324°C), and render it completely liquid. On the other hand, it has been calculated that the amount of heat generated by the rapid passage of small bodies of glass through the atmosphere, taken in conjunction with conductivity of the glass, would be insufficient to melt them to the extent that tektites were melted prior to their arrival on earth (cf. Watson, 1935; Opik, 1937; Spencer, 1937).

THEORY OF AUSTRALITE SHAPE AND SCULPTURE FORMATION UNDER THE CONTROL EXERTED BY AERODYNAMICAL PHENOMENA

In view of the foregoing remarks, it becomes necessary to accept as a starting point:

a) the evidence indicating that australites entered the final phase of earthward flight—the atmospheric phase—as cold, non-rotating objects of glass having pre-formed (primary) shapes, and

b) the fact that if australites are in reality glass meteorites, then they traversed the earth's atmosphere at very high speeds.

The following observations and deductions are regarded as providing satisfactory arguments in favor of the above items.

In possession of small primary shapes (spheres, spheroids, apioids and dumbbells), and a few accidental or aberrant shapes, each with a pre-determined complex internal flow pattern ("schlieren") in natural glass with slight inhomogeneities from limited miscibility in the original silicate melt, resulting from very rapid fusion followed by equally rapid cooling, there arrived a time in the Recent epoch of the earth's geological history, when a cloud of australites approached and was drawn into the earth's sphere of attraction. The cluster was limited in size, as evidenced by the confinement of australites to Australia and Tasmania only.

By analogy with meteorites, the speed of transit through the earth's atmosphere was considerable at the onset. Some distance above the earth's surface, the australite cloud was slowed down and thereafter its individual components fell to earth along parabolic paths. Even with the decreased rate of infall, the australites would have traversed the remaining distance to earth in seconds or very few minutes at most. During this short time, but at such high speeds, primary shapes were subjected to considerable front surface modification through heating by air friction. The motion involved during transit was one of forward propagation only, with no rotation whatsoever, although a few forms occasionally provide evidence of having developed a slight wobble due to buffeting at high speeds as a result of form drag.

A short distance above the earth's surface, the australites rapidly lost speed, and descended through the denser, lower layers of the atmosphere at such reduced speeds that they landed without being smashed to pieces on contacting hard ground, and probably without becoming too deeply buried on striking soft earth.

At no time during the atmospheric phase of earthward flight had australites become completely fluid throughout, hence there was no flattening possible, as has been previously suggested, of any of the forms on impact with the ground.

Throughout the zone of the atmosphere where secondary modification took place, the australites were travelling at ultra-supersonic speeds. Under the terms of gas dynamics at such high speed flow, the effects of airflow over their front surfaces played the dominant role in determining their ultimate modified shapes, structures and sculpture patterns. Secondary thin film fusion arose from the heat generated by intensive friction with the atmosphere. It is apparent that during the formative stages of secondary shape development, the nature of the airflow over and past the forwardly directed surface of each australite, was undoubtedly maintained as a steady flow while ultra-supersonic speeds prevailed, but must have changed considerably with the decreased speeds of earthward approach when much nearer the earth's surface.

In experiments relating to fluidal flow past bodies, fluids are made to flow at varying speeds around the bodies, because it has so far proved more practicable to produce and note the effects of flow in moving fluid, the velocity of which can be more readily controlled. With australites, it is required to know the effects of the external streaming of the fluid medium (air) upon the fast-moving objects themselves. Since they are small bodies that travelled at such high speeds through the atmosphere, great stream velocities were generated, and boundary layers in the fluid medium (air) would of necessity be thin layers. Within these layers, all frictional effects between the front surfaces of the australites and the atmosphere took place, and considerable temperature rises resulted.

Density and temperature variations in the earth's atmospheric envelope, throughout the secondary shape formative zone in particular, would have had but very minor effects, if any, on the ultimate character of the newly forming front surfaces of the australites, because of the high speeds of transit. Any very minor density variations existing in the tektite glass itself, are far too insignificant to have effected the issue in any way. The coefficient of thermal conductivity of the glass being low (between 0.002 and 0.0005 cal/cm/°C), the rate of heat transference from the front surface inwards would be of negligible account—the temperature gradient is very steep—so that no heat was conducted through the body of australite glass to the rear surface. Moreover, since the australites traversed the atmosphere so rapidly, there would be insufficient time for frictional heat to raise the interior portions and the rear surface to a temperature (1,324°C) at which a completely molten state could arise. It becomes apparent from studies of their internal flow patterns (Baker, 1944; 1955; 1956) that during the atmospheric phase of flight, the front surfaces were subjected to superficial melting only, at any particular instant. Such melted films did not exceed 0.01 mm in thickness at any one time, under the influence of frictional heat (cf. Dodwell's calculations (*in* Fenner, 1938)) based on Opik's (1937) work on the mechanics of meteor phenomena).

The different character of the rear surfaces of australites, taken in conjunction with the low rate of heat transference, and the fact that regions of "dead air" with low temperatures and pressures developed immediately behind fast-moving objects (see fig. 1), lead to the conclusion that these rear surfaces remained at low temperatures throughout atmospheric flight, and were in no way ablated.

SHAPE ORIGIN OF BUTTON, LENS AND ROUND-CORE FORMS OF AUSTRALITES FROM PRIMARY SPHERES

On the most common of the primary australite forms (spheres), pressure would have been greatest at the front pole, least at the rear during fall to earth. In the initial stages, frictional forces would have been negligible or non-operative at the poles of the sphere, greatest at the periphery (i.e. equatorial regions). They would have been operative to varying degrees between the front polar and the equatorial regions, but absent at the rear pole during sustained steady flight. The effects of frictional forces were thus confined to front surfaces, where they varied somewhat with various arcs of curvature of individual spheres having different radii of curvature, the range of which was limited from 2 mm up to 50 mm.

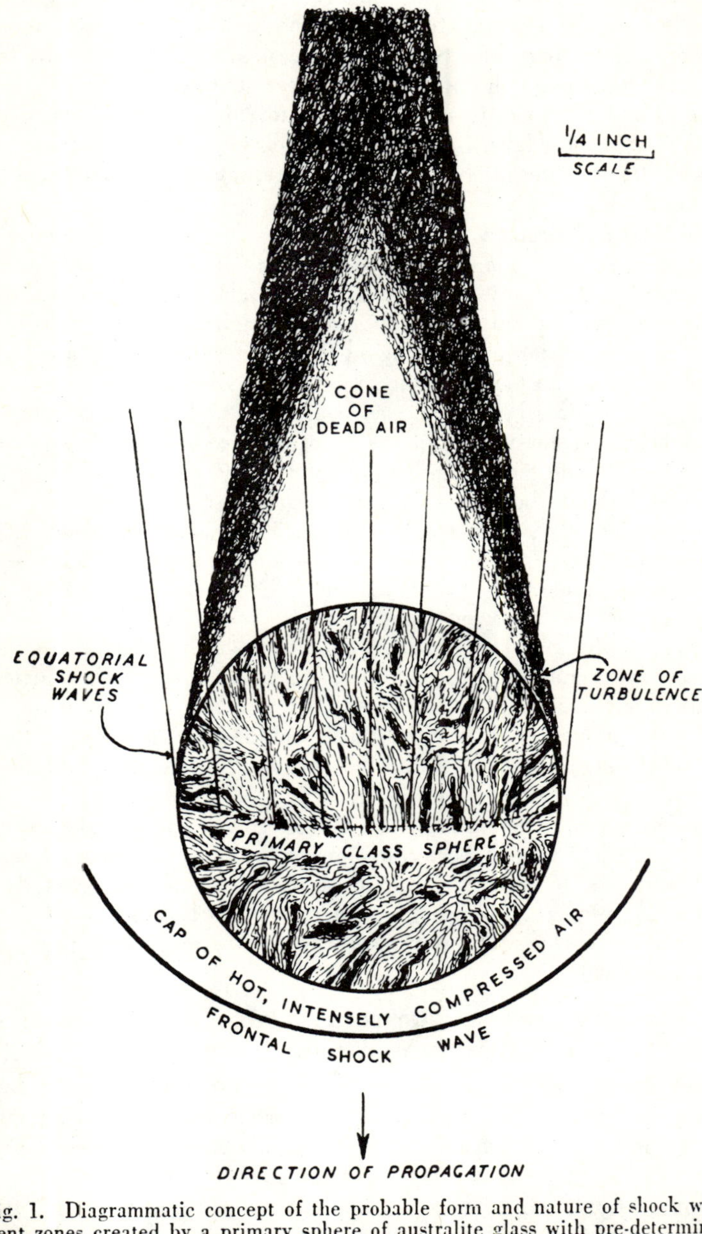

Fig. 1. Diagrammatic concept of the probable form and nature of shock waves and turbulent zones created by a primary sphere of australite glass with pre-determined complex internal flow-line pattern, travelling earthwards in its initial stages of atmospheric flight at ultra-supersonic velocity.

A combination of these pressure and frictional effects on the different sized spheres created an optimum condition of heat stability for each; the degree of viscosity attained by the heated front film was controlled by the temperature developed from frictional resistance of the atmosphere during flight, and hence by the velocity of fall of australites.

Travelling at ultra-supersonic speeds through the "formative zone" of the atmosphere, certain aerodynamical factors operated upon the front surfaces of the initial primary forms, to produce the modified shapes now possessed by australites.

Since the australites would have had high Mach Numbers[1] of approximately 20 to 25 at a height of some 70 to 80 miles above the earth's surface, a state of disturbance of a permanent type was set up in shock waves produced in the air piled up ahead of each form. The shock waves (cf. Durand, 1935) are regarded as sheets where there is a discontinuity of velocity, i.e. they are places where rate of change in velocity and density of the fluid medium (air) become infinite. These shock waves travelled ahead of the fast-moving australites producing them, at the same speed and in the same direction.

The frontal shock wave had an arc of curvature little different from that of the front surface of the australite, and its radius of curvature was but slightly greater. At all other points, such as at the equator of the sphere of australite glass and from any irregularities existing between the front pole and the equatorial regions, subsidiary shock waves moved obliquely to the direction of air flow (see fig. 1).

The shock wave was actually a narrow, almost hemispherical zone of intense compression where greatly increased temperatures arose. Thus, any air that flowed over the front surface of the primary australite sphere, did so only after penetrating the narrow arcuate region of the frontal shock wave. In the broader zone behind the frontal shock wave and ahead of the front surface of the sphere of australite glass, the air expanded to some extent, but high pressures and consequently high temperatures persisted. Due to the viscosity and conductivity of the air in the expansion zone, the mechanical energy produced at ultra-supersonic speeds, was converted into heat. While sufficiently high speeds prevailed, a cap of highly heated compressed air thus travelled ahead of each australite.

During the initial stages of the processes promoting the development of the ultimate secondary shapes of australites, this cap of hot, compressed air supplied the temperature rise necessary to induce superficial sheet fusion of thin films of the tektite glass on the front surfaces, more especially at this stage in the immediate polar regions. Here, pressure effects were at their greatest and effectively removed the fused film as soon as it formed—partly by fusion stripping, partly as a result of accompanying ablation which involved liquefaction followed by some vaporisation and immediate removal from the heated surface sheet. With the generation of another fused film at the next exposed level, and the progress of further fusion stripping and abla-

[1] The Mach Number (M) is the ratio of the speed of supersonic flow to the speed of sound, so that if $M = 1.0$, the speed of supersonic flow equals the speed of sound (760 m.p.h at the standard sea level temperature of 15°C).

tion, the arc of curvature of the front surface of the sphere was modified, becoming less steep, so that its radius of curvature increased in value; the front polar regions migrated rearwards and a rim was formed separating two now distinct surfaces—the back and the front surfaces. At this stage, the rim was merely a relatively sharp line of demarkation between the two surfaces with their now different arcs of curvature.

The part played by the equatorial skirt of subsidiary shock waves (see fig. 1), would not be expected to be of significant importance at this stage. In these subsidiary shock waves, compression, although high, would not be as intense as in the frontal shock wave, and they would be arranged very obliquely backwards and somewhat detached from the now slightly modified sphere, because of thickening of the boundary layer of air in these regions as induced by the primary australites having such a high Mach Number. Such subsidiary shock waves, however, could have had marked effects in subsequent fusion plucking around the equatorial regions of the larger forms of australites—they could have been responsible for the generation of such secondary shapes as are possessed by the core-like australites, which are forms that seldom developed, or if developed, rarely retained flanges throughout the later stages of earthward flight, but which usually, though not universally, possess characteristic flaked equatorial zones not to be entirely ascribed to subsequent tertiary processes of weathering.

While ablation of the front surface proceeded, turbulence was created behind the fast-moving sphere, arising where the main airflow separated from the equatorial regions. This arose from the action of reverse, secondary reverse, and tertiary laminar flow producing vorticity in the thin boundary layers of air in contact with the front surface. Within the thin boundary layers, arose all frictional effects between the outermost skin of the front surface of the australite and the innermost layer of the air through which it had its trajectory. Stresses produced in this layer of air gave rise to skin friction as a tangential component. Stresses produced in the positions where turbulent flow was produced in the equatorial regions, gave rise to form drag (cf. Whitlock, 1943, chapter V). Within the zone of turbulence and situated behind the rear surface of each fast-moving australite sphere undergoing modification, a cone of virtually "dead air" (see fig. 1), ensured maintenance of the rear surface at temperatures below the fusion temperature of australite glass. Thus, such of the rear surface that remained after the reduction of the original sphere, was unaltered throughout flight, and is left as testimony to the arc of curvature of the original sphere; from it, the radius of the primary sphere can be determined by reconstruction (Baker, 1955; 1956). Moreover, the remnant portion of the rear surface provides evidence of the originally bubble-pitted and flow-swirled character of the sculpture pattern of the primary spheres of glass of extraterrestrial origin.

Continued removal of the front surface films by ablation, reduced the front surfaces of the australite spheres to different extents, as evidenced by studies of curvature—size relationships of the ultimate secondary shapes (Baker, 1955a; 1956). The larger of the spheres (25 to 50 mm across) had become modified to form the normally flangeless core-like types in which equatorial plucking was dominant. In them, the subsidiary equatorial shock waves provided the necessary temperature rise, while the effects of the separation of the main airflow became important in shaping the flaked equatorial zone, for in the equatorial regions occurred the transition zone in which laminar flow in the thin boundary layers ended, and turbulence supervened.

The medium size spheres were progressively reduced in bulk all over their front surfaces, and at a stage when approximately 60 to 65 percent of the original sphere had been ablated, the front pole had passed beyond the position of the original centre of the primary sphere, the arc of curvature of the front surface had become considerably flatter with an attendant increase in radius of curvature, and the equatorial edge of the form had progressively receded (in the form of a rim) beyond the position of the original equatorial zone of the primary sphere. As a consequence, the frontal shock wave became less oblique to the airstream than the original shock wave ahead of the primary spherical form. Pressures are greater in less oblique shock waves, so that increased drag would be expected along the secondarily developing front surface.

Throughout these alterations in the shape and character of the front surface, the radius of curvature of the rear surface remained unchanged, so that the arc of curvature of the rear surface was preserved in character although not in amount, for a stage was now reached where the rim of the form had migrated rearwards to such an extent that only some 30 to 50 percent of the original hemispherical rear surface remained. As the modified form approached the closing phases of its period of ultra-supersonic flight, a situation was reached when glass from the front surface was not entirely whipped away under the influence of skin friction or by turbulent friction in equatorial regions. A condition had been created favorable to flange building, for now equatorial regions (peripheral edge) of the rear surface were less steeply sloping than when the rim was situated in positions agreeing with the equatorial regions of the primary sphere. Moreover, they now slope away from the airstream compared with a slope into the airstream when the rim was migrating from the original front polar regions of the primary sphere towards its original equatorial regions. Under the new conditions, therefore, a more stable position is available for the retention and accumulation of such melted glass as reached the edge of the new shape, although some was still no doubt lost in the turbulent airstream.

Some of the glass migrating rapidly under pressure from the polar regions of the newly created shape, was, in the fused state, stretched backwards in the airstream on reaching the equatorial regions. It was immediately caught up in eddy currents associated with the turbulent zone (see fig. 3), and a proportion (7 percent of the bulk of the original sphere) became inrolled towards the equatorial edge of the rear surface, without spreading out (in the majority of forms) for very far over that surface. There it chilled and was moulded into a circumferential flange composed of secondarily fused glass with plane spiral coiling of the internal flow structures (fig. 2). Contortion of some of the flow structures resulted from the jamming effects produced by warmer glass being forced against cooler glass in the developing flange.

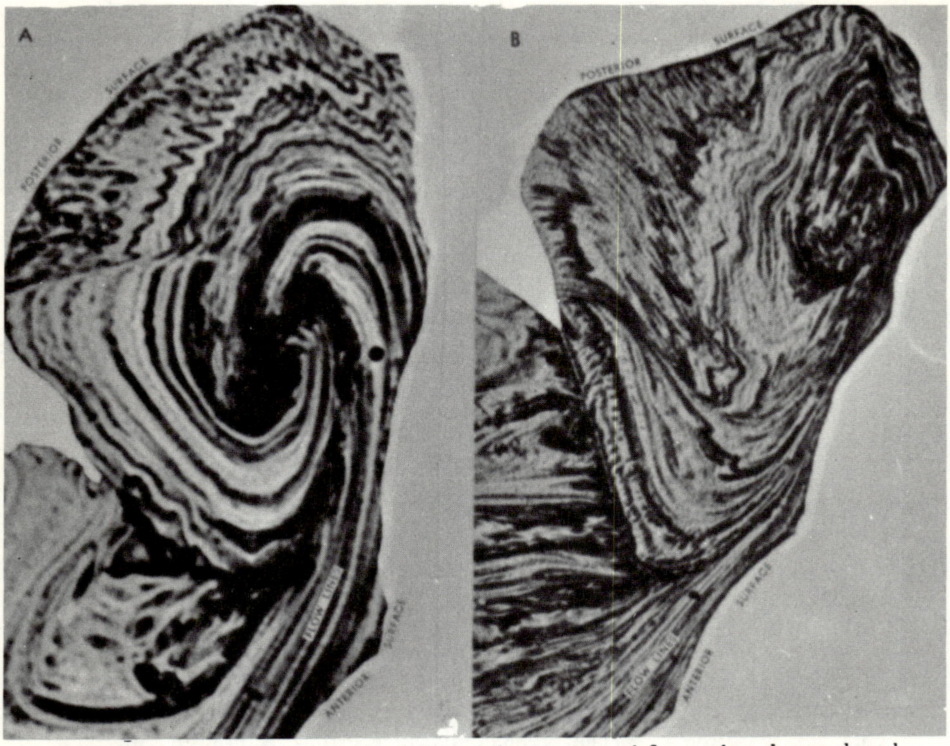

Fig. 2. Enlarged photographs (x20) of thin sections of flanges from button-shaped australites. Showing spirally coiled flow lines contorted in "chin" region and flow lines near anterior surface truncated in the trough structures between ridges. Both specimens from Port Campbell, Victoria.

As an outcome of the turbulence in the separating main airstream, there thus arose a newly created feature in the secondary australite shape, a feature delimiting the newly formed front surface from the remnant portions of the primary rear surface. In the uncoiled state, it is found by measurement along a prominent flow line that the glass of the flange would have been stretched back from the equatorial edge of the secondary shape for a distance equal to three times the present distance across the plane spiral structure, before the process of inrolling occurred. Inrolling was evenly developed all around the circumference of the secondary shape, as evidenced from the examination of a series of radial thin sections cut along the direction of the polar axis of flanged, button-shaped australites.

Approaching the final phases of high speed flight through the earth's atmosphere, the velocity of the newly shaped australite considerably declined. Much lower supersonic speeds prevailed, and the passage through transonic into subsonic phases was close at hand. Small amounts of drag were experienced by the front surface at this stage, with resultant whipping away of thin films of somewhat softened glass from certain areas, thus producing the "troughs" separated by the "ridges" which are so typical of the front surfaces of well-preserved button-shaped australites. This final, smaller-scale sculpturing effect is best evident from thin section studies (Baker, 1944), which reveal that the trough structures cut across the last-developed secondary flow line structures revealed immediately below the front surface (see fig. 2). These effects were rather more pronounced nearer the equatorial edges of the secondary form, more particularly on the front surfaces of the flange structures, for here the frictional forces created by the separation of boundary layer flow into the main airstream, were still operative to a lesser degree in the final phases of supersonic flight. Such forces were responsible for the complex "wrinkling" effects shown by the ridges on anterior surfaces of flanges near their equatorial edges.

At this stage of secondary shape development, aerodynamical flow phenomena had become far less effective in removing fused glass from front surfaces; temperatures were evidently not nearly as high as at earlier stages, and the work of the aerodynamical flow phenomena was largely confined to putting the finishing touches to the sculpture features. The equatorial skirt of subsidiary shock waves possessed by the original sphere at the onset of conflict with the atmosphere, had now been replaced by less oblique subsidiary shock waves attached to the small projections provided by the ridge structures on the front surface (fig. 3).

Behind each of the subsidiary shock waves attached to the ridges, existed narrow zones of expansion where it was possible for the trough-like structures to be partially scoured out during the very limited time of existence of such subsidiary shock waves. It is not yet clear what the nature of the aerodynamical flow phenomena would have been when considered in terms of the fact that ridges and troughs are concentrically arranged on some front surfaces, while on others, they are spirally arranged in sometimes clockwise, sometimes anti-clockwise fashion (Baker, 1956).

On passing from the supersonic through the transonic into subsonic phases of earthward flight, some of the flanged button-shaped australites evidently lost their flanges, more particularly those with less stable planes of union between flange and central body portion which had been subjected to rather more extensive scouring out of glass from trough regions neighboring the planes of union (see fig. 2). Such a process yielded the relatively common non-flanged, lens-shaped australites which are fundamentally identical with the central body portions of the flanged button-shaped australites.

SHAPE ORIGIN OF THE ELONGATED FORMS OF AUSTRALITES

The operation and effects of aerodynamical flow phenomena at ultra-supersonic speeds, followed by short-duration lesser supersonic and transonic speeds as outlined for the generation of secondary shapes from primary spheres, apply equally as well to the other known primary shapes of australite glass such as the ellipsoidal, dumbbell and apioidal primary forms. The general conditions, however, were possibly rather more complicated for dumbbell-shaped forms which presented bi-polar front surfaces to the shock wave

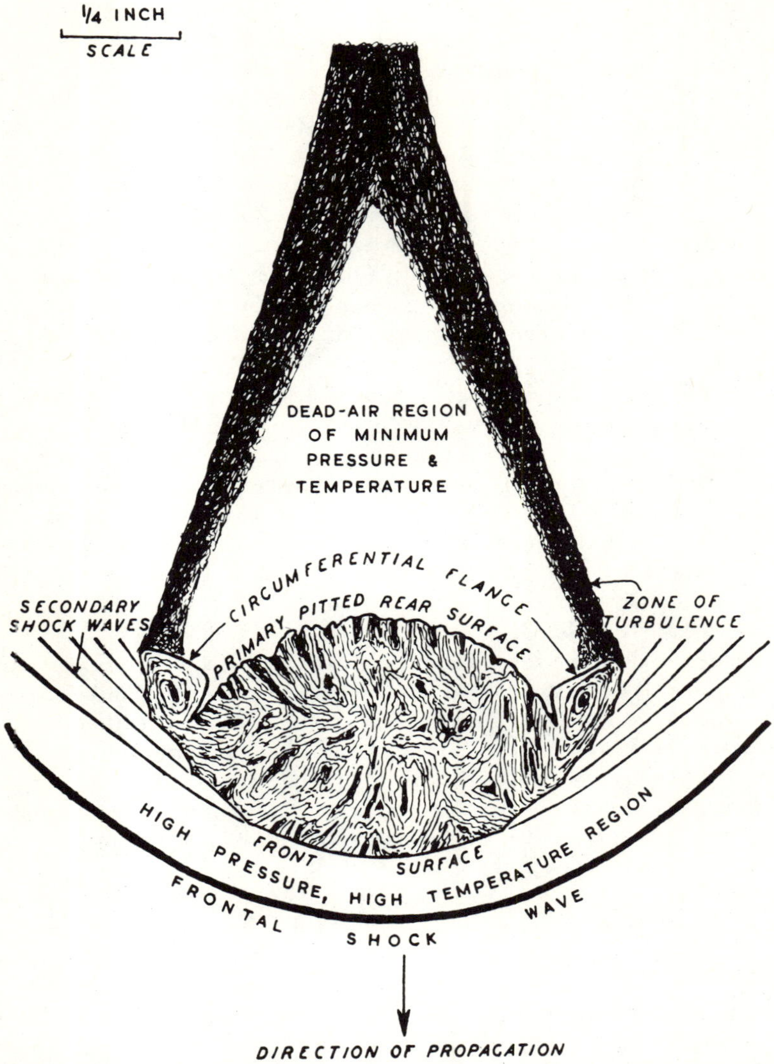

Fig. 3. Radial section of button-shaped australite with circumferential flange, showing complex internal flow pattern and relationship of the newly developed secondary shape to the surmised aerodynamical phenomena created during the end phases of supersonic flight through the earth's atmosphere.

and boundary layer flow phenomena. Nevertheless, identical features such as (a) flanges with plane spiral inrolling structures, (b) ridges and (c) troughs intervening between the ridges, were developed on their front surfaces during the phase of atmospheric flight earthwards, in much the same way as on the button-shaped australites which were modified from primary spheres. Moreover, the development of these features in comparable positions on the modified elongated primary forms of australites, testifies to their stable positions of flight through the atmosphere—nearly all travelled in such a position that their longer axes were maintained normal to the direction of earthward propagation. A very small number of types allied to the apioids, provide evidence of probably having fallen with their long axes parallel with the direction of earthward flight, such forms being of the "aerial-bomb" type.

For all the australite shape types, therefore, shock waves were produced at high speeds of flight, and only in shock waves can a state of steady flow of a permanent type be set up in the medium through which they have their passage, when the motion of the objects is limited to one direction. Under such circumstances, it becomes apparent that the australites in the earthward-falling cluster, behaved more or less in unison, each maintaining its own particular path of flight through the atmosphere, without tumbling about or shooting in all directions. Rotation of the forms was entirely unnecessary about an axis parallel with the direction of flight, and there is evidence to indicate that rotation did not occur; the ultimate secondary shapes and their structures all find satisfactory explanation in terms of the effects of aerodynamical flow phenomena arising at high speeds of flight.

Among experiments designed to emulate the secondary shapes possessed by australites, perhaps the most significant is one carried out by Dr. E. S. Hills some fifteen years or so ago. Small spheres of paraffin wax, when held under a controlled stream of hot water, developed shapes and structures simulating the shapes and structures of australites as they are found on the earth's surface. No rotation was imparted to the small spheres. The conclusion is that any rotation that did occur during the history of australite formation, was entirely pre-atmospheric phase, when the smaller percentage comprised of ellipsoidal forms, dumbbells and apioids, were produced as the common figures of revolution at their source in some unknown extraterrestrial environment. The rare aberrant forms evidently arose by accident prior to entry into the earth's atmosphere, for where not badly weathered subsequently, some of them show evidence of having been subjected to similar treatment by high speed flow during rapid atmospheric flight, as for the more normal symmetrically shaped australites.

SHAPE ORIGIN OF ROUND DISC-LIKE AND ELONGATED PLATE-LIKE AUSTRALITES

The formation of the rare and special disc-like and elongated plate-like types of australites, which are small, thin forms (1.0 to 1.5 mm thick), and are usually flat or sometimes bowl-like, is difficult of interpretation under the terms of any of the theories of shape development so far propounded. They can be more readily envisaged as resulting from special circumstances of high speed flow, commencing with the very smallest primary forms present in the cloud of australites that entered and survived the effects of rapid traverse through the earth's atmosphere. Thus the round discs derive from the smaller spheres, while the elongated plate-like forms were produced from the smallest spheroids of which we have evidence.

Reduction of the smaller spheres by ablation and fusion stripping while moving at ultra-supersonic speeds through the atmosphere, under the control of the aerodynamical flow phenomena outlined above for other australite shapes, yielded small, flanged button-shaped secondary forms in the first place. These had become reduced to a thickness of from 2 to 4 mm from spheres originally 4 to 10 mm across as determined from the radii of curvature of their remnant rear (primary) surfaces. Moreover, they possessed flanges which were more "flattened" compared with flanges on larger australites, and which constituted a greater bulk of each complete specimen than did the remnant body portions.

During ultra-supersonic flight, it is but a stage further for sustained aerodynamical phenomena to reduce the smallest formed button-shaped objects of australite glass to flat, round (in plan aspect) disc-shaped forms, which consist approximately of 90 percent broad, flat flange with internal flow lines coiled inwards in plane spiral fashion after the manner of flanges on larger australites. The remaining 10 percent of each of these flat forms consists of a minute body portion, and even this is wanting in some disc- and plate-like australites, which are thus fundamentally "flattened" flanges only.

Some of the smallest of the original spheres existing on initial entry into the earth's atmosphere, were probably completely consumed during ultra-supersonic flight, having been dissipated by the processes of ablation and fusion stripping, for ablation depends essentially upon the size of the surface, and with diminishing volume, the relative size of the surface increases—in a somewhat similar, but not identical way, raindrops that reach the earth's surface, have a certain minimum size. For these reasons, australites of microscopic dimensions are unknown.

During formation of the small disc- and plate-like forms of australites, it is evident that as the small primary spheres became smaller and "flattened" by ablation, the nature of the airflow at ultra-supersonic speeds changed considerably, and rather more special aerodynamical conditions than hitherto, were created at a certain stage. The major difference concerns the nature of the frontal shock wave and the consequent pressure forces associated therewith. Whereas the frontal shock wave ahead of the larger forms of australites and ahead of the smaller primary spheres and spheroids, lay obliquely down the airstream, because of their arcuate, convex front surfaces, the shock wave ahead of an ablation-reduced, "flatter" disc- or plate-like secondary modification, would approach normality to the airstream. The shock wave would thus be parallel to and ahead of the gradually "flattening" front surface, and thus more or less normal to the direction of propagation of the object. The shock wave pattern therefore changed from a fundamentally oblique type to a normal type during the all-important formative stages of these special, small secondary shapes. In the oblique shock wave, pressures, which depend upon deflection, were seldom more than 50 percent of the pressures generated when perpendicular (i.e. normal to the airstream) shock waves formed ahead of the "flattened" secondary forms, which presented a perpendicular reflecting surface to airflow.

"Flattening" of thinned, softened small forms occurred near the final phases of atmospheric flight, when the effects of supersonic speeds were diminishing and the transonic region was nearly reached. Frontal shock waves were then perpendicular, and pressures consequently up to 50 percent greater than hitherto, causing increased drag on the front surface. The thinnest forms (1.0 mm) were sufficiently softened to result in bending backward into bowl-shaped forms.

Since the bowl-like forms have small (5 to 6 mm) front surface radii of curvature, oblique frontal shock waves were re-established and pressures thus much reduced for the final stages of supersonic flight. After passing through the transonic region, these thin, disc-, plate- and bowl-like forms of secondary shapes, were no longer heated by frictional resistance at any point, and as cold, no further modified fragile bodies, they traversed the subsonic region to reach the ground undamaged. They are so light and thin that they were evidently wafted down through the lower, denser layers of earth atmosphere, and gently deposited upon the ground. Only in this way can the exceptionally well-preserved disc- and plate-like forms have escaped breakage, and thus account for the preservation of such lightweight forms as the smallest known complete australite—an oval, plate-like form of 1.0 mm thickness, weighing only 0.0645 grams, from the Port Campbell strewnfield, S. W. Victoria.

Since such forms are symmetrical and reveal no signs of damage on landing, it canot be accepted that they resulted from the flattening of plastic glass on impact with the ground, as advocated by some earlier theories.

SHAPE ORIGIN OF HOLLOW AUSTRALITES

The original shapes of the rare hollow australites were formed at some extraterrestrial source in company with the original forms from which the other shape groups of australites were produced. The presence within them of large internal cavities, had no influence on their external configuration.

Preservation of some hollow forms as complete entities throughout the process of frontal ablation during high speed earthward flight, demands the original development of primary hollow forms (most of which were spheres) possessing an eccentrically disposed internal cavity, so arranged within, that the walls of australite glass were thicker at one pole than at the other.

All the known hollow australites provide evidence that the stable position during earthward flight, was one in which the thickest wall was at the front. Those with sufficiently thick front surfaces, withstood ablation to the same extent as some of the larger of the solid australites. Similar sculpture features were produced upon their front surfaces, and a few examples developed small flange structures (Baker, 1956). In some, the front walls collapsed because they were thinner or were subjected to rather more ablation. Such examples sometimes reveal inrolled edges on remnant portions of the collapsed walls and sometimes show signs of minor secondary flowage of glass over the collapsed edges for short distances on to the typically smooth, "hot-polished" walls of the internal cavity.

CONCLUSIONS

This newly advanced "Aerodynamical Control Theory" of australite shape development, is not only capable of explaining satisfactorily all the characteristic shapes of secondary origin attained by flanged and non-flanged australites, but it also supplies a means for logical explanation of (a) plane spirally coiled circumferential flange structures, (b) relationships between the arcs and radii of curvature of front and rear surfaces of forms of different original size (Baker, 1955a), (c) variations in specific gravity as between cores and flanges of australites (Baker, 1956), (d) the rare colour banding in some flanges and its absence from central body portions (Baker, 1956), (e) the existence of occasional etch pits but no true bubble pits on the front surfaces of australites (Baker, 1956), and (f) the nature of such sculpture features as ridges, troughs and radial flow lines on front surfaces.

The theory thus provides reasons for the existence of two such sharply contrasted surfaces on australites—front surfaces which are virtually free of bubble pits and flow swirls, but which possess such features as ridges, troughs, radial flow lines and occasional etch pits, features which clearly distinguish them from rear surfaces (in well-preserved specimens), and which are of secondary origin. Rear surfaces only possess such initially developed sculptural elements as bubble pits (caused by escaping gas) and flow swirls (caused by flow streaming at the surface of primary shapes, and confined to much less bubble-pitted areas); they represent remnants of primary shapes possessing limited original sculptural elements, produced in an extraterrestrial birthplace.

References

Baker, G., 1944, The flanges of australites: Nat. Mus. Victoria (Melbourne), Mem. 14 (1), p. 7-22.

————, 1955a, Curvature—size relationships of Port Campbell Australites, Victoria: Roy, Soc. Victoria, Proc. 67, p. 165-219.

————, 1955b, Australites from Harrow, Victoria: Mineralog. Mag., v. 30, p. 596-603.

————, 1956, Nirranda Strewnfield australites, southeast of Warrnambool, Western Victoria: Nat. Mus. Victoria (Melbourne), Mem. 20, p. 59-172.

Black, J., 1953, The advance to supersonic flight: Discovery (Magazine of Scientific Progress), v. 14, p. 250-255.

Bonney, E. A., 1950, Engineering supersonic aerodynamics: New York, McGraw-Hill Book Co.

Clarke, W. B., 1855, On the occurrence of obsidian bombs in the auriferous alluvia of New South Wales: Geol. Soc. London, Quart. Jour., v. 11, p. 403.

Durand, W. E., 1935, Aerodynamical theory: Berlin, Julius Springer, vol. III, p. 212-243.

Fenner, C., 1934, Australites, part I. Classification of the W. H. C. Shaw Collection: Roy. Soc. South Australia Trans. and Proc., v. 58, p. 62-79.

————, 1938, Australites, part III. A contribution to the problem of the origin of tektites: Roy. Soc. South Australia, Trans. and Proc. 62, p. 192-216.

Grant, K., 1909, Obsidianites—their origin from a physical standpoint: Roy. Soc. Victoria, Proc., v. 21 (2), p. 444-448.

Opik, E., 1937, Researches on the physical theory of meteor phenomena, III: Observatoire Astronomique Université Tartu Publ. 29 (5), 69 p.

Pope, A., 1950, Aerodynamics of supersonic flight: New York, Pitman Aeronautical Publications, Pitman Publishing Corp.

Walcott, R. H., 1898, The occurrence of so-called obsidian bombs in Australia: Roy. Soc. Victoria, Proc. 11 (1), p. 23-53.

Watson, F., Jr., 1935, Origin of tectites: Nature, v. 136, p. 105-106.

Whitlock, T. O., 1943, Elementary applied aerodynamics: Oxford University Press, Humphrey Milford.

Editors' Comments on Papers 30 and 31

That the intricate sculpture found on tektites is caused by etching has been conceded for many years; however, Nininger and Huss,* supported by O'Keefe,* returned to the explanation given by Suess (1898a, b; 1900)† that this sculpture is caused by ablation. They illustrate indochinites "that suffered surface breaks and interior stretching during plastic bending," proposing that the surface sculpturing took place immediately before the tektites arrived, while they were passing through the atmosphere.

The cause of tektite sculpture has been a source of controversy for three-quarters of a century, since Suess compared the surfaces of freshly fallen meteorites with the sculpture of tektites and decided that both were caused by ablation as they passed through the atmosphere. Previously Stelzner (1893a) had correctly interpreted the intricate surface sculpture of moldavites as due to corrosion. Subsequently Merrill (1911) supported the corrosion theory, as did many others including Wright (1915), Van der Veen (1923), Michel (1925), Lacroix (1931a), Martin (1934b), Rosicky (1934), Beyer (1934), Fenner (1935a), Barnes (1940a), Krinov (1958), Baker (1959), and Rost (1964). Chapman et al. (1967) were able to develop by etching all the sculpture usually attributed to etching on models of natural tektite glass which had been ablated by aerodynamic heating in an arc-jet.

We have also collected specimens in South Viet Nam similar to those illustrated by Nininger and Huss* in which the truncated bottoms of etch pits are preserved within the area of the scars. These pits could not have formed either by ablation or by etching after the scar developed without affecting the rest of the surface of the scar. Furthermore, tektites are under stress, and these scars occur in the area of greatest curvature where failure by spalling would be most apt to take place. Nininger and Huss cite the lack of conchoidal fracture as evidence against these scars being caused by spalling. Conchoidal fractures result from blows, whereas spalling is spontaneous, with failure starting at the weakest point; the resulting surface is smooth.

---

†For references prior to 1959, the reader is referred to Baker's* bibliography.

O'Keefe is correct in stating that tektites recovered from the sea bottom were found to be sculptured in a way which was not greatly different from land tektites of adjoining areas. He implies that this sculpturing is therefore aerodynamical, whereas an argument can just as well be made that the sculpturing results from etching by chemical means. The specimen that Barnes (1971)* studied, which is one of those that O'Keefe refers to, had etch pits on it in which fossils had grown a million years ago and no apparent etching had taken place since. The specimen very likely was etched on land before it was washed into the sea or was submerged by the rise of sea level.

The plate in O'Keefe's* paper duplicates one in the paper by Nininger and Huss* and is omitted to conserve space.

Dr. Harvey H. Nininger for half a century has been collecting, describing, and publishing on meteorites and tektites. The first recorded mention of tektites by Dr. Nininger was in connection with the 1940 Annual Meeting of the Geological Society of America, held in Austin, Texas. He gave a paper on this occasion entitled "The Moon as a Source of Tektites." Barnes also gave his first paper on tektites at this meeting, "Distribution and Origin of Tektites," in which he announced the discovery of tektites in North America. Dr. Nininger's present address is Meteorite Investigations, P.O. Box 446, Sedona, Arizona 86336. His son-in-law, Glenn I. Huss, is Director, American Meteorite Laboratory, P.O. Box 2098, Denver, Colorado 80201.

Dr. John A. O'Keefe is Assistant Chief for Planetary Studies in the Laboratory of Space Physics at NASA's Goddard Space Flight Center. His current duties include the theoretical study of the moon from the standpoint of planetology.

References

*Barnes, V. E. (1971). Age of Asian tektites. *Geol. Soc. Amer. Bull.,* **82,** 1995–1996.

Chapman, D. R., Zimmerman, N. B., Centolanzi, F. J., and Scheiber, L. C. (1967). Effects of etching and of earth landing on aerodynamically ablated tektites. Meteoritical Society, 30th Ann. Meeting, Ames Research Center, Moffett Field, California.

Krinov, E. L. (1958). Some considerations on tektites. *Geochim. Cosmochim. Acta,* **14,** 259–266.

Rost, R. (1964). Surfaces of and inclusions in moldavites. *Geochim. Cosmochim. Acta,* **28,** 931–936.

30

Reprinted from Science, July 7, 1967, Vol. 157, No. 3784, pages 61-62

Tektites That Were Partially Plastic after Completion of Surface Sculpturing

Abstract. *Among the 50,000 tektites collected over an 8-year period as part of a representative collection of the indochinities in an area near Dalat, South Vietnam, several individual ones have been found that show evidence of having been internally plastic after surface sculpturing was essentially completed. Two drops, which were bent after having formed a thin exterior skin or crust, exhibit surface breaks and stretching of their plastic interiors within the breaks. The lack of deep sculpturing in this interior stretched area, coupled with twisting within the break on one of them, indicates that the surface features on these tektites were formed in the atmosphere and not by etching by soil acids, as had been widely believed.*

The external features of tektites from southeast Asia have generally been considered to be the result of deep etching by soil acids. Evidence found in the surface features of several indochinites from an area near Dalat, South Vietnam, now indicates that this concept may be in error and that the general surface features are the result of aerial ablation. This evidence was discovered during the examination of field collections, totaling 50,000 specimens, made during an 8-year period beginning in 1958.

In inspecting the first 40,000 specimens, many shapes were found that indicated stretching: elongate drops, rods and straps that appeared to represent portions of the tails of drops that had been elongated to perhaps 6 inches (15 cm) or more, dumbbells, and discoidal shapes that showed whirlpool or swirling effects that had resulted from spinning. A large percentage of the elongate specimens also showed twisting and bending. Sixty-six percent of all specimens showed breakage, and an even larger percentage showed spalling. However, it was only during examination of the next 10,000 specimens that the full significance of these forms began to be realized.

As long ago as 1900, Suess (*1*) concluded that the surface sculpturing of tektites was due to aerial ablation and was "not due to some sort of corrosion process," because corrosion causes a matte appearance that is lacking on the surfaces of most tektites and that fails to account for the grooves and cupules found on tektite surfaces. Most students of the subject disagreed with this hypothesis, insisting that the present surfaces of tektites, particularly those of indochinites, were the result of deep etching by acids in soil.

Serious difficulties for the theory of attack by ground chemicals come from a study of the broken indochinites. Many indochinites have flattish surfaces that cut sharply across the rounded surfaces that seem to be part of the original shape, and these flattish surfaces are logically interpreted as breakage surfaces. The degree of pitting and sculpturing is much less on the broken surfaces than it is on the rounder original surfaces. In terms of the theory of attack by ground acids, this clearly implied that breakage had taken place after more than half the life of the tektites on the ground and that it had occurred simultaneously throughout the area strewn with indochinites. It was suggested (*2*) that this breakage might have resulted from an enormous tidal wave.

Doubts about these solutions to the problem were enhanced by the discovery of a number of onion-shaped tektites (Fig. 1) that look as though they had encountered either the surface of the ground or else the denser layers of the atmosphere while still in a molten condition.

The really decisive evidence was furnished, however, by two tektites (Fig. 2) that seem to have suffered sharp bending while the interior was still plastic. The exteriors of these had actually broken. Close inspection showed that breakage had occurred after cooling from the outside had produced a thin skin or crust over a plastic interior. The broken skin of the tektites had pulled apart, much as the skin of a cut finger would, while the plastic interior stretched but did not break. Thus, we conclude that although more than two-thirds of the indochinites were broken after more or less complete cooling had taken place, not all of them were cold rigid bodies at this point.

The internal plasticity of these tektites this late in flight is explained by the conclusion that tektites did not arrive at the earth's atmosphere as a

Fig. 1. Bulbous drops that show plastic flattening and external spallation after completion of aerial sculpturing.

Fig. 2. Tektites that suffered surface breaks and interior stretching during plastic bending. Note lack of sculpturing within breaks.

Fig. 3 Dorsal view of larger drop shown in Fig. 2, demonstrating twisting that accompanied surface separation and stretching.

swarm, but were spawned by a larger, semisatellite parent body (*3*) rotating rapidly within the confines of the atmosphere (*2*). Tektites were thrown off in all directions, up as well as down, and a very small percentage of them cooled in the act of plastic bending. Had they arrived at the atmosphere as individuals, there would have been little or no internal heating or plastic shaping and bending, but only surface ablation from atmospheric friction.

Some will argue that this bending occurred before the tektites encountered the atmosphere and that the breaks are not true breaks, but are only the result of much later spalling induced by the stress of bending, coupled with the action of soil acids. This argument might carry considerable weight were it not for the following considerations.

The interiors exposed by surface separations in the bent drops show only very slight etching, similar to the etching noted in the freshest of australites. These surfaces show no evidence of flight markings. They must, therefore, have been exposed, at the earliest, only after major aerial ablation had ceased to operate upon the exterior of the tektites.

The breaks thus exposed cannot be the products of spalling for they show no concoidal fracturing as do the normal spall markings. Many of the markings on either side of the separations would match if placed back together, showing that the specimen was completely formed and marked before bending and separation took place; and, most important, there is considerable twisting within the break in the larger specimen (Fig. 3). There is no conceivable manner in which such twisting could have occurred had not the specimen first been broken while in a partially or internally plastic state. The specimen must have been broken upon encountering strong aerodynamic resistance, or the soil, or by a collision with another individual very near the end of its terminal flight.

It seems evident, therefore, that the major features on the exterior of the indochinites from Dalat, Vietnam, were not formed by soil etching. Had they been, the areas of these breaks would have been as deeply etched as the exteriors of the specimens are widely thought to be. The major features had to be formed during their atmospheric flight. It is also evident that much of the breakage and the so-called spalling, which is widespread and shows the same etching age as the surfaces exposed by the skin fractures, also dates from the time of original fall and not from later breakage.

It is suggested that the forms and surface features of the indochinites be reevaluated in light of this new information and its relation to other known facts and to tektites of other areas.

H. H. NININGER

Meteorite Investigations, Sedona, Arizona

GLENN I HUSS

American Meteorite Laboratory, Denver, Colorado

References and Notes

1. F. E. Suess, *Jahrb. Geol. Reichsanst.* **50**, 193 (1900).
2. H. H. Nininger, *J. Ariz. Acad. Sci.* **1**, 125 (1961).
3. J. A. O'Keefe and B. E. Shute, *Aerospace Eng.* **20**, 26 (1961).
4. We thank Dr. John A. O'Keefe for reading the manuscript and making valuable suggestions for its improvement.

28 April 1967

31

Geochimica et Cosmochimica Acta, 1967, Vol. 31, pp. 1931 to 1933. Pergamon Press Ltd. Printed in Northern Ireland

Tektite sculpturing

JOHN A. O'KEEFE
Laboratory for Theoretical Studies, Goddard Space Flight Center,
National Aeronautics and Space Administration, Greenbelt, Maryland 20771

(*Received* 22 *May* 1967)

Abstract—Tektites are divided into three types on the basis of external form: Type A (Muong Nong), Type B (spheres, teardrops, dumbbells, etc.), and Type C (derived from Type B forms but with evidence of aerodynamic ablation).

The surface sculpturing of Type B tektites (grooves, pits, notches) is ascribed to aerodynamic processes rather than to chemical etching after arrival on the surface of the earth.

TEKTITES may be classified by their forms into three groups: Type A (the Muong Nong type of BARNES, 1963) which is chunky or, often, platy in external form, but generally shows indications of layering; Type B, which usually has the form of spheres, ellipsoids, teardrops, rods, dumbbells, or else of thin scales, often bent or rolled; and type C, which is derived from bodies having the form of spheres, ellipsoids, teardrops, rods, or dumbbells, etc., but have been subjected to attack by heat on one side (called the anterior side) so that a part of the mass has been lost. Australian tektites, which are nearly all of Type C, show a flange, in about 10% of the specimens, evidently derived from the melted material of the anterior side.

CHAPMAN and LARSON (1963) by a series of experiments with high-temperature wind-tunnels have shown convincingly that the anterior surfaces of Type C tektites are the result of aerodynamic ablation acting on spherical, ellipsoidal, etc., bodies moving at hypersonic speeds. The observed amount of ablation could be produced either by bodies moving at 7–8 km/sec at low angles, or by velocities of 11 km/sec at steep angles. E. W. ADAMS and his associates (1963) favor the former alternative, CHAPMAN and his group the latter. Both groups consider that origin from the ground is not reconcilable with the data; both support the hypothesis of NININGER (1943) that tektites result from meteorite impact on the moon.

The posterior surfaces of Type C tektites, and all surfaces of Type B tektites, usually show a system of pits, notches and grooves which is very characteristic. The pits are typically hemispherical; they very rarely show any trace of overhanging sides as would be expected for bubble-pits. The grooves are of two types: relatively short and straight furrows, and meandering marks like worm-tracks. Notches are typically flat-sided and occur chiefly on the edges of tektites. All markings tend to have a lacquered appearance which, on worn tektites, may contrast with the dull appearance of the majority of the tektite.

The vitally important question for the tektite problem is: What is the origin of the sculpturing on Type B tektites and on the posterior surfaces of Type C tektites? Is it due to aerodynamic causes like that on the anterior surfaces of Type C tektites or is it due to attack by ground chemicals? The latter view has, until recently, had the support of most students of tektites.

SUESS (1900) in his classic memoir on the European tektites gave powerful reasons for believing that the sculpture was due to aerodynamic causes. He pointed out that tektites contain, in their interior, a system of contorted fluidal structures now called schlieren and defined by differences in density or transparency. These are expressed at the surface as delicate stripes, which he calls fluidal striping. When a tektite is attacked by hydrofluoric acid for example, the attack follows the lines of the fluidal striping. On the other hand, the natural tektite sculpture, especially the system of furrows, is usually completely independent of the fluidal striping. This is significant evidence against an origin of the surface sculpture by chemical corrosion.

SUESS noticed some significant relations between overall shape and the pattern of sculpture. On tektites whose general form was an oblate ellipsoid he found that grooves near the equator tended to be perpendicular to the equator. Grooves near the poles, on the other hand, tended to radiate from the poles, especially if the body was highly flattened On convex surfaces, the grooves ran in the direction of strongest curvature, while on concave surfaces they ran in the direction of minimum curvature.

The significance of these observations lies in the fact that aerodynamic forces often depend on the overall size and shape of a body, while it is hard to see why action by ground chemicals should be governed by anything except the local properties of the surface.

BAKER (1944) pointed out that the pits on the posterior surfaces of australites which he called bubble-pits, are occasionally covered by flange glass or filled by it. These pits clearly antedate arrival on the ground. It is not usual to regard them as genuine bubble-pits, because they rarely overhang and because they are so common relative to the true bubbles in the interior.

When glass, whether artificial (BRILL, 1961) or natural (WRIGHT, 1915) is attacked by ground solutions, the effect is to remove the metal oxides, and leave behind the silica skeleton of the glass microstructure. These residues are not found on tektites.

KING (1964) has reported some characteristic Type C forms, without flanges, from Texas. These have not been altered significantly since arrival at the earth; yet they are found in the same sites as some deeply sculptured tektites. It is easier to believe that these tektites encountered different aerodynamic conditions than that they encountered different ground chemistries.

GLASS (1967) reports the discovery of tektite-like glasses from the sea bottom at levels which indicate an age of 700,000 yr. The cores in which these tektites were found contain at least 50% water. The glass bodies range from 40 μ to 800 μ in size; most are in the small range. Clearly sea-water does not remove as much as 20 μ of glass in 700,000 yr. But a few tektites recovered from the sea bottom (Millies-Lacroix, quoted by O'KEEFE, 1963) were found to be sculptured in a way which was not greatly different from land tektites of adjoining areas.

JOHNSON (1965), using cold helium gas on low-melting point materials, has shown that, under certain conditions, aerodynamic flow can produce cavities. ZOTIKOV (1959) found he could simulate the markings on certain meteorites using hot gases. J. GOLDEN (quoted by O'KEEFE, 1966) found that tektite glass could be ablated in a simulated rocket exhaust to produce forms not unlike the equatorial grooves on moldavites. The common feature of these experiments was the use of a relatively dense gas; the Reynolds number was high enough to produce turbulence; no such grooves or pits have been found in the wind-tunnel experiments of CHAPMAN's group, which employed lower Reynolds numbers and laminar flow.

NININGER and HUSS (1967) have recently provided powerful evidence that the attack is aerodynamic. From the investigation of some 50,000 Dalat tektites, they found a number which appeared to have been deformed plastically after the formation of pits and grooves on the surface. Their two most significant specimens are shown in Fig. 1. Such specimens are very rare; but a large proportion of the tektites from Indochina show evidence of a clean break suffered after most of the pitting and grooving had occurred, but before the end of the period of pitting and grooving. To explain this result in terms of chemical attack means supposing an event which caused tektites to break throughout Indochina, all at the same time. It is much easier to understand in terms of stresses in flight, as was pointed out by the Czechoslovakian students of moldavites, (KASPAR, 1938).

Despite this evidence, the great majority of students of the problem regard the question as settled in favor of origin by chemical attack. BAKER (1963) and BARNES (1963) refer to etching as the mode of formation of grooves and pits in tektites. They argue that the anterior surfaces of Type C tektites show what aerodynamic attack will do. They argue that Type B tektites are younger than Type C; in particular that the australites are only a few thousand years old. On the other hand, the small glassy objects found by GLASS are of the same age off the coast of Australia, where Type C tektites are found, as off the coast of the Philippines where Type B tektites are found. Hence the australites are probably of the same age as the other tektites.

It is concluded, therefore, that there are strong reasons for thinking that the grooves and pits in Type B and Type C tektites were produced by turbulent gas flow in the earth's atmosphere.

REFERENCES

ADAMS E. W. (1963) Aerodynamic analysis of tektites and their hypothetical parent bodies. In *Tektites* (editor J. A. O'Keefe), p. 150. University of Chicago Press.

BAKER G. (1944) Flanges of australites. *Mem. Nat. Museum Victoria* **14,** Part 1.

BAKER G. (1963) Form and sculpture of tektites. In *Tektites* (editor J. A. O'Keefe), p. 1. University of Chicago Press.

BARNES V. E. (1963) Tektite strewn fields. In *Tektites* (editor J. A. O'Keefe), p. 25. University of Chicago Press.

BRILL R. H. (1961) The record of time in weathered glass. *Archaeology* **14,** No. 1.

CHAPMAN D. R. and LARSON H. K. (1963) On the lunar origin of tektites. *J. Geophys. Res.* **68,** 4305–4358.

GLASS B. (1967) Microtektites in deep sea sediments. *Nature* **214,** 372–374.

GOLDEN J. (1966) Quoted on p. 207 of The Origin of Tektites. *Space Sci. Rev.* **6.**

JOHNSON T. H. (1965) Flow instabilities relating to the surface markings of tektites. *J. Geophys. Res.* **71,** 945–949.

KASPAR J. (1938) Czechoslovakian tektites and the problem of their origin. *Pop. Astr.* **46,** 47–51.

KING E. A. (1964) An aerodynamically sculptured bediasite. *J. Geophys. Res.* **69,** 4731–4733.

NININGER H. H. (1943) The Moon as the source of tektites. *Sky and Telescope* Part I, **2,** No. 4; Part II, **2,** No. 5.

NININGER H. H. and HUSS G. I. (1967) Tektites that were partially plastic after completion of surface sculpturing. *Science,* **157,** 61–62.

O'KEEFE J. A. (1963) Origin of tektites. In *Tektites* (editor J. A. O'Keefe), p. 167. University of Chicago Press.

SUESS F. E. (1900) Die Herkunft der Moldavite. *Jahrbuch der kk Reichsanstalt. Vienna* **50,** 193–382.

WRIGHT F. E. (1915) Obsidian from Iceland. *Bull. Geol. Soc. Amer.* **26,** 255–286.

Editors' Comments on Papers 32 and 33

Of the numerous papers concerning etching as the medium for producing intricate tektite sculpture, two of the most recent have been chosen.

Rost* presents compelling evidence that moldavite sculpture is caused by etching. Suess (1900), also working with moldavites, arrived at the conclusion that the sculpturing was caused by ablation.

The paper by Margolis, Barnes, Cloud, and Fisher* treats two important features of tektites: (1) the thin inhomogeneous surface layer of australites preserved beneath flanges; and (2) tektite sculpture, compared with igneous analogs.

Although one manner in which the inhomogeneous surface layer preserved beneath flanges might have formed is described in this paper, it is necessary to add here a discussion of possible source materials. Kurat and Keil (1972) suggest that differences in composition that they observed between the surfaces and interiors of some lunar spherules may have been caused by vapor condensation on the spherules immediately following the impacting event that formed the spherules. It is likely that similar vapor condensation on australites helped produce the surficial layer preserved beneath flanges. However, from the wide variation in composition of this surficial layer, it is judged that the source material probably was fine-grained and polymineralic. With such source material the larger drops of melt, composed of large numbers of the various grains, would tend toward similar compositions; whereas small droplets, representing only a few grains or even individual grains of the source material, could vary as much in composition as do the grains in the source material. The impingement of droplets on other drops indicates variation in speed of the various drops. Chao (1963, Plate I*d*) illustrates a droplet that deeply penetrated an indochinite, breaking up as it went. King and Bouška (1968) described a two-colored moldavite which could have been two different moldavites that joined during flight.

Figures 14 and 15 clearly illustrate the difference in etching of two Miocene basaltic glass spherules recovered in the Deep Sea Drilling Program. One spherule is deeply etched and the other is only faintly etched and retains spatter on its surface. This finding is at variance with the belief of O'Keefe (1967) concerning the sculpture on bediasites. He states: "It is easier to believe that these tektites encountered different aerodynamic conditions than that they encountered different ground chemistries." Obviously the igneous spherules were not subjected to aerodynamical ablation, yet they show wide differences in etching, possibly because some specimens were protected by a coating, such as clay, and others were not.

Dr. Rudolf Rost is Professor and Head, Department of Mineralogy, Geochemistry, and Crystallography at Charles University, Prague, Czechoslovakia.

Dr. Stanley V. Margolis is now with the Department of Oceanography, University of Hawaii, 2525 Correa Road, Honolulu, Hawaii 96822. He is also a research geologist in the Hawaii Institute of Geophysics, where he is the director of the scanning electron microscope and microanalysis facility. The address of Barnes should have been listed as University of Texas at Austin.

References

Chao, E. C. T. (1963). The petrographic and chemical characteristics of tektites. *In* "Tektites" (J. A. O'Keefe, ed.), pp. 51–94. Univ. Chicago Press, Chicago, Illinois.

King, E. A., Jr., and Bouška, V. (1968). Electron microprobe analysis of a two-colored moldavite from Slavče-Lipí (Bohemia), Czechoslovakia. Rept. 23rd Internat. Geol. Congress, Prague, Abstr. Vol., p. 341.

Kurat, G., and Keil, K. (1972). Effects of vaporization and condensation on Apollo 11 glass spherules: implications for cooling rates. *Earth Planet. Sci. Lett.* **14,** 7–13.

O'Keefe, J. A. (1967). Tektite sculpturing. *Geochim. Cosmochim. Acta,* **31,** 1931–1933.

Suess, F. E. (1900). Die Herkunft der Moldavite und verwandter Gläser. *Jahrb. Königl Geol. Reichsanstalt, Wien,* **50** (2), 193–382.

32

Journal of Geophysical Research Vol. 74, No. 27, December 15, 1969

Sculpturing of Moldavites and the Problem of Micromoldavites

Rudolf Rost

Department of Geochemistry, Mineralogy and Crystallography
Charles University, Prague, Czechoslovakia

Nininger and Huss stated in 1967 that the sculpturing of the surfaces of two very rare indochinites from South Vietnam occurred in the atmosphere and was not due to etching by soil agencies. However, very important and convincing features observed in moldavites and reported in this paper support the opinion that moldavite sculpturing is due to chemical corrosion in soils and gravels. It is concluded that when moldavites fell on the earth's surface they were already cool solid bodies. Many of the falling moldavites, however, were fragmented on landing. All of them, emplaced in the soil, were strongly corroded. Pure (hyaline) and translucent glassy drops, a few tenths of a millimeter in diameter, have been separated by washing the material encountered by an 80-m-deep bore through Pannonian sediments in southern Moravia. These glassy drops may have been produced as slag spherules from steam engines; they may have been in the volcanic dust from Slovakian rhyolites; or they may be micromoldavites. To solve this problem, a more detailed investigation would be necessary.

Introduction

Nininger and Huss [1967] have shown on two rare specimens of indochinites from the Dalat area, South Vietnam, that these pieces were still plastic during their flight through the atmosphere of the earth. The typical pitted sculpturing and a solid crust on the surface of these indochinites had evidently been formed as early as before they landed on the earth's surface. The two specimens under consideration were chosen from among 40,000 indochinites found in the Dalat area. Although more than two-thirds of the indochinites were broken after a more or less complete cooling had taken place, not all of them were cold rigid bodies at that time. For example, a number of onion-shaped indochinites discovered there appear as if they had encountered the surface of the ground or the denser layers of the atmosphere while still in a molten state. From their studies Nininger and Huss concluded that the general surface sculpturing of the indochinites examined was due to aerial action. These authors suggested that the forms and surface features of indochinites be re-evaluated in the light of this new information and its relation to other known facts and to knowledge about tektites from other areas.

I support Nininger and Huss's interpretation of the two rare indochinites from the Dalat area. Studying a perfect black plastic cast of the specimens, I have established that the angle of bending of the drops corresponds to the angle of the opening of the break.

However, the theory that the present-day sculptured surface of moldavites has resulted from intensive etching by soil agencies cannot be rejected.

Recently, several interesting moldavites whose surface sculpturing can be interpreted only by chemical etching in sands were dug from still intact sand pits in southern Bohemia.

In the sand pit near Chlum nad Malší, southwest of Ločenice and about 25 km south of České Budějovice, at a depth of about 1 meter beneath the surface, a small flat moldavite was found. Part of it is shown in Figure 1. This moldavite is green, of a tint typical of moldavites of southern Bohemia. At one of its ends a lenticular bubble parallel to the surface is ellipsoidal in outline, its longer axis being 7.4 mm. As the moldavite is relatively translucent, the outline of the bubble is distinctly visible, as shown in the figure. The upper wall of the lens-shaped bubble is not complete, as almost half of the wall had disappeared owing to chemical corrosion. No trace of mechanical damage of the upper wall is visible. The margins of the remaining part of the upper wall are very sharp, thin, and finely pitted. The usual deeply pitted sculpturing can be observed over the whole surface of the moldavite. The subsurface bubble has been opened by chemical corrosion, but opening could certainly not have taken place during the flight of the moldavite through the atmosphere. The sharp points at the margin of the remaining upper wall of the bubble could not be due to any factor other than chemical corrosion affecting the moldavite at the site of its occurrence. Figure 1 shows deep pits on the sculptured surface of the specimen which agree both with those on the outer part of the wall of the bubble and with those on the rest of the surface of the specimen. Therefore, the present-day appearance and the deep surficial sculpturing of this flat moldavite are the results of a long-term chemical corrosion. The whole lower inner wall of the lenticular bubble is smooth and only very slightly etched, and this observation is consistent with my conclusion [*Rost,* 1964] that the intact inner walls of bubbles within moldavites (and also the fracture surfaces) are much more resistant to etching than any other surfaces on moldavites. This also agrees with the fact that the lower inner surface of the bubble of the moldavite was accessible to an etching attack in sand as late as after the upper wall of the bubble had been partly opened, which must have occurred much later.

Supporters of the air sculpturing theory admit shallow etching by groundwater but claim that the main deep sculpturing originated during flight. Other convincing evidence of strong chemical corrosion of moldavites is furnished by the moldavites shown in Figure 2*E, F*. These specimens have been dug from the sand pit near Ločenice at a depth of about 1.5 meter. Both specimens are very thin, about 1–3.5 mm thick. Originally, they were thin smooth spalls (splinters) only because of the shattering of major moldavites during their fall. After their emplacement in the sediment, advancing chemical corrosion produced their present-day appearance. Photographs (Figure 2) clearly show that in the moldavite matter less resistant portions have been more intensively attacked by chemical corrosion. It is impossible to believe that the sharp points visible in Figures 1 and 2*C, F* were formed during flight through the atmosphere (in either a plastic or a solid state). Further evidence of strong chemical corrosion is furnished by projecting points of the 'schlieren' of lechatelierite, on some moldavites, especially those from some sand pits in southern Bohemia: points of strongly siliceous glass just out as much as 1 mm above the level of the surrounding surface of the moldavite. Moldavite glass melts at about 1400°C, and pure lechatelierite melts at about 1700°C. The difference in the melting points of the two components of moldavite matter being only about 300°C, and the difference between the viscosity grades therefore being little, the temperature and pressure occurring during flight would hardly make it possible for such fine and sharp points to be formed.

Their sharp forms can be best explained by chemical corrosion. From this it follows that around the resistant lechatelierite points, a layer at least 1 mm thick of moldavite groundmass has disappeared. Figure 3 shows a projecting point of more siliceous glass in a small groove on the surface of a moldavite excavated from the sand pit near Chlum nad Malší. The whole specimen weighs 11.7 g. From Figure 3 the sharpness and fineness of the projecting points can be judged. I saw several moldavites from the Stilec and Hrozinka sand pits near České Budějovice on whose surface points of schlieren stand out at least 1 mm above the surrounding level of the surface. Because of incautious washing and manipulation of moldavites, most of the points are broken off. In one case I established on one such point the refractive index $n = 1.460$. It was an optically quite isotropic lechatelierite which, however, mostly occurs in moldavites as isometrical grains up to 1 mm in size. Many of the points are tips of schlieren ranging in silica content between the values of the moldavite groundmass and those of a pure lechatelierite (=100% SiO_2).

In many observations published in earlier papers [*Rost,* 1964, 1966*a*, 1966*b*, 1967] I have shown evidence that chemical corrosion affects the surfaces of moldavites. From this point of view, two additional moldavites are interesting, those from Ločenice and Bukovec near České Budějovice. The Ločenice specimen was dug from a sand pit by F. Maroušek (Figure 1). After emplacement in sand, a marginal portion of this moldavite spontaneously broke off. In 1968, both portions fitting to each other were

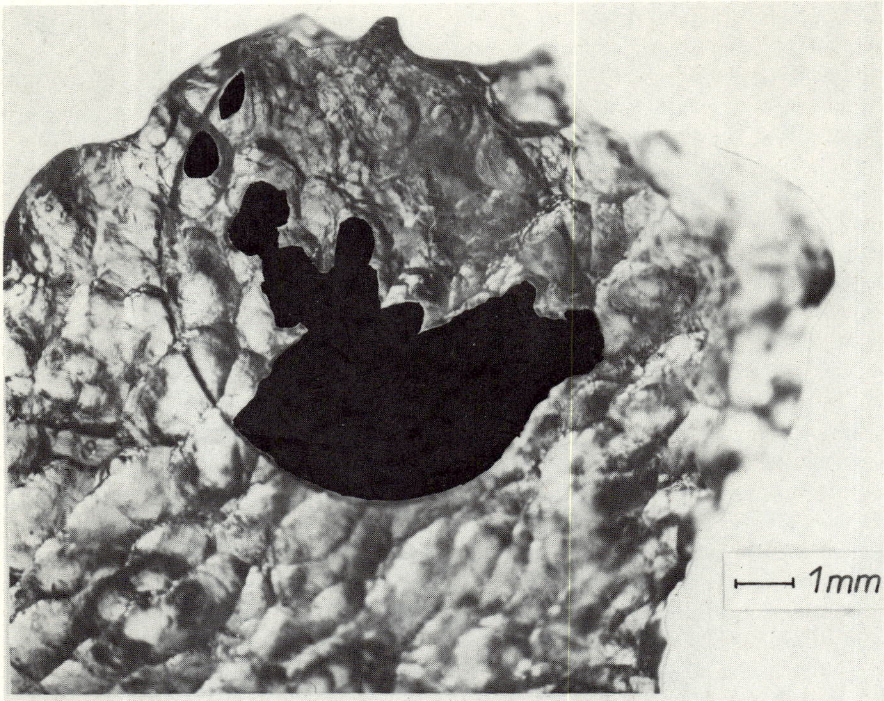
1mm

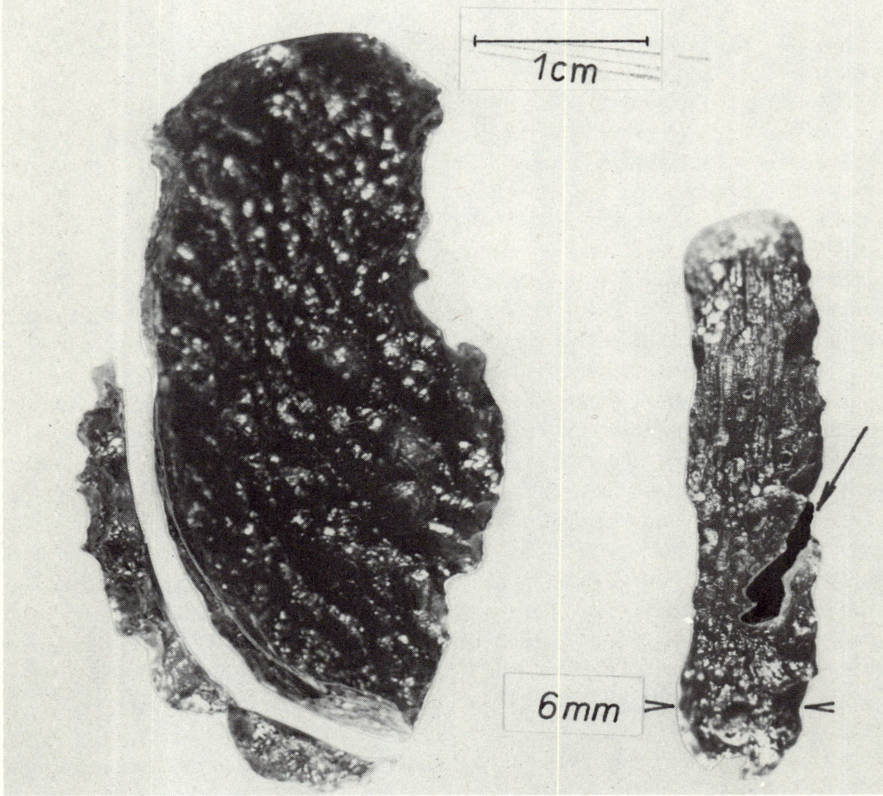
1cm
6mm

found by F. Maroušek, lying next to each other and separated by 2 mm of sand. From the shape of the two parts shown in the photograph, it is evident that originally they formed one piece. The outer surfaces of both parts of the moldavite display a normal deep sculpturing; the fracture surfaces are finely sculptured only. The fine sculpturing is striking; it is distinctly visible in the photograph, lower right. A supporter of the aerodynamic sculpturing theory could interpret both these pieces of the same moldavite in the terms of his opinion, for instance, as follows: Flying through the atmosphere, this specimen was intensively sculptured over its whole surface. After the infall into sand the marginal portion spalled off and the fracture surface was only slightly corroded in the sand. This moldavite from Ločenice has been described by *Bouška and Maroušek* [1969].

In examining the slightly etched fracture surfaces of the larger piece, we see a large deep groove (Figure 1, the place designated by an arrow) which was produced by chemical corrosion from the time of spallation of the minor part. It is clear that in this region a less resistant portion of moldavite glass must have existed, which later was etched off. A counterpart of the deep groove is not visible on the minor chipped piece. On the major part of the specimen, the deep groove continues toward the side of this major piece. It can be observed on the specimen that chemical corrosion was responsible for the origin of the deep groove and that before separation of the minor piece neither groove nor bubble existed there.

Several examples of two pieces of a moldavite fitting together to form a larger one are known from the literature and collections of southern Bohemia. Such pairs always have a distinctly finer sculpturing on the fracture surfaces. However, there are numerous pieces of moldavites whose fitting parts could not be found. In the Ločenice area, a number of such spalls have been encountered; I assess them to be 5% of the total amount of moldavites coming from this area and assume that they will not exceed 10% at this site of occurrence. To illustrate this idea, I refer the reader to Figure 2, *A–D*, which shows four moldavites recently dug from the sand pit. The sculpturing of all of them is different. The specimen shown in *A* is an example: the corner of the spall is sharp and the fracture surface forming the corner displays a slight sculpturing only, resembling that shown in Figure 1, below right, and in Figure 2*B*, *C*. *B* and *C* illustrate rounded fragments without sharp corners, where even the edges and all surfaces exhibit fairly deep sculpturing, almost as deep as that on the main surfaces of these moldavites. According to *Zebera* [1967], the Ločenice area belongs to the fluviolacustrine type of moldavite-bearing sediments. In this site of occurrence, moldavites suffered a not very long transport by water. Most of them were somewhat worn, and, having been emplaced in sand for a long time, they have been so solution-etched as to assume the present-day appearance. Since such moldavites as shown in Figure 2*A* and Figure 4*A* were not subjected to long transport by water after the second part separated, two subsequent histories can be considered: either at a time of flooding the moldavites were washed down from their nearest environment into the lake and in this way also into the moldavite-bearing sediments, or the sediment, carrying moldavites that were already settled and consolidated, was in later geological time transported again by water and redeposited; the youngest fracture surfaces of moldavites then occurred during this last redeposition of moldavites.

The same type of surface phenomena can also be observed on a small flat moldavite dug from the sand pit at Chlum nad Malší. Its weight is 5.5 g and its size 8 × 19 × 37 mm (Figure 4*A*). Its surface was already somewhat

Fig. 1. Above: Part of the moldavite from the sand pit near Chlum nad Malší displaying a lenticular bubble under the surface. About half of the upper wall of the lens-shaped bubble has been removed by chemical corrosion. The missing part of the upper wall is represented in black. Below: A moldavite from the sand pit near Ločenice. A small fragment was found at a distance of 2 mm from the larger piece. The total weight of the two pieces together is 8.5 g. Dimensions of the larger piece: 6 × 20 × 41 mm. To the right the fracture surface of the larger (main) piece shows the fine sculpturing of the surface. An arrow designates a deep groove due to corrosion.

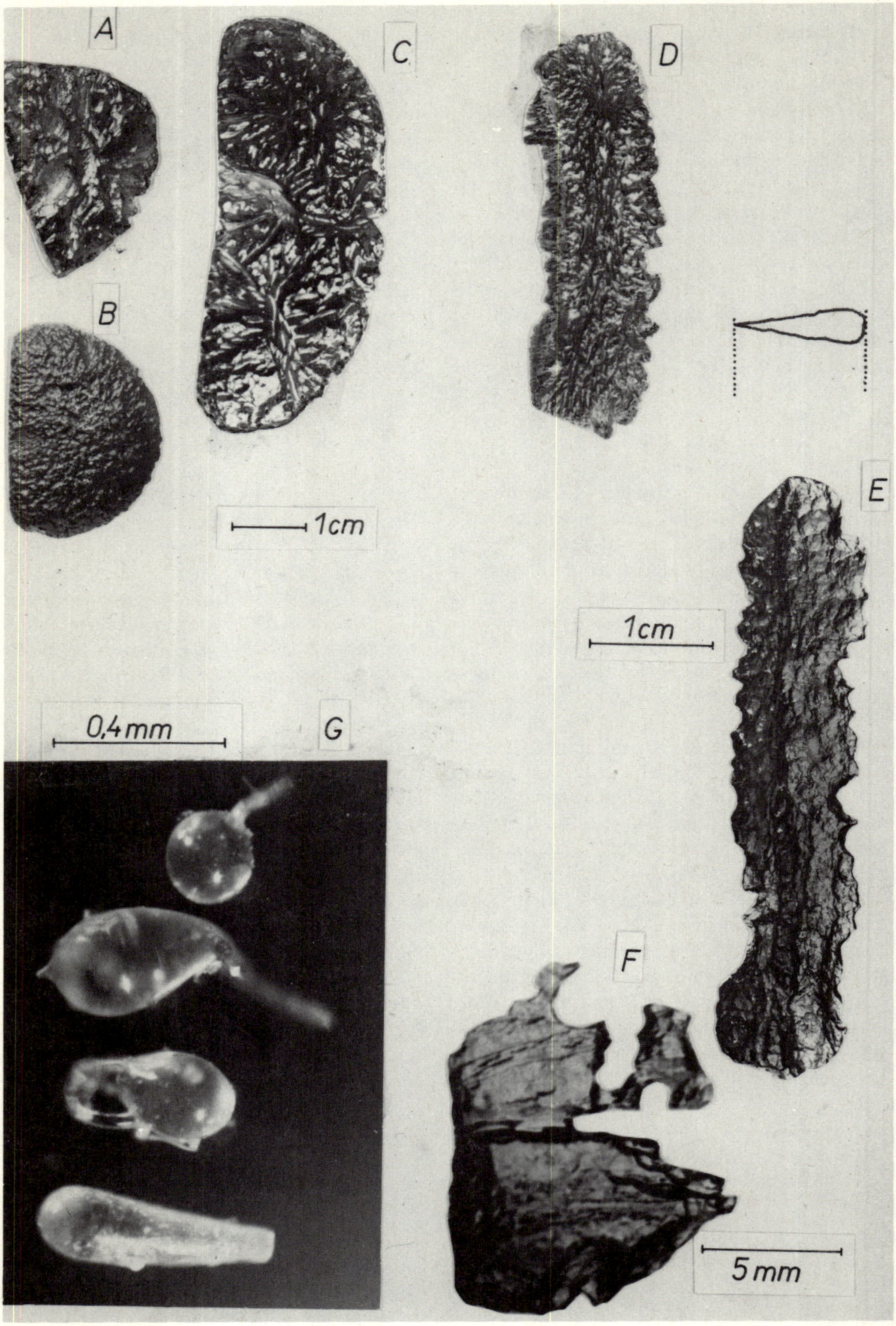
A
B
C
D
E
F
G
1cm
1cm
0,4mm
5mm

sculptured when a part of it spalled at its right upper end. A conchoidal fracture is still visible on the fracture surface. The bottom of a small vesicle incidentally opened by this conchoidal fracturing has a high vitreous luster. Sporadic microscopic pits due to the beginning of corrosion are visible on the bottom of the vesicle. This is consistent with my earlier observations [*Rost,* 1964]. Even in the photograph one can perceive a finer and therefore younger sculpturing in the place of spallation. This is a case analogous to that illustrated in Figure 2*A*. As already stated, similar moldavites derived from the Ločenice area make up about 5% of the total moldavites found there. This percentage is certainly not high, and it is very likely that these fragments of larger pieces spalled a relatively long time after the infall of the moldavites. What evidence of this can be furnished, irrespective of other facts, even for this small number of spalls? At the time the moldavite shower landed on the earth, the impact caused fragmentation of most of the moldavites. For instance, *Nininger and Huss* [1967] reported that about 66% of the indochinites fallen in the Dalat area had been shattered. Moldavites are a similar case. But how can we explain the fact that moldavite spalls with slightly corroded fracture surfaces are so few in number? The most probable explanation is that the spalls with lustrous surfaces, together with unbroken complete forms, have been etched since their fall upon the strewn field and were then washed down into the moldavite-bearing sediments. It is assumed that the moldavites landed about 14.8 m.y. ago, when in southern Bohemia there were several opportunities for their transport and redeposition. I do not wish to discuss purely geological aspects, but it must be considered that there exist several independent proofs of the fact that, with the uplift of southern Bohemia during Miocene time, layers of rocks at least tens

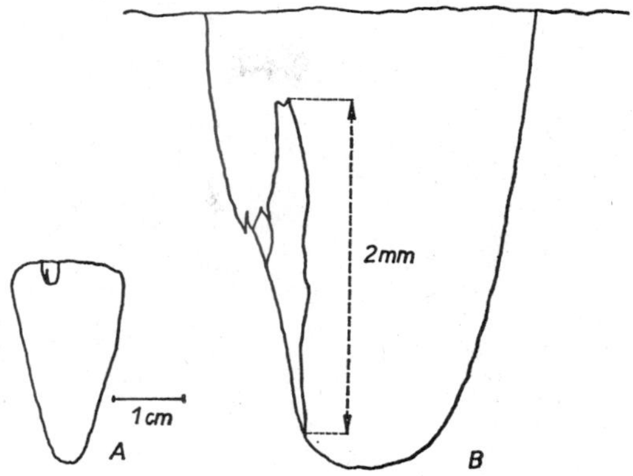

Fig. 3. A needle of lechatelierite etched out from the surrounding surface by chemical corrosion, or schlieren, on the surface of a moldavite from the sand pit near Chlum nad Malší. Weight of the moldavite 11.7 g, length 25 mm. Total length of the needle 2 mm. *A*, outline of the moldavite; *B*, pit with projecting point (i.e., the etched-out needle) of a more resistant siliceous glass.

of meters thick (in some places in Bohemia up to 200 meters) were removed by erosion. Under such conditions in Bohemia and Moravia only relics of the total of fallen moldavites have been preserved, and practically no specimen of a moldavite can lie in the place where it landed on the earth's surface. This also explains why a corresponding quantity of small spalls is missing; they were destroyed during transport. Accordingly, in Moravia the transport of moldavites was, on the whole, longer than in southern Bohemia. In any case, here also a more or less long transport by water and redeposition of moldavites occurred (a minimum distance of tens of meters, a maximum of tens of kilometers).

All the moldavites dealt with here have been dug from sand pits, and geologically they are the oldest ones. *Zebera* [1967] attributes these moldavite-bearing sands and sandy gravels to

Fig. 2. *A–F*, moldavites dug from the sand pit near Ločenice. *A*, moldavite with fracture surface to the left. Sharp edges indicate a short transport. *B*, a moldavite with fracture surface to the left. Rounded edges indicate a rather long transport by water. *C*, a fragment of moldavite showing deep sculpturing. The fragment's edges are rounded. *D*, a moldavite showing deep sculpturing on its sides. *E*, a thin moldavite, considerably corroded, wedge-shaped in cross section. The upper half is 1 mm, the lower half up to 3.7 mm thick. *G*, four glassy drops washed out by *Pokorný* [1950] from drill cores from a depth of 80 meters (southern Moravia).

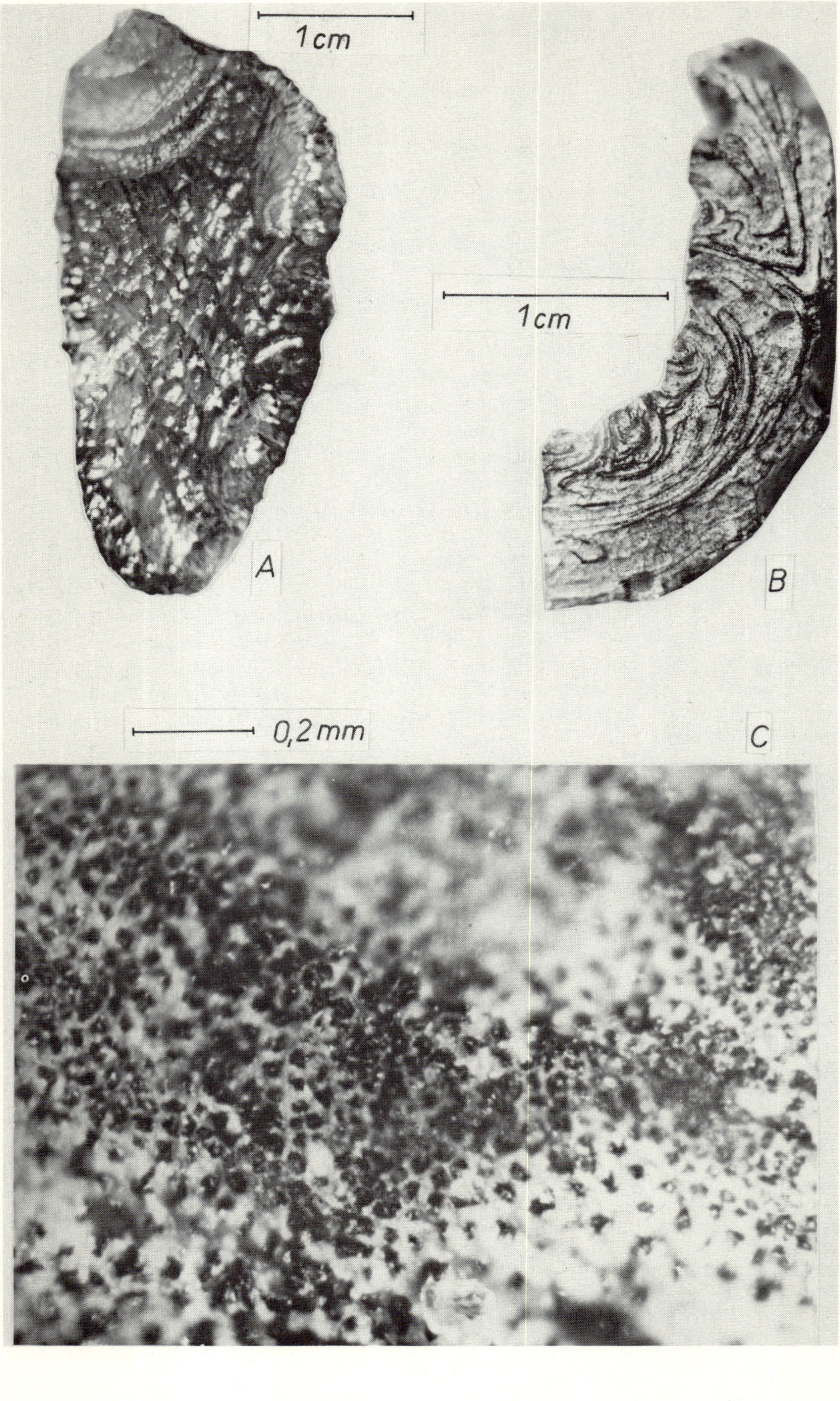
1 cm
A
1 cm
B
0,2 mm
C

the Middle Eocene (Tortonian); *Malecha* [1968] regards them as Late Pliocene (Pontian). The problem of the geological age of moldavites has not yet been solved satisfactorily. As the boundary between the end of the Tertiary and the beginning of the Quaternary has recently been shifted from 1 m.y. to 2.5 to 3 m.y., the age of the moldavite-bearing sands is estimated to be between 4 and 15 m.y. This also represents a time range for the action of chemical corrosion upon moldavites in sediments occurring today as much as 10 meters beneath the present-day surface of the earth.

If the age of the moldavite-bearing sediments could be proved to be low, i.e., about 3–4 m.y., the less strongly etched fracture surfaces (e.g., those shown in Figure 2*A* and Figure 4*A*, *B*, as well as others) would significantly support the above age determination.

The difference in quality and sculpturing of moldavites from one site of occurrence (Figure 2*A*–*D*) can be explained by variable conditions of the individual localities as, for instance, a rapidly varying permeability of sands, presence of clayey intercalations, nests, or admixtures, presence of ferric compounds, varying degree of solidification of sands, etc. The agencies affecting the speed and progress of chemical corrosion are, however, much more numerous and have not so far been fully recognized. A considerable difference is perceptible, even to the naked eye, between the moldavites dug from sand pits and those recovered from the clay of the brickyard at Besednice in southern Bohemia. The moldavites from sand pits are fairly lustrous and show a deep sculpturing. The moldavites from the Besednice clay are rather dull and somewhat rough to the touch.

In examining the surface of sculptured moldavites under the microscope or under a strong magnifying glass, we can usually see very fine pits which cover the whole deep sculpturing. *Störr and Konta* [1964] called attention to a still more rare type of moldavite—with microscopic pyramids on the surface. I have seen such a pyramidal type on the whole surface of a matted moldavite found by J. Prokopec in a field at Bukovec near České Budějovice (Figure 4*B*). The latter is an angular splinter (spall), weight 5.5 g, dimensions 8 × 14 × 27 mm. It is so finely etched that the course of schlieren is distinctly visible. This is also supported by a light gray patina. In Figure 4*B*, *C* the course of the schlieren is shown. A similar schlieren pattern can be observed on the opposite surface of the moldavite (breadth 14 mm). The general etching of this moldavite was weak.

In my paper in 1964 I mentioned that obsidians embedded in loess in eastern Slovakia became deeply sculptured during approximately 5000 years. A further examination of the material from this area, however, revealed that their sculpturing (sculpture pattern) probably dates from the Tertiary, having already occurred at the primary localities. According to an oral communication by K. Zebera, the sculpturing of these obsidians was due to more intensive weathering during the Tertiary period. I assume, however, that the sculpturing in question originated early in the post-eruptive phase as a result of etching by hot post-magmatic solutions.

Problem of Micromoldavites

I did not succeed in finding micromoldavites in southern Bohemia, as moldavite-bearing sediments were mostly eroded away. However, another problem concerning southern Moravia should be mentioned. In 1950, minute glassy spherules and drops resembling microtektites in size, shape, translucency, and pellucid color were found by micropaleontologists in the 80-m-deep bores penetrating the sediments of southern Moravia in an area far from that where the Moravian moldavites occur. However, in the

Fig. 4. Surfaces of two moldavites from southern Bohemia. *A*, a moldavite excavated from the sand pit near Chlum nad Malší. Weight 5.5 g. The upper part of the moldavite is broken, showing a conchoidal fracture surface; it has a finer sculpturing than the main part of the moldavite. *B*, a fragment of moldavite found in a field near Bukovec. The fracture surface is slightly etched. The more resistant schlieren stand out distinctly. *C*, photomicrograph of the surface of moldavite *B* displaying a fine pyramidal sculpturing. In the photomicrograph, minute black pyramids jut out above the surrounding light coating (patina) which consists of settled secondary compounds derived from soil.

same year, the Czech micropaleontologist V. Pokorný established that the so-called micrometeorites from boreholes greatly resemble glassy spherules gushed forth by steam engines. According to *Pokorný* [1950], the drilling equipment transported on railway cars was contaminated by the slag spherules from steam engines and in this way the spherules got into the deep core samples. The discovery of microtektites brings this problem back to mind in spite of a slight difference between the geological age of the beds in southern Moravia yielding glassy micrometeorites and the age of the Moravian moldavite-bearing sediments. It should also be taken into account that a rather powerful superficial rhyolitic volcanism took place in central and eastern Slovakia and that very small, pellucid volcanic 'bomblets' from the eruption of Slovakian rhyolites could have been transported by high air currents to Moravia and into the flysch areas of Poland.

The photomicrographs of four spherules isolated in 1950 by V. Pokorný from the drill cores drawn up from a depth of 80 m are shown in Figure 2*G*. The morphological conformity with the microtektites described by *Glass and Heezen* [1967] is considerable. The parent rocks of the glassy spherules coming from a depth of 80 m in southern Moravia are approximately of the same geological age (Sarmatian, Pannonian).

Acknowledgments. I wish to express my thanks to Karel Zebera, Geological Survey of Czechoslovakia, Prague, for the kind loan of the moldavites shown in Figures 1 and 2*A*, to F. Maroušek from České Budějovice for the moldavites illustrated in Figure 1*B–D*, and to J. Prokopec for the moldavite shown in Figure 4*B*. I am also very much obliged to Vl. Pokorný of the Charles University, Prague, for the material for photograph *G* of Figure 2.

References

Bouška, V., and F. Maroušek, Příspěvek k poznání vzniku skulptace u vltavínů, *Časopis Národního Musea, Prague,* oddíl přírodovědný, in press, 1969.

Glass, B., Microtektites in deep-sea sediments, *Nature, 214*(5086), 372–374, 1967.

Glass, B., and B. C. Heezen, Tektites and geomagnetic reversals, *Nature, 214*(5086), 372, 1967.

Malecha, A., K zastoupení pliocénu v jihočeských pánvích, *Zprávy o geol. výzkumech v r. 1966,* vol. 1, pp. 305–308, ÚÚG-Academia, Prague, 1968.

Nininger H. H., and G. I. Huss, Tektites that were partially plastic after completion of surface sculpturing, *Science, 157*(3784), 61–62, 1967.

Pokorný, V., O původu skleněných a magnetických kuliček ve výplavech z vrtních vzorků, *Věstn. SGÚ ČSR, 25,* 77–80, 1950.

Rost, R., Surfaces of and inclusions in moldavites, *Geochim. Cosmochim. Acta, 28,* 931–936, 1964.

Rost, R., Příspěvek k morfologii vltavínů I, *Sb. Národ. Musea Praze, 22B*(5), 169–192, 1966*a*.

Rost, R., A Muong Nong-type moldavite from Lhenice in Bohemia, *Acta Univ. Carolinae, Geol., 4,* 235–242, 1966*b*.

Rost, R., Behaviour of moldavites in polarized light, *Acta Univ. Carolinae, Geol., 2,* 95–112, 1967.

Störr, M., and J. Konta, Petrographische Untersuchungen moldavit-führender Sedimente von Besednice in Südböhmen, *Chem. Erde, 23,* 259–278, 1964.

Zebera, K., Moldavite bearing sediments between Koroseky and Holkov in South Bohemia, *Věstn. Ústředního Ústavu Geol., 42,* 327–337, 1967.

(Received June 27, 1969.)

Correction: Dr. B. P. Glass, using an electron microprobe, ascertained that one among the spherules in Fig. 2G (page 260) is not of tektite composition. Also according to Glass, the index of refraction of this spherule is substantially higher than that of a tektite. (Written communication, 1970.)

33

Proceedings of the Second Lunar Science Conference, Vol. 1, pp. 909–921.
The M.I.T. Press, 1971.

Surface micrography of lunar fines compared with tektites and terrestrial volcanic analogs*

STANLEY V. MARGOLIS,† VIRGIL BARNES, PRESTON CLOUD, and RICHARD V. FISHER
Department of Geological Sciences, University of California, Santa Barbara, California 93106

(*Received* 20 *February* 1971; *accepted in revised form* 5 *April* 1971)

Abstract—Optical and scanning electron microscopy of the surfaces of lunar glassy objects and mineral grains returned by Apollo 11 and Apollo 12 exhibit a variety of micron-sized features indicating processes responsible for syn- and postdepositional modification. (1) Blebs of adhering glassy matter are believed to be produced by spattering of semi-molten particles. (2) Pitted and abraded surfaces probably result from postsolidification impact by small bypassing objects. (3) Dome-shaped blisters reflect solidification of outgassing bubble surfaces. (4) Circular depressions and microcraters include those caused by solidification concurrent with outgassing, as well as others with frothy glass surfaces and fractured haloes caused by high velocity impact by small projectiles. (5) Fractures are caused by impact and thermal effects. (6) Rims, flanges (CLOUD *et al.*, 1970), and flow structures in the glasses imply partial melting and shaping during high speed flight.

The unfractured surfaces of lunar glass, terrestrial volcanic glasses and the primary surfaces beneath the appressed flanges of australites all exhibit domal gas blisters bubble pits, ablation phenomena and perhaps spatter—indicative of exposure to similar melting, outgassing, and ablative processes on their respective planetary bodies.

INTRODUCTION

SAMPLES OF LUNAR DUST, microbreccia, and igneous rock returned by Apollo 11 and Apollo 12 were examined by optical and electron microscopy for evidence of past or present microorganisms, and for processes active in modifying rocks and minerals at the lunar surface. In addition, lunar glass found in the dust and microbreccia at these Apollo sites was compared with terrestrial volcanic glass and tektite glass to see what, if any, similarities and differences might be discerned. No evidence was found to support the present or past existence of life on the moon, a notion that now seems more improbable than ever. This report, therefore, focuses on comparisons and contrasts between lunar and terrestrial glasses, and on observations that bear on lunar surface processes.

SURFACE FEATURES OF LUNAR GLASSES COMPARED WITH TEKTITES

The superficial similarity between shaped glassy objects found in the lunar soil, and terrestrial microtektites has recently revived the interest of proponents of a lunar origin for tektites. Chemical data, however, reveal great differences between tektite and lunar glasses (GOPALAN *et al.*, 1970; TATSUMOTO, 1970; PHILPOTTS and SCHNETZLER, 1970; TAYLOR and EPSTEIN, 1970; FRONDEL *et al.*, 1970; KEIL *et al.*,

* Contribution No. 17 from the Biogeology Clean Laboratory Department of Geological Sciences University of California Santa Barbara California 93106.

† Present address: Department of Geology University of Georgia, Athens, Georgia 30601.

1970; FREY *et al.*, 1970). Comparative surface micromorphology offers an independent basis for comparison.

Morphologic similarities between lunar glass spheres and microtektites extend beyond shape and size range (CLOUD *et al.*, 1970). Some lunar spheres, when examined with a scanning electron microscope (SEM), exhibit a surface that is covered with a fine spattering of glassy material of varying composition and shape (Fig. 1). Optical examination and electron microprobe study has shown that this spatter ranges from glass resembling melted lunar dust, to material extremely rich in iron, possibly derived from meteorites (CARTER and MACGREGOR, 1970). Other features found on these glass surfaces are hollow dome shaped blisters (Fig. 2) caused by outgassing as a result of melting, flow structure, and surfaces with irregular small-scale fracture plates (Figs. 3 and 4) similar to those found on eolian sand grains and described by KRINSLEY and MARGOLIS (1969) as being the result of crack propagation from low velocity impacts.

Surface textures presently found on tektites do not represent their original pre-weathered morphology. All tektites have been subjected to varying degrees of chemical weathering and are covered with etch pits (Fig. 5) that in part reflect internal variations in composition (STELZNER, 1893; BARNES, 1940). Fractured tektites, when etched for one hour in a 20% solution of HF, produced pits virtually identical to those found on normally weathered tektite surfaces (Fig. 6).

The only place where the original tektite surface is preserved is beneath flanges welded to the bodies of many australites (BARNES, 1962). These flanges are appressed to the margin of the tektite body, thereby sealing and protecting the original surface and crust from weathering. Such flanges, accepted by all as evidence of secondary melting during high-speed flight through an atmosphere, are taken by some as evidence for an extraterrestrial origin. Others, however, have suggested that the flanges may have formed as the australites plunged back through the terrestrial atmosphere after first being ejected from the earth and partially orbiting the globe (BARNES, 1961).

The boundary between the flange and the body of australites was examined optically in millimeter-thick polished plates cut across it for study. The inhomogeneous crust-like layer beneath the australite flange was found to be composed of glass having a wide range of refractive index. It includes areas of high-index glass with diffuse borders, and numerous sharply bounded hemispherical bodies of siliceous low-index glass (Figs. 7 and 8). The wide range in refractive index of crustal components and

---

Fig. 1. Amber sphere from 12001 lunar dust sample. Sphere is covered with iron-rich spatter of varying morphologies. Dark circular pits are the result of outgassing. Fig. 2. Surface of sphere from 12001 lunar dust showing outgassing blisters and craters from burst blisters. Fig. 3. Dark gray glassy sphere embedded in microbreccia of lunar rock 12034. Fig. 4. Surface of sphere shown in Fig. 3 showing crack-propagation plates similar to those found on eolian quartz-sand grains. Fig. 5. Surface of indochinite from Chaumpae, Thailand, showing natural etched surface. Fig. 6. Surface of indochinite from same locality as Fig. 5, after fracturing and etching with a 20% solution of HF for 1 hour. All photographs except Figs. 7 and 8 are made with the electron scanning microscope.

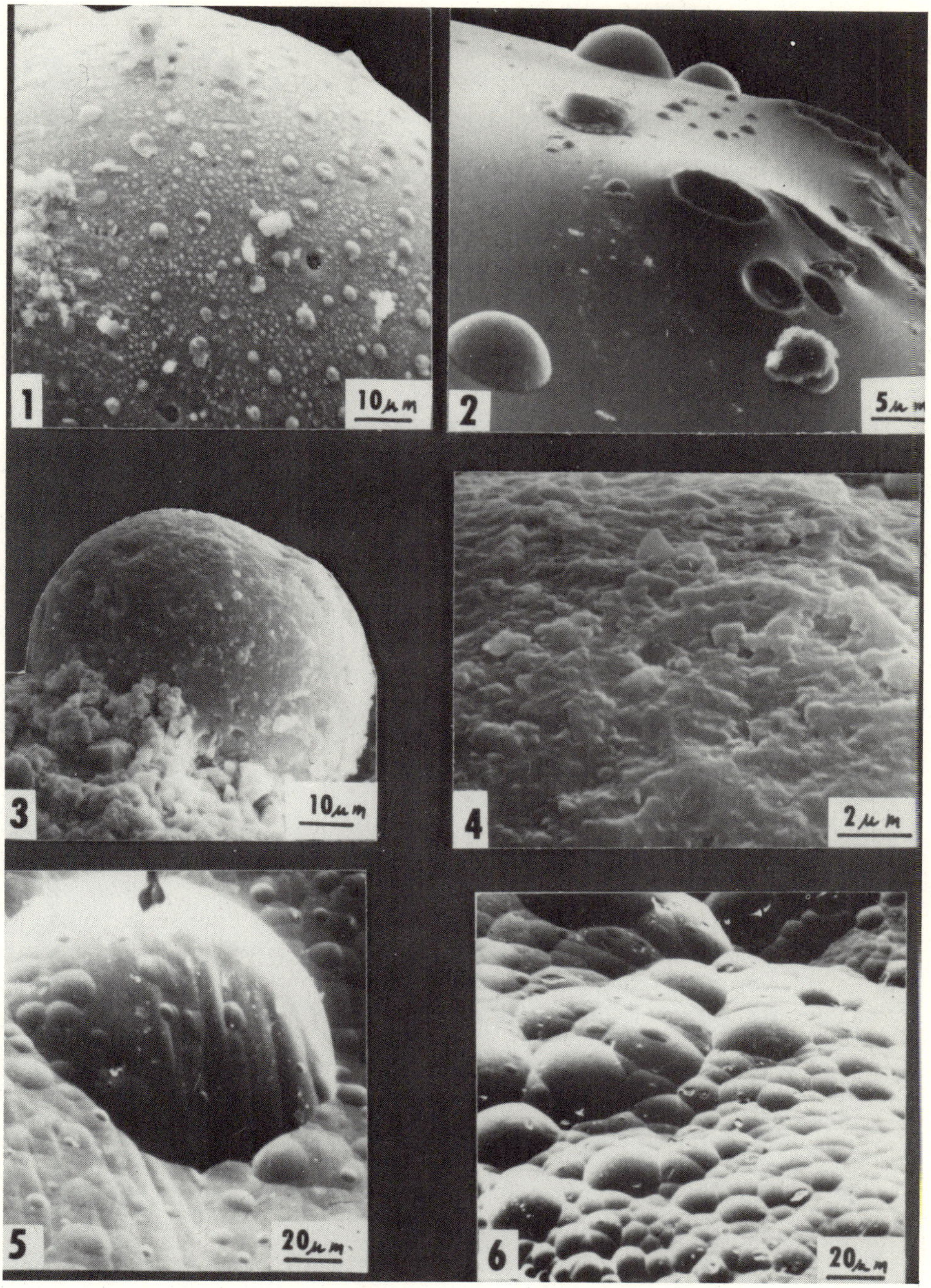
1
10μm
2
5μm
3
10μm
4
2μm
5
20μm
6
20μm

strong internal strain shows that the original tektite crust varied considerably in composition, and that it is much less homogeneous than the body of the australite. During weathering, the high-index, less siliceous glass in the original crust, would have dissolved faster than the low-index, more siliceous glass, thus accounting for the more intricate patterns of surface etching observed in naturally weathered tektites as contrasted with the artificially etched internal portions.

On some australites the flange was fairly easily detached from the body, and optical examination showed that the surface of detachment was coincident with the original outer surface of the tektite. Examination of such surfaces with the SEM showed that they (Fig. 9) are locally covered with minute blebs of material that display an apparent orientation, as well as with circular bubble-shaped depressions. These blebs are about an order of magnitude smaller, and of a different shape, than the hemispheres identified using optical methods. Also, the flat surfaces of the hemispheres coincide with the outer surface of the australite, and are neither raised nor depressed, while the minute blebs are raised above the surface. It is possible that these blebs may be artifacts, produced by detaching the flange from the body. If not, they can be likened to the spatter found on the spheres of lunar glass (Fig. 10). They could have been formed, as the australite plowed back into the atmosphere, from a spray of droplets of molten glass produced by frictional melting. The orientation of the blebs (Fig. 9) could have been produced by such a spray of droplets sweeping across its surface.

Impacting blebs of material of varying composition may perhaps explain the inhomogeneities in the original tektite surfaces. The larger hemispheres could have resulted from viscous droplets which impacted viscous tektites, and interaction of the surface tensions of the two melts could have drawn the drops into hemispheres with their flat sides parallel to the surface of the australite.

To check this hypothesis, we examined natural spheres of tektite-glass having ablation flanges that were produced experimentally by a hypervelocity arc-jet (Chapman and Larson, 1963, Sample P–1192). Examination of the ring-wave flow ridges on the circumferential flange with a scanning electron microscope (Fig. 11) reveals that the crests of the ridges exhibit rough surfaces formed by flowing and ablation under the hypervelocity air blast. They also show outgas bubbles due to the frictional heating of the thin outer layer.

Fig. 7. Light-micrograph of contact between flange and body of australite from Beltana, Australia. Flange is in lower portion. Note hemispherical-shaped inclusions in contact with surface, and "fingers" of higher index glass. Fig. 8. Higher magnification optical photograph of central portion of Fig. 7 showing hemispheres. Fig. 9. Surface of australite, beneath flange, showing spatter-like blebs of material. Fig. 10. Higher magnification image of Fig. 1 showing details of spatter. Some of the smaller iron-rich blebs may have exsolved from the sphere during cooling. Fig. 11. Surface of simulated flanged tektite (Chapman and Larson, 1963, Sample P-1192) produced by subjecting phillipinite glass sphere to hypervelocity arc-jet. Note rough surface at crest of waves, produced by ablation. Direction of air flow was from top of photo to bottom. Fig. 12. Dumbbell-shaped glassy object on surface of sphere from lunar dust sample 12042. Note raised rims around circular pits, probably outgassing structures.

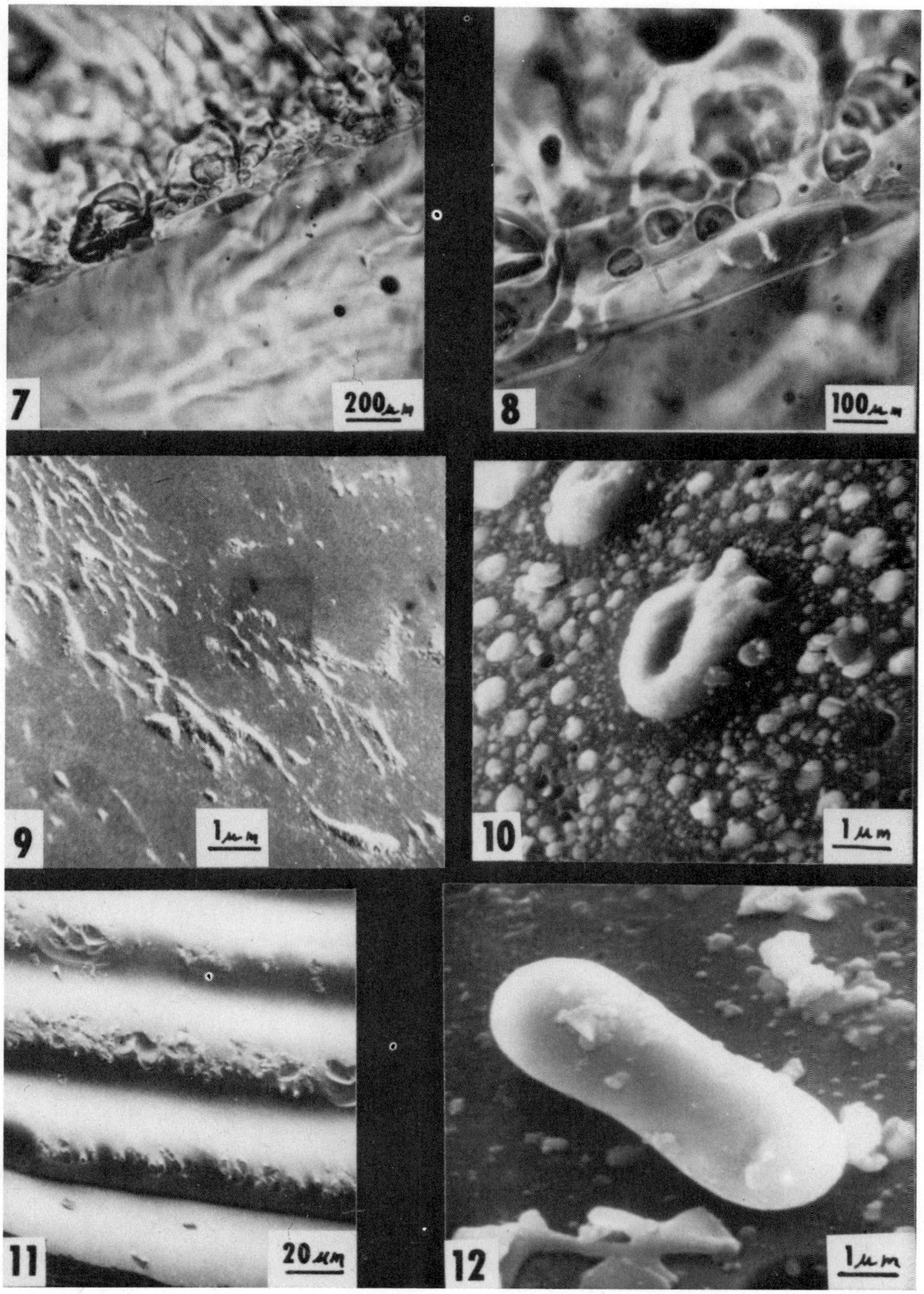
7
200 μm
8
100 μm
9
1 μm
10
1 μm
11
20 μm
12
1 μm

When the experimentally formed flange was detached from the body, however, no observable spatter was found on the contact, although high concentrations of bubble pits were found at the surface of the body. The absence of spatter on the surface beneath the experimentally produced flange casts some doubt on the interpretation of the blebs in Fig. 9 as spatter, despite the potential source in flow-wave ablation. Further work, therefore, is needed to resolve this question.

Morphologic similarities between tektites and shaped particles of lunar glass prove only that similar mechanical and chemical processes were responsible for their initial formation. The cometary impact theory (UREY, 1957; BARNES, 1961) could explain both the origin of tektites on the earth, and shaped glassy objects on the moon, if cometary (or other meteoritic) bodies impacting on each planet could melt and fuse local surficial material, and eject it as spray which quickly cools as variously shaped glassy objects.

LUNAR SPHERES COMPARED WITH TERRESTRIAL VOLCANIC GLASSES

Pele's tears and other microlapilli from basaltic volcanic glasses can assume the same shapes and size distribution as lunar glasses and microtektites (Figs. 12–15). The surfaces of Pele's tears also show hollow dome-shaped blisters and spherical bubble pits from outgassing (Fig. 14). Spheres produced by melting lunar dust in a vacuum of 10^{-7} torr at between 1200 and 1400°C (CLOUD *et al.*, 1970) show similar dome-shaped blisters. Other volcanic spheres show flow structures and spatter as well (Figs. 15–16). When basaltic volcanic glass is weathered in terrestrial or marine environments, it assumes the same etched surface sculpture as glass spheres from deep-sea cores identified by GLASS (1968) as microtektites (Figs. 13, 14, and 17). Vesicles within the glass are exposed by the etching (Fig. 14). In questionable cases, however, chemical analysis must be performed in order to distinguish microtektites from etched spheres of volcanic glass—but refractive indices give sufficient indication of geochemistry for differentiation in most instances. Glass spheres and grains from volcanic base-surge deposits also show spatter and impact abrasion features similar to those on lunar spheres, although hypervelocity pitting is lacking (Fig. 18). The lunar dust contains pieces of pumiceous and scoriaceous glass shards similar to terrestrial pyroclastics (Figs. 19 and 20).

Fig. 13. Glass sphere from deep-sea core near the Ivory Coast described by GLASS (1968). Surface sculpture is caused by chemical etching in sea-water. Object identified by Glass as a microtektite based on chemical similarity to Ivory Coast tektites. Note morphologic similarity to Fig. 14. Fig. 14. Volcanic basaltic glass sphere from North-east Pacific deep-sea sediments of Miocene age collected from a depth of 28 cm in core no. 10 from Site 32 on Leg 5 of the Deep Sea Drilling Program (VON DER BORCH, in press). Note etched surface sculpture and vesicle. Fig. 15. Basaltic glass sphere from same locality as Fig. 14. Curiously this sphere is only slightly etched and has retained its original surface with flow lines and spatter. Fig. 16. Surface of basaltic volcanic sphere from same locality as Fig. 14, showing spattered, slightly etched surface. Spatter is similar to Fig. 1. Fig. 17. Etched, teardrop-shaped, basaltic glass object from same locality as Fig. 14. Fig. 18. Dumbbell-shaped Pele's tear from recent eruption of Kilauea, Hawaii. Note vesicles on surface due to outgassing.

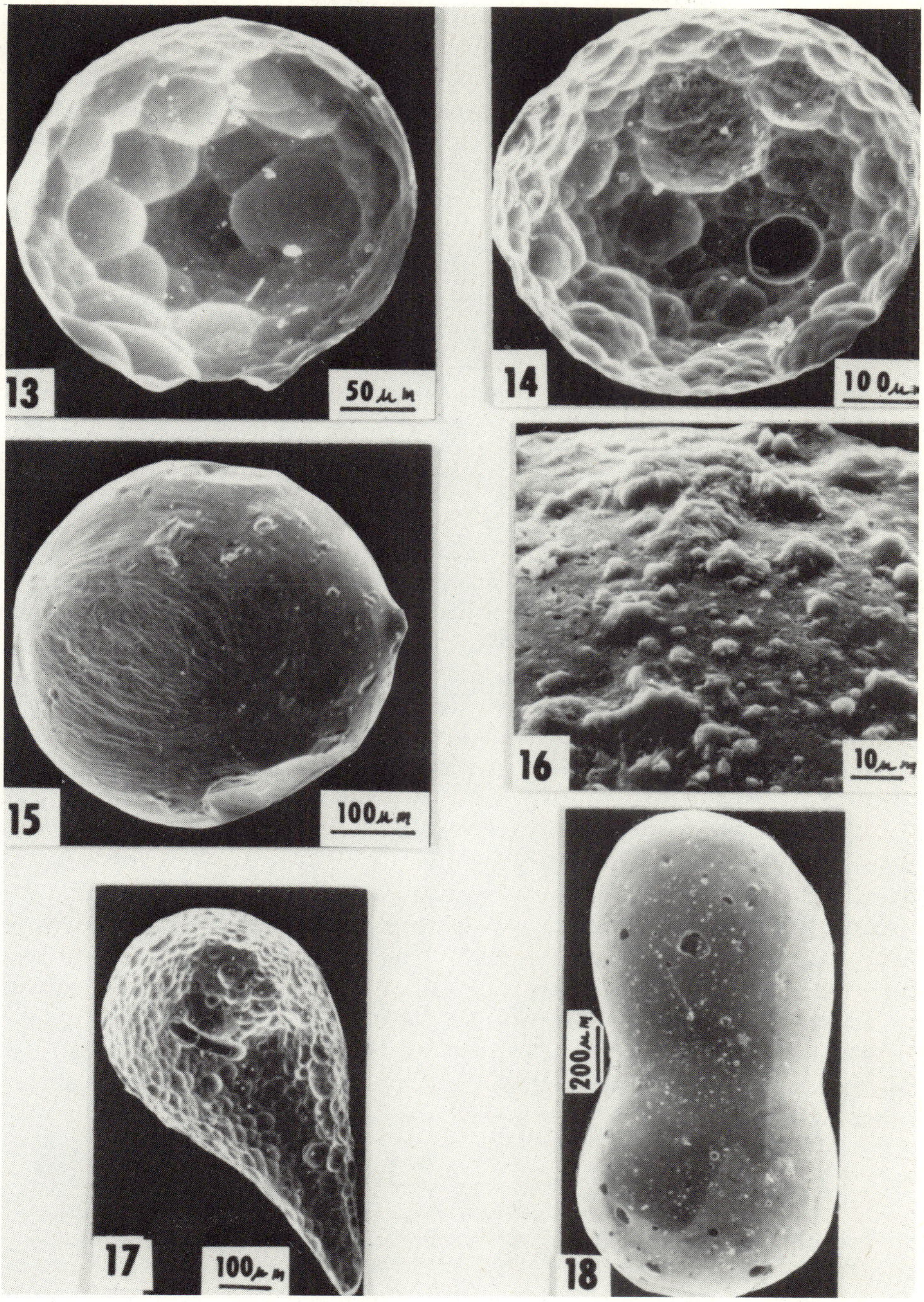
13
50 μm
14
100 μm
15
100 μm
16
10 μm
17
100 μm
18
200 μm

Surface features on lunar glasses are similar enough to features of terrestrial pyroclastic fragments to suggest that a volcanic origin is possible for some components of the lunar regolith, although a means of distinguishing between impact melted and volcanically derived components based on surface micromorphology alone has not as yet been established. Petrographic data presented by many investigators, however, strongly favors an impact origin for most lunar glasses.

HYPERVELOCITY IMPACT PITS ON LUNAR PARTICLES

As reported by several other authors, external surfaces of lunar rocks and mineral fragments contain numerous hypervelocity impact pits or craters that are glass lined, and vary in morphology, diameter, and surface density. The morphology of the pits and the nature of their surrounding fracture halo seems to be a function of the composition, size, and velocity of the impacting projectile, as well as the nature of the object being hit. Typical impact pits on lunar microbreccia surfaces are lined with frothy glass (CLOUD *et al.*, 1970, Fig. 3J), while pits on glass surfaces and igneous rocks have smooth linings and highly fractured haloes which extend for a distance several times that of the crater diameter (Fig. 21). Blebs of the lining material are often spattered onto the fractured margins (Fig. 22). Other impact pits lack glass linings and contain micropulverized centers surrounded by fractures (Fig. 23). Fractured haloes are often small or missing on pits on the microbreccia, and the glazed central portion of the pit is raised above the surface of the rock.

A possible explanation for this condition could be that the impact spalled off the surrounding matrix, leaving behind only the glass-welded portion. Alternatively, the highly fractured halo could have been removed after the impact by long-term lunar weathering, leaving the more resistant glazed central portion behind. The weaker intergranular bonding forces in the microbreccia could explain the differences in the fracture haloes in these rock types. Some small impact pits are occasionally superimposed on the glass lining of a larger pit, indicating several episodes of impacting. Overlapping and closely spaced craters occur more commonly on the surface of igneous rocks than on microbreccia, possibly because of the tendency of their fracture haloes to spall. A new impact pit, therefore, would tend to destroy a previously existing one close by, resulting in an equilibrium density of craters after an unknown but probably long time interval (HARTUNG *et al.*, 1970). The amount of material added to the rock or mineral surface by the hypervelocity impact event is far exceeded by the material removed by the fracturing, so that impacts tend to reduce the size of rocks on the lunar surface.

Fig. 19. Pumice-like fragment from light-gray lunar dust sample 12033, which has been described as a vitric ash. Fig. 20. Vesicular, scoriaceous glass with flow structure from lunar sample 12033. Fig. 21. Microcrater with fracture-halo superimposed on the margin of a larger glazed impact pit on the surface of lunar sample 12018 type B, a medium-grained crystalline rock. Fig. 22. Details of Fig. 21 showing spatter or fractured margin of pit. Fig. 23. Microcrater on lunar sample 12018 showing micropulverized central portion. Fig. 24. Microfracture on mineral grain from external surface of 12063 microgabbro.

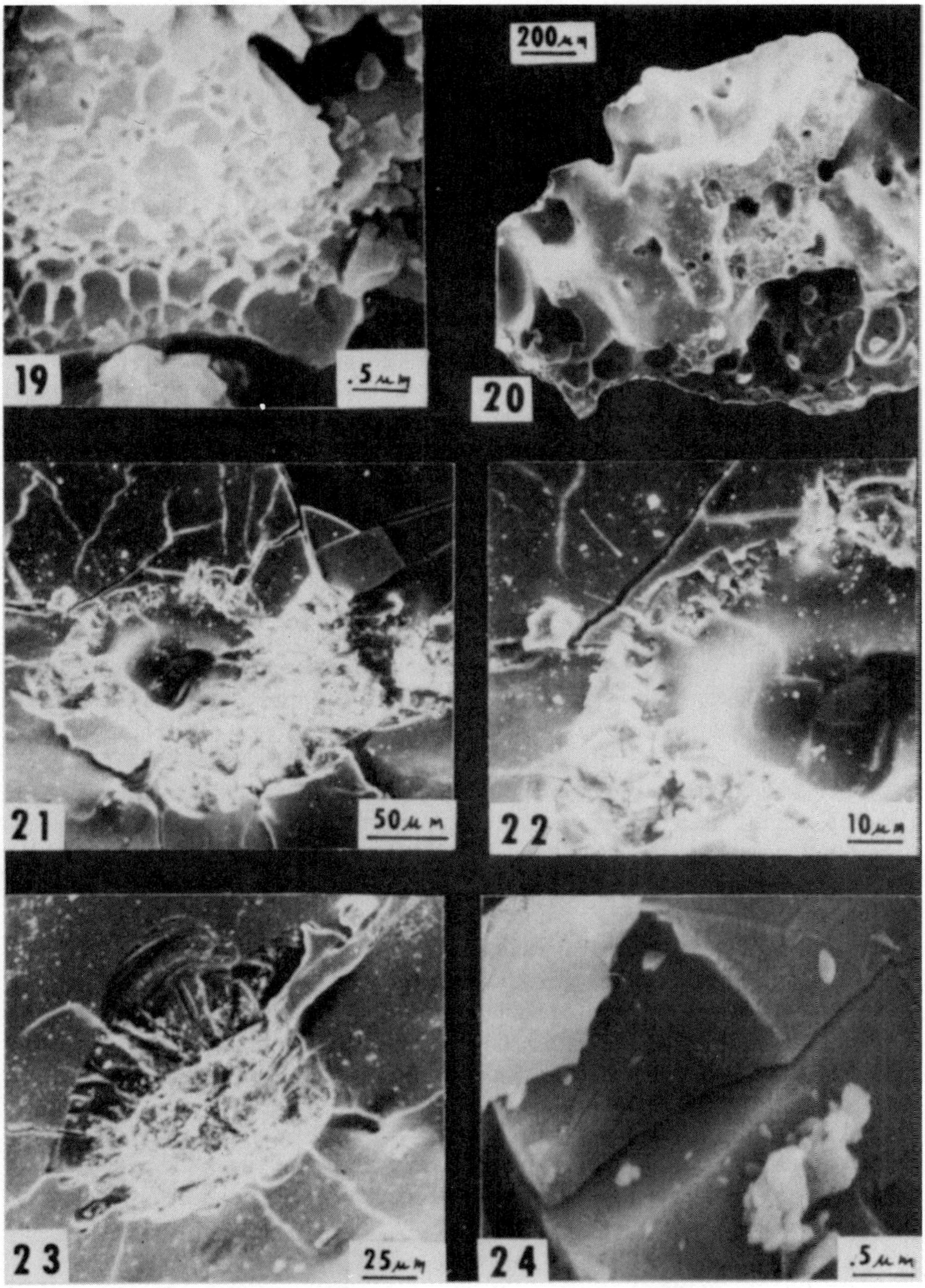
19
.5 μm
200 μm
20
21
50 μm
22
10 μm
23
25 μm
24
.5 μm

Fractures are also found within the margins of the glazed pits of lunar samples (Fig. 22), possibly due to contraction upon cooling from the molten state.

Thermal Fractures

Microfractures not associated with any impact pits were also found on the surfaces of microbreccia and igneous rocks (Figs. 24–27). It is hypothesized that this fracturing was produced by thermal change.

Simple experiments were performed to check this hypothesis. Samples of terrestrial basaltic glass and minerals similar to those observed in the Apollo 11 and Apollo 12 samples (augite, ilmenite, bytownite feldspar) were first broken out of larger samples and examined with the SEM. These fragments were then heated with a bunsen burner and quenched by immersion in liquid nitrogen through eight cycles to approximate the lunar thermal cycle. Although there was no obvious optical difference before and after this treatment, striking physical changes were observed on examination with the SEM. The plagioclase feldspar was the most severely affected by this treatment (Fig. 28). For example, after repeated heating and cooling, the previously smooth surface between cleavage fractures of the feldspar broke into many small subrectangular plates separated by cracks with displacements in some areas. Where such plates were plucked out, a conchoidal fracture was seen on the fresh surface below (Fig. 28). The ilmenite and augite were similarly affected although not as severely as the feldspar. They splintered into small prisms and plates (Figs. 29 and 30) reflecting their characteristic cleavages. This pattern was not found anywhere on the original fracture surface before heating and cooling. Similar patterns were found on mineral grains of like composition within the lunar dust and in the matrix of the lunar breccia.

These experiments, to be sure, do not duplicate lunar surface conditions. They were performed in the terrestrial atmosphere and frost which formed on the minerals after removal from the liquid nitrogen accentuated the shattering process. The range of temperature was also different (moon $+130°$ to $-153°C$, experiment $+300°$ to $-195°C$) and the heat was conductive rather than radiative. Furthermore, materials on the lunar surface gradually go from maximum to minimum temperature, and are not quenched as these in the experiments. Nevertheless, the observed analogies suggest that differential thermal expansion could possibly account for some or much lunar surface weathering. Crystal breakage and disruption due to day-to-night temperature variation may well contribute to the production of the minute angular mineral grains and glass particles in the dust, and to slow inplace rock disintegration.

Summary

Surface features on lunar glassy spheres are generally similar to those found on terrestrial pyroclastics and also to those on the original outer surface of australite

---

Fig. 25. Microfracturing on augite grain from external surface of 10091 lunar microbreccia. Fig. 26. Fractured surface of ilmenite grain picked from lunar dust sample 12001. Fig. 27. Fractured surface of feldspar grain picked from lunar dust sample 12001. Fig. 28. Grain of feldspar (bytownite) after similar heating and cooling (Fig. 29). Fig. 29. Grain of augite after 8 cycles of alternate heating and cooling ($+300°C$ to $-195°C$). Fig. 30. Grain of ilmenite after similar treatment to that described in Fig. 29.

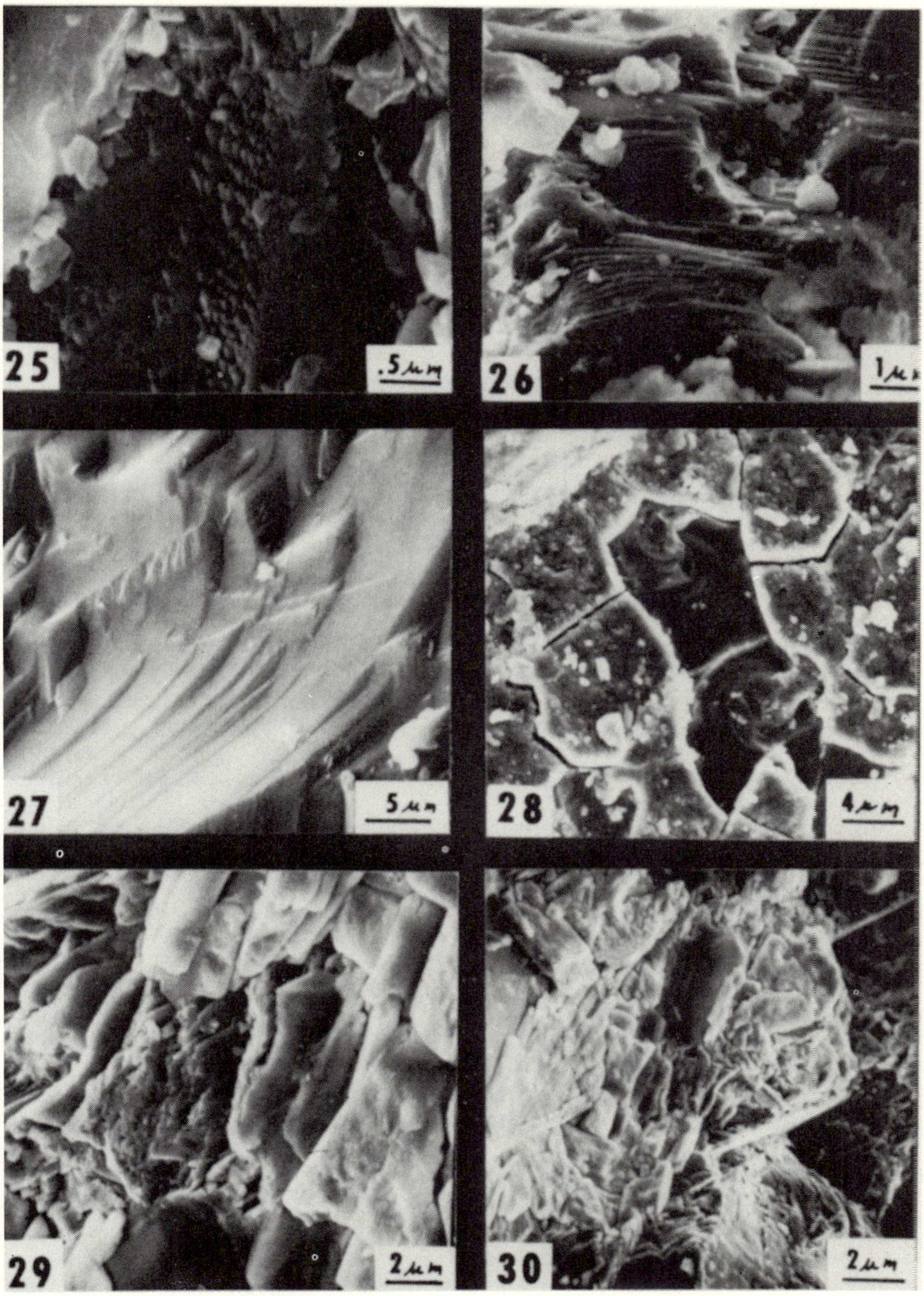
25
.5 µm
26
1 µm
27
5 µm
28
4 µm
29
2 µm
30
2 µm

tektites, preserved beneath flanges. Evidence of destructive lunar degradation processes include glass-lined and spallation hypervelocity impact pits caused by micrometeorite or cometary impact, microfracturing caused by diurnal temperature changes, and abrasional rounding on rocks and mineral grains, caused perhaps by a type of low-velocity "sand blasting" by clouds of fine-grained material set in motion either by large scale meteor impacts, or volcanic activity.

Acknowledgments—Supported by NASA Contract No. 9-7882 for major equipment services, and travel, NASA Grant No. NGR 05-010-035 for collateral operations, University of California for facilities and partial services, NSF Grant No. GB-7851 for equipment and general support of the Biogeology Clean Laboratory, and NSF Grant No. GA-914 for tektite research. We thank Dr. Dean Chapman, Ames Research Center, for the loan of the arc-jet ablated natural tektite, Dr. Billy Glass for loan of Ivory Coast microtektites, Dr. Brian Mason, U.S. National Museum, for flanged australites, and Dr. Chris Von der Borch for volcanic microlapilli.

REFERENCES

BARNES V. E. (1940) North American Tektites. Univ. of Texas Publ. 3945, 447–582.

BARNES V. E. (1961) Tektites. *Sci. Amer.* Nov., 58–65.

BARNES V. E. (1962) Tektites: Original surfaces preserved beneath flanges, rayed bubbles, and oriented strain. *Trans. Amer. Geophys. Union* **43,** 466.

CARTER J. L. and MACGREGOR I. (1970) Mineralogy, petrology, and surface features of some Apollo 11 samples. *Proc. Apollo 11 Lunar Sci. Conf., Geochim. Cosmochim. Acta* Suppl. 1, Vol. 1, pp. 247–265. Pergamon.

CHAPMAN D. R. and LARSON H. K. (1963) On the lunar origin of tektites. *J. Geophys. Res.* **68,** 4305–4358.

CLOUD P. E., MARGOLIS S. V., MOORMAN M., BARKER J. M., LICARI G. R., KRINSLEY D., and BARNES V. E. (1970) Micromorphology and surface characteristics of lunar dust and breccia. *Science* **167,** 776–778.

FREY F. A., SPOONER C. M., and BAEDECKER P. A. (1970) Microtektites and tektites: A chemical comparison. *Science* **170,** 845–847.

FRONDEL C., KLEIN C., JR., ITO J., and DRAKE J. C. (1970) Mineralogical and chemical studies of Apollo 11 lunar fines and selected rocks. *Proc. Apollo 11 Lunar Sci. Conf., Geochim. Cosmochim. Acta* Suppl. 1, Vol. 1, pp. 445–474. Pergamon.

GLASS B. P. (1968) Glassy objects (microtektites?) from deep-sea sediments near Ivory Coast. *Science* **161,** 891–893.

GOPALAN K., KAUSHAL S., LEE-HU C., and WETHERILL G. W. (1970) Rb–Sr and U, Th–Pb ages of lunar materials. *Proc. Apollo 11 Lunar Sci. Conf., Geochim. Cosmochim. Acta* Suppl. 1, Vol. 2, pp. 1195–1205. Pergamon.

HARTUNG J. B., HORZ F., and GAULT D. (typescript, 1970) Microcraters on lunar rock surfaces. NASA Publ.

KEIL K., BUNCH T. E., and PRINZ M. (1970) Mineralogy and composition of Apollo 11 lunar samples. *Proc. Apollo 11 Lunar Sci. Conf., Geochim. Cosmochim. Acta* Suppl. 1, Vol. 1, pp. 561–598. Pergamon.

KRINSLEY D. and MARGOLIS S. V. (1969) A study of quartz sand grain surface textures with a scanning electron microscope. *N.Y. Acad. Sci. Trans.* Ser. 11, **31**, no. 5, 457–477.

MARGOLIS S. V. (1968) Electron microscopy of chemical solution and mechanical abrasion features on quartz sand grains. *Sediment Geology* **2,** 243–256.

NININGER H. H. and HUSS G. I. (1967) Tektites that were partially plastic after completion of surface sculpturing. *Science* **157,** 61–62.

O'KEEFE J. A. and SCHUTE B. E. (1963) Origin of tektites. *Science* **139,** 1288–1290.

PHILPOTTS J. A. and SCHNETZLER C. C. (1970) Apollo 11 lunar samples: K, Rb, Sr, Ba, and rare-earth concentrations in some rocks and separated phases. *Proc. Apollo 11 Lunar Sci. Conf., Geochim. Cosmochim. Acta* Suppl. 1, Vol. 2, pp. 1471–1486. Pergamon.

STELZNER A. W. (1893) Uebereigenthumliche Obsidianbomben aus Australien. *Deutsche geol. Gesell. Z.* **45,** 299.

TATSUMOTO M. (1970) Age of the moon: An isotopic study of U–Th–Pb systematics of Apollo 11 lunar samples. II. *Proc. Apollo 11 Lunar Sci. Conf., Geochim. Cosmochim. Acta* Suppl. 1, Vol. 2, pp. 1595–1612. Pergamon.

TAYLOR H. P., JR., and EPSTEIN S. (1970) O^{18}/O^{16} ratios of Apollo 11 lunar rocks and minerals. *Proc. Apollo 11 Lunar Sci. Conf., Geochim. Cosmochim. Acta* Suppl. 1, Vol. 2, pp. 1613–1626. Pergamon.

UREY H. C. (1957) Origin of tektites. *Nature* **179,** 556–557.

VON DER BORCH C. C. (1971) *Basaltic Glass Spherules in North-East Pacific Deep Sea Drilling Cores.* Marine Geology (in press).

Microtektites

V

Editors' Comments on Papers 34–37

The discovery of microtektites in deep-sea cores in the Australasian area of tektite occurrence, described by Glass (1967) and Glass and Heezen (1967), has added another segment to the puzzle of tektite origin. The story of microtektites is a fascinating one, and as more is learned about them and their distribution the last pieces of the tektite puzzle may fall into place. Four papers have been chosen to cover present knowledge of microtektites: where and how they occur, their physical and chemical properties, and a comparison of their range in composition with the range in composition of one australite.

The discovery of microtektites in association with the Ivory Coast strewn-field, described in the first paper, following rapidly after the original discovery of microtektites, indicates that it will be only a matter of time before microtektites related to other strewn-fields will be discovered in ocean sediments. Indeed, Storzer and Gentner (1970) have already reported micromoldavites, but in Bavarian molasse, not in ocean sediments.

The bottle-green objects of low silica and high magnesia content found by Cassidy, Glass, and Heezen* associated with normal microtektites add still another dimension to the tektite puzzle. These objects do not correspond in composition to any known stony meteorite, lunar sample, or rock outcrop on earth; and if from the earth, probably represent rather deep-seated mantle material. Because of space limitations and the fact that this article is in a readily available journal, the section on "Discussion of Data" and "Table 5. Description of Analyzed Samples" are omitted. In the "Discussion of Data" numerous variation diagrams show that the normal microtektites and tektites form a continuous compositional series if the bottle-green "microtektites" are excluded. The authors also make a conservative estimation that the total mass of microtektite material present in the strewn field amounts to 10^{10} g which is equivalent to a sphere about 200 m in diameter.

The chemical data in the third paper by Glass,* confirms the relationship between Ivory Coast tektites and normal microtektites recovered from a core off the Ivory Coast. Again, as in the Australasian area, bottle-green "microtektites" are associated with normal microtektites. If Bosumtwi Crater is entirely within the kind of rock exposed at the surface, which is likely, no source for the bottle-green specimens would have been available, as Bosumtwi Crater is far too shallow to reach the mantle. Additional studies should be made to determine if these bottle-green objects are artifacts, or if they actually do represent mantle material from some locality other than Bosumtwi Crater.

The fourth paper, by Glass* is important not only in showing that the range of microtektite composition is similar to that found in one australite flange, but indirectly in giving analytical data concerning the inhomogeneous surficial layer of australites preserved beneath flanges, pictured by Margolis et al. (1971, Figs. 7 and 8).* The character of this surficial layer was first described by Barnes (1962); however, Baker (1944, p. 13) previously had noted "hemispherical bubble pits in the surfaces of australites beneath flanges filled with flange glass." These features described by Baker are probably chiefly the low-refractive-index glass hemispheres of Margolis et al.* Wisps and ghostly forms of high-refractive-index glass are interspersed with the hemispheres. In general this thin layer, preserved in its entire thickness only beneath flanges, is far more inhomogeneous than the body of the tektite as a whole. In examination of plates cut across bodies of australites and their flanges, it is seen that a thin layer of tektite softens and moves laterally beneath the surface of ablation and that the more highly siliceous inhomogeneities remain as distinct particles as they become incorporated in the flange; also that the low-silica material is drawn out in streaks. The silica-rich and low-silica intrusions within the flange, analyzed by Glass, probably represent such material.

We interpreted the significance of this surficial layer of australites in the discussion of the paper by Margolis et al.* We had hoped to have analytical data to support the interpretation; however, this paper by Glass* furnishes two relevant analyses giving some basis for the interpretation. In 1964 and again in 1969, Corning Glass Works offered to make electron microprobe analyses of these inhomogeneities beneath flanges. The first group of specimens was lost, and as yet no results from the second group of specimens have been received.

Dr. Billy P. Glass, as previously noted, is now with the Department of Geology, University of Delaware, Newark, Delaware 19711.

References

Baker, G. (1944). Flanges of australites (tektites). *Mem. Natl. Museum Victoria, Melbourne,* No. 14, Part. 1, pp. 7–22.

Barnes, V. E. (1962). Tektites—original surfaces preserved beneath flanges, rayed bubbles, and reoriented strain (abstract). *Trans. Amer. Geophys. Union,* **43,** 466.

Glass, B. P. (1967). Microtektites in deep-sea sediments. *Nature,* **214,** 372–374.

Glass, B. P., and Heezen, B. C. (1967). Tektites and geomagnetic reversals. *Nature,* **214,** 372; also, *Scient. Amer.* **217,** 32–38.
*Margolis, S. V., Barnes, V., Cloud, P., and Fisher, R. V. (1971). Surface micrography of lunar fines compared with tektites and terrestrial volcanic analogs. Proc. Second Lunar Sci. Conf., Vol. 1, pp. 909–921, M. I. T. Press, Cambridge, Massachussetts.
Storzer, D., and Gentner, W. (1970). Micromoldavites from the Bavarian molasse (abstract). *Meteoritics,* **5,** 225.

Reprinted from
30 August 1968, Volume 161, pp. 891-893

SCIENCE

Glassy Objects (Microtektites?) from Deep-Sea Sediments near the Ivory Coast

Abstract. *Glassy objects of spherical, oval, dumbbell, teardrop, and irregular shapes have been found in a deep-sea sediment core taken off the Ivory Coast. They occur in a layer of sediment that was apparently deposited about 800,000 years ago. Their geographic location, appearance, and physical properties suggest that they are microtektites and that they are related to the tektite-strewn field of the Ivory Coast.*

Several hundred small (<1 mm) glassy objects have been found in deep-sea sediments from north of the Philippines, south of Sumatra and Australia, and southeast of Madagascar. On the basis of their geographic location, age of deposition, and physical properties, it has been suggested that these glassy objects are microtektites and that they represent a portion of the Australasian tektite-strewn field (*1*). The chemical composition of 60 of the glassy objects was determined by electron-microprobe analysis at the Goddard Space Flight Center (*2*). In general, they appear to be quite similar in their compositional trends to Southeast Asian and Australian tektites.

Recently, a search was undertaken to see if any glassy objects could be found in the deep-sea sediments near the Ivory Coast of Africa. A few hundred glassy objects were found in one core (V19-297) taken at 2°37′N and 12°00′ W. The appearance and physical properties of these glassy objects are similar to that of the Australasian microtektites.

Out of the 277 glassy objects recovered, approximately 90 percent are spherical in shape, and several hollow hemispheres were found. Most of the remainder are either oval or disc- or dumbbell-shaped (Fig. 1). Only two teardrop-shaped objects were found. In addition, fragments are very abundant. Like the Australasian microtektites, these glassy objects are mostly less than 1 mm in diameter; however, a few dumbbell-shaped objects are as long as 1.2 mm. They are brittle and break with a conchoidal fracture.

In transmitted light most (~70 percent) of the glassy objects are olive brown to olive green in color. A few are colorless, and the remainder are yellowish-green or dark brown to black. Thus, the color of these glassy objects is in contrast to the Australasian microtektites which are mostly yellowish-brown (from the Wharton Basin) or colorless to yellowish-green; however, they are similar in color to the Ivory Coast tektites which, according to Vand (*3*), are brownish in thin section.

The surface texture of the glassy objects varies from glassy smooth to very irregular and badly pitted. Unlike the Australasian microtektites, these glassy objects appear to have undergone extensive solution which has resulted in surface markings that are not observed on the Australasian microtektites. Elongated bubble or solution pits, grooves, and flow lines are among the most abundant surface markings (Fig. 1). Like the Australasian microtektites and the Ivory Coast tektites, the glassy objects contain spherical bubble cavities, but no crystalline inclusions. Several of the glassy objects have elevated circular areas. In some instances, these elevated areas are seen to be hemispherical in shape, with the rounded surface directed in toward the center of the glassy object, and are attached to a short pedestal. These caplike protrusions (Fig. 1) are probably the results of differential solution due to variations in silica content.

The refractive indices of 27 of the Ivory Coast glassy objects were determined. Three of them have rather high refractive indices ranging from 1.581 to 1.592. They are bottle green in color and have a deeply corroded and irregular, but smooth, surface and are thus similar to the bottle-green glassy objects with high refractive indices (1.582 to 1.620) that were found associated with the Australasian microtektites in two of the cores. The remainder of the glassy objects have a range in refractive index from 1.514 to 1.542 with an average of 1.525. In addition, one spherule has a refractive index of 1.461 ± 0.002. Electron-microprobe analysis of this spherule indicates that it is nearly pure SiO_2. Thus the spherule is probably a lechatelierite particle. Other lechatelierite particles were found as inclusions in several of the other glassy objects. The caplike protrusions found have lower refractive indices than the remainder of the glassy object from which they protrude.

From simultaneous paleomagnetic and radiometric studies of Pliocene and recent lavas on land it has been established that the earth's magnetic field has reversed polarity several times in the last 4 million years (*4*). Zones of normally and reversely magnetized sediments can be found in many deep-sea sediment cores by determining the polarity of closely spaced samples with a magnetometer (*5*). The resulting magnetic stratigraphy can then be used indirectly to date certain layers in the cores by correlating the reversal boundaries with those dated on land by the K-Ar method (*5, 6*).

The magnetic stratigraphy of the core containing the glassy objects found off the Ivory Coast was determined by the use of a spinner magnetometer designed by Foster (*7*). This core was taken near the magnetic equator, and therefore changes in polarity can be detected only by shifts in declination of 180°, since the inclination is of little use on the equator. Unfortunately, the core was cut into sections, and each section was split in half with a random orientation with respect to the sections above and below it. Therefore, the location of the boundaries between the various

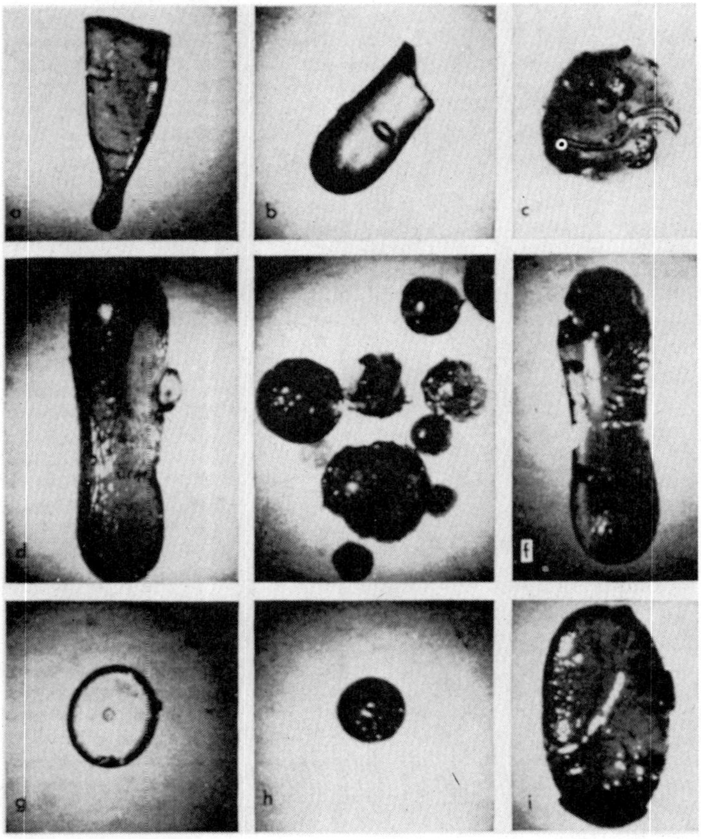

Fig. 1. Glassy objects found in deep-sea sediments from core V19-297 taken near the Ivory Coast. (a) Irregular teardrop-shaped glassy object; length 794 μ. (b) Fragment of a dumbbell with large bubble cavity; length 350 μ. (c) Sphere with pitted and grooved surface; diameter 280 μ. (d) Dumbbell with circular caplike protrusion on right side and flow lines parallel to length of dumbbell; length 1.24 mm. (e) Group of glassy objects; largest diameter, 400 μ. (f) Dumbbell with elongated grooves and pits; length 1.14 mm. (g) Disc-shaped glassy object with circular caplike protrusion on right side; maximum diameter, 180 μ. (h) Opaque sphere; diameter 160 μ. (i) Flattened oval-shaped glassy object with flow structure and large caplike protrusions; length 862 μ.

sections had to be considered in determining the magnetic stratigraphy of the core. The only 180° shift in declination not associated with one of these boundaries was observed at a depth of 490 cm. That this is indeed a reversal and, in fact, is the last reversal, is indicated by a drop in abundance of the planktonic Foraminifera *Sphaeroidinella dehiscens* just above this depth. Such a decrease in abundance of this species occurs at or just above the last reversal (*8*). Closely spaced samples taken below the reversal boundary at 490 cm do not reveal the presence of a Jaramillo event, however. This may be due to magnetic instability (although there is no indication of such) or to a hiatus in the sedimentary record.

Although scattered glassy objects were found as high as 40 cm above the reversal, they are most abundant in a layer 15 cm thick that occurs about 40 cm below the reversal. Extrapolation from this reversal indicates that the glassy objects were deposited about 800,000 years ago. This age is somewhat less than the K-Ar age determination of 1.2 ± 0.2 million years (*9*) given for the Ivory Coast tektites, but it is greater than the age of deposition of the Australasian microtektites (*1*). The glassy objects were most abundant in a sample taken at 540 to 543 cm. None was found below that depth. Because of the absence of a Jaramillo event, the exact age of deposition is uncertain, but seems to be about 800,000 years.

Although the chemical composition has not yet been determined, it is suggested that the glassy objects are microtektites and that they represent a portion of the tektite-strewn field of the Ivory Coast because of (i) their geographic location near the Ivory Coast strewn field; (ii) their apparent age of deposition which is similar to, but somewhat less than, the presently accepted K-Ar age determination of the Ivory Coast tektites; and (iii) their similarity in appearance and physical properties to both the Australasian microtektites and the Ivory Coast tektites. Preliminary chemical analyses, performed with the electron-microprobe at Goddard Space Flight Center, Greenbelt, Maryland, indicate that the glassy objects are also similar to the Ivory Coast tektites in composition.

BILLY P. GLASS*

Lamont Geological Observatory, Columbia University, Palisades, New York 10964

References and Notes

1. B. Glass, *Nature* **214**, 372 (1967).
2. W. A. Cassidy, B. Glass, B. C. Heezen, "Physical and chemical properties of Australasian microtektites," in preparation.
3. V. Vand, *Advan. Geophys.* **2**, 1 (1965).
4. A. Cox, R. R. Doell, G. B. Dalrymple, *Nature* **198**, 1049 (1963); *Science* **142**, 382 (1963); R. R. Doell, G. B. Dalrymple, A. Cox, *J. Geophys. Res.* **71**, 531 (1966); I. McDougall and D. H. Tarling, *Nature* **200**, 54 (1963); F. H. Chamalaun and I. McDougall, *Nature* **210**, 1212 (1966). The present normal period of the earth's magnetic history began 700,000 years ago. From 0.7 to about 2.5 million years ago the earth's magnetic field had a polarity opposite to that of the present field, except for two short periods (approximately 50 to 150 thousand years) about 0.95 and 1.9 million years ago, called the Jaramillo and Olduvai events, respectively.
5. N. D. Opdyke, B. Glass, J. D. Hays, J. Foster, *Science* **154**, 349 (1966).
6. D. Ninkovich, N. Opdyke, B. C. Heezen, J. H. Foster, *Earth Planet. Sci. Lett.* **1**, 476 (1966).
7. J. H. Foster, *ibid.*, p. 463.
8. B. Glass, D. B. Ericson, B. C. Heezen, N. D. Opdyke, J. A. Glass, *Nature* **216**, 437 (1967).
9. J. Zähringer, in *Tektites*, J. A. O'Keefe, Ed. (Univ. of Chicago Press, Chicago, Ill., 1963), p. 137; W. Gentner, H. J. Lippolt, O. Muller, *Z. Naturf.* **19a**, 150 (1964); R. L. Fleischer, B. P. Price, R. M. Walker, *Geochim. Cosmochim. Acta* **29**, 161 (1965).
10. I thank Dr. M. Ewing, director of Lamont Geological Observatory, for having taken cores off the Ivory Coast; Dr. N. D. Opdyke, for the use of the facilities of Lamont's paleomagnetic laboratory which were used to determine the magnetic stratigraphy of core V19-297; Drs. Burckle and A. Kaneps, for help with the photographic equipment used to photograph the glassy objects; Drs. B. C. Heezen, J. A. O'Keefe, O. L. Anderson, D. B. Ericson, and J. A. Glass, for discussions and help with the manuscript; F. Wood and Dr. Walter, for help with microprobe analyses. Supported by NSF grant GA 580 and the U.S. Navy, Office of Naval Research. Lamont Geological Observatory Contribution No. 1228.
* Present address: U.S. Army Corps of Engineers at the Goddard Space Flight Center, Laboratory for Theoretical Studies, National Aeronautics and Space Administration, Greenbelt, Maryland 20771.

27 April 1968; revised 1 July 1968

35

JOURNAL OF GEOPHYSICAL RESEARCH VOL. 74, NO. 4, FEBRUARY 15, 1969

Physical and Chemical Properties of Australasian Microtektites[1]

W. A. CASSIDY,[2] B. GLASS,[3] AND B. C. HEEZEN

Lamont Geological Observatory of Columbia University
Palisades, New York 10964

A group of glassy particles found in ocean sediments that were deposited about 700,000 years ago appear to be related to the Australasian tektites by their age, area of occurrence, physical properties, and chemical compositions. They are called microtektites because of their small sizes (<1 mm in diameter). A second group of microscopic glassy particles has been found associated with microtektites. They are also similar to tektites in age and area of occurrence but are quite different in their refractive indices, specific gravities, and compositions, except that some of them are close in composition to two unusual tektites from Java. This second group has been differentiated from the 'normal' microtektites by calling them 'bottle-green' microtektites. Whereas composition trends in the 'normal' microtektites can be explained by high-temperature vapor fractionation, composition variations among bottle-green microtektites can be explained by crystal fractionation in a basic magma. Compositions of bottle-green microtektites are suggestive of planetary mantle material. A minimum estimate of the total amount of microtektite material in the Australasian strewn field is 10^{10} kg.

INTRODUCTION

During the investigation of deep-sea sediment cores taken south of Australia, many small (<1 mm) glassy objects were discovered [*Glass*, 1967]. They are mostly spherical in shape, although dumbbell, teardrop, pear, and irregular shapes are also present. They are isotropic and contain no crystalline inclusions, although bubble cavities are common. They do not have the appearance of typically angular, glassy volcanic shards.

Approximately 45 cores have now been investigated for the presence of microtektites (Table 1). Glassy particles were found in 10 of them; 1 from south of the Japanese Islands, 1 from south of Sumatra, 5 from south and southwest of Australia, 1 from the central Indian Ocean, and 2 from southeast of Madagascar (Figure 1). The magnetic stratigraphy had been determined for these cores with the cooperation of Neil Opdyke before they were searched for glassy particles. The particles were found in a zone 30–50 cm thick associated in time with the last geomagnetic reversal.

In all but one core (RC9-137) scattered microtektites were found below as well as above the reversal. In this core the microtektites were confined to a layer about 20 cm above the reversal. In two cores (V20-138 and V16-70) approximately equal numbers of microtektites were found below and above the reversal. In the remainder of the cores the microtektites were most abundant above the reversal. Although the microtektites were most abundant in sediments occurring about 10–30 cm above the reversal, which is from 2 to 5 meters below the top of the core, there is no constant relationship between the height of this concentrated layer above the reversal and sedimentation rates, as would be expected if the microtektites had fallen some time after the reversal. Whether or not the microtektites fell exactly at the time of the reversal, the close association of the microtektite layer with the last reversal of the earth's magnetic field indicates that the microtektites were deposited on the ocean floor approximately 700,000 years ago. The fact that they occur in a zone 30–50 cm thick does not rule out the

[1] Contribution 1283 of Lamont Geological Observatory.

To be included in a volume of reprints in memory of the late Craig M. Merrihue.

[2] Now at Department of Earth and Planetary Sciences, University of Pittsburgh, Pittsburgh, Pennsylvania 15213.

[3] Now at Laboratory for Theoretical Studies, Goddard Space Flight Center, Greenbelt, Maryland 20771.

TABLE 1. List of Cores Searched and Their Locations

Lamont Core No.	Micro-Tektites	Lat.	Long.	Description
RC8-49		53°16'S	76°55'E	Silty diatomaceous lutite
RC8-52	X	41°06'S	101°22'E	Foraminiferal lutite
RC8-53	X	39°23'S	104°22'E	Foraminiferal lutite
RC8-63*		51°05'S	129°58'E	Diatomaceous lutite
RC8-93		29°22'S	105°14'W	Foraminiferal lutite
RC9-115		31°24'S	163°36'W	Brown lutite
RC9-117		24°30'S	168°42'W	Pale brown lutite
RC9-137	X	45°01'S	132°45'E	Foraminiferal lutite grading into lutite
RC9-142	X	42°43'S	116°54'E	Foraminiferal lutite
RC9-143	X	41°22'S	114°08'E	Foraminiferal lutite
RC10-95		03°31'N	129°45'W	Foraminiferal lutite
RC10-159		31°13'N	162°18'E	Brown lutite
V12-18		28°42'S	34°30'W	Foraminiferal lutite
V16-39		24°44'S	04°46'W	Foraminiferal lutite
V16-57		45°14'S	24°29'E	Siliceous lutite
V16-60		42°39'S	45°40'E	Siliceous lutite
V16-69		37°35'S	56°44'E	Siliceous lutite
V16-70	X	32°06'S	55°51'E	Foraminiferal lutite
V16-75		22°13'S	58°23'E	Silty lutite
V16-76	X	25°09'S	59°54'E	Siliceous lutite
V16-84		29°44'S	79°34'E	Silty lutite with some Foraminifera
V16-134		61°54'S	91°15'W	Siliceous lutite
V18-188		37°38'S	37°52'E	Silty lutite
V19-116		11°14'N	130°10'E	Pale brown lutite
V19-153	X	08°51'S	102°07'E	Silty lutite
V19-169		10°13'S	81°37'E	'Red clay'
V19-170		07°54'S	81°25'E	Brown lutite
V19-171	X	07°04'S	80°46'E	Brown lutite
V20-109		47°19'N	174°39'W	Siliceous lutite
V20-138	X	28°52'N	135°33'E	Brown lutite
V20-142		17°11'N	133°16'E	Brown lutite
V20-156		14°26'S	116°19'E	alternate layers of lutite and diatomaceous lutite
V20-167		21°03'S	72°30'E	Calcareous lutite
V20-169		21°38'S	69°55'E	Foraminiferal ooze
V20-184		29°48'S	53°41'E	Brown lutite
V21-48		09°31'S	126°22'W	Foraminiferal lutite
V21-67		24°58'N	176°16'E	Brown silty lutite
V21-73		29°28'N	154°36'E	Brown lutite
V21-74		29°51'N	150°58'E	Brown silty lutite
V21-89		23°35'N	145°39'E	Foraminiferal ooze
V21-118		15°53'N	137°23'E	Diatomaceous ooze to brown lutite
V21-125		13°41'N	128°29'E	Yellowish brown lutite
V21-145		34°03'N	164°50'E	Brown lutite

* Magnetic stratigraphy confusing.

possibility that they were all deposited in a thin layer during a single event, since mixing by bottom-dwelling scavengers and local erosion and redeposition could easily result in a vertical distribution this thick [*Glass and Heezen*, 1967]. The fact that the glassy particles are found in greater concentrations above the reversal than below it is also consistent with a distribution by bottom dwellers since an initial thin layer would be distributed by them only to a limited depth. The upper zones of this region, however, would be continually reworked as later sediments were added, and material would therefore be carried upward and incorporated in later sediments.

The area in which these glassy particles have been found is shown in Figure 1. They were not found in all the cores examined from this area. This could be because of chemical solution, hiatuses in the sediment column, or an original patchy distribution of the particles. Their total area of occurrence, however, is surprisingly large, and there are only a few ways of explaining this wide horizontal distribution. We do not think it likely that they were spread this widely by winds or ocean currents because sand particles of the same size are not. Except for some apparently ice-rafted mineral grains scattered through the more southerly of the ten cores, the only particles of equal size in the sediments are the tests of microscopic pelagic animals.

We do not feel that they were shot out radially from a single large terrestrial disturbance such as a volcanic eruption because they do not show adequate signs of increasing toward any patricular center. Distribution studies [*Ninkovich and Heezen*, 1965; *Ninkovich*, 1968] show that if a thin layer of volcanic particles were found in deep-sea sediments at the limits of the area outlined in Figure 1, a deposit hundreds of meters thick would be expected at a postulated source near the center of the area.

Although we offer no proofs of the terrestrial or extraterrestrial origin of these glassy particles, we do find support for the suggestion [*Glass*, 1967] that they are microtektites and that they represent a portion of the Australasian tektite strewn field. In the following material we present additional data on the physical properties of the glassy objects and chemical analyses of sixty of them.

Physical Properties

To determine the size distribution of the microtektites, a sample weighing approximately 10 grams from core RC9-143 was washed and then sieved with a 38-μ sieve. The coarse fraction ($>38\ \mu$) was then mounted in caedax on a glass slide, and the diameter of each of the microtektites (54 in all) was measured by using an ocular micrometer. Classes were defined on

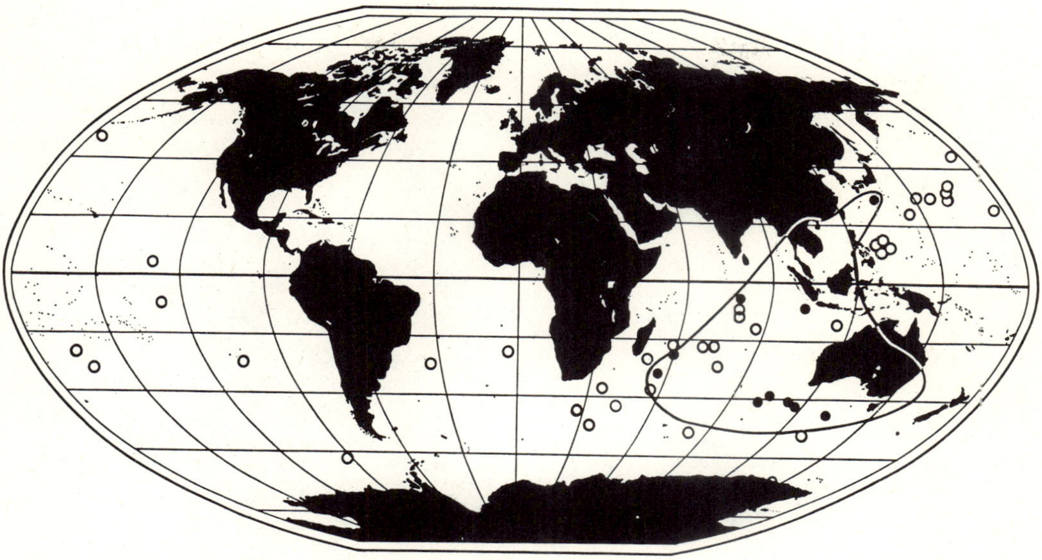

Fig. 1. Map of microtektite occurrence. Boundaries are provisional. Open circles indicate cores without microtektites; closed circles, with microtektites.

the basis of size range, and the per cent of microtektites occurring in each size range was then calculated. By using the average size within each class, the weight per cent of each size was also calculated (Table 2). Although the largest number of microtektites occurs in the smallest size range studied, the largest per cent by mass occurs in the largest size range. We have not found microtektites larger than about 1 mm, however.

The refractive indices of 40 microtektites were determined, and they were found to fall into 2, or possibly 3, different groups. The particles ranged most commonly between 1.497 and 1.540 but others were found in the ranges 1.560–1.570 and 1.580–1.620. Representatives of the highest index group were found in only 3 of the 10 cores (V19-153, V19-171, and RC9-143). The low-index microtektites are generally some shade of yellow, brown, or yellowish-green with the smaller ones (<200 μ) apparently colorless. The high-index group, however, is quite different in general appearance, having a distinct bottle-green color and apparent deep corrosion pits that in some cases have destroyed any original spheroidal shape. However, both the high- and low-index groups are glassy and are found in the same sedimentary layer. The microtektites with refractive indices between 1.560 and 1.570 (see Figure 2 and Table 5) are yellowish-green in color and represent microtektites with compositions intermediate between the high index (bottle-green) group and the low-index group (see section on discussion of data).

TABLE 2. Size Distribution of Microtektites

Size Range, mm	Abundance, %	Weight % (Cal.)
0.038–0.062	47.0	1.7
0.062–0.088	26.0	2.9
0.088–0.124	12.0	3.6
0.124–0.175	6.0	5.3
0.175–0.246	5.0	11.7
0.246–0.351	2.5	19.6
0.351–0.495	1.0	23.4
0.495–0.707	0.5	31.6

The per cent abundance of microtektites according to refractive index is shown in Figure 2. A maximum occurs around 1.515, and the high-index group could be either the tail of an assymetric distribution or a second mode. Not seen in the graph is the fact that a number of specimens had very-low-index inclusions that are apparently lechatelierite particles (see next section).

The specific gravities of 7 of the larger low-index microtektites were determined by using the sink-or-float method in heavy liquids. The range is from 2.37 to 2.55. Six of these have

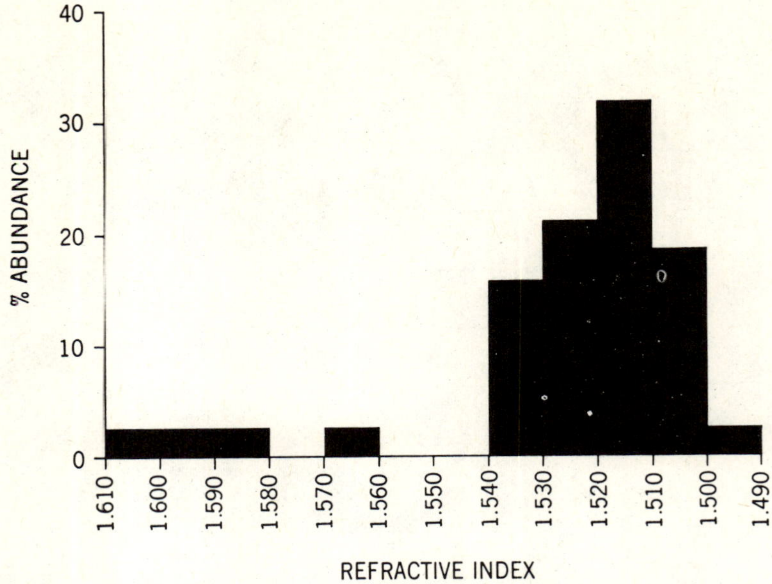

Fig. 2. Distribution of microtektites according to refractive index. The average value is about 1.520. Total number of determinations: 38.

been plotted on a diagram (Figure 3) showing specific gravity versus refractive index [*Barnes*, 1940]. The low-index microtektites agree well with the trend previously established for tektites.

Chemical Composition

We selected specimens for electron microprobe analyses in a nonrandom manner by their appearance to get a reasonable number of representatives of all varieties. This resulted in a disproportionate number of what appeared to be exceptional types, and ensured that samples with a wide variety of shapes, colors, and other characteristics, such as inclusions, surface texture, and scoriaceous structure, were represented. Among the specimens prepared for analysis were several where the surface to be analyzed intersected siliceous inclusions resembling lechatelierite particles. Figure 4 shows a concentration map of the analyzed surface of one of these, made with Si Kα radiation. An

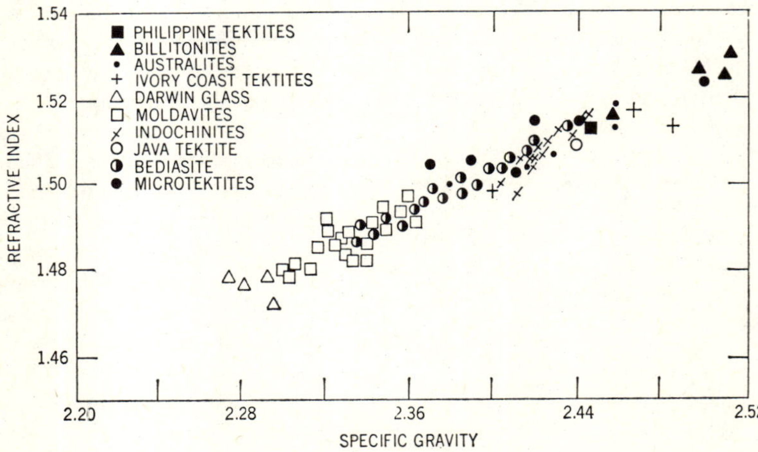

Fig. 3. Specific gravity versus refractive index for microtektites and tektites. A seventh microtektite measurement at specific gravity = 2.55 lay on the same trend line.

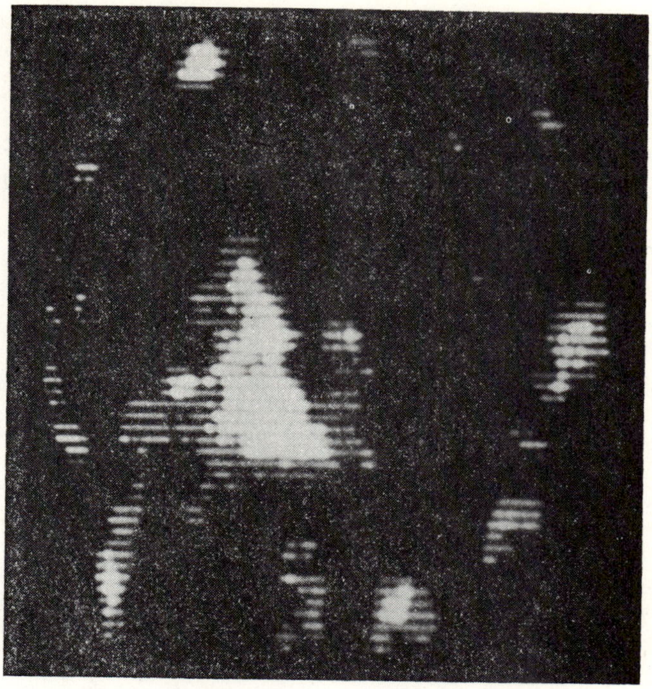

Fig. 4. Concentration map showing lechatelierite inclusions in sample 19. The bright spots are almost pure SiO_2. The polished section has a diameter of 130 μ. Most of the smaller inclusions were near the surface of the spherule and one, in the lower left corner, projected above the surface.

analysis of the largest silicate inclusion is given in Table 3.

The samples were mounted in epoxy, ground to expose a section, and polished. In most cases they were analyzed for 8 oxides in the following 3 groups of 3: Mn-Si-Mg, Fe-K-Na, and Ca-Si-Al. For one group of samples Mn-Ti-Mg replaced Mn-Si-Mg. The work was done on the ARL instrument at the Theoretical Division,

TABLE 3. Electron Microprobe Analysis of Lechatelierite Inclusion in Sample 6

Component	Wt. %
SiO_2	98.8
Al_2O_3	1.6
FeO	0.4
MgO	0.1
CaO	0.0
Na_2O	0.7
K_2O	0.6
TiO_2	0.1
MnO	0.0
Total	102.3

Goddard Space Flight Center, Greenbelt, Maryland.

For most samples, 5 different points on the surface were analyzed for each group of 3 oxides. Where composition differences were obviously not great, the data from the 5 spots were averaged to get the average composition of the spherule in those 3 oxides. Where significant composition differences were found to exist between the 5 points, composition profiles were run across the section by step-scanning, and those results were averaged. The data count for each point was accumulated to a level where the statistical variation was much lower than the other uncertainties in the determination.

A number of standards of known composition were run for comparison including a suite of multicomponent artificial glasses made up to resemble tektites.

The raw data were analyzed by computer according to an existing program at Goddard in which data from the standard samples were used to establish a first-order and a second-order curve by the method of least squares for each oxide. In most cases, 1 of the 2 standard curves looked better than the other for a given concentration of the oxide, so we were able to choose which of 2 concentrations derived for the unknown sample fit the data better. Derived oxide concentrations were then combined to form a complete major-element analysis.

The described method relies on the use of standards as close as possible in composition to the unknowns. This method has obvious drawbacks, particularly in multicomponent systems, but in our opinion the results obtained in this work justify the method: most of the analyses summed to 100 + 2.5%. Bulk analyses are given in Table 4.

* * * * * * *

TABLE 4. Major-Element Analyses of Glassy Spherules by Electron Microprobe (in Weight Per Cent)

Sample No.	1	2	3*	4	6†	8	10	11*	12	14	15	16
Analysis No.	1	2	3	4	5	6	7	8	9	10	11	12
SiO_2	68.7	70.7	56.8	65.8	72.9	67.9	68.9	59.7	62.6	67.1	67.2	72.1
Al_2O_3	12.2	14.4	8.0	16.3	13.1	15.2	13.7	11.7	19.3	15.4	15.5	13.8
FeO	5.2	5.5	8.4	5.2	4.5	5.7	5.1	7.7	5.3	5.4	5.9	4.7
MgO	2.8	2.9	20.6	2.5	2.2	2.7	2.5	14.6	3.1	4.0	4.5	2.3
CaO	1.3	2.8	2.9	2.6	2.1	2.9	2.9	3.9	2.2	2.6	2.8	2.4
Na_2O	1.9	0.9	0.8	1.4	0.6	1.7	1.4	1.2	1.3	1.2	1.7	1.2
K_2O	3.7	3.0	0.3	3.2	3.0	3.2	3.1	0.6	3.1	2.7	2.3	3.0
TiO_2‡	0.7	0.8	0.6	n.d.	0.7	0.8	0.8	0.7	1.0	0.9	0.8	0.8
MnO	0.1	0.1	0.1	0.1	0.1	0.1	0.1	0.1	0.1	0.1	0.1	0.1
Totals	96.6	101.1	98.5	97.1	99.2	100.2	98.5	100.2	98.0	99.4	100.8	100.4
Sample No.	17	19§	21*	22	25	28	29	30	31	32	33	34
Analysis No.	13	14	15	16	17	18	19	20	21	22	23	24
SiO_2	75.7	66.2	64.2	62.7	71.6	68.6	72.3	70.0	71.6	65.6	70.0	74.9
Al_2O_3	15.5	14.5	11.0	16.6	8.4	13.6	11.2	13.1	13.0	14.5	10.6	9.7
FeO	5.3	6.0	9.6	7.4	8.1	6.2	5.7	5.6	5.9	6.6	5.7	5.0
MgO	2.5	2.7	7.2	7.0	4.5	8.7	4.6	6.5	2.7	4.4	6.4	3.4
CaO	2.4	3.0	3.3	4.8	2.2	4.4	2.7	3.4	3.1	4.4	2.9	2.2
Na_2O	1.2	1.5	1.8	1.6	2.8	1.2	1.1	1.0	0.6	1.6	1.2	1.0
K_2O	3.0	2.9	1.5	1.4	2.0	0.6	1.2	1.2	2.7	2.4	0.9	1.9
TiO_2	0.8	0.9	0.6	0.8	n.d.	0.8	0.6	0.6	0.7	0.7	0.7	0.6
MnO	0.1	0.1	0.1	0.1	0.1	0.1	0.1	0.1	0.1	0.1	0.1	0.1
Totals	106.5	97.8	99.3	102.4	99.7	104.2	99.5	101.5	100.4	100.3	98.5	98.8
Sample No.	35	36	38	39	40	41*	43	45	46	47§	48	51
Analysis No.	25	26	27	28	29	30	31	32	33	34	35	36
SiO_2	75.9	60.5	67.5	49.6	69.5	58.0	66.2	70.8	56.3	57.8	71.7	69.9
Al_2O_3	7.5	21.7	14.8	18.3	16.0	11.5	17.1	14.4	17.8	19.7	13.3	15.2
FeO	5.4	4.0	5.4	4.9	4.8	8.1	5.0	5.1	7.1	7.1	5.5	5.2
MgO	3.4	4.2	2.7	17.1	3.1	15.6	4.1	2.5	11.3	11.1	3.2	2.6
CaO	1.4	5.4	4.7	5.8	3.2	4.2	5.8	3.0	3.8	3.5	2.9	3.1
Na_2O	1.7	1.0	1.4	0.2	0.8	0.5	1.0	0.4	1.5	0.6	0.4	0.8
K_2O	2.6	0.6	2.5	0.1	1.3	0.2	1.1	2.1	0.4	0.3	1.6	1.7
TiO_2	0.5	1.0	n.d.	0.9	0.8	n.d.	0.8	n.d.	n.d.	n.d.	n.d.	n.d.
MnO	0.1	0.1	0.1	0.1	0.1	0.1	0.1	0.1	0.1	0.1	0.1	0.1
Totals	98.5	98.5	99.1	97.0	99.6	98.2	101.2	98.4	98.3	100.2	98.7	98.6

TABLE 4 (*continued*)

Sample No.	53	54	55	57	58§	60	61	62	63*	64*	65	67
Analysis No.	37	38	39	40	41	42	43	44	45	46	47	48
SiO_2	70.2	68.6	67.0	60.7	64.9	59.9	59.2	71.8	48.1	54.8	69.7	69.7
Al_2O_3	15.0	15.5	17.7	18.4	16.7	19.3	22.1	13.5	11.3	9.6	12.2	11.3
FeO	5.6	5.7	4.7	5.5	5.0	4.5	3.5	4.9	5.5	8.5	5.6	6.7
MgO	2.6	2.7	3.2	6.6	5.4	7.1	5.8	2.6	27.3	20.4	3.4	4.2
CaO	3.0	3.0	4.1	4.9	4.0	5.4	4.5	3.0	5.3	3.8	3.5	3.1
Na_2O	0.8	1.1	1.0	0.9	0.8	0.6	0.7	0.8	0.5	0.6	1.2	1.5
K_2O	1.6	2.0	1.4	0.4	0.8	0.3	0.3	2.0	0.1	0.3	2.3	2.4
TiO_2	n.d.	n.d.	n.d.	n.d.	n.d.	n.d.	n.d.	n.d.	n.d.	n.d.	n.d.	n.d.
MnO	0.1	0.1	0.1	0.1	0.1	0.1	0.1	0.1	0.1	0.1	0.1	0.1
Totals	98.9	98.7	99.2	97.5	97.7	97.2	96.2	98.7	98.2	98.1	98.0	99.0
Sample No.	68	70	71	72	73	74	75	76	77	79	81	82*
Analysis No.	49	50	51	52	53	54	55	56	57	58	59	60
SiO_2	63.2	60.8	76.2	73.3	63.8	77.0	72.2	70.0	71.1	68.9	64.6	52.5
Al_2O_3	19.2	21.2	13.8	15.5	20.9	10.2	13.7	16.0	13.5	15.3	17.6	12.3
FeO	7.6	5.6	4.9	5.2	3.0	4.8	5.2	3.9	3.5	5.5	5.4	6.4
MgO	2.9	4.7	2.3	2.6	3.7	5.1	2.6	3.1	1.9	2.4	3.5	21.4
CaO	1.0	3.0	3.5	4.3	4.4	3.0	3.4	4.8	2.6	1.5	4.2	4.6
Na_2O	1.3	2.1	0.8	0.8	0.6	0.6	0.5	0.6	0.7	0.7	1.0	0.2
K_2O	3.3	2.2	1.8	1.5	0.8	1.6	1.8	1.0	2.6	3.5	1.8	0.1
TiO_2	n.d.	n.d.	n.d.	n.d.	n.d.	n.d.	n.d.	n.d.	n.d.	n.d.	n.d.	n.d.
MnO	0.1	0.1	0.1	0.1	0.1	0.1	0.1	0.1	0.0	0.1	0.1	0.1
Totals	98.6	99.7	103.4	103.3	97.3	102.4	99.4	99.5	95.9	97.9	98.2	97.6

* Indicates 'bottle-green' glassy object with high refractive index.

‡ TiO_2 was determined in only 27 samples. In these it ranged between 0.5 and 1.0% by weight, with most values 0.7 or 0.8. TiO_2 can therefore be estimated at around 0.7 or 0.8% in those samples where the determination was not made.

† Analysis of matrix only. See Table 3 for analysis of lechatelierite inclusion.

§ Analysis of matrix only.

References

Barnes, V. E., North American tektites, *Texas Univ. Publ. 3945,* 477–582, 1940.

Glass, B., Microtektites in deep-sea sediments, *Nature, 214,* 372–374, 1967.

Glass, B., and B. C. Heezen, Tektites and geomagnetic reversals, *Nature, 214,* 372, 1967.

Ninkovich, D., Pleistocene volcanic eruptions in New Zealand recorded in deep-sea sediments, *Earth Planetary Sci. Letters, 4,* 89–102, 1968.

Ninkovich, D., and B. C. Heezen, Santorini Tephra, *Submarine Geology and Geophysics, Colston Papers, 17,* 413–453, 1965.

Schmidt, R. A., Microscopic extraterrestrial particles from the Antarctic Peninsula traverse, *Res. Rept. Ser., G3-3,* 65 pp., Geophysical and Polar Research Center, University of Wisconsin, 1963.

Schnetzler, C. C., and W. H. Pinson, Jr., The chemical composition of tektites, in *Tektites,* edited by J. A. O'Keefe, pp. 95–127, University of Chicago Press, Chicago, 1963.

Shima, M., Glassy spherules (microtektites?) found in ice at Scott base, Antarctica, *J. Geophys. Res., 71*(14), 3595–3596, 1966.

Walter, L. S., Tektite compositional trends and experimental vapor fractionation of silicates, *Geochim. Cosmochim. Acta, 31,* 2043–2063, 1967.

(Received June 21, 1968;
revised September 16, 1968.)

36

eochimica et Cosmochimica Acta, 1969, Vol. 33, pp. 1135 to 1147. Pergamon Press. Printed in Northern Ireland

Chemical composition of Ivory Coast microtektites

BILLY P. GLASS*

Planetology Branch, Goddard Space Flight Center, Greenbelt, Maryland 20771

(*Received* 1 *March* 1969; *accepted in revised form* 28 *April* 1969)

Abstract—The chemical composition of twenty-eight microscopic glassy objects, found in a deep-sea sediment core taken near the Ivory Coast, has been determined by electron microprobe analysis. Like the Australasian microtektites, these glassy objects can be divided into two main groups. One group is composed of individuals with chemical compositions similar to tektites. These are referred to as "normal" microtektites. The other group is composed of bottle-green glassy objects with low silica contents (<53%) and high MgO contents (16–21%). The "normal" Ivory Coast microtektites are similar in composition to Ivory Coast tektites and like Ivory Coast tektites, they differ from the Australasian tektites and microtektites by having a high MgO/CaO ratio (>3) and an average Na_2O/K_2O ratio greater than unity. Although the bottle-green objects differ from the "normal" microtektites in their appearance, physical properties and chemical compositions, they are nevertheless found in association with them and are probably therefore related to them. Both the Ivory Coast and Australasian microtektites have higher MgO and lower Na_2O contents than igneous glasses with comparable SiO_2 contents. The chemical data presented in this paper are interpreted as confirming a previous suggestion that the glassy objects found in deep-sea sediments near the Ivory Coast are microtektites and that they represent a portion of the Ivory Coast tektite strewn field.

INTRODUCTION

SEVERAL hundred microscopic (<1 mm) spherical, dumbbell, teardrop, disc and irregular shaped glassy objects have been found in ten deep-sea sediment cores taken adjacent to the Australasian tektite strewn field (Fig. 1). These glassy objects occur in a rather thin layer (20–50 cm thick) associated with the Matuyama-Brunhes reversal, indicating that they fell on the ocean floor approximately 700,000 years ago (GLASS, 1967). On the basis of their geographical location, age of deposition (700,000 yr), appearance, physical properties and chemical composition, it has been concluded that these glassy objects are microtektites and that they represent a portion of the Australasian tektite strewn field (GLASS, 1967; CASSIDY *et al.*, 1969).

Recently, several hundred microscopic glassy objects were found in a single deep-sea core (V19-297) taken near the Ivory Coast (Fig. 1). On the basis of their similarity in appearance to the Australasian microtektites, their geographical location and apparent age of deposition (~0·8–1·0 million years) it has been suggested that these glassy objects are also microtektites and that they represent a portion of the Ivory Coast tektite strewn field (GLASS, 1968). The purpose of this paper is to present the composition of twenty-eight of the Ivory Coast microtektites (determined by electron microprobe analysis) and to compare the Ivory Coast microtektites with the Australasian microtektites and Ivory Coast tektites. In addition, both the Ivory Coast and Australasian microtektites are compared with igneous glass and hay-silica glass.

* U.S. Army Corps of Engineers assigned to Goddard Space Flight Center.

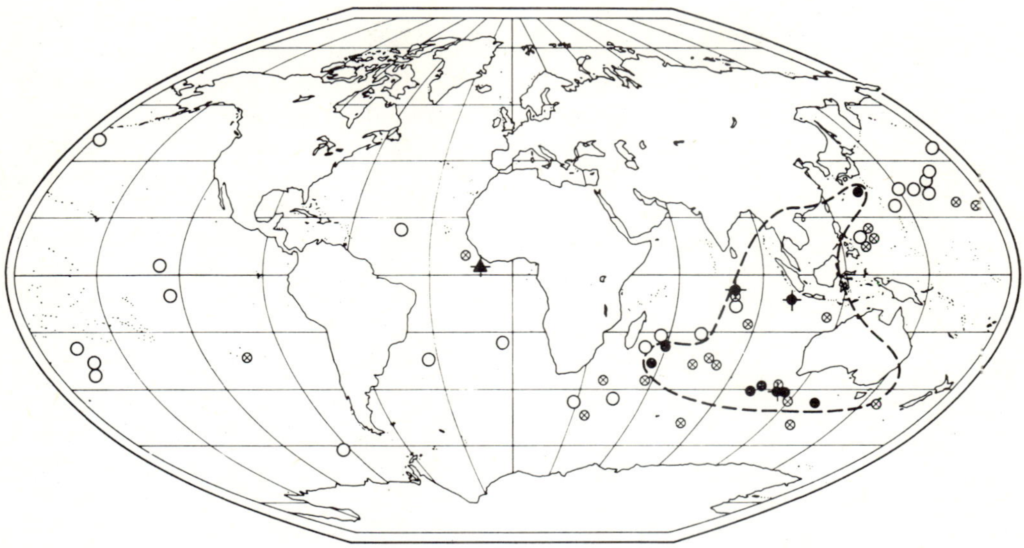

Fig. 1. Geographical distribution of microtektite-bearing cores.

METHOD OF ANALYSIS

Thirty specimens were selected for microprobe analysis. The samples were individually mounted in epoxy inside $\frac{1}{4}$ in. dia. aluminum cylinders. They were ground to expose a section and polished with successively finer abrasives to yield a flat, smooth surface. Two of the specimens were lost during preparation. The remaining samples were then vacuum coated, at the same time as the standards, with a thin layer of carbon. Each of the specimens was then analyzed for nine elements in the following groups of three: Fe–K–Na, Ca–Si–Al and Mn–Ti–Mg. The analyses were performed with an Applied Research Laboratories electron microprobe X-ray analyzer. The measurements were made at 20 kV and 0·03 μA. Ten to forty seconds integration time was used.

For most samples, five different points on the surface were analyzed for each group of three oxides; but as many as twelve points were analyzed if the sample seemed heterogeneous. The data from the different points were averaged to obtain the average composition of the specimen.

The method of microprobe analysis used in this work involved direct interpolation (and limited extrapolation) using homogeneous synthetic glass standards of tektite composition, SiO_2 (quartz) and Al_2O_3.

The raw data were analyzed by computer according to an existing program at the Goddard Space Flight Center (WALTER, 1967). Most of the analyses summed to 100 $\pm$ 2·5% (Table 1).

DISCUSSION OF DATA

Sample No. 12 is a small, badly pitted spheroid with an isotropic colorless core of glassy material having a refractive index of 1·461 $\pm$ 0·002 (measurements of refractive index were made by the oil immersion method using a petrographic microscope). The core is surrounded, except for one small irregular area, by a thin yellowish outer layer with a refractive index of 1·514 $\pm$ 0·002. Microprobe analysis indicates that the core of the spherule is composed of almost pure SiO_2

(Table 1); thus the spherule is probably lechatelierite. The yellowish outside layer is variable in composition with an average silica content of about 80%.

Four of the spherules (Nos. 9, 10, 21 and 26) have a bottle-green color and high refractive index (1·584–1·592), and are badly corroded. These spherules have lower silica contents and higher MgO contents than the remainder of the spherules (Table 1). They contain less silica and up to eight times as much magnesia as the standards. The composition of one of the spherules was corrected using the procedures described by BENCE and ALBEE (1968). A comparison of the uncorrected values with the corrected values indicates that the silica per cent derived from the uncorrected analysis may be too high by about 4% of the amount present and the alumina may be low by about 6% of the amount present. The remainder of the oxides remained unchanged.

The remaining twenty-three spherules have compositions similar to tektites but with somewhat lower average silica contents. Three of these (Nos. 8, 14 and 17) were found, during microprobe analysis, to contain lechatelierite particles (for the composition of one of the lechatelierite particles, see 8b in Table 1). This means that lechatelierite was found at three of the 145 points analyzed for silica. Thus the glassy objects may contain as much as 2% lechatelierite by volume.

Samples No. 14 and 17 are angular fragments with a smooth convex surface on one side which suggests that they are fragments of a fairly large microtektite. The fact that they are the same color and have the same refractive index (1·522 ± 0·002) and similar chemical compositions (Table 1) indicates that they may be fragments of the same microtektite which was probably greater than 1 mm in diameter.

Comparison with Australasian microtektites

Using the same methods, standards and instrument as were used to determine the composition of the spherules reported in this paper, CASSIDY *et al.* (1969) determined the chemical composition of sixty of the Australasian microtektites. On the basis of appearance, physical properties and chemical composition, CASSIDY *et al.*, were able to divide the Australasian microtektites into two extreme groups with some intermediate specimens. One of the extreme groups consists of microtektites that have physical and chemical compositions similar to the Australasian tektites, but with a somewhat lower average silica content. These constitute about 65% of the microtektite population and are referred to as "normal" microtektites. The other extreme group is composed of badly corroded, bottle-green glassy objects with low silica contents and high MgO contents. These are referred to as "bottle-green" microtektites because of their quite distinctive bottle-green color. The bottle-green objects constitute approximately 5% of the total Australasian microtektite population. The intermediate specimens, which constitute the remainder of the Australasian population, are usually yellowish-green in color. They have intermediate physical properties and chemical compositions.

Like the Australasian microtektites, the Ivory Coast microtektites can also be divided into two distinct groups on the basis of their appearance, physical properties and chemical compositions: the "normal" microtektites with physical properties and compositions similar to tektites; and the "bottle-green" objects

9

with high refractive index, low silica contents (50–53%) and extremely high MgO contents (up to ~21%). In core V19-297 the "normal" microtektites make up approximately 84% of the Ivory Coast microtektite population. The bottle-green objects constitute approximately 5% of the total microtektite population. The remainder are yellowish-green in color and many of these may have intermediate physical properties and compositions.

The "normal" Ivory Coast microtektites seem to have a much narrower range in composition than the "normal" Australasian microtektites. This is apparently not the result of a fewer number of analyses, since an equal number of analyses chosen at random from the forty "normal" Australasian microtektite analyses still shows a greater range in composition. Nor does it seem to be due to the fact that the forty "normal" Australasian microtektites were picked from seven cores from widely separated geographical areas, since eighteen "normal" microtektites analyzed from core V19-153 from the Australasian strewn field have a greater range in composition than the twenty-three "normal" Ivory Coast microtektites.

As is indicated by their higher average refractive index (1·525 vs. 1·515 for the Australasian microtektites) the Ivory Coast microtektites have a somewhat lower average silica content than the Australasian microtektites (Table 2). The Ivory Coast microtektites are distinguished from the Australasian microtektites by their lower CaO content (Fig. 2). In addition, the average K_2O content of the Ivory Coast microtektites is somewhat lower and the average FeO (total iron) and Na_2O contents are somewhat greater than for the Australasian microtektites. The major oxide vs. SiO_2 trends are similar for the two groups except for the MgO vs. SiO_2 trend which is much steeper for the Ivory Coast microtektites than for the Australasian microtektites (Fig. 2).

Of the more than fifty cores (from various parts of the world's oceans) examined by the author, the bottle-green glassy objects have only been found in four (Fig. 1). Three of the cores are from the Australasian area and the other is the core containing the Ivory Coast microtektites. In these four cores the bottle-green objects are found only in association with the "normal" microtektites and make up from three to approximately 36% of the total population of glassy spherules. It may be of significance that three of the cores in which the bottle-green objects were found contained a larger number of microtektites than the remaining microtektite-bearing cores.

The bottle-green objects generally are roughly spheroidal in shape with a badly corroded surface. Only a few have a good spherical shape with a rather smooth surface. With the exception of one specimen which appeared to be a fragment of a dumbbell, primary shapes such as teardrop, dumbbell, disc and others have not been observed among the bottle-green objects. Like the "normal" microtektites, the bottle-green objects do not contain any crystalline inclusions; yet unlike the "normal" microtektites, none of the bottle-green objects have been observed to contain bubble cavities, siliceous inclusions or lechatelierite particles.

Based on four analyses, the Ivory Coast bottle-green objects do not seem to show as well defined trends in composition as the Australasian bottle-green objects. The Ivory Coast bottle-green objects seem to have higher Al_2O_3 contents, and lower MgO contents than the eight Australasian bottle-green objects (Table 2,

Table 1. Chemical composition and description of Ivory

Sample No.	SiO_2	Al_2O_3	FeO	MgO	CaO	Na_2O	K_2O	TiO_2
1	66·2 ± 0·6	15·0 ± 0·2	6·2 ± 0·1	4·3 ± 0·1	1·5 ± 0·1	2·0 ± 0·3	1·8 ± 0·1	0·9 ± 0·1
2	65·1 ± 0·4	15·7 ± 0·4	7·1 ± 0·2	4·7 ± 0·1	1·3 ± 0·1	1·9 ± 0·4	1·9 ± 0·4	0·8 ± 0·1
3	66·3 ± 1·5	16·5 ± 0·5	6·8 ± 0·1	4·1 ± 0·1	1·7 ± 0·0	1·9 ± 0·2	1·9 ± 0·2	0·8 ± 0·1
4	66·1 ± 1·0	15·2 ± 0·2	5·7 ± 0·2	4·2 ± 0·1	1·5 ± 0·1	2·0 ± 0·3	1·8 ± 0·1	0·6 ± 0·0
5	64·2 ± 0·5	14·0 ± 0·5	6·8 ± 0·2	5·4 ± 0·1	1·3 ± 0·1	1·7 ± 0·2	1·5 ± 0·1	0·5 ± 0·1
6	66·5 ± 0·3	16·9 ± 0·2	6·5 ± 0·3	3·1 ± 0·1	1·1 ± 0·1	2·0 ± 0·2	2·1 ± 0·1	0·6 ± 0·0
7	63·5 ± 0·2	14·5 ± 0·5	7·7 ± 0·1	5·8 ± 0·2	1·6 ± 0·1	2·2 ± 0·2	1·5 ± 0·1	0·7 ± 0·0
8a	67·4 ± 0·6	15·1 ± 0·5	6·0 ± 0·1	3·7 ± 0·2	1·6 ± 0·1	2·1 ± 0·1	1·7 ± 0·1	0·5 ± 0·1
8b	98·1	0·9	0·1	0·0	0·0	0·1	0·0	n.d.
9	52·0 ± 1·5	18·5 ± 0·6	7·8 ± 0·1	16·4 ± 0·4	6·3 ± 0·4	0·4 ± 0·1	0·2 ± 0·1	1·0 ± 0·0
10	50·5 ± 1·0	12·7 ± 0·5	9·1 ± 0·1	17·4 ± 0·4	2·3 ± 0·1	0·5 ± 0·1	0·2 ± 0·1	0·6 ± 0·1
11	63·4 ± 0·3	16·6 ± 0·5	7·1 ± 0·3	5·5 ± 0·3	1·6 ± 0·1	1·8 ± 0·4	1·5 ± 0·2	0·7 ± 0·1
12	98·6 ± 0·2	1·1 ± 0·0	0·1 ± 0·1	0·0	0·0	0·0	0·0	0·0
13	64·8 ± 0·3	15·6 ± 0·2	7·2 ± 0·2	6·0 ± 0·1	1·6 ± 0·1	1·9 ± 0·2	1·5 ± 0·1	0·7 ± 0·1
14	67·3 ± 2·5	17·3 ± 1·5	6.2 ± 0·6	2·3 ± 0·8	1·0 ± 0·1	2·0 ± 0·5	2·3 ± 0·2	0·7 ± 0·0
15	62·4 ± 0·9	15·5 ± 0·4	8·7 ± 0·2	7·7 ± 0·2	1·8 ± 0·1	1·4 ± 0·2	1·0 ± 0·1	0·7 ± 0·1
17	67·3 ± 0·4	16·4 ± 0·4	6·4 ± 0·1	2·8 ± 0·2	0·9 ± 0·1	2·0 ± 0·2	2·4 ± 0·3	0·7 ± 0·1
18	65·8 ± 0·7	15·4 ± 0·3	6·8 ± 0·7	4·5 ± 0·1	1·3 ± 0·1	2·1 ± 0·2	2·1 ± 0·1	0·6 ± 0·1
19	66·2 ± 1·3	14·8 ± 0·2	7·2 ± 0·6	5·5 ± 0·2	1·1 ± 0·1	1·8 ± 0·3	2·0 ± 0·3	0·6 ± 0·1
20	65·8 ± 0·5	16·4 ± 0·3	6·4 ± 0·2	4·1 ± 0·1	1·3 ± 0·1	1·6 ± 0·3	1·8 ± 0·1	0·7 ± 0·0
21	53·1 ± 2·0	16·1 ± 0·7	10·7 ± 0·1	16·3 ± 0·2	4·6 ± 0·2	0·5 ± 0·2	0·3 ± 0·2	0·7 ± 0·0
22	64·7 ± 0·4	16·9 ± 0·3	6·9 ± 0·2	5·3 ± 0·2	1·8 ± 0·1	1·9 ± 0·2	1·3 ± 0·1	0·7 ± 0·1
23	66·9 ± 1·1	15·7 ± 0·8	6·6 ± 0·5	4·3 ± 0·2	1·7 ± 0·1	1·9 ± 0·2	1·8 ± 0·1	0·7 ± 0·1
24	68·2 ± 0·3	16·1 ± 0·2	7·9 ± 0·2	2·0 ± 0·2	0·2 ± 0·0	1·4 ± 0·1	2·9 ± 0·1	0·8 ± 0·0
25	62·6 ± 0·8	16·2 ± 0·1	8·1 ± 0·3	5·7 ± 0·2	1·3 ± 0·1	1·7 ± 0·1	1·6 ± 0·1	0·9 ± 0·0
26	49·7 ± 0·5	16·6 ± 0·4	6·4 ± 0·1	20·9 ± 0·7	3·6 ± 0·2	0·4 ± 0·2	0·2 ± 0·1	0·8 ± 0·1
28	66·5 ± 1·3	13·4 ± 0·4	6·5 ± 0·2	4·4 ± 0·1	1·3 ± 0·1	2·3 ± 0·5	2·9 ± 0·2	0·6 ± 0·0
29	64·2 ± 0·6	12·8 ± 0·2	7·9 ± 0·2	6·7 ± 0·1	1·3 ± 0·1	2·3 ± 0·2	1·9 ± 0·1	0·7 ± 0·0
30	65·9 ± 0·3	17·1 ± 0·3	6·7 ± 0·2	3·2 ± 0·1	1·1 ± 0·1	1·6 ± 0·1	2·0 ± 0·1	0·7 ± 0·1

Coast microtektites. Range of values indicated by ± sign

Sample No.	MnO	Total	Refractive index	Shape	Dia. (μ)	Color	Comments
1	0·06 ± 0·02	98·0	1·521 ± 0·001	Spheroid	170 × 263	Olive-Green	1 large bubble cavity
2	0·06 ± 0·01	98·6	1·526 ± 0·001	Sphere	416	Olive-Brown	2 bubble cavities
3	0·09 ± 0·00	100·1	n.d.	Sphere	800	Olive-Brown	
4	0·09 ± 0·01	97·2	1·523 ± 0·002	Spheroid	655	Brown	~25 bubble cavities
5	0·06 ± 0·01	95·5	n.d.	Dumbbell	n.d.	Greenish-Yellow	Badly corroded
6	0·06 ± 0·00	98·9	n.d.	Sphere	n.d.	Dark Olive-Green	Slightly pitted
7	0·08 ± 0·02	97·6	n.d.	Sphere	n.d.	Dark Olive-Brown	Several protrusions
8a	0·06 ± 0·01	98·2	n.d.	Sphere	850	Dark Olive-Brown	Matrix
8b	0·01	99·2	n.d.	—	—	—	Inclusion
9	0·13 ± 0·01	102·7	n.d.	Fragment	n.d.	Bottle-Green	Badly corroded
10	0·13 ± 0·01	93·4	n.d.	Fragment	n.d.	Bottle-Green	Corroded
11	0·11 ± 0·01	98·3	1·528 ± 0·003	Fragment	n.d.	Deep Yellow	
12	0·0	99·8	1·461 ± 0·002	Sphere	n.d.	Colorless	
13	0·11 ± 0·01	99·4	1·534 ± 0·001	Spheroid	n.d.	Dark Brown	Badly pitted
14	0·09 ± 0·01	99·2	1·522 ± 0·002	Fragment	n.d.	Yellow-Brown	Numerous bubble cavities
15	0·13 ± 0·00	99·3	1·542 ± 0·002	Fragment	n.d.	Yellow-Brown	Clear protrusions
17	0·05 ± 0·01	99·0	1·522 ± 0·002	Fragment	n.d.	Yellow-Brown	Similar to No. 14
18	0·10 ± 0·01	98·7	1·525 ± 0·002	Spheroid	n.d.	Very Dark Brown	Several bubble cavities
19	0·11 ± 0·01	99·3	1·528 – 1·532	Spheroid	n.d.	Dark Brown	Badly pitted
20	0·10 ± 0·01	98·2	1·524 ± 0·002	Fragment	n.d.	Yellow-Brown	
21	0·15 ± 0·01	102·4	1·584 ± 0·002	Spheroid	n.d.	Bottle-Green	Corroded
22	0·11 ± 0·01	99·6	1·524 ± 0·002	Spheroid	n.d.	Greenish-Yellow	Badly pitted
23	0·10 ± 0·01	99·7	1·523 ± 0·002	Fragment	n.d.	Yellow-Brown	Badly pitted
24	0·09 ± 0·01	99·6	n.d.	Spheroid	~100	Very Dark Brown	Badly pitted
25	0·12 ± 0·01	98·2	1·540 ± 0·002	Fragment	n.d.	Brown	
26	0·10 ± 0·01	98·7	1·592 ± 0·002	Fragment	n.d.	Bottle-Green	Badly corroded
28	0·08 ± 0·01	98·0	1·522 ± 0·001?	Cylindrical	n.d.	Yellow-Brown	Pitted
29	0·10 ± 0·01	97·9	1·534 ± 0·002	Spheroid	n.d.	Dark Brown	Circular protrusion
30	0·06 ± 0·01	98·4	1·522 ± 0·002	Cylindrical	n.d.	Yellow-Brown	1 small bubble cavity

Table 2. Comparison between Ivory Coast and Australasian microtektites

	Australasian microtektites				Ivory Coast microtektites		
	A	B	C	D	E	F	G
SiO_2	66·4	57·0	65·8	69·2	64·7	51·3	65·6
Al_2O_3	14·7	12·0	15·2	14·9	15·1	16·0	15·6
FeO	5·7	7·0	5·4	5·3	6·9	8·5	6·9
MgO	5·9	18·0	6·1	3·2	6·3	17·8	4·6
CaO	3·4	4·0	3·9	3·1	1·7	4·2	1·4
Na_2O	1·0	0·8	1·1	1·1	1·6	0·4	1·9
K_2O	1·7	0·4	1·0	2·3	1·6	0·2	1·8
TiO_2	0·7	0·6	0·7	0·8	0·7	0·8	0·7
MnO	0·1	0·1	0·1	0·1	0·1	0·1	0·1
Total	99·6	99·9	99·3	100·0	98·7	99·3	98·6

A. Average of all 60 analyses.
B. Average of 8 bottle-green objects.
C. Average of 14 yellowish-green microtektites.
D. Average of 38 "normal" microtektites.
E. Average of all 28 analyses.
F. Average of 4 bottle-green objects.
G. Average of 23 "normal" microtektites; for range see Table 3.

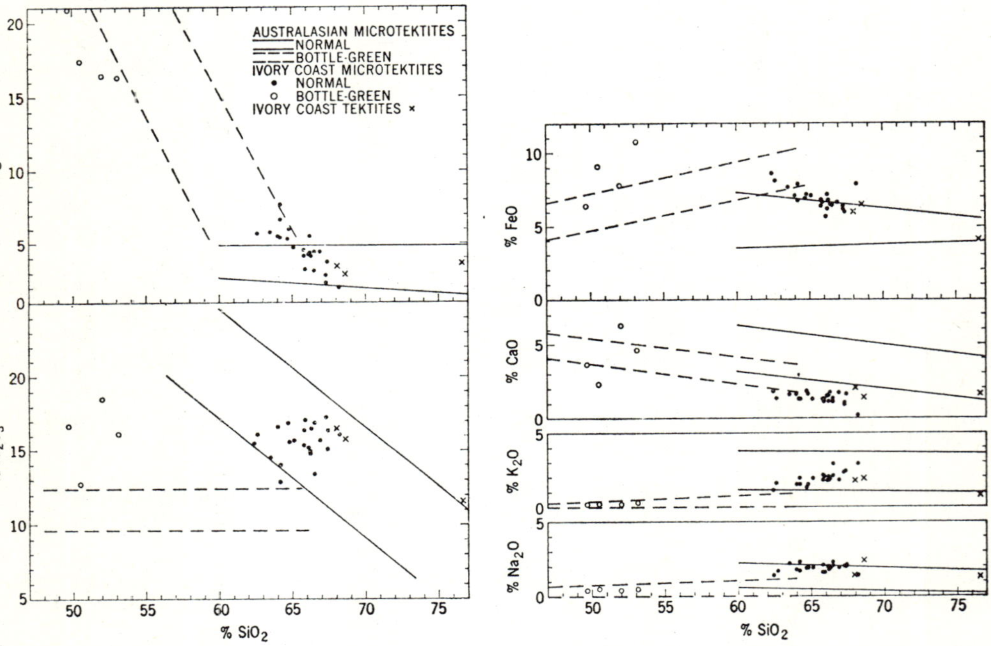

Fig. 2. Variation diagrams of MgO, Al_2O_3, FeO (total iron), CaO, Na_2O and K_2O vs. SiO_2 for Ivory Coast microtektites, Australasian microtektites and Ivory Coast tektites. Australasian microtektite data from CASSIDY *et al.* (1969). Ivory Coast tektite data from BARNES (1940).

Fig. 2). Otherwise, the Ivory Coast bottle-green objects seem to be similar to the Australasian bottle-green objects.

Using the electron microprobe the spectrum of ten of the glassy objects (six normal and four bottle-green microtektites) was scanned from 1·1 to 9·9 Å. The only peaks observed for the six normal microtektites were those for Si, Al, Fe, Mg, Ca, Na, K, Ti and Mn. In addition to the above elements, the four bottle-green objects also gave a peak for chromium. Comparison of the height of these chromium peaks to chromium peaks in pyroxene standards of approximately the same composition as the bottle-green objects indicates that the bottle-green objects contain approximately 0·15% Cr_2O_3.

Yellowish-green glassy objects of probable intermediate compositions make up only about ten percent of the Ivory Coast microtektite population. Only two (Nos. 5 and 22) yellowish-green spherules were analyzed and these did not appear to be much different from "normal" microtektites (Table 1).

Comparison with Ivory Coast tektites

Like Ivory Coast tektites, the Ivory Coast microtektites are usually spherical or ovoid in shape with pits and corrosion grooves. Likewise, the glass is usually brownish in thin section and contains spherical bubble cavities. Grooves forming an annular depression around elevated central portions have been observed on a few of the larger microtektites. These may be similar to the navel-like "cupule" observed on an Ivory Coast tektite specimen described by BAKER (1959).

Unfortunately only three analyses of Ivory Coast tektites have been published (BARNES, 1940). Two of these have low silica contents (~68%) and the other has a high silica content (75·6%). The two analyses with low silica contents fall at roughly the same position on Harker diagrams as the Ivory Coast microtektites (Fig. 2). CHAO *et al.* (1965) have analyzed seven Ivory Coast tektites, which, on the basis of physical properties, show the widest range in composition among a suite of fourteen Ivory Coast tektites. Although the analyses are not published, CHAO *et al.* state that the Ivory Coast tektites are characterized by comparatively low SiO_2 (67·2–69·1%), relatively high Al_2O_3 (16·1–16·8%), unusually high MgO/CaO ratios (about 3·1), and by Na_2O being slightly greater than K_2O. Likewise, the "normal" Ivory Coast microtektites have an unusually high average MgO/CaO ratio of about 3·6 and an average Na_2O/K_2O ratio of about 1·1, in contrast to the "normal" Australasian microtektites which have an average MgO/CaO ratio of approximately 1·2 and an Na_2O/K_2O ratio less than one (~0·5). In addition, the Ivory Coast microtektites have slightly lower average silica contents than the Australasian microtektites. Thus the Ivory Coast microtektites are similar to the Ivory Coast tektites in many respects (Table 3) and differ in composition from Australasian microtektites in the same manner in which Ivory Coast tektites differ from the Australasian tektites.

Comparison with igneous glass

Before the chemical composition of the Australasian microtektites had been published, several authors (MARTIN, 1967; VON KOENIGSWALD, 1968; BAKER, 1968) suggested that the Australasian microtektites, or at least a portion of them, might be volcanic glass analogous to Pele's Tears. However, a comparison between the chemical composition of the Australasian and Ivory Coast microtektites and the composition of igneous glasses shows that although there is little or no difference in the alumina content between microtektites and igneous glasses, the microtektites definitely have higher MgO and lower Na_2O contents than igneous glasses with comparable SiO_2 contents (Fig. 3). In addition, the microtektites have higher

Table 3. Comparison between Ivory Coast tektites and Ivory Coast microtektites

	Ivory Coast tektites BARNES (1940) (3)	Ivory Coast tektites CHAO *et al.* (1965) (7)	Ivory Coast microtektites (23)
SiO_2	68·00–76·56	67·2–69·1	62·41–68·20
Al_2O_3	11·54–16·46	16·1–16·8	12·75–17·33
FeO	6·08–6·62		5·70–8·61
MgO	2·88–3·60		2·01–7·69
CaO	1·40–2·00		0·22–1·82
Na_2O	1·32–2·35		1·41–2·30
K_2O	0·82–1·92		1·12–2·87
TiO_2	0·60–0·80		0·45–0·93
MnO	0·06–0·09		0·05–0·13
$\frac{Na_2O}{K_2O}$	1·2	>1	1·1
$\frac{MgO}{CaO}$	2·0	3·1	3·6

( ) Refers to number of analyses.

average iron and CaO contents and lower K_2O contents than igneous glasses (Fig. 3). An exception is the CaO content of the Ivory Coast microtektites which is somewhat lower than igneous glasses. Thus, the normal Australasian and Ivory Coast microtektites differ from igneous glasses in the same manner that the Australasian and Ivory Coast tektites differ from igneous glasses (BARNES, 1940). This is to be expected since the Australasian and Ivory Coast microtektites have compositions similar to the Australasian and Ivory Coast tektites, respectively.

Like the "normal" microtektites, the bottle-green objects have Al_2O_3 contents that cannot be distinguished from igneous glasses. Likewise the bottle-green objects have lower Na_2O and higher MgO contents than igneous glasses. Also, the K_2O content of the bottle-green objects is less than for igneous glasses of comparable silica content. However, unlike the "normal" microtektites, the bottle-green objects also have lower iron, CaO and TiO_2 contents than igneous glasses. This change in relationship means that many of the yellowish-green objects, with intermediate compositions, are similar to igneous glasses in FeO and CaO as well as Al_2O_3 contents (Fig. 3). Yet even these microtektites have distinctly higher MgO contents and lower Na_2O contents than igneous glass.

VON KOENIGSWALD (1968) suggested that some of the glassy objects from the Australasian strewn field might have been derived from volcanism in the Indochina or New Zealand area, and BAKER (1968) suggested that the bottle-green, high specific gravity glassy objects (which he thought were found only in the Wharton Basin) could have had their origin in Javanese volcanic eruptions. The chemical composition (determined by microprobe analysis) of two ash samples from deep-sea sediments taken south of Sumatra and Java, and one sample from a core taken east of New Zealand shows no similarity to the microtektites. (One of the samples, from core V19-153, taken south of Sumatra, was taken immediately (~5 cm) below sediment containing abundant microtektites.)

NINKOVICH (1968) has made an extensive study of volcanic ash deposits in deep-sea sediment cores taken around New Zealand. Five distinct layers of

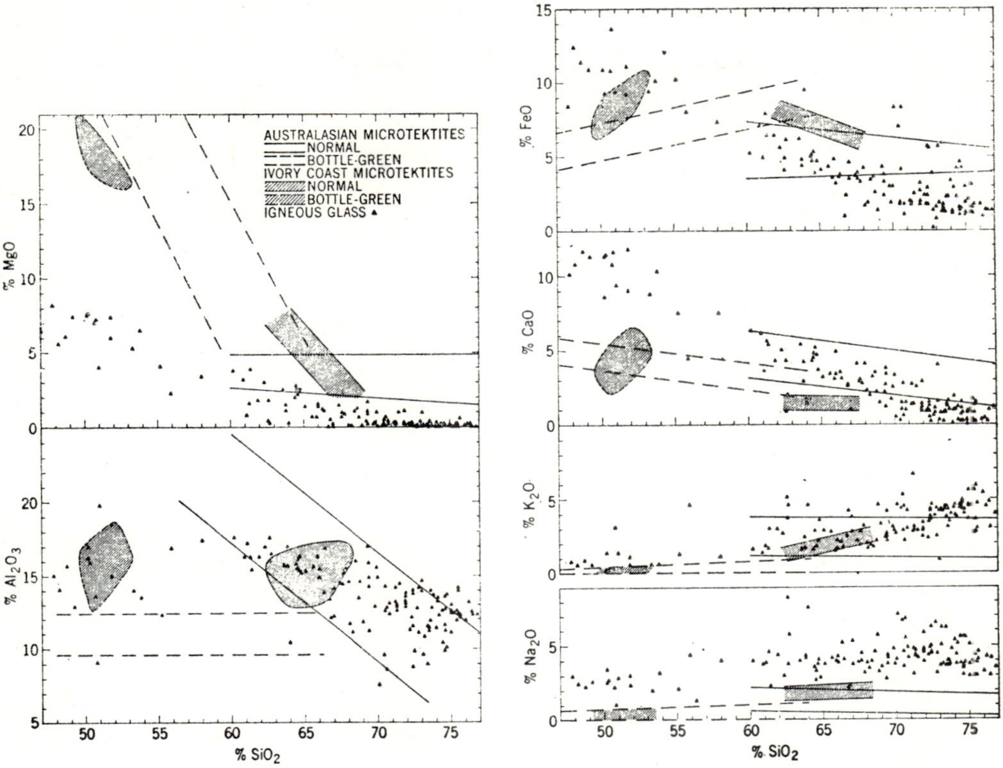

Fig. 3. Variation diagrams of MgO, Al_2O_3, FeO (total iron), CaO, Na_2O and K_2O vs. SiO_2 for Ivory Coast and Australasian microtektites and igneous glasses. Igneous glass data is from 120 analyses of igneous glasses (obsidian, perlite, pitchstone, pumice, tuff, volcanic ash, residual glass, etc.) collected at random from the literature. Three analyses of "Pele's Tears" (SiO_2 contents of 47·0, 49·1 and 50·8%) are included as well as analyses of three volcanic ashes from deep-sea sediments taken near or from within the Australasian strewn field.

windborne rhyolitic ash were identified only in cores taken east of New Zealand. According to the paleomagnetic stratigraphies of the cores, the ash layers range in age from 0·3 to 0·9 m.y. The ash layers have been correlated with ignimbrite deposits on New Zealand and their distribution indicates that the ash was transported by westerly winds. The ash in all five layers has a shard-like shape typical of volcanic ashes (NINKOVICH, 1968). A study was made of two of the cores investigated by NINKOVICH. Both cores contain ash layers and penetrate the Matuyama–Brunhes reversal boundary. No microscopic glassy objects ($>125\ \mu$ in dia.) with rounded shapes were observed in either core near the Matuyama–Brunhes boundary. Thus there is no evidence to indicate that the glassy objects (microtektites) found in deep-sea sediments adjacent to the Australasian tektite strewn field are the result of New Zealand volcanism.

Although very fine volcanic ashes can be carried several times around the globe by winds before being deposited, the sand size ash fragments always fall in the immediate neighborhood of the eruption (RITTMAN, 1962). Thus, because of their large size (up to 0·8-mm dia.) it is unlikely that the distribution of the Australasian microtektites is a result of a single giant volcanic eruption. In addition, windborne ashes are sorted by the wind with the mean grain size decreasing away from the source. No such sorting is observed among the microtektite-bearing cores; nor

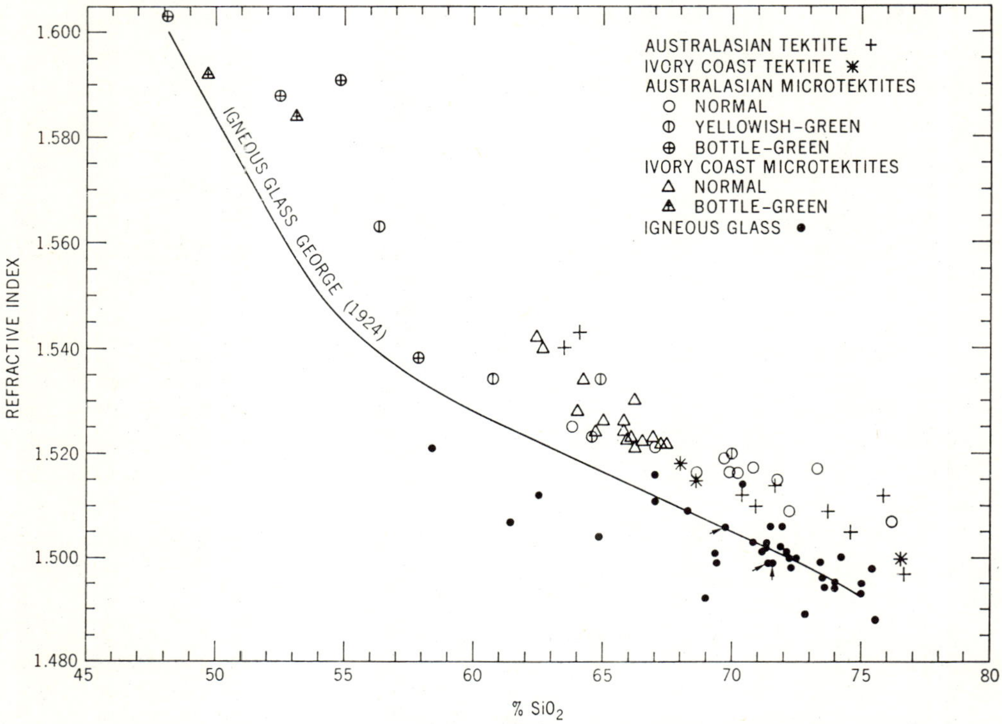

Fig. 4. Refractive index vs. SiO_2 content for Australasian and Ivory Coast tektites, Australasian and Ivory Coast microtektites and igneous glass. Tektite data from BARNES (1940). Australasian microtektite data from CASSIDY *et al.* (1969). The curve for igneous glass is from GEORGE (1924). The igneous glass data were taken from various sources and seems to confirm the high silica end of GEORGE's curve. The arrows indicate volcanic glass from deep-sea sediments from the Australasian tektite strewn field and near New Zealand (SiO_2 content determined by microprobe analysis).

is there a systematic increase in numbers toward any hypothetical source area as would be expected from a volcanic origin (CASSIDY *et al.*, 1969).

BARNES (1940) showed that igneous glasses can be distinguished from tektites on a plot of silica content vs. refractive index. The same relationship exists for microtektites and igneous glasses (Fig. 4) illustrating once again that the microtektites are similar to tektites and distinct from igneous glasses.

Comparison with hay-silica glass

BAKER (1968) suggested that the low specific gravity microtektites, described by GLASS (1967) from the Australian and Wharton Basins, might be hay-silica glass produced in a prehistoric forest, bush or grass fire and dispersed into the Wharton and Australian Basins by the southeast Trade and Brickfielder Winds. He gives the following composition for a sample of hay-silica glass: SiO_2 61·7%, TiO_2 0·19, Al_2O_3 1·16, Fe_2O_3 0·83, FeO 0·28, MgO 4·88, MnO 0·30, CaO 6·77, Na_2O 8·73, K_2O 8·53, P_2O_5 5·66, CO_2 nil, SO_3 tr., Cl 0·08, H_2O+ 0·20. The low Al_2O_3 content, which is characteristic of glass produced by the incineration of opal-bearing vegetation, is in contrast to the high Al_2O_3 content of the microtektites which contain approximately ten times as much Al_2O_3. In addition, glass produced in the above manner, contains more CaO, Na_2O and K_2O (approximately five times as much) and less iron than the microtektites.

In his description of the micro-forms of hay-silica glass, BAKER states that some have tapering ends which sometimes represent the proximal ends of long, drawnout, filamentous glass that connected one lenticular form with another. Such lenticular and filamentous shapes have not been observed among the microtektites.

CONCLUSIONS

It is the opinion of the author that the data presented in this paper confirm the suggestion (GLASS, 1968) that the microscopic glassy objects found in deep-sea sediments near the Ivory Coast are microtektites and that they represent a portion of the Ivory Coast tektite strewn field. The following observations support this conclusion:

1. Both the Ivory Coast and Australasian microtektites contain too much Al_2O_3 and not enough potash to be glass formed by the incineration of opal-bearing vegetation.

2. Like the Australasian microtektites, the Ivory Coast glassy objects are distinct from igneous glasses in that for a given silica content they have a higher MgO and lower Na_2O content. In addition, the widespread geographical extent and lack of sorting characteristic of windborne material of the Australasian microtektites and the rounded shapes, absence of crystalline inclusions and presence of lechatelierite particles in both the Australasian and Ivory Coast microtektites all suggest that both groups of deep-sea microscopic glassy objects are related to tektites rather than to terrestrial volcanic glasses.

3. The composition of the Ivory Coast microtektites is similar to the composition of the Ivory Coast tektites. And like the Ivory Coast tektites the Ivory Coast microtektites are distinguished from the Australasian microtektites and tektites by having a higher MgO/CaO ratio and an average Na_2O/K_2O ratio greater than one. (The comparison between the Ivory Coast microtektites and Ivory Coast tektites is not as strong as it could be, however, because of the lack of published analyses of Ivory Coast tektites. The author feels that additional analyses of Ivory Coast tektites are badly needed.)

4. In addition to the above, it has already been shown (GLASS, 1968) that the age of deposition of the Ivory Coast deep-sea glassy objects is similar to the K–Ar age of the Ivory Coast tektites.

Badly corroded bottle-green glassy spherules with high refractive index (1·584–1·630), low silica content and high MgO content have been found associated with both the Australasian and Ivory Coast microtektites. These bottle-green objects, although different from microtektites (and thus tektites) in appearance, physical properties and chemical composition, have nevertheless been found only in association with them, and therefore are most likely related to them.

Acknowledgments—I thank Dr. L. S. Walter and F. Wood for help with microprobe analyses. Drs. J. A. O'Keefe, C. C. Schnetzler, B. Mason, D. R. Chapman and W. A. Cassidy critically reviewed the manuscript. Discussions with D. Ninkovich were helpful. I am indebted to Dr. M. Ewing, director of Lamont-Doherty Geological Observatory, for taking the core (in which the Ivory Coast microtektites were discovered) in 1963 for the joint purpose of making heat flow measurement and searching for cosmic particles.

References

Baker G. (1959) Tektites. *Mem. Nat. Mus. Vict.* **23,** 1–313.

Baker G. (1968) Micro-forms of hay–silica glass and of volcanic glass. *Mineral. Mag.* **36,** 1012–1023.

Barnes V. E. (1940) North American tektites. Contributions to Geology, Part 2. *Univ. Texas Publ.* No. 3945, 477–582.

Bence A. E. and Albee A. L. (1968) Empirical correction factors for the electron microanalysis of silicates and oxides. *J. Geol.* **76,** 382–403.

Cassidy W. A., Glass B. and Heezen B. C. (1969) Physical and chemical properties of Australasian microtektites. *J. Geophys. Res.* **74,** 1008–1025.

Chao E. C. T., Cuttitta F., Carron M. K., Annell C. and Mount P. (1965) New data on some Ivory Coast tektites. *Trans. Amer. Geophys. Union* **46,** 427 (abstract).

George W. O. (1924) The relation of the physical properties of natural glasses to their chemical composition. *J. Geol.* **32,** 353–372.

Glass B. (1967) Microtektites in deep-sea sediments. *Nature* **214,** 372–374.

Glass B. (1968) Glassy objects (microtektites?) from deep-sea sediments near the Ivory Coast. *Science* **161,** 891–893.

Martin R. (1967) Tektites, impactites and volcanic glass. 30th Annual Meeting of the Meteoritical Society (abstract).

Ninkovich D. (1968) Pleistocene volcanic eruptions in New Zealand recorded in deep-sea sediments. *Earth Planet. Sci. Lett.* **4,** 89–102.

Rittman A. (1962) *Volcanoes and Their Activity*, pp. 74–75. John Wiley.

von Koenigswald G. H. R. (1968) The relationshipof shape, size and texture in Asiatic tektites. Reprinted from *Kon. Ned. Akad. Wetensch.* Proc. **B61,** No. 1 (Tektite Studies X).

Walter L. S. (1967) Tektite compositional trends and experimental vapor fractionation of slicates. *Geochim. Cosmochim. Acta.* **31,** 2043–2063.

37

RTH AND PLANETARY SCIENCE LETTERS 9 (1970) 240–246. NORTH-HOLLAND PUBLISHING COMPANY

COMPARISON OF THE CHEMICAL VARIATION IN A FLANGED AUSTRALITE WITH THE CHEMICAL VARIATION AMONG "NORMAL" AUSTRALASIAN MICROTEKTITES

Billy P. GLASS

Planetology Branch Goddard Space Flight Center, Greenbelt, Md. 20771, USA

Received 30 April 1970

Revised version received 27 June 1970

A single flanged australite was found by electron microprobe analysis to have a wide range in chemistry. The silica content varies from 60 to 85%. The Al, Fe, Mg, Ca, Na, Ti, Mn and Ba contents vary inversely, and K and P vary directly with the silica content. The chemical variations and trends for the australite are similar to those found among the Australian Basin deep-sea microtektites.

ntroduction

Aicrotektites have been found in deep-sea sedi-
its adjacent to the Australasian and Ivory Coast
tite strewn fields [1,2]. Based on the physical
chemical properties of sixty Australasian micro-
tites, Cassidy et al. [3] were able to divide them
two groups. One group is composed of micro-
tites which have a characteristic corroded appear-
e and bottle-green color. These microtektites,
ed bottle-green microtektites because of their
or, have low silica contents and high magnesium
tent. The second group is composed of micro-
ites with physical and chemical properties similar
Australasian tektites and are thus called normal
rotektites. The chemical compositions of the nor-
microtektites are similar to the compositions of
Australasian tektites except that about one-third
e silica contents less than 68% which was the lower
t of silica content most often quoted for Austral-
n tektites before 1969.

Cassidy et al. [3] concluded from their study of
tralasian microtektites that Australasian tektites
t have a wider range in chemistry than previously
ight and that there must be a high Mg (bottle-
n) variety of Australasian tektite. Both of these
clusions have been at least partially confirmed by the work of Chapman and Scheiber [4] who have determined the chemical composition of 530 Australasian tektites. They define a high Mg variety of Australasian tektites which seem to be related to the high Mg microtektites. In addition, they found individual specimens with silica contents as low as 62%. If 62% is used as a lower limit for the silica content of Australasian tektites, then all but four microtektite specimens, designated as normal microtektites by Cassidy et al., would have silica contents within the silica range for Australasian tektites.

In order to obtain further evidence that the Australasian tektites have a greater range in chemistry than is generally given, the chemical variation of a single flanged australite was studied with an X-ray microprobe analyzer.

2. Method of analysis

The tektite used for this investigation was an australite button given to Dr. O'Keefe by Dr. Baker. Two thin sections were cut from the tektite. One included only the flange, while the other one included both the flange and the core (fig. 1). Both were cut at right angles to the posterior and anterior surfaces and to the outside edge of the flange. During a preliminary

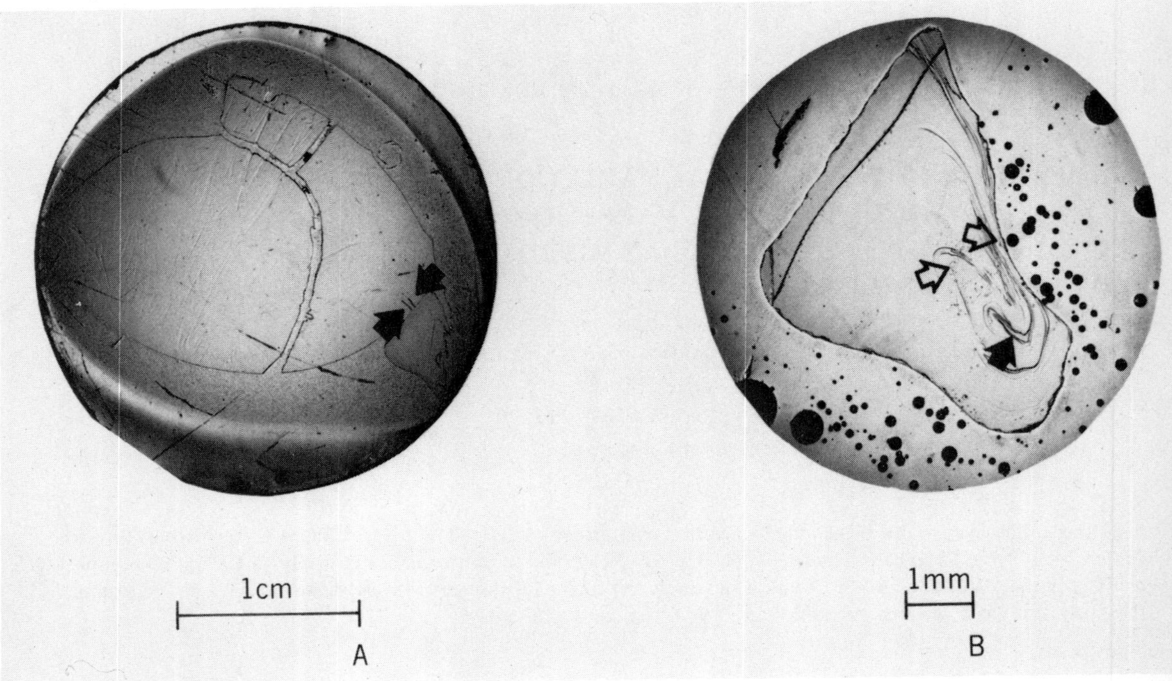

Fig. 1. Photomicrographs of australite thin sections. (A) Thin section of flange and core. Solid arrows indicate the location of two low silica inclusions. (B) Thin section of flange. Solid arrow indicates location of low silica inclusion. Open arrow indicate location of silica-rich inclusions.

investigation the thin sections were scanned in order to locate the areas with the highest and lowest silica content. In order to obtain a continuous range in composition, traverses were made across the low and high silica inclusions. The percent abundance of the following elements was determined in groups of three: Si, Al, Fe, Mg, Ca, Na, K, Ti, Mn, P and Ba. Silicon was analyzed along with two other elements in each set of analyses, so that the variation of each element could be compared directly with the silica content.

The standards, method of analysis and computer program for converting the raw data to percent oxides were the same as those used by Cassidy et al. [3] for determining the chemical composition of the Australasian microtektites.

3. Presentation of data

A total of three low silica and two high silica inclusions were found in the two thin sections (fig. 1). Two of the low silica inclusions are in the core area. The greatest range found in the silica content was approximately 60 to 85% (table 1), however the lc silica and high silica areas represent less than one p cent of the thin section. The bulk of the glass has average composition of about 76% SiO_2 (table 1).

Table 1

Range of composition in a single flanged australite as dete mined by electron microprobe analysis.

	Low silica inclusion (%)	High silica inclusion (%)	Avera (%)
SiO_2	60	85	76
Al_2O_3	23	7.5	12
FeO	6.0	2.6	3.9
MgO	4.0	0.7	1.9
CaO	4.9	0.9	2.5
Na_2O	0.7	0.2	0.7
K_2O	1.0	1.7	1.8
TiO_2	1.0	0.4	0.6
MnO	0.10	0.04	[illegible]
	100.7	99.0	99.5

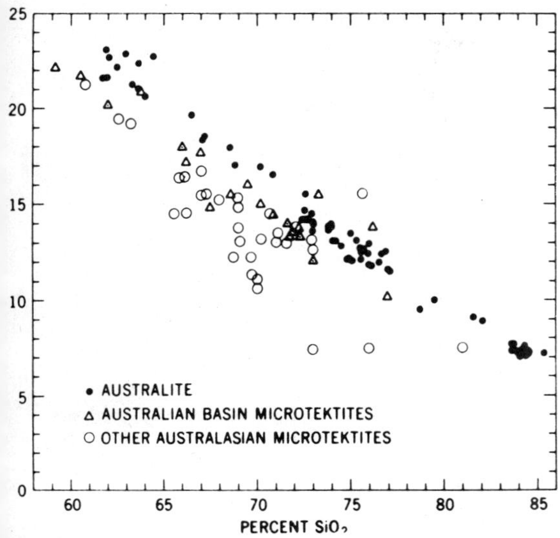

ɡ. 2. Al_2O_3 vs. SiO_2 for Australasian microtektites and a ıgle australite. Microtektite data from Cassidy et al. [3] pen symbols) and this paper (open symbols with a dot in the center).

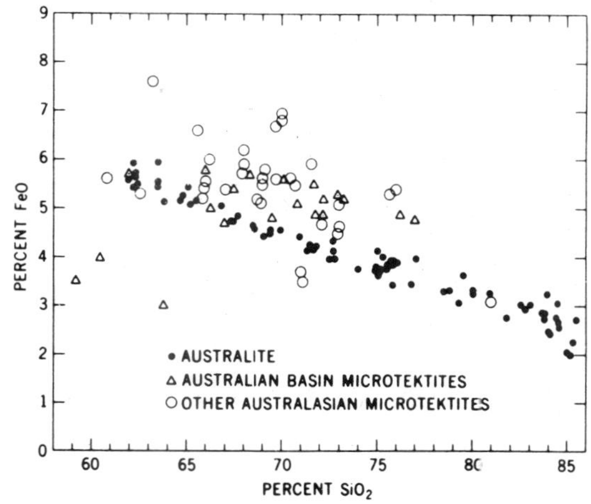

Fig. 3. FeO vs. SiO_2. See fig. 2 for further explanation.

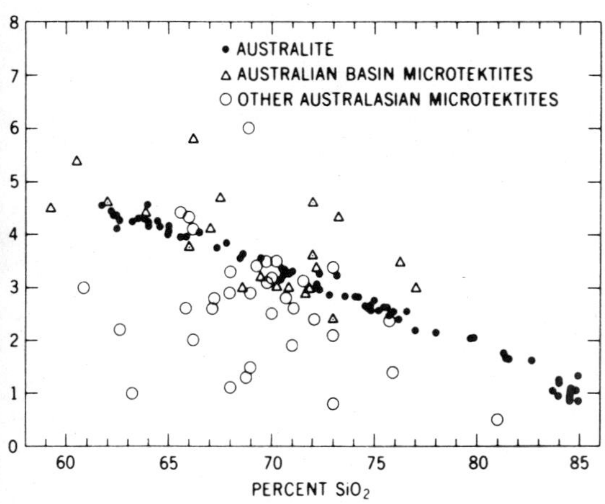

Fig. 4. CaO vs. SiO_2. See fig. 2 for further explanation.

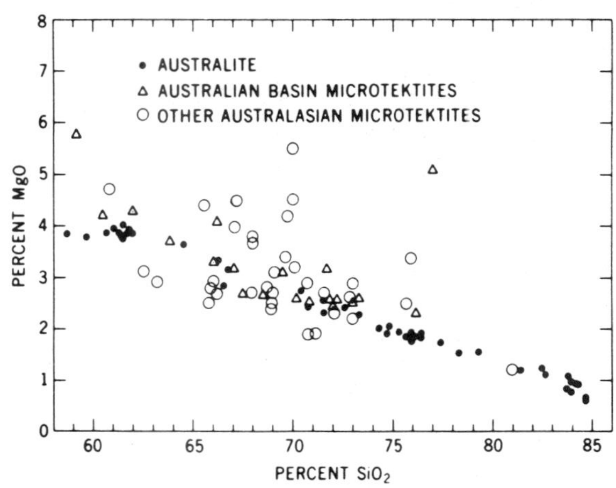

Fig. 5. MgO vs. SiO_2. See fig. 2 for further explanation.

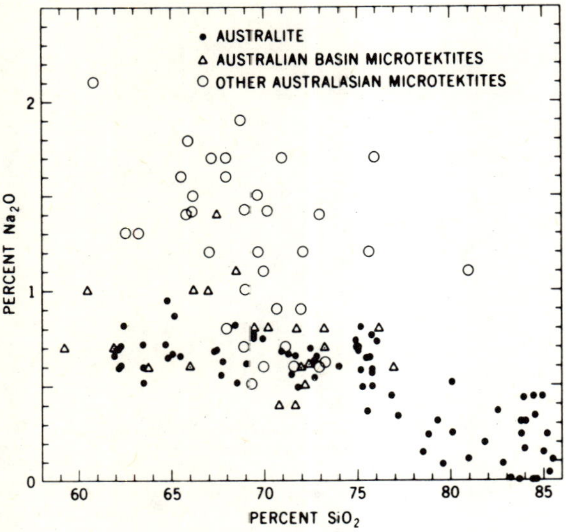

Fig. 6. Na_2O vs. SiO_2. See fig. 2 for further explanation.

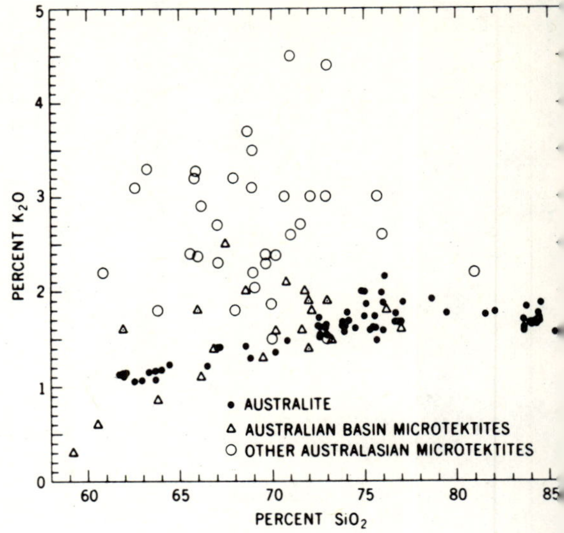

Fig. 7. K_2O vs. SiO_2. See fig. 2 for further explanation.

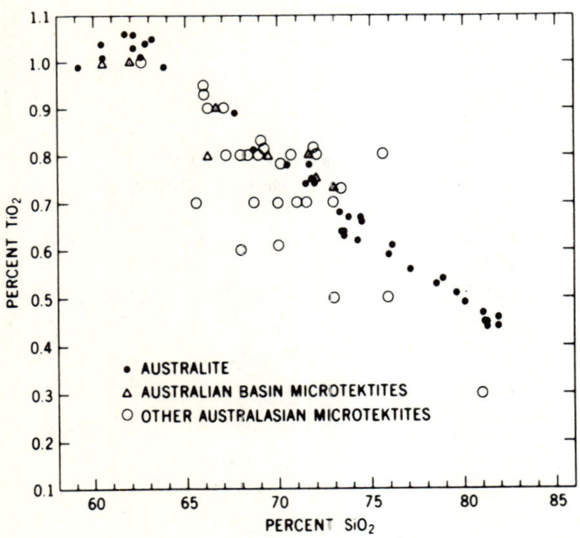

Fig. 8. TiO_2 vs. SiO_2. See fig. 2 for further explanation.

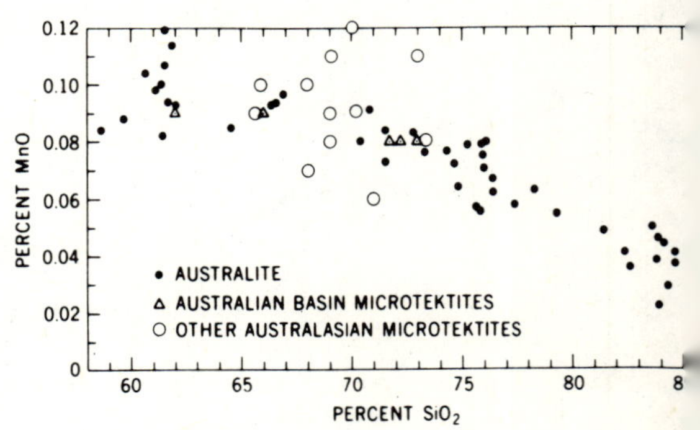

Fig. 9. MnO vs. SiO_2. See fig. 2 for further explanation.

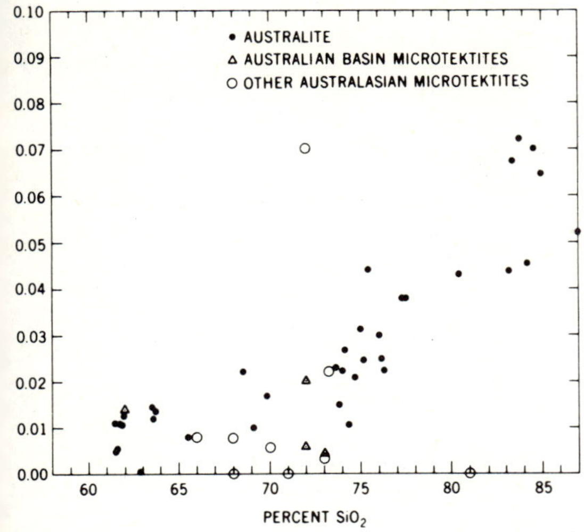

'ig. 10. P_2O_5 vs. SiO_2. See fig. 2 for further explanation.

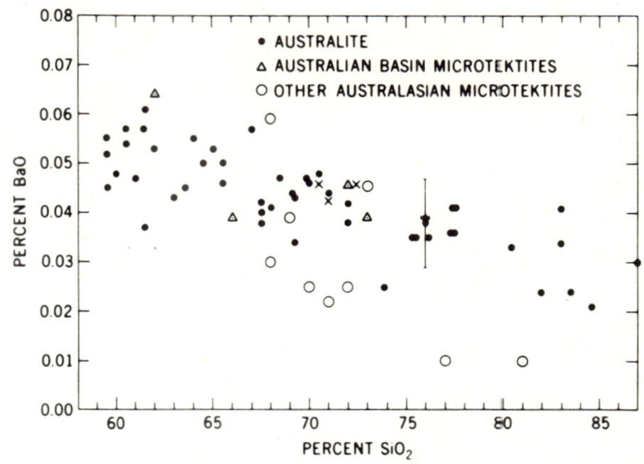

Fig. 11. BaO vs. SiO_2. The bar represents the variation in BaO observed in one small area with a homogeneous SiO_2 content of 76% which is the average silica content of the australite. The X's represent normal australite analyses from Chapman and Scheiber [4].

systematic difference was observed between the re and the flange areas. As can be seen in table 1, average composition of the tektite is what one uld expect for an australite except for the someat lower Na and K values.

Harker diagrams of the various oxides versus silica gs. 2–11) show that there is a strong inverse relanship between silica and Al_2O_3, iron (as FeO), O, CaO, TiO_2, MnO and BaO. Soda also seems to y inversely with silica, especially between 76 and % silica, but below 76% silica the Na_2O content ms to remain almost constant. However, scatter of Na_2O data is quite large. Phosphorus, on the ner hand, varies directly with the silica content. e potash also varies directly with the silica content to approximately 76% (which is the average comsition of the tektite). Above 76% silica, there is a ght inverse correlation between K_2O and SiO_2.

Australasian normal microtektite data from Casy et al. [3] and more recent unpublished Australan normal microtektite data are plotted on the rker diagrams for comparison with the australite ta. Cassidy et al. did not determine the Ba or P ntent of the Australasian microtektites, thus only ta obtained during this investigation are plotted the Harker diagrams for BaO vs. SiO_2 and P_2O_5 SiO_2. In addition, only Mn data obtained for this estigation are plotted on the MnO vs. SiO_2 diagram since Cassidy et al. rounded all the MnO values to 0.1%.

Although there is greater scatter in the microtektite data, the similarity between the Australian Basin microtektites and the australite data is obvious. (The FeO vs. SiO_2 diagram, fig. 3, shows the most disagreement because of three low FeO values at the low silica end of the diagram.) As would be expected, microtektites from other parts of the strewn field have compositions which do not agree as well with the australite data. This is especially obvious in the Na_2O vs. SiO_2 and K_2O vs. SiO_2 plots (figs 6 and 7).

4. Discussion

The similarity between the normal Australian Basin microtektites and the australite is even more striking when it is considered that the australite data are from a single specimen and the microtektite data are of individual microtektites from four different locations. Analysis of different australite specimens would certainly result in a greater range of values and thus in better agreement with the microtektite data.

Several tektite investigators have questioned the identification of the Australasian microtektites because of their greater range in chemistry than reported for Australasian tektites. However, the ranges

Table 2
Comparison of the chemical variation among normal Australian Basin microtektites and previously reported australites with t[illegible] variation in a single australite determined by electron microprobe analysis.

	Normal Australian Basin microtektites [21] Cassidy et al. [3]	Australites [23] Schnetzler and Pinson [7]	Australite thin section
SiO_2	59.2 - 76.2	69.0 - 78.8	59 - 85
Al_2O_3	12.1 - 22.1	9.4 - 15.3	7 - 23
FeO	3.0 - 5.8	3.7 - 5.7	2 - 5.9
MgO	2.3 - 5.8	1.4 - 2.5	0.6 - 4.1
CaO	2.4 - 5.8	2.1 - 5.1	0.8 - 4.6
Na_2O	0.4 - 1.4	0.9 - 1.8	0.0 - 0.96
K_2O	0.3 - 2.5	1.7 - 2.6	1.0 - 2.2
TiO_2	0.7 - 1.0	0.5 - 0.8	0.43 - 1.07
MnO	0.08 - 0.1	0.08 - 0.12	0.02 - 0.12

in the major oxide contents in a single flanged australite, reported on in this paper, are not only greater than the ranges reported among 23 australite analyses, but are even greater than the ranges among 21 normal Australian Basin microtektites (table 2).

The K_2O concentration in australites shows little variation with SiO_2 content and for most published data there seems to be a slight inverse relationship between K_2O and SiO_2. This is in contrast to the K_2O-SiO_2 trend for the australite specimen reported on in this paper. Otherwise, however, within the same SiO_2 range the chemical trends obtained for the single australite specimen are similar to those among australites and other Australasian tektites.

Some of the chemical variations observed for Australasian tektites, especially between the various groups (australites, indochinites, philippinites, etc.), is undoubtedly due to heterogeneity of the parent material. However, it is difficult to understand how incomplete mixing of a heterogeneous parent material could produce such well-defined trends, over such a wide range in concentrations, as is found for the single australite specimen reported on in this paper. On the other hand, the trends observed for this australite are similar to those produced by vapor fractionation in an oxidizing environment as proposed by Walter [5] and such a process is capable of producing well-defined trends.

Unlike those of the other oxides, the Na_2O-SiO_2 and K_2O-SiO_2 plots do not appear to be straight lines. Below approximately 76% SiO_2, the Na_2O-SiO_2 slope decreases and is almost horizontal and above approximately 76% SiO_2, K_2O appears to b[e] inversely correlated with SiO_2 rather than directly correlated. These changes in slope may be due to [a]n increase in volatility of the alkalis at lower SiO_2 c[on]tents. (In fact, Walter and Giutronich's [6] vapor fractionation data can be interpreted as showing t[he] same changes in slope for Na_2O-SiO_2 and K_2O-Si[O_2] at approximately 76% SiO_2.) An alternative expla[na]tion for the inverse correlation between Na and Si [and] K and Si above 76% SiO_2 may be that the trends [are] the result of mixing with lechatelierite (SiO_2 glass). However, the extrapolated trends do not seem to pass close to 100% SiO_2 at 0% Na_2O or 0% K_2O.

5. Summary and conclusions

Microprobe analysis of thin sections taken fro[m a] flanged australite shows that the australite has a c[on]tinuous range in SiO_2 content from at least 60% [up] to 85% (excluding lechatelierite). This range in sil[ica] content is greater than that observed among all th[e] normal Australian Basin microtektites. No system[atic] difference in composition was observed between [the] core and the flange of the australite. Aluminum, [Fe,] Mg, Ca, Ti, Mn, Ba, and to a lesser extent, Na, var[y] inversely with the silica content, while P and K va[ry] directly with the silica content. The chemical vari[a]tions and trends observed for the australite are si[milar] to those observed among Australasian microtekti[tes] specially the Australian Basin microtektites – and chemical trends for both the australite and microtites are similar to those produced by vapor fracıation in an oxidizing environment.

The data presented in this paper are interpreted as ther evidence that the Australasian tektites and rotektites have a common origin.

knowledgements

am grateful to Dr. J.A. O'Keefe of Goddard ce Flight Center for suggesting that I look for mical variations in a thin section of an australite ge which was prepared for him by W. Kouns. Vood, Jr. helped with the microprobe analysis discussions with Dr. L.S. Walter were helpful. s research was performed while the author was on ve duty with U.S. Army Corps of Engineers ased to Goddard Space Flight Center. The author resently a National Research Council Postdoctoral Resident Research Associate At Goddard Space Flight Center.

References

[1] B.P. Glass, Microtektites in deep-sea sediments, Nature 214 (1967) 372.

[2] B.P. Glass, Glassy objects (microtektites?) from deep-sea sediments near the Ivory Coast, Science 161 (1968) 891.

[3] W.A. Cassidy, B.P. Glass and B.C. Heezen, Physical and chemical properties of Australasian microtektites, J. Geophys. Res. 74 (1969) 1008.

[4] D.R. Chapman and L.C. Scheiber, Chemical investigation of Australasian tektites, J. Geophys. Res. 74 (1969) 6737.

[5] L.S. Walter, Tektite compositional trends and experimental vapor fractionation of silicates, Geochim. Cosmochim. Acta 31 (1967) 2043.

[6] L.S. Walter and J.F. Giutronich, Vapor fractionation of silicate melts at high temperatures and atmospheric pressure, Solar Energy J. 11 (1967) 163.

[7] C.C. Schnetzler and W.H. Pinson, Jr., The chemical composition of tektites, in: Tektites, edited by J.A. O'Keefe (University of Chicago Press, Chicago, 1963) p. 95.

Origin of Tektites

VI

Editors' Comments on Papers 38 and 39

We had hoped to include Lyttleton's (1964) stimulating paper on the cometary origin of tektites as a third paper in the group of papers on the extraterrestrial origin of tektites, but a reply had not been received to our request at the time this volume went to press. However, the essence of Lyttleton's paper is included in our introductory chapter.

Hardcastle (1926) and Hanus (1928) postulated that tektites are "plastic sweepings" from meteorites passing through the atmosphere. An objection to this postulate is the absence of meteorites with tektite composition from which the sweepings could be derived. The same objection applies to the hypothetical parent body of tektites postulated by Adams (1963) and Adams and Huffaker.* Although some workers have postulated that Muong Nong-type tektites are meteorites, instead they almost surely are puddles of tektite melt formed on earth. The analysis by Adams and Huffaker* starts with the entry of the hypothetical parent body of tektites into the atmosphere. The surface area of this body necessarily would have been tremendous to have provided the Australasian microtektites alone, estimated by Cassidy et al. (1969)* to be equivalent to a sphere 200 m in diameter.

The vapor-pressure data used by Adams and Huffaker* has been questioned by Centolanzi and Chapman (1966), who measured the vapor pressure for tektite melt and concluded that the vapor-pressure data by Walter and Carron (1964) is in error. If this conclusion is correct, the use of Walter and Carron's data by Adams and Huffaker* invalidates their aerodynamic calculations of very shallow, grazing tektite trajectories, as well as their theory as a whole. However, Walter and Adams (1967) argue that the method that Walter and Carron used to determine vapor pressure is valid.

Chapman* analyzes much of the data accumulated on tektites during ten years or more research at Ames Research Center. The effort in chemistry alone for the Australasian area of tektite occurrence surpasses that of all workers combined for any other group of tektites. The determination of the distribution of the chemical types of Australasian tektites on earth and the matching of the pattern found with ejection along one ray from Tycho was a superb accomplishment. However, is this beautiful theory, based on prodigious research and so logically presented, consigned to oblivion because the Apollo Missions have failed to recover material of tektite composition? The reader is advised to read Chapman's paper and make up his own mind. He is particularly asked to read the acknowledgments.

Nearly everyone will agree that Dean Chapman is a superior aerodynamicist. It is now recognized that tektites are not from the moon, and a first-rate aerodynamicist, such as Chapman is, should be capable equally well of demonstrating how tektites could arrive, for example, from Mars or from an even more likely place, Earth.

Dr. Ernst W. Adams is now Professor, Department of Applied Mathematics, University of Karlsruhe, Karlsruhe, West Germany.

References

Adams, E. W. (1963). Aerodynamic analysis of tektites and their hypothetical parent body. *In* "Tektites" (J. A. O'Keefe, ed.), pp. 150–166. Univ. Chicago Press, Chicago, Illinois.

*Cassidy, W. A., Glass, B., and Heezen, B. C. (1969). Physical and chemical properties of Australasian microtektites. *J. Geophys. Res.,* **74,** 1008–1025.

Centolanzi, F. J., and Chapman, D. R. (1966). Vapor pressure of tektite glass and its bearing on tektite trajectories determined from aerodynamic analysis. *J. Geophys. Res.,* **71,** 1735–1749.

Hanus, F. (1928). Les moldavites (tectites) de la Bohême et de la Moravie. *Rozpravy Český Akad. II,* **37,** 1–83 (*Bull. Intern. Acad. Sci. Bohême, Prague*).

Hardcastle, H. (1926). The origin of australites. Plastic sweepings of a meteorite. *New Zealand J. Sci. Technol.,* **8** (2), 65–75.

Lyttleton, R. A. (1964). A cometary mechanism for the formation of tektites. *Geochim. Cosmochim. Acta,* **28,** 807–820.

Walter, L. S., and Adams, E. W. (1967). Vapor pressure of natural tektite melts at high temperatures and its application to aerodynamic analysis. *J. Geophys. Res.,* **72,** 3717–3728.

Walter L. S., and Carron, M. K. (1964). Vapor pressure and vapor fractionation of silicate melts of tektite composition. *Geochim. Cosmochim. Acta,* **28,** 937–951.

Geochimica et Cosmochimica Acta 1964, Vol. 28, pp. 881 to 892. Pergamon Press Ltd. Printed in Northern Ireland

Aerodynamic analysis of the tektite problem

ERNST W. ADAMS* and R. MILTON HUFFAKER
G. Marshall Space Flight Center, Huntsville, Ala.

Abstract—If a recently measured vapor pressure function of tektite material is used, trajectory and ablation analysis gives a band less than 2° wide, and adjacent to the skipping limit, for the entry angles of button-type australites. Problems of shape, distribution and atmospheric flight of tektites can be solved if tektites are assumed to have ablated from a hypothetical skipping parent body. In particular, this hypothesis explains the narrow range of shallow entry angles of australites. Aerodynamic evidence precludes terrestrial origin of tektites or tektite material.

NOMENCLATURE

c_p kcal/kg°K	Specific heat.
H km	Flight altitude above sea level.
p_{v1} kg/m^2	Vapor pressure function of tektite material assumed by ADAMS and HUFFAKER (1962a, b; 1963a, b) and ADAMS (1963), see Fig. 2.
p_{v2} kg/m^2	Vapor pressure function of fused quartz (SiO_2), see Fig. 2.
p_{v3} kg/m^2	Preliminary empirical vapor pressure function (ref. 22), see equation (1) and Fig. 2.
p_{v4} kg/m^2	Empirical vapor pressure function (WALTER, private communication) see Fig. 2.
p_{v5} kg/m^2	Correction of p_{v4} for effects of surface tension, see Fig. 2.
p_{vc} kg/m^2	Semi-empirical vapor pressure function (ref. 9), see Fig. 2.
r m	Radius of spheres from which button-type australites were derived.
R km	Horizontal range covered by australites outside of atmosphere according to parent body hypothesis.
t sec	Time; $t = 0$ at $H = 150$ km.
T °K	Temperature.
V km/sec	Flight speed.
γ kg/m^3	Specific weight.
ε	Emissivity constant with regard to thermal radiation.
θ	Angle between flight trajectory and local horizon; $\theta > 0$ for descending flight.
$\sigma = 1{\cdot}38 \times 10^{-11} \dfrac{\text{kcal}}{°\text{K}^4\text{m}^2\,\text{sec}}$	Stefan–Boltzmann constant.

* Now at: Deutsches Versuchsanstalt fur Luft- und Raumfahrt, Freiburg i. Br., Germany.

Subscripts

a australite
d droplet
e exit from atmosphere at $H = 150$ km
i entry into atmosphere at $H = 150$ km
p parent body

1. INTRODUCTION

A LARGE portion of the approximately 1000 published papers on tektites deals with the problem of terrestrial or extraterrestrial origin. It has become evident by now that this problem can be solved only by a critical survey of the limitations furnished by individual disciplines with regard to the variation of pertinent parameters. Quite significant limitations on the origin of tektites follow from trajectory and ablation calculations for the final descent of button-type australites. A typical cross section of these buttons is shown in Fig. 1. Because of their regular geometry,

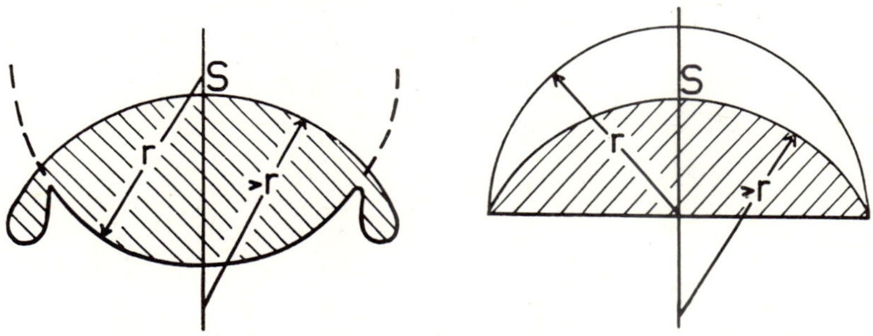

Fig. 1(a). Button-type australite. Fig. 1(b). Hemispheric model of button-type australites.

aerodynamic analysis of these objects is possible. The fairly uniform occurrence of button-type specimens in the Australian strewn field renders the results presented in this paper generally valid for all the australites.

2. FLIGHT OF BUTTON-TYPE AUSTRALITES IN THE ATMOSPHERE

2.1 Calculation model for button-type australites

CHAPMAN *et al.* (1962) have presented experimental evidence that solid spheres made of tektite material become similar to button-type australites when placed in an electric-arc jet-tunnel and exposed to heating rates of the order of those experiences in hypersonic flight. Because the initial spherical shape of these objects at the outer edge of the atmosphere is unknown, the subsequent descending trajectories and ablation processes can be calculated rigorously by use of the large amount of aerodynamic information available on high-speed flight. The experimental results on the formation of button-shapes can be used to decrease the amount of calculation work involved in the aerodynamic analysis of button flight. The button model analysed in CHAPMAN *et al.* (1962), ADAMS and HUFFAKER (1963a, 1963b) and ADAMS (1963) has been established in accordance with this consideration. This model, which is shown in Fig. 1, is hemispheric before ablation begins. The ablation process is calculated as a function of time on the stagnation point axis only. The model retains its circular rear face while ablation takes place and can be represented by segments of spheres with continuously increasing radius r. This model has been selected in order to simulate the forefront of the shape depicted in Fig. 1(a). A comparison of Figs. 1(a) and 1(b) reveals that this initially hemispheric model does not simulate the remainder of the original spherical shape at the rear face of the button in Fig. 1(a). The model, therefore, overestimates the mass loss of the sphere of radius r from which the button in Fig. 1(a) was derived. A similar button model has been used in CHAPMAN and LARSON (1963), whereas the ablation process on calculated trajectories has only been estimated in CHAPMAN (1960).

2.2 Vapor pressure of tektite material

The accuracy of quantitative ablation studies depends on the degree of reliability to which pertinent material properties of tektites are known as functions of temperature T. Except for the gas pressure, to be called vapor pressure p in the following,

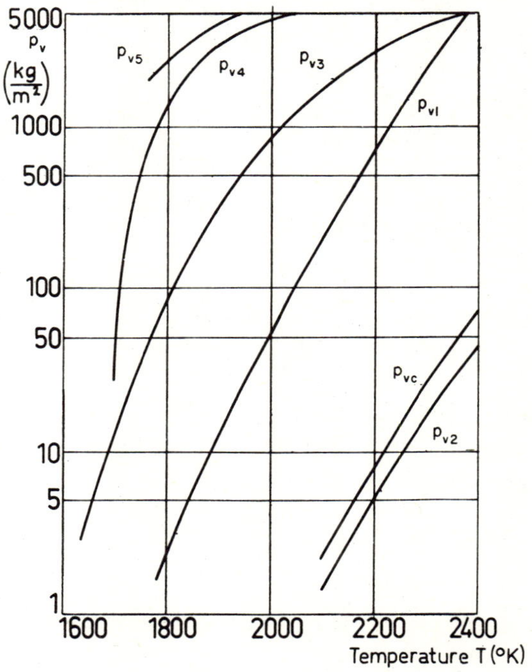

Fig. 2. Vapor pressure functions p_v.

these properties have been fairly well known for a long time. Besides 65–75% of silica, tektite material has a large content of volatile constituents. The vapor pressure of a heated test specimen, therefore, must exceed considerably the one of silica until the specimen has given away all the volatile components. For this reason, the vapor pressure function $p_{v1}(T)$ employed in ADAMS and HUFFAKER (1962a, b; 1963a, b) and ADAMS (1963) has been estimated considerably larger than the one of fused quartz, $p_{v2}(T)$, see Fig. 2. Recently, Dr. L. Walter (private communication) has measured the vapor pressure of indochinite material by observing pressure and temperature at the incidence of boiling in a heated crucible filled with ground test material. Dr. Walter's final results are represented by $p_{v4}(T)$ in Fig. 2. According to Walter, the margin of experimental error is $t \pm 75°K$. By eliminating the influence of the surface tension on the formation of vapor bubbles, Dr. Walter has calculated a corrected function, $p_{v5}(T)$ in Fig. 2. Since the final functions p_{v4} and p_{v5} were not available at the time the trajectory and ablation calculations presented here were carried out, a preliminary vapor pressure function, $p_{v3}(T)$ from Walter, was used. This function is also shown in Fig. 2; it can be curve-fitted by the expression

$$p_{v3} = \exp\left[(-148\,028\,820/T^2) + (115\,260/T) - 13{\cdot}607\right] (\mathrm{kg/m^2}) \tag{1}$$

It is explained in Section 2.3 of this paper why the difference between p_{v3} and p_{v4} has only negligible influence on the calculated ablative mass losses. By substituting mass loss measurements into a calculation method for the ablation rate at the stagnation point of a heated test specimen, Chapman and Larson (1962) have obtained a semi-empirical vapor pressure function, $p_{vc}(T)$ shown in Fig. 2. This indirect method of determining the vapor pressure function $p_v(T)$ is less reliable than Dr. Walter's direct empirical method. In particular, it is very difficult to determine theoretically both scattering and reflection of internal radiant heat exchange in the strongly non-homogeneous tektite material. This problem is much more severe in the case of test specimens than in the case of the considerably smaller natural button-type australites. It is irrelevant that the semi-empirical calculation method for p_{vc} (Chapman and Larson, 1962) gives the correct answer when applied and compared to experimental results for commercial clear silica.

2.3 The mass loss of button-type australites

The results presented in this section have been obtained by use of a calculation method outlined by Adams and Huffaker (1962a, 1963a). According to Fig. 20 of Baker (1959), average-size natural button-type australites were produced by mass losses between 70 and 90% from spheres of 11 g initial weight. Figure 3 presents calculated mass losses of the hemispheric model with 11 g initial weight and the vapor pressure function $p_{v3}(T)$. The significance of the initial temperature T_{ai} in Fig. 3 is discussed in Section 3.4. Calculations for a hemispheric model with 1·38 g initial weight and p_{v3}, entering at $V_{ai} = 11{\cdot}2$ km/sec, give between 20 and 100% mass loss depending on the entry angle. For any entry speed V_{ai} in Fig. 3, the possible range of entry angles θ_{ai} is bounded by the overshoot limit and by the curve representing the observed minimum mass loss of 70% of average-size natural buttons. The vapor pressure function p_{v2} ($< p_{v3}$) of silica yields a wide range of entry angles θ_{ai} according to Adams and Huffaker (1962a; 1963a,b) and Adams (1963); this is in agreement with results presented in Chapman and Larson (1963) by use of the semi-empirical vapor pressure function $p_{vc}(T)$ which is very nearly coincident with $p_{v2}(T)$ according to Fig. 2. The preliminary empirical function $p_{v3}(T)$ gives a narrow band of entry angles θ_{ai} adjacent to the overshoot limit, see Fig. 3. This band actually is even narrower since the initially hemispheric model of Fig. 3 overestimates the mass loss and since the final empirical vapor pressure function $p_{v4}(T) > p_{v3}(T)$ according to Fig. 2. Because of the small width of the band of entry conditions in Fig. 3, the change of this band due to the substitution of p_{v4} for p_{v3} must be negligible.

A test specimen is considered which ablates due to convective heating in a given high-enthalpy flow. If this specimen is replaced by one with a higher vapor pressure function, both the convective heat transfer rate and the surface temperature decrease in the region of laminar boundary layer flow due to the shielding mass transfer effect. The lower surface temperature explains why the mass loss by

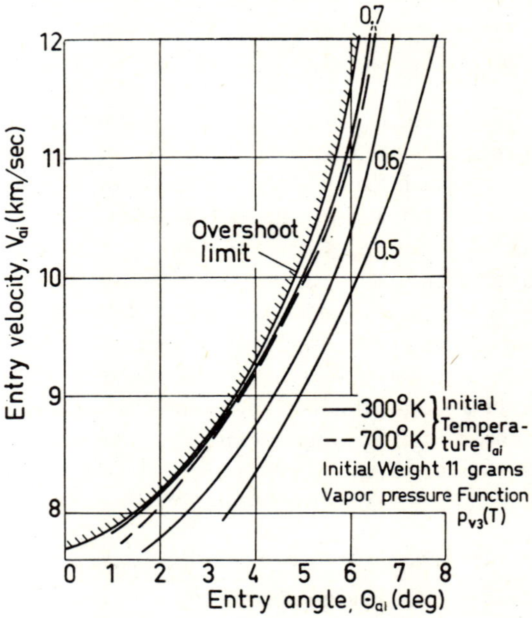

Fig. 3. Relative mass loss of the hemispheric model.

evaporation and (in general) also by melt flow drops. To reach a given mass loss, the duration of the convective heating pulse in the atmosphere has to increase if a higher vapor pressure function is introduced for a given one. Correspondingly, average-size buttons with the relatively high vapor pressure functions p_{v3}, p_{v4}, or p_{v5} (see Section 2.2) reach the observed minimum mass loss of 70% only in case of extremely long-lasting entry flight, i.e. as the overshoot limit is approached at decreasing entry angle θ_{ai} relative to the Earth's horizon.

Natural button-type australites exhibit a system of ring waves on the ablated front face, e.g. plate VIII in BAKER (1959) or Fig. 1 in CHAPMAN and LARSON (1963). The static jet tests carried out by CHAPMAN and LARSON (1963) and CHAPMAN, LARSON and ANDERSON (1962) show that the stagnation pressure must exceed 0·7 atm if these ring waves are to form on the forefront of artificial buttons. According to the stability analysis (CHENG, 1959), however, the molten film on a meteor surface is less stable in descending flight than under static conditions. According to trajectory and ablation calculations, australites experience decelerations of more than 20 g at the end of their ablation period when these ring waves solidify. In the static tests of CHAPMAN *et al.*, this deceleration is simulated by a difference of 2 g in gravity force due to the jet bowling upward or downward vertically. CHAPMAN *et al.* observed only negligible changes with regard to the ring wave formation upon the change of the jet from one vertical direction to the other. This is hardly surprising since stability phenomena of liquid films are nonlinear functions of the relevant forces. It is also pointed out (CHAPMAN and LARSON, 1963) with regard to ring wave formation on buttons that mechanical similarity can be achieved between static test and flight conditions by a proper decrease of aerodynamic forces to compensate for the difference of the g-forces. In view of the large number of similarity parameters governing this viscous high speed and high enthalpy problem, however, it is impossible to obtain the necessary complete dynamical and thermodynamical similarity in this way. According to CHENG (1959), the ring wave pattern observed on natural button-type australites, therefore, must have formed at stagnation pressures smaller than the lower bound of 0·7 atm following from the tests described by CHAPMAN, LARSON and ANDERSON (1962). Calculations for the initially hemispheric button model entering in the range $5° < \theta_{ai} < 8°$ give stagnation pressures of the order of 0·1 atm at the end of the ablation period.

2.4 *Consequences with regard to the origin of australites*

The preceding discussion for button-type australites yields the conclusion that they must have entered in a surprisingly narrow range of shallow entry angles θ_{ai} adjacent to the overshoot limit of entry. The observed smallest mass loss of 70% of average-size natural buttons shows that this range is bounded by a minimum entry speed of 7·7 km/sec (Fig. 3). The negligible Al^{26} content of tektites established by VISTE and ANDERS (1962) renders the escape speed of the Earth–Moon system, 11·2 km/sec, highly probable as an upper limit of the possible range of entry conditions of button-type australites. The question arises: how could these objects have entered the Earth's atmosphere under conditions of approximately glancing flight? To answer this question, a hypothetical parent body mechanism as origin of tektites is discussed in Section 3 of this paper.

3. THE PARENT BODY HYPOTHESIS FOR ORIGIN OF TEKTITES

3.1 *Impossibility of terrestrial origin of tektites*

A review of papers proposing terrestrial origin of tektite showers is given in pp. 123–126 of BAKER (1959). Experimental and theoretical evidence in Section 2 of this paper shows that the button-type australites must have entered the Earth's atmosphere as solid spheres, at very shallow angles, and at more than 7·7 km/sec. According to aerodynamic arguments by ADAMS (1963) and CHAPMAN and LARSON (1963), re-entry under these conditions is impossible for objects of any size and mass departing from the Earth and traversing the atmosphere without the aid of a propulsion system. This statement also negates the formation of australite droplets from parent bodies with terrestrial origin and ablating in hypersonic flight. UREY (1957; 1963) has proposed that a comet–Earth collision blew out the atmosphere so that tektites could have originated from the Earth. The same argument has been pursued by BARNES (1962), COHEN (1963), HAWKINS (1962), and other authors. By use of the approximate empirical scaling law of crater volume proportional to expended energy and the equations of radically symmetric blast wave propagation, however, it has been shown by CHAPMAN and LARSON (1963) that liquid ejecta of tektite size could not have been thrown up from the Earth to re-enter the atmosphere as has been determined in Section 2 even if the atmosphere was blown out over a large area, e.g. by meteor impact. It should be considered here that all tektites have experienced complete melting at some time in their past before they finally descended to the Earth.

3.2 Extraterrestrial origin of individual tektites

The negligible Al^{26} content in tektites (VISTE and ANDERS, 1962) precludes extended flight times in space and renders the Moon the most likely place of origin of tektite material. Hypervelocity impact experiments simulating a meteor impact reveal that lunar ejecta can reach velocities sufficiently large to escape the Moon (CHAPMAN, 1960). In general, these ejecta do not arrive in a dense and short-lasting shower at the Earth. A uniform tektite distribution on the Earth's surface then is to be expected, and the existence of a few and relatively small strewn fields can not be explained.

SUESS (1951), O'KEEFE *et al.* (1962) and ZÄHRINGER (1962) have studied the gas content in tektite bubbles. ZÄHRINGER points out: "The fact that the A^{40}/A^{36} value as well as the A:K:Xe ratio agrees very closely with the atmospheric values gives conclusive evidence that these gases are of terrestrial origin . . . The air pressure of 40 mm found in tektite bubbles occurs at an altitude of 20 mm. Since it has been shown on aerodynamic grounds that tektites or tektite material must have extraterrestrial origin, the trapping of atmospheric gases in tektites requires that these objects be strongly deformed in the liquid state during their flight in the atmosphere. Australites, however, have experienced only superficial melting during their final descent, so that atmospheric gases could not have been trapped in their unmolten cores during this flight of australites in the atmosphere. The presence of atmospheric gases in tektites then demands another preceding flight phase in the atmosphere and precludes the extraterrestrial origin of individual tektites; this statement does not relate to the origin of tektite material.

3.3 Parent body flight in the atmosphere

Figure 3 refers to average-size natural button-type australites of 11 g initial weight with the vapor pressure function $p_{v3}(T)$. Results for buttons of 11 g and 1·38 g initial weight both with $p_{v1}(T) < p_{v3}(T)$ are presented by ADAMS and HUFFAKER (1962a; 1963a, b) and ADAMS (1963). A comparison of all these data shows that the observed minimum mass loss of 20% (BAKER 1959) in the case of natural buttons of 1·38 g initial weight and p_{v3} requires an entry speed $V_{ai} \cong 5$ km/sec if the entry angle is close to the skipping limit; these shallow angles are necessary to obtain the observed minimum mass loss of 70% for average-size specimens. Extraterrestrial objects, however, in general arrive at the Earth with a considerably higher speed than 5 km/sec.

The parent body hypotheses for tektite origin is able to explain the four problems mentioned above with regard to the origin of individual tektites: (1) the existence of small button-type australites with mass losses of only 20%, (2) the very shallow entry angles of button-type australites, (3) the trapping of atmospheric gases in tektites, and (4) the existence of only a few and relatively small tektite strewn fields. The parent body hypothesis stipulates that tektites had been released as liquid fragments from the ablating surface of a hypothetical parent body which traversed the atmosphere in glancing hypersonic flight. Terrestrial origin of this parent body is impossible according to Section 3.1. The negligible Al^{26} content in tektites reveals the Moon as the most likely place of origin of the parent body. It is not attempted here to discuss by which physical processes the parent body could have been formed and could have departed from the Moon or any other celestial body.

According to detailed discussions by ADAMS and HUFFAKER (1962b; 1963a, b) and ADAMS (1963), a heavy blunt-nosed parent body is predominantly heated by radiation from the gas cap behind the shock wave. The incident radiation rates of as much as 10^4–10^6 kcal/m^2 sec cause a layer adjacent to the surface to boil unless the material is opaque with regard to thermal radiation. Liquid masses of widely varying size and shape then are stripped off the parent body. These pieces of melt subsequently are decelerated relative to the parent body and continuously deform and break up. The deformations, which can be seen by observing the striae pattern in cross sections of any tektites, explain the trapping of atmospheric gases discussed in Section 3.2. By far the majority of the pieces of melt descends directly while breaking up. This situation is simulated by an experiment described on p. 11 of CHAPMAN and LARSON (1963) "... a mass of highly viscous glass was forcibly disrupted by aerodynamic pressure, a number of forms were observed which were rather similar to some indochinites. For example, flattened discs and flattened tear drops ...". The parent body hypothesis assumes correspondingly that the tektites found in the Indomalayan strewn fields were released from the parent body to descend directly while continuously breaking up.

A small percentage of the residual pieces of melt, whose parent masses were stripped off at lower altitudes during the parent body's ascending flight phase, escaped from the atmosphere. The exit angles θ_{de} and exit velocities V_{de} of these escaping droplets are bounded, $0 \leqq |\theta_{de}| < |\theta_{pe}|$ and $0 \leqq V_{de} < V_{pe}$, with small values of $|\theta_{de}|$ and V_{de} enormously more likely than larger ones in the range bounded by the parent body's exit angle θ_{pe} and exit speed V_{pe}. Obviously, a coherent droplet shape can be attained and preserved in a non-rigid state of the material at very high altitudes in the atmosphere only. According to CHAPMAN and LARSON (1963), australite-size droplets could not have been stripped directly from the parent body at these high altitudes (>60 km) where there is not enough aerodynamic heating available. The assertion that these droplets resulted from a sequence of intermediate liquid masses stripped off the parent body at low altitude is confirmed by the total pressure corresponding to an altitude of 40 km in tektite bubbles (Section 3.2).

According to the parent body hypothesis, exiting liquid droplets subsequently re-entered to form the australites. This re-entry is determined by $\theta_{ai} = -\theta_{de}$ and $V_{ai} = V_{de}$. After having escaped from the atmosphere, the droplets cooled down by radiation and took spherical shape due to surface tension if they were sufficiently small and if their exit parameters $|\theta_{de}|$ and V_{de} were large enough. Droplets whose flight time outside of the atmosphere, as compared to their size, was not long enough to attain a spherical shape may account for the australites derived from general bodies of revolution, e.g. the canoe-shaped forms described by BAKER (1959). The small spherical javanites can be explained similarly.

Shallow entry angles θ_{ai} adjacent to the horizontal direction of the overshoot limit follow from both the analysis of australite descent in the atmosphere (Section 2.3) and from the discussion of the preceding skipping flight phase of the hypothetical parent body (Section 3.3). The effect of atmospheric perturbations on ballistic tektite trajectories increases as the flight path becomes shallower. Since small entry angles θ_{ai} are much more likely than larger ones in the range $0 \leqslant \theta_{ai} < |\theta_{pe}|$, it is seen that australites found in one location may come from different trajectories and that the strewn corridor may have a considerable lateral width. It is also seen that only in exceptional cases the exiting droplets were able to escape with large enough exit parameters $|\theta_{de}|$ and V_{de} to reach Australia, whereas by far the majority of the liquid masses stripped off the parent body descended directly or traveled only short distances outside the atmosphere so that no particular shape could be formed by surface tension.

3.4 *Relationships between parent body and australite flight*

For a skipping and non-ablating spherical parent body of 9·67 m radius and 10^4 tons weight, Fig. 4 presents exit angle θ_{pe} versus entry angle θ_{pi}, both measured at

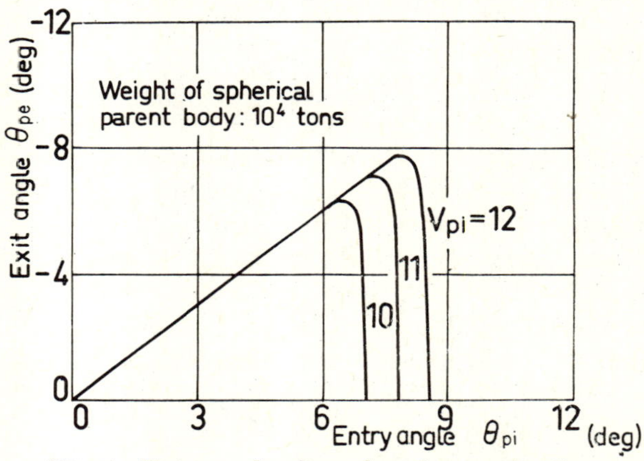

Fig. 4. Entry and exit angles of parent bodies.

the altitude $H = 150$ km above sea level relative to the Earth's horizon. This sphere experiences significant deceleration, and thus ablation, only if the entry angle θ_{pi} exceeds a certain value θ_{pi}^* which depends on the entry speed V_{pi} (Fig. 4). The exit angle θ_{pe} decreases rapidly within a range of approximately 2° width once the entry angle θ_{pi} has surpassed θ_{pi}^*. Figure 4 shows that a parent body must enter within this narrow range of $\sim$2° width if both skipping and ablation are to take place. Correspondingly, it is unlikely that a parent body carries out more than one pass through the atmosphere if these two conditions are simultaneously satisfied. A second entry in the atmosphere with ablation taking place then terminates in general with the impact of the parent body. The droplets escaping from the atmosphere, consequently re-enter at angles larger than their skipping limit.

Keplerian elliptic arcs with end points at $H = 150$ km sea level have been used to calculate the horizontal range R of spheres with 1·3 cm radius and 22 g weight travelling outside of the atmosphere. In accordance with the parent body hypothesis, these small spheres are residuals, contracted by surface tension, from liquid masses stripped off the parent body at low altitudes. Figure 5 presents the horizontal

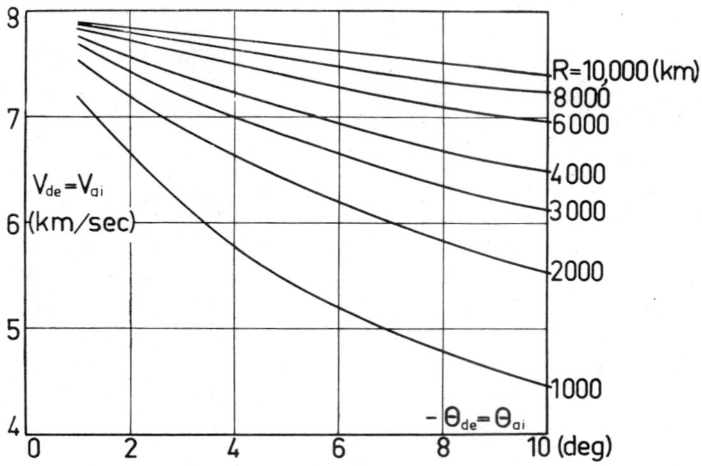

Fig. 5. Horizontal range of exiting droplets outside of atmosphere.

flight range R of these spheres along Keplerian arcs as a function of initial angle θ_{de} and initial speed V_{de}, both at exit from the atmosphere. Figure 5 shows that the influence of V_{de} on the horizontal range R goes to zero together with θ_{de}.

Under the assumption that a sphere of radius r possesses a uniform but time-dependent temperature $T(t)$, radiation cooling by use of the Stefan–Boltzmann law gives the following formula for the temperature gradient:

$$\frac{dT(t)}{dt} = -\frac{3\sigma\varepsilon}{\gamma c_p r} T^4(t) \tag{2}$$

Here t is the time, ε stands for the emissivity constant of the sphere, σ is the Stefan–Boltzmann constant, c_p is the specific heat, and γ the specific weight. Equation (2) is integrated by

$$\frac{T^3(t)}{T^3(0)} = \frac{1}{1 + (9\sigma\varepsilon/\gamma c_p r)T^3(0)t} \tag{3}$$

The temperature $T(0)$ at the begining ($H = 150$ km) of the elliptic arc is estimated to be 3000°K. Figure 6 presents evaluations of equation(3) for $T(0) = 3000$°K, $\varepsilon = 0{\cdot}1$, 0·5 and 0·9, $r = 1{\cdot}3$ cm, $c_p = 0{\cdot}21$ kcal/kg°K, and $\gamma = 2400$ kg/m^3. A comparison of Figs. 5 and 6 shows for $\varepsilon = 0{\cdot}1$ and $\theta_{ai} = 4°$ that spheres of 1·3 cm radius have to travel approximately 3000 km outside of the atmosphere before $T(t)$ reaches 1000°K and 6000 km for $T(t)$ to take the value 840°K. Even for a value of ε as low as 0·1, however, radiation cooling brings $T(t)$ down to a level of 1000°K within a flight distance comparable to the geographic extension of the Indomalayan and Australian strewn fields. At the time of re-entry, the australites then were in the solid state, so that aerodynamic heating and not aerodynamic pressure acting on soft material formed the final australite shapes. Figure 3 shows that approximately the same mass loss of average-size button-type australites results from initial temperatures $T_{ai} = 300°K$ and $700°K$ at the beginning, $H = 150$ km, of their final descent. The same conclusion should be very nearly correct if initial temperatures T_{ai} of 300°K and 1000°K are compared.

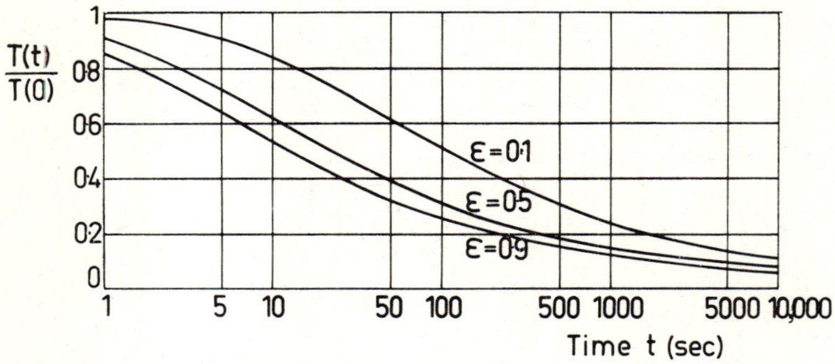

Fig. 6. Cooling time for exiting droplets flying outside of atmosphere.

4. Conclusions

Terrestrial origin of tektites or tektite material is shown to be impossible on aerodynamic grounds whether the atmosphere is present or has been removed locally by, say, meteor impact. Trajectory and ablatic analysis gives a band of shallow entry angles less than 2° wide and adjacent to the overshoot limit if a new empirical vapor pressure function of tektite material is used. (1) This narrow range of entry angles, (2) the presence of atmospheric gases in bubbles in tektites, (3) the limited extension of tektite strewn-fields, and (4) the existence of small button-type australites with only 20% mass loss can only be explained if tektites are assumed to have ablated from a hypothetical parent body which carried out a skipping flight in the Earth's atmosphere. Aerodynamic discussions of parent body flight show that the australites escaped from the atmosphere together with the parent body to re-enter subsequently as solid bodies of revolution. Re-entry angles close to the horizontal direction or the skipping limit, respectively, are enormously more likely than larger ones in the narrow range of shallow angles determined by the skipping parent body, which experiences deceleration and thus ablation only within a band of entry angles approximately 2° wide. In general, a parent body then can not carry out more than one pass through the atmosphere under simultaneous conditions of skipping and ablation. It is seen that the parent body hypothesis for tektite origin reconciles all the relevant observed, measured, and calculated aspects of tektite flight in the Earth's atmosphere. It has not been attempted in this paper, however, to present speculations on the formation of parent bodies.

References

Adams E. W. and Huffaker R. M. (1962a) *Nature, Lond.* **193,** 1249–1251.
Adams E. W. and Huffaker R. M. (1962b) *Nature, Lond.* **195,** 681–684.
Adams E. W. and Huffaker R. M. (1963a) NASA Tech. Rep. R-149.

ADAMS E. W. and HUFFAKER R. M. (1963b) *Proceedings of the Third International Space Science Symposium, Washington 1962.* North Holland, Amsterdam.

ADAMS E. W. (1963) Chapter 7 in *Tektites* (Ed. J. A. O'Keefe). Univ. of Chicago Press, Chicago.

BAKER G. (1959) *Mem. Nat. Mus. Vict.* No. 23, 1–313.

BARNES V. E. (1962) *Proc. Nat. Acad. Sci.* **48,** 947–955.

CHAPMAN D. R. (1960) *Nature, Lond.* **188,** 353–355.

CHAPMAN D. R. and LARSON H. K. (1963) Paper presented at the XIII International Astronomical Congress, 1962 and NASA Tech. Note D-1556.

CHAPMAN D. R., LARSON H. K. and ANDERSON L. A. (1962) NASA Tech. Rep. R-134.

CHENG S. I. (1959) *J. Amer. Rocket Soc.* **29,** 579–589.

COHEN A. J. (1963) *Proceedings of the Third International Space Science Symposium, Washington 1962.* North Holland, Amsterdam.

GRANT K. (1909) *Proc. Roy. Soc. Vict.* **21** (2) 444–448.

HAWKINS G. S. (1962) Boston Univ. Rep., No. 14.

O'KEEFE J. A., DUNNING K. L. and LOWMAN P. D. JR., (1962) NASA Tech. Note D-1342.

SUESS H. E. (1951) *Geochim. et Cosmochim. Acta* **2,** 76–79.

UREY H. C. (1957) *Nature, Lond.* **179,** 556.

UREY H. C. (1963) *Nature, Lond.* **179,** 228–230.

VISTE E. and ANDERS E. (1962) *J. Geophys. Res.* **67,** 2913–2919.

VOLAROVICH M. P. and LENOTIEVA A. A. (1939) *C.R. Acad. Sci., Paris* **22,** 589–591.

ZÄHRINGER J. (1962) *Symposium on Radioactive Dating.* SM-33-37, Athens, Greece.

VOL. 76, NO. 26 JOURNAL OF GEOPHYSICAL RESEARCH SEPTEMBER 10, 1971

Australasian Tektite Geographic Pattern, Crater and Ray of Origin, and Theory of Tektite Events

DEAN R. CHAPMAN

Ames Research Center, NASA, Moffett Field, California 94035

Chemical data recently obtained on 18 major and minor elements in 507 tektites from 205 localities are used to map the geographic distribution pattern of Australasian tektites. Areas of distinct chemical type outline a coarse structure to the pattern: an elongate zone of HCa tektites stretching northwest across Australia, a crescent zone of HMg tektites curving from Australia to Indonesia to the Philippines, and a teardrop-shaped zone of normal indochinites arcing northeast over Southeast Asia. Various sets of matching polygons of specific gravity and matching chemical analyses of individual specimens define a fine structure compatible with this coarse structure. The over-all pattern is not radial, as would be required for a terrestrial origin, but is systematically curved. Numerous moon-to-earth trajectories were computed for ejecta leaving various lunar craters in a search for compatible places of origin. It was found that the complex tektite distribution pattern is matched by the trajectory landing pattern for ejecta leaving Tycho, and that the required heading direction for this ejecta coincides with one of Tycho's most prominent rays. From a study of visible elements comprising this and other similar rays, a 'connate crater' theory of tektite events is formulated which offers a reconciliation of certain tektite observations previously considered contradictory. Some implications to selenology, and comparisons with Apollo data (especially with rock 12013), are briefly discussed.

INTRODUCTION

Among four separate tektite groups presently known, the Australasian group is the youngest, most numerous, and most widely strewn. Tektites from this group have been discussed in the scientific literature for 126 years, beginning with the description of an oval australite button by *Darwin* [1844]. He suggested an origin as a volcanic bomb that burst after spinning in flight through the air (Figure 1). Subsequent evidence, however, has focused on a type of natural event much rarer than earth volcanism. Concordant age datings by the K-Ar method [*Gentner and Zähringer*, 1960] and the fission-track method [*Fleischer et al.*, 1965; *Gentner et al.*, 1967] indicate that the Australasian tektites formed in a single event about 700,000 years ago. Approximately three million of these tektites have been recovered to date at numerous localities spread from Tasmania to South China, from Thailand to the Philippines. Moreover, Australasian microtektites have been recovered in deep-sea cores from the Indian and Pacific Oceans at localities separated by as much as 10,000 km [*Glass*, 1967]. On the basis of the observed microtektite abundance, the mass of glass estimated to have fallen in this tektite event is of the order of 100 million tons [*Cassidy et al.*, 1969].

At present there is reasonable agreement as to how tektites formed, but rife contention as to where. After the discovery of metallic spherules of meteoritic Fe-Ni within phillippinites and indochinites [*Chao et al.*, 1964], and of grains of coesite in thailandites [*Walter*, 1965], it has been widely though not universally accepted that tektites originated as splash from a large meteoritic impact crater. In order to allow the small tektites to disperse over the vast distances observed, it would have been necessary, if this impact occurred on earth, for the atmosphere above the terrestrial crater to be removed. Such a removal, however, requires the order of 10^{32} erg of energy, an amount sufficient to excavate an earth crater several hundred kilometers in diameter [*Lin*, 1966; *Chapman and Gault*, 1967]. No crater this size of Australasian tektite age has yet been identified on earth. If the impact occurred on the at-

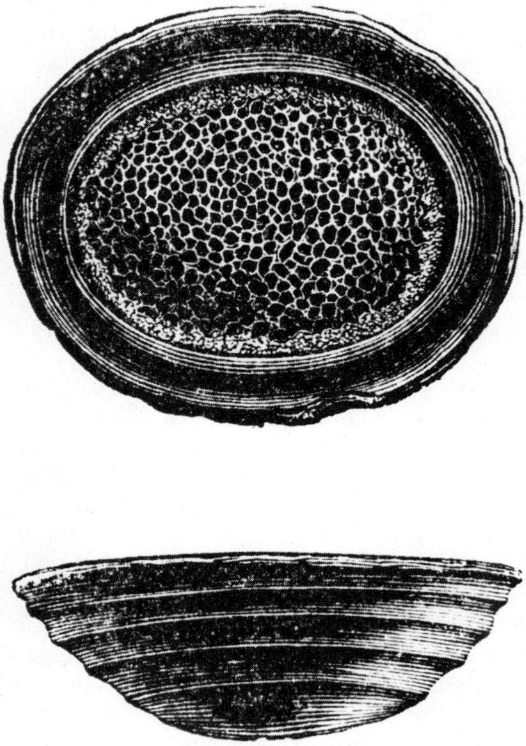

Fig. 1. First illustration known of a tektite [from *Darwin,* 1844], thought by Charles Darwin to be a volcanic bomb.

mosphereless moon, a smaller source crater, several tens of kilometers to a hundred kilometers in diameter, could account for 10^8 tons of glass on earth. Tektite form and surface sculpture have been closely reproduced in the aerodynamic laboratory, and an ablation analysis, checked by laboratory experiments, has indicated that the amount of ablation on australites is compatible with a lunar origin [*Chapman and Larson,* 1963]. A preliminary trajectory study of material originating from ten large, young, lunar craters has pointed to Tycho, in the lunar uplands, as a prime suspect [*Chapman,* 1964]. At present, however, age values for Tycho and other lunar craters are unknown; and Apollo missions have not yet returned samples from landing sites in the uplands.

The present paper is based on a 7-year program of tektite collection and chemical analysis conducted for the specific purpose of delineating the geographic distribution pattern of the Australasian strewnfield. It was hoped that from such a pattern the place of origin could be pin-pointed. During this period, more than a million tektites were inspected, about 47,000 individually measured for specific gravity, and 507 tektites from 205 different localities were chemically analyzed for major and minor elements by a single method of good precision. An additional 200 tektites were analyzed for major elements only. A summary of the chemical data, without a discussion of the geographic distribution pattern, has been presented recently [*Chapman and Scheiber,* 1969]. Complete tables of the chemical analyses are now being prepared for publication. The objective of the present paper is to present and analyze the data on earth distribution pattern of the Australasian tektites.

A fundamental thesis of this paper is that the tektite geographic distribution pattern establishes the trajectory landing pattern; and that this pattern, in turn, provides a basis for determining from where in space the shower originated. The landing pattern for earth and moon origin would differ greatly. Objects of earthly origin would travel in nearly planar trajectories for which the over-all distribution pattern, as delineated by various chemically distinct components within the shower, would form a set of radial elements, like the spokes of a wheel, projecting toward a common hub where the terrestrial crater of origin would be located. Objects of lunar origin, however, would travel to earth along nonplanar trajectories curved in three dimensions, so that their distribution on earth, owing to the combined effects of trajectory curvature and of earth rotation about an inclined axis, would form a nonradial pattern of curved elements. Such elements, together with computer calculations of moon-to-earth trajectories, would provide a means of identifying the particular lunar crater of tektite origin.

Previous evidence greatly simplifies the task of finding the place of origin. The aerodynamic ablation conditions and the initial tektite temperatures required to reproduce the common spallation-type core shapes demonstrate that before atmosphere entry the tektites were individual pieces of rigid glass. The absence of cosmic-ray exposure evidence, e.g., Al^{26}, cosmogenic neon, and cosmic-ray tracks like those present in meteorites, indicates that tektites have neither traveled far nor long in space

[*Reynolds,* 1960; *Viste and Anders,* 1962; *Fleischer et al.,* 1965]. Hence, they came as a shower from an impact either on the moon or someplace on earth. Since only a very small percentage of glass is formed during a hypervelocity impact, it follows that the crater of genesis was large indeed. The current age of 0.7 m.y. for the Australasian shower corresponds to formation in geologically recent times. Consequently, the task reduces to one of finding a large, geologically young crater, on the earth or moon, so positioned that ejecta spurted from it would shower on earth in precisely the tektite landing pattern.

It is the design of the sections that follow, first, to outline various sysematic features in the distribution pattern of the Australasian tektites; second, to illustrate a few of many computed landing patterns of moon-to-earth trajectories that do not match this strewnfield distribution pattern; third, to document the one computed landing pattern that does match; fourth, to note some consequences to selenology that are believed to follow from the main conclusion of this paper; and finally, to sketch a theory of tektite events based on a 'connate crater' concept, which offers a resolution of certain tektite evidence that heretofore has appeared to be contradictory.

Geographic Distribution Pattern

Two assumptions underlie the present method of mapping the Australasian tektite strewnfield: (1) that the target crust material at the impact site was chemically inhomogeneous, and (2) that lunar impact ejecta upon reaching earth is highly strung out. The first assumption is based on the diverse specific gravity populations and multifarious chemical variations observed among Australasian tektites. The second assumption, as will be illustrated later, is a basic property of the mechanics of moon-to-earth trajectories. The stringing out of ejecta is also believed to be a natural consequence of the fluid dynamics of streaming material at high velocity along narrow rays from a crater. These two mechanisms acting together would tend to convert any irregular pattern of chemical variations, which may have existed within a ray as it left the moon, into an elongated or streak-like pattern by the time it reached earth.

The tektite chemical data obtained specifically for mapping the distribution pattern has revealed several distinct chemical groups and many discernible chemical types within these groups [*Chapman and Scheiber,* 1969]. A classification nomenclature evolved along two lines: some groups, like certain commercial products, were named after the chief chemical ingredient that sets them apart from the multitude (e.g., HCa, HMg, LCaHAl, HCu,B, wherein H denotes 'high' and L 'low'); other groups and types, like French wines, were named after the area from which the varietal is chiefly renowned (e.g., Dalat type, Chiang-Rai type, Serpentine Lakes type, normal indochinite, normal australite-philippinite). From recent investigations of the isotopes of oxygen [*Taylor and Epstein,* 1969] and of rubidium-strontium [*Compston and Chapman,* 1969], the distinctness of these principal groupings made on the basis of chemistry have been independently corroborated.

In attempting to map out details of the earth distribution pattern, considerable attention is given to chemical varieties that are relatively uncommon or rare. Very common types, such as the normal australites-philippinites and the normal indochinites, which are found in great numbers at many different localities across the strewnfield, presumably represent rock types that were relatively widespread at the impact site. On the basis of chemistry alone, it would be difficult to trace out within the strewnfield the landing path of tektites that are so widespread and mutually similar. Hence both specific-gravity polygons and chemistry are used to map the landing path of these common types. But the landing path of any rock mass of uncommon chemistry which happened to be at the impact site, or of any chemically unique portion of an inhomogeneous rock mass, could be readily traced out within the strewnfield simply by noting where tektites of this uncommon or unique chemistry are found. In the rest of this section, a variety of chemical characteristics are used in mapping the distribution pattern.

Coarse structure from chemical groups. The individual geographic zones for each distinct chemical group are demarked on the map of Figure 2. These zones outline a coarse structure to the distribution pattern. Prominent in Australia is an LSG-HCa 'streak,' about 300 km wide, stretching northwest for 2200 km from Tasmania to central Australia. Found only within this streak, and nowhere else in the strewnfield, are HCa populations of low modal SG (mode $\leq$2.41). Arcing from west-central Australia northwest across Indonesia, then northwest over the South China Sea, is an HMg 'crescent.' Only within this curved zone have tektites been found with HMg chemistry (the large square symbols in this zone denote HMg specimens with MgO $>$ 3.4%, Ni $\geq$ 210 ppm, and Cr $\geq$ 210 ppm). On the fringes of this crescent, and also at some localities within this zone, are 'nearly HMg' tektites, small square symbols in Figure 2, denoting MgO $>$ 2.8%, Ni $\geq$ 200 ppm, and Cr $\geq$ 190 ppm. Within a teardrop-shaped zone stretching northeast over Southeast Asia are the normal indochinites, with CaO/MgO = 1 $\pm$ 0.2, Ni $<$ 35 ppm, and Cu $<$ 5 ppm. Situated to the east, and partly overlapping the normal indochinites, is the zone of HCu, B indochinites, tektites of Muong-Nong type with Cu $\geq$ 10 ppm and B $\geq$ 30 ppm. Other chemical types for which geographic zones are not demarked, but which are identified by different symbols in Figure 2, are HAl australites ($Al_2O_3 \geq$ 14.9%) found only in eastern Australia; Chiang-Rai type (indochinites with CaO/MgO

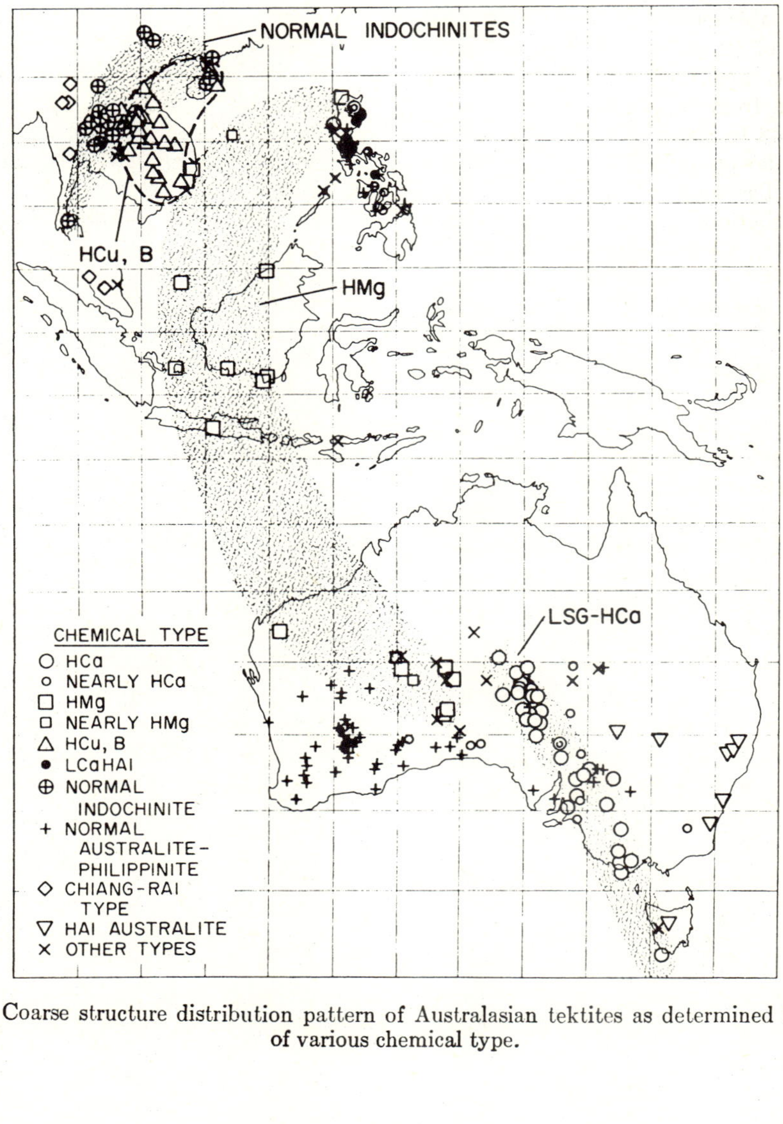

Fig. 2. Coarse structure distribution pattern of Australasian tektites as determined by zones of various chemical type.

greater than or nearly equal to unity, and with 100 < Ni < 170 ppm) found in Southeast Asia only in western Thailand and in Malaya; normal australite-philippinite ($CaO > MgO$ and $Ni \leq 41$ ppm) found essentially throughout the Philippines and southern Australia; 'nearly HCa' tektites (CaO near lower boundary of the HCa domain on a CaO versus MgO plot, and with $Na_2O > 1.25\%$ instead of $Na_2O < 1.25\%$ as in the HCa group), which are found within and near the HCa zones in Australia and the Philippines.

The principal coarse structure features in the strewnfield distribution pattern can also be demarked from the range in abundance of individual elements. Useful for this purpose is the most variable element, Ni, which ranges over a factor of 50, from 11 ppm to 615 ppm. This range is arbitrarily divided into four parts: low Ni, from 11 to 41 ppm; intermediate Ni, from 42 to 100 ppm; medium-high Ni, from 100 to 290 ppm; and high Ni, from 290 to 615 ppm. The map in Figure 3 shows the regional distribution of Ni abundance. It is seen that low Ni tektites are found throughout the strewnfield, in Australia, Southeast Asia, and the Philippines. Hence, low Ni does not define a distinct pattern. This circumstance arises because several different chemical types each are low in Ni (e.g., HCa group, LCaHAl group, normal indochinites, and normal australite-philippinites). The high Ni tektites, however, are found only in the HMg group, and demark the same crescent zone as that demarked by the HMg tektites. High values of Co and Cr provide still another independent method of delineating precisely the same crescent.

As another example, tektites that are both uncommonly low in Ni (below about 25 ppm Ni) and high in silica (above about 77% SiO_2) define two different geographic zones: one that is the same as the LSG-HCa streak in Australia, and another that is the same as the HCu,B zone of Muong-Nong type tektites in Southeast Asia. Australasian tektites both low in Ni and high in SiO_2 have not been found elsewhere.

Still another independent source of data that is most helpful in defining the distribution pattern is provided by measurements of specific gravity. Such measurements were made on about 47,000 individual tektites of the normal, nonspongy variety. A wide variety of SG polygons was found, as is illustrated in Figure 4. Some populations are very homogeneous (Wiang Papao, Figure 4*c*, others are heterogeneous (Sangiran II, Figure 4*b*); some are bimodal (Figure 4*a*), other trimodal or even quadrimodal. These variations in SG polygon correlate well with chemical variations: HMg chemical populations have heterogeneous SG polygons always with high-SG specimens above 2.47 (Figure 4*b*); normal australite-philippinites, in contrast, have relatively homogeneous SG polygons without specimens above 2.47 (Figure 4*d*); while HCa populations are distinguished by bimodal or multimodal SG polygons (Figure 4*a*). Population polygons with low modal SG, less than 2.42+, are found only in two zones: within the LSG-HCa streak in Australia, and within the normal indochinite zone in Southeast Asia. Apart from a few obvious exceptions, an SG polygon reflects closely the true material SG polygon rather than variations in bubble content [*Scheiber*, 1970]. Inasmuch as SG varies inversely with SiO_2, it follows that an SG population polygon for a given locality represents the spectrum of SiO_2 variation. Hence it should not be surprising that the coarse structure pattern of chemical variations and the SG data are mutually consistent.

Several early attempts to define a geometric pattern within the strewnfield distribution of Australasian tektites each have led to patterns different from that constructed herein. As is explained in Appendix 1, these differences are attributed to the relatively meager quantity of data available in earlier investigations.

Fine structure from SG-polygon matches and chemical matches. Four separate conditions are employed to define additional detail within the strewnfield pattern. Fine structure is delineated by tracing loci of 'matching' localities. The criterion selected for a match depends upon how common or rare is a given chemical type. For a match between localities of the very common types (e.g., normal australites-philippinites, normal indochinites, Dalat-type indochinites), it was required that either (1) SG polygons match for the same type of tektite chemistry or (2) chemical analyses match for the same type of SG polygon having the same modal SG. For less common types (e.g., HCa,

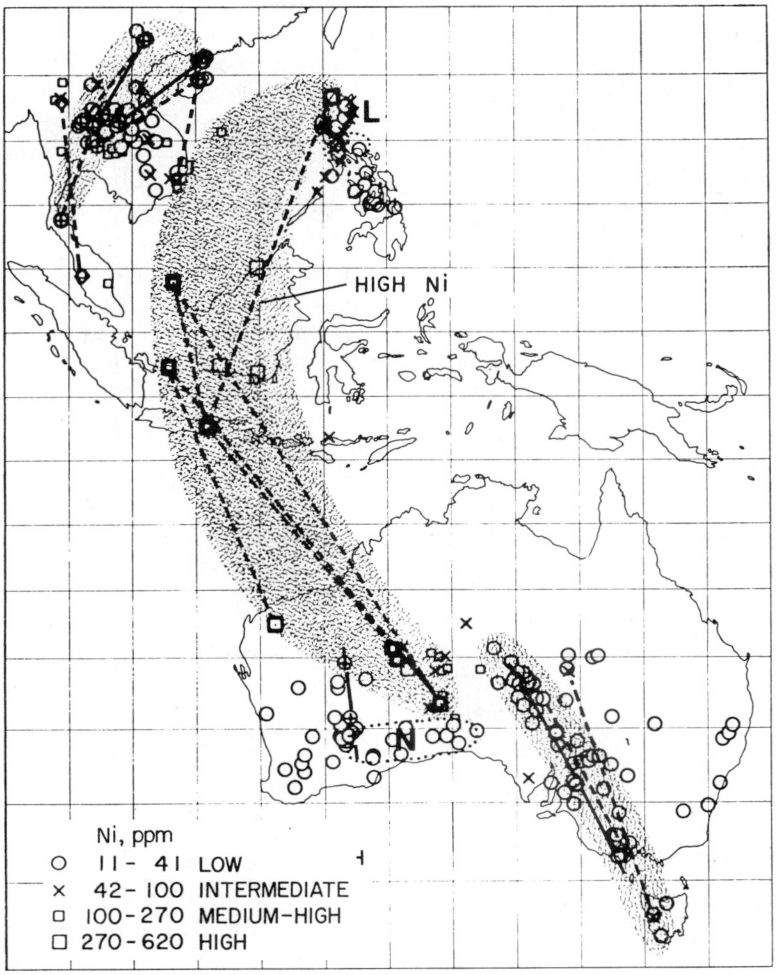

Fig. 3. Distribution pattern of Australasian tektites as determined by Ni abundance.

HMg, HCu,B, Serpentine Lakes type), it was required that (3) chemical analyses match for the same type of SG polygon having the same modal SG to within 0.01. For the uncommon or rare types (e.g., Sangiran II, which combines HMg with SG > 2.5), it was required only that (4) chemical analyses match. The first two matching conditions, which are applied to very common types, are envisioned as trailing the splash from a part of a rock mass that was relatively widespread at the impact site. Match condition 4, in contrast, is envisioned as trailing smaller masses of rarer composition, or of unique chemistry, that happened to be swept into the fused ejecta as it streamed from the source crater.

It is emphasized that the identification of strewnfield fine structure for all but the rare or uncommon chemical varieties involves consideration of *both* the chemistry of individual specimens and the polygons of specific gravity. Various examples illustrating the importance of considering both chemistry and SG polygons are presented in Appendix 2.

Inasmuch as a broad spectrum of SG polygons is observed within the strewnfield (Figure 4), a match between populations from two or more different localities is significant. Six sets of matching SG populations are shown in Figure 5. Elementary computations of probability indicate that the chance of accidentally reproducing two population polygons that match to this degree over their entire spectrum of SG is of the order of one in a thousand. Each match illustrated is for the same type of tektite chemistry. From left to right in Figure 5*a*, the four-locality match is for the low-SG mode of HCa populations found along the LSG-HCa streak in Australia; the match of Pia Oac (North Vietnam) and Kasetsomboon (Thailand) is for normal indochinites; and the match of Ooldea (South Australia) and Coco Grove (Luzon) is for the normal australite-philippinite type. In Figure 5*b* the match of Kuchinari (Thailand) and Fort Bayard (South China) is for normal indochinites; that of West Kalgoorlie (West Australia) and Kubao (Manila) is for the normal australite-philippinite type, as is the three-locality match of Boyce Creek, Lake Lapage, and Earaheedy (all in West Australia). Since both Ooldea and Kalgoorlie are from the Nullabor Plains of Australia, and both Kubao and Coco Grove are from Luzon, these two matches are referred to subsequently as the Nullarbor-Luzon match. Such a match has been observed previously [*Chapman et al.*, 1964]. In Figure 6 the Nullarbor area is designated by the encircled *N*, the matching Luzon area by an *L*. Other localities of matching SG polygons are shown connected together by solid lines in this figure.

A computer was used to search for all possible matching analyses within the chemical data. The 507 tektite analyses for 18 major and minor elements were digitized on punch cards. Altogether about 125,000 different combinations of pairs of tektite analyses were then tested by the computer for possible chemical matches. The quantitative criteria used for a match were based on the experimental precision of determining a given element. The mean deviation σ between replicate determinations of that element from a given sample of tektite powder was taken as a measure of the precision. For a chemical match it was required that all major elements with at most one exception be the same within 2σ, that the one excepted major element (if any) be the same within $\pm 2\sigma$, that all minor elements with at most one exception likewise be the same within 2σ, and that the one excepted minor element (if any) be the same within $\pm 2\sigma$. The 2σ precision was taken as 6% for Mg, Na, Al; 7% for Ca, K, Fe; 20% for Ti, Ni, Mn, V; 25% for Ba, Co, Cr, Y, Zr; and 30% for B,Cu. The 2σ precision for silicon, which was determined by

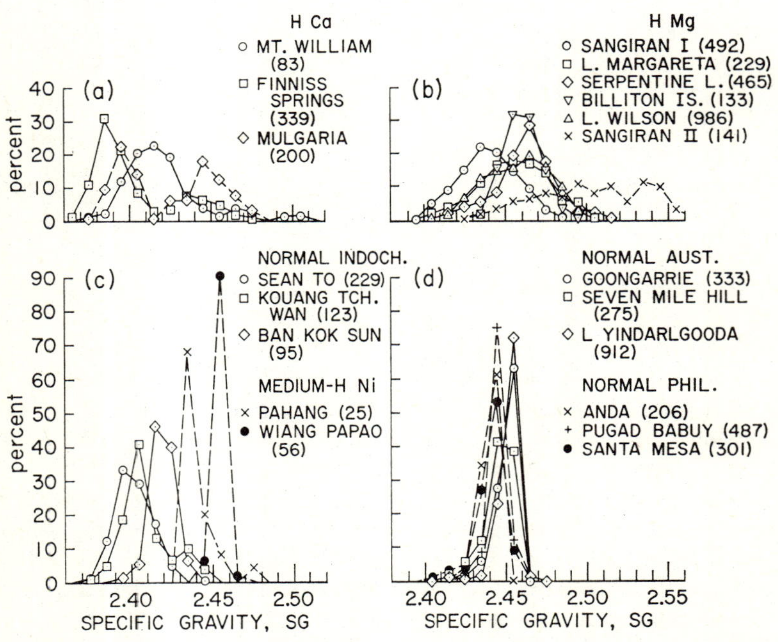

Fig. 4. Specific-gravity population polygons for various chemical types: (*a*) HCa australites, (*b*) HMg tektites, (*c*) normal indochinites and localities with medium-high Ni, (*d*) normal australites and normal philippinites.

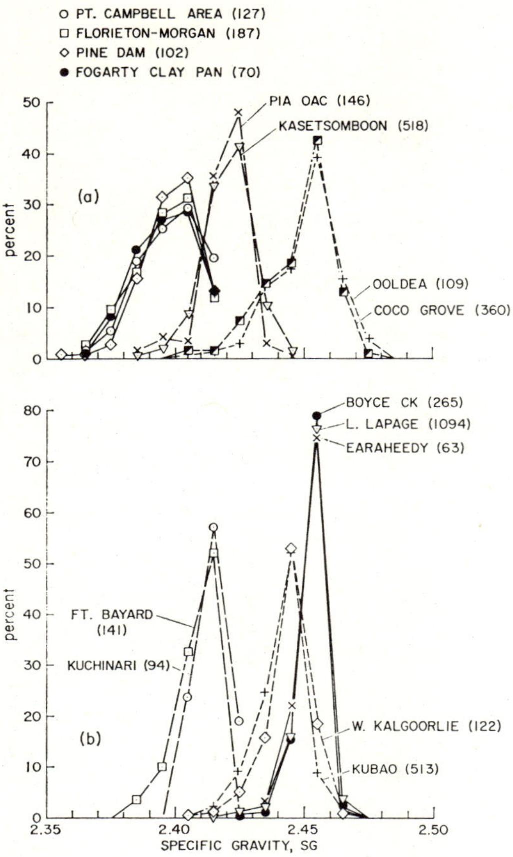

Fig. 5. Six sets of matching polygons of specific gravity: (*a*) left to right, matches along LSG-HCa australite streak, North Vietnam-Thailand match, and Nullarbor Plains-Luzon match; (*b*) left to right, South China-Thailand match, West Australia-Luzon match, and West Australia matches.

difference from the total of the other elements, was taken as 1.5%. In a few cases where an element was uncommonly low in abundance, the 2σ precision was taken as an appropriate increment rather than as a percentage.

Not all matching analyses identified by the computer, nor all locality pairs at which the SG polygons match, can be considered a significant feature of the original shower pattern. Obviously, a match between two localities only several tens of kilometers apart may reflect some geological transport process rather than a characteristic of the tektite trajectory landing pattern. Also, locality data for some specimens are uncertain by as much as 50 km. More important, the elongate zone of a given type is at least several hundred kilometers across. Consequently, only those locality pairs of chemical matches separated more than about 1000 km would provide relevant evidence on the 'streak paths' within the over-all trajectory landing pattern.

Various examples of matching analyses as determined by the computer search are presented in Table 1. These include matches between normal australites and philippinites (matches 1 and 2), between Dalat type tektites (match 3), between normal indochinites (match 4, a tri-locality match), between HMg tektites (matches 5, 6, 7), between other less common types (matches 8, 9), and between rare types (matches 10, 11). In Figure 6 the locality pairs for matching analyses are shown connected together by dashed lines. It is evident from this figure that the fine structure, as represented by these lines of matching chemistry, is quite consistent with the coarse structure (shaded zones) and with the fine structure of matching SG polygons (solid lines).

Directional conditions and relative temperature of formation. In comparison to australites, the internal flow structures of tektites from Southeast Asia are indicative of formation in a much more viscous condition. Within indochinites, the lechaterlierite inclusions appear less softened and contorted, the internal bubbles are commonly elongate instead of spherical as in australites, and minute grains of unfused minerals are found [*Barnes*, 1963*a*, *b*; *Glass*, 1970]. Moreover, their external shapes are larger and much more irregular than australites. All these are signs that the indochinites formed at a higher viscosity than the australites. Higher viscosity implies lower temperature, a lower internal energy increase ΔE, and a lower particle velocity V_p of shock acceleration ($\Delta E = (1/2)V_p^2$ according to the Rankine-Hugoniot conservation equations for strong shock waves). The irregularly shaped tektites from Southeast Asia, therefore, were ejected from their crater of origin at a lower velocity than were the australites. This condition of decreasing ejection velocity in the direction from Australia to Thailand is an important directional condition to be imposed on the trajectory landing pattern of the Australasian tektite shower.

From study of the geographical variation of

tektite shapes and sizes within a given chemical group, a number of other directional conditions have been deduced. Within the low-SG component population along the LSG-HCa streak, for example, the average size of substantially whole tektites (all but the broken ones) increases systematically in the northwest direction, varying from about 2 grams in Victoria to 2.6 grams at Florieton, 4 grams in the area of Mulgaria-Witchelina-Pine Dam, about 6 grams at Charlotte Waters, and about 10 grams at Henbury. Also, 100% of the cores in Victoria have a smoothly curved base, whereas about 1/3 of the cores at Charlotte Waters have bases that are irregularly curved. Clearly, the direction of increasing viscosity of formation, and hence of decreasing velocity of ejection, runs northwest from Victoria to Henbury. Another example is provided by the HMg javanites and australites. The billitonites, averaging 19 grams, and the javanites (Sangiran I with 2.38 < SG < 2.48) are considerably larger and more irregularly shaped than the HMg australites of similar SG range. This indicates a decreasing velocity of ejection in the direction running NNW from the HMg area in Australia to Sangiran in Java. Still another example of a directional condition is provided by the indochinites at Pia Oac in North Vietnam and at Kuchinari in northeast Thailand, which have matching SG polygons: the Pia Oac tektites average about 30 grams versus about

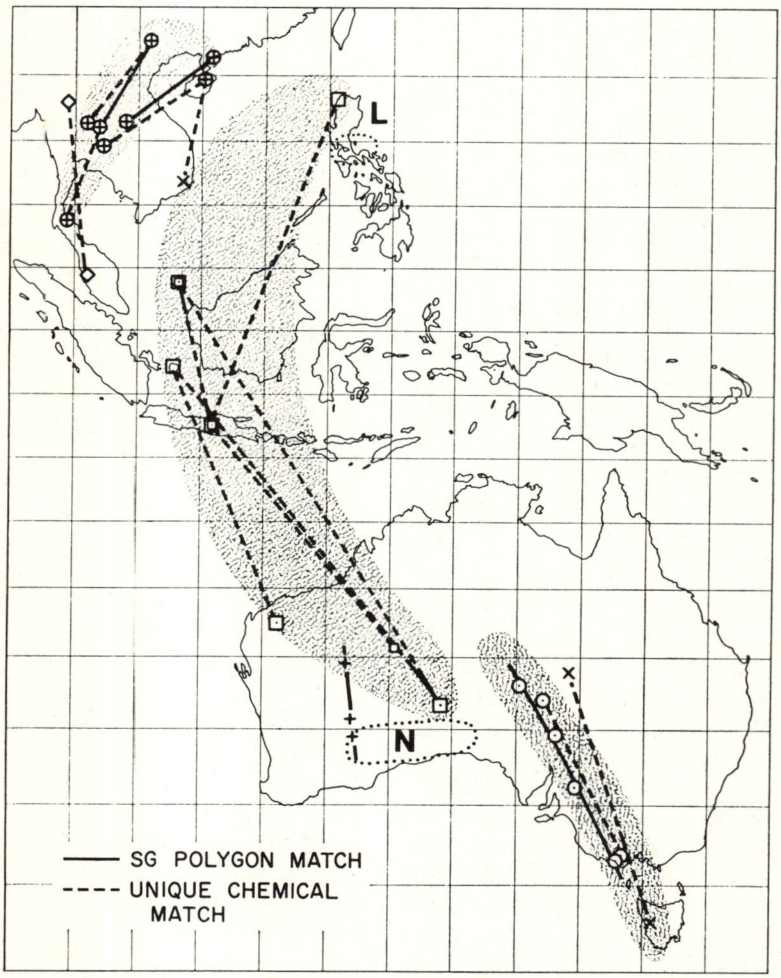

Fig. 6. Fine structure distribution pattern of Australasian tektites as determined by matching chemical analyses and matching polygons of specific gravity.

TABLE 1. Examples of Matching Chemical Analyses (Oxides in weight per cent, elements in parts per million.)

	Match 1		Match 2		Match 3		Match 4			Match 5		Match 6	
	P420 Bato-balani C. Grove	AN58 27 mi. so. of Cook	AN315 Yandal	P314M Kubau	I125 N. Suoi Vang, Dalat	SCN22 Kouang Tcheou Wan	T241 Surat Thani	T254 Ban Kok Sun	I110 Pia Oac	AN59 Giles Creek	J83 Sangi-ran Java	AN226 Ser-pentine Lake	BN5 Billiton Is.
SG	2.435+	2.445+	2.455		2.410+	2.425+	2.420+	2.400+	2.426	2.445+	2.467	2.477	2.486
RI	1.509	1.516	1.513	1.513	1.504	1.507	1.507	1.506	1.508	1.515	1.518	1.516	1.519
SiO_2	72.5	72.1	70.7	71.6	74.2	73.3	73.1	73.4	72.6	73.0	72.1	70.5	71.0
Al_2O_3	12.8	12.9	13.4	13.0	12.0	12.5	13.3	13.1	13.8	11.2	11.9	12.4	11.7
FeO	4.52	4.71	4.96	4.55	4.77	4.89	4.66	4.56	4.60	5.50	5.65	5.66	6.04
MgO	2.15	2.15	2.30	2.42	2.24	2.35	2.00	1.97	1.99	3.42	3.48	3.42	3.61
CaO	2.71	2.89	3.36	3.31	1.94	2.02	2.22	2.18	2.37	2.82	2.71	3.21	3.16
Na_2O	1.54	1.50	1.58	1.50	1.31	1.36	1.23	1.31	1.17	1.08	1.07	1.42	1.30
K_2O	2.61	2.53	2.54	2.47	2.47	2.45	2.36	2.39	2.39	2.00	2.03	2.25	2.12
TiO_2	.75	.82	.73	.78	.72	.75	.76	.67	.73	.61	.68	.78	.67
B	19	20	22	39	31	27	11	13	18	22	20	23	28
Ba	440	430	440	400	350	340	395	445	380	315	400	360	380
Co	12	14	16	13	18	20	11	11	12	33	33	37	36
Cr	78	100	92	105	120	130	58	61	56	265	240	225	280
Cu	4	5	6	8	5	4	-4	-4	-4	5	5	5	6
Mn	800	780	700	670	800	920	735	740	880	760	790	825	830
Ni	27	29	31	35	100	110	16	15	15	250	290	295	370
V	89	89	87	87	65	71	63	63	65	65	73	66	70
Y	33	30	27	32	26	28	30	27	29	26	32	29	27
Zr	240	270	280	270	280	280	300	300	280	245	280	280	280

	Match 7			Match 8		Match 9		Match 10		Match 11	
	AN122 near Young Range	J71 Sangi-ran Java	NAI1 Natuna Is.	T237 Ban Mae Jong	MYN1 Batu Gajah Malay	AN338 Mort-lake	AN258 E. of Div. Bluff	AN254 Mount Darwin Tas.	AN198 Alton Downs	P192 Pasuquin Luzon	J59 Sangi-ran Java
SG	2.455+	2.457	2.455	2.445+	2.442	2.375+	2.380+	2.390+	2.385	2.507	2.511
RI	1.514	1.513	1.512	1.511	1.511	1.496	1.497	1.500	1.497	1.525	1.527
SiO_2	71.7	72.0	71.9	72.7	72.9	78.1	77.9	76.8	76.8	69.4	69.6
Al_2O_3	12.5	12.0	12.2	12.5	12.3	10.5	10.4	11.6	11.7	12.3	12.2
FeO	5.20	5.32	5.41	4.85	4.74	3.69	3.90	3.97	3.92	6.70	6.94
MgO	2.99	3.19	3.04	2.41	2.59	1.46	1.52	1.59	1.53	5.01	5.06
CaO	2.84	2.85	2.54	2.79	2.73	2.01	2.08	1.75	1.72	2.78	2.72
Na_2O	1.36	1.31	1.40	1.33	1.28	1.10	1.16	1.04	1.07	.88	.75
K_2O	2.34	2.26	2.42	2.31	2.29	2.22	2.16	2.12	2.18	1.75	1.67
TiO_2	.72	.68	.68	.75	.73	.48	.53	.73	.66	.76	.68
B	28	34	44	19	20	16	22	20	26	25	16
Ba	380	370	400	440	395	410	360	400	425	400	490
Co	31	26	27	18	23	8	10	9	10	41	40
Cr	210	210	195	120	140	54	59	76	59	310	300
Cu	5	5	6	5	4	5	-4	-4	-4	4	4
Mn	840	840	805	830	900	610	610	680	680	990	900
Ni	230	200	215	120	165	16	17	24	20	325	300
V	73	64	80	76	69	46	54	46	45	69	63
Y	26	24	30	29	29	31	28	27	24	29	27
Zr	260	260	290	310	305	350	290	310	320	315	265

19 grams in northeast Thailand. In this case the direction of decreasing velocity of ejection (increasing size) runs northeast. At three different localities in northern Luzon where tektites of intermediate Ni content are found (42 ppm $\leq$ Ni $\leq$ 53 ppm), the average weight is between 6 and 14 grams, whereas it is a little less than 6 grams at Busuanga in the southwest Philippines. In this case the inferred direction of decreasing ejection velocity is NNE. The Sangiran II tektites of high SG average less than 1 gm weight; and, among 141 examined, the largest weighed 6.1 grams, whereas all the rest weighed less than 2.6 grams. In contrast, the tektite with a matching analysis in northern Luzon (P192, match 11 in Table 1) weighs 8.3 grams. Hence this inferred direction of decreasing ejection velocity is from Java NNE to the Philippines.

Composite distribution pattern. Assembled in Figure 7 is a composite map showing the localities of microtektites, the zones of various chemical type, the lines of fine structure, and arrows indicating the directional condition of decreasing ejection velocity (V_E). This map summarizes the principal elements known about the earth landing pattern of the Australasian tektite shower. That the pattern is nonradial, without a center of emanation, is evidence that the shower was not of terrestrial origin. In essence, Figure 7 provides a rough Rosetta stone for deducing from which lunar crater the tektites came.

Determination of Origin

The method used to determine which lunar crater could have distributed ejecta on earth in the particular pattern of the tektites is the time-honored method of trial and error. Evidence that the source crater was both large and geologically young greatly limits the number of trial craters; the fact that the moon presents substantially the same face toward earth greatly limits the variety of distribution patterns for a given crater. For each trial crater, a determination was made first of the azimuthal heading direction δ (measured in degrees north of lunar east) required to launch material toward earth, then of the elevational angle β (measured from the local lunar zenith) required to hit the Australasian part of the globe. Numerous moon-to-earth trajectories were computed for a 2° spread in both δ and β, and for the complete range of lunar ejection velocity V_E that sends material directly to earth. These computed patterns and the tektite distribution pattern were then compared.

The computer program used in the present calculations of moon-to-earth trajectories is a much refined version of a program used earlier [*Chapman*, 1964]. In this previous work the lunar equator and the lunar orbit plane were assumed to coincide with the ecliptic plane, and the lunar orbit was taken to be circular. The present program takes into account the 5°9′ mean inclination of the lunar orbit plane to the ecliptic, the 1°32′ angle between the lunar equator and the ecliptic, the elliptical orbit of the moon, and the librations in latitude and longitude as prescribed by the laws of Cassini. It also takes into account the regression of nodes, the advance of apsides, as well as

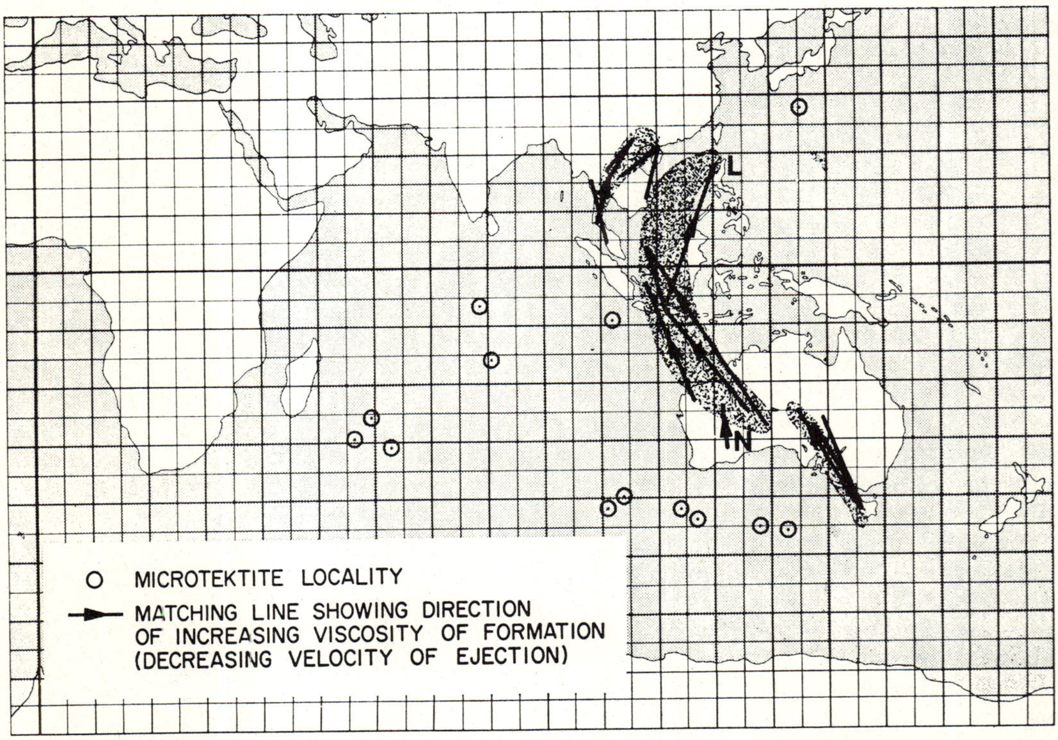

Fig. 7. Australasian tektite distribution pattern: Microtektite distribution and principal features of distribution pattern for land tektites.

perturbations of the sun according to an approximate theory worked out by Dr. William Mersman. In order to obtain a check on accuracy, seven test cases were run both on the present program and on the Apollo program of the Real Time Computer Complex at the Manned Spacecraft Center, Houston. The agreement in all cases was satisfactory.

As was noted above, a 2° spread in both azimuth (δ) and zenith (β) angle at lunar ejection was used for most calculations. An azimuthal dispersion $\Delta\delta = 2°$ was selected from observations of the width of lunar rays. Most rays are less than 2° wide. An equal elevational dispersion $\Delta\beta = 2°$ was selected partly on the basis of a theoretical interpretation (described later in this paper) of the mechanism by which lunar ray elements are ejected from their parent crater. The laboratory impact experiments of *Gault et al.* [1962] would suggest that a value somewhat larger than $\Delta\beta = 2°$ might be more appropriate. For the relatively narrow range in velocity of 2.5 to 2.9 km/sec, which represents the velocity range for 90% of the material that can hit the earth from the moon, these laboratory experiments would suggest values for $\Delta\beta$ up to about 3° or 4°. Consequently, in order to investigate what effect a greater elevational dispersion would have, some landing patterns were computed for $\Delta\beta = 4°$ (these are illustrated later). In order to investigate what effect a smaller dispersion would have, such as is observed for some ray elements, landing patterns were computed for a dispersion $\Delta\beta = \Delta\delta = 1°$ (also illustrated later).

The earth distribution patterns for each trial crater were found to depend chiefly on the lunar celestial longitude at the time of lunar ejection. Only weak dependence was found on such parameters as libration angles, sun position, and the moon's position relative to perigee. Increments in lunar longitude were taken small enough to safely interpolate between computed patterns. No two patterns were the same for different craters. Thus each crater can be characterized by a unique set of earth landing patterns, analogous to a set of fingerprints, any one of which may serve for the purpose of identification.

From the family of trajectory landing patterns for a given crater, the only pattern illustrated herein is the one judged as coming closest to meeting the basic requirement of spreading ejecta across at least the 74° range in earth latitude (from 46°S to 28°N) presently known for Australasian tektites. Among the many trial craters that failed to match the tektite strewnfield pattern, only a few will be mentioned. Throughout the illustrations, the direction of decreasing lunar ejection velocity V_E is indicated on the earth trail lines by an arrow. Each trail line represents constant values of δ and β, and variable values of V_E ranging over the full spectrum from the highest down to the lowest velocity that will land material on earth.

Landing patterns for three trial craters are shown in Figure 8. Strabo, situated in the northern hemisphere of the moon's near side, meets the requirement of spreading material over a 74° latitude range, but is incompatible with the tektite streak pattern, and with the directional requirement of V_E decreasing *northward* (Figure 8*a*). Strabo clearly provides no match for the tektite distribution. Another example of a trial crater that failed, as Figure 8*b* shows, is one situated on the moon's far side (latitude 10°N, longitude 118°E). As V_E decreases, material from this crater would sweep eastward. Other craters on the far side are similarly incompatible because of their generally eastward trails on earth. Still another example of a crater that failed is Copernicus (Figure 8*c*). In this case the latitude spread is comparable to that of the Australasian strewnfield, but the looped pattern does not match that of the tektites.

It would be pointless to document further various trial craters that failed. Six years ago, when relatively little was known about the earth landing pattern of the tektites, a number of moon-to-earth trajectories were investigated for 10 large, young, lunar craters. It was found that of these only Tycho was so situated as to be able both to spread ejecta generally northward as V_E decreased, and to spread it over the required latitude range [*Chapman*, 1964]. Two years ago, when the present chemical data were partially complete, Tycho still appeared to be the prime suspect [*Chapman*, 1968]. With many details of the landing pattern now delineated, this prime suspect can be put to a severe test.

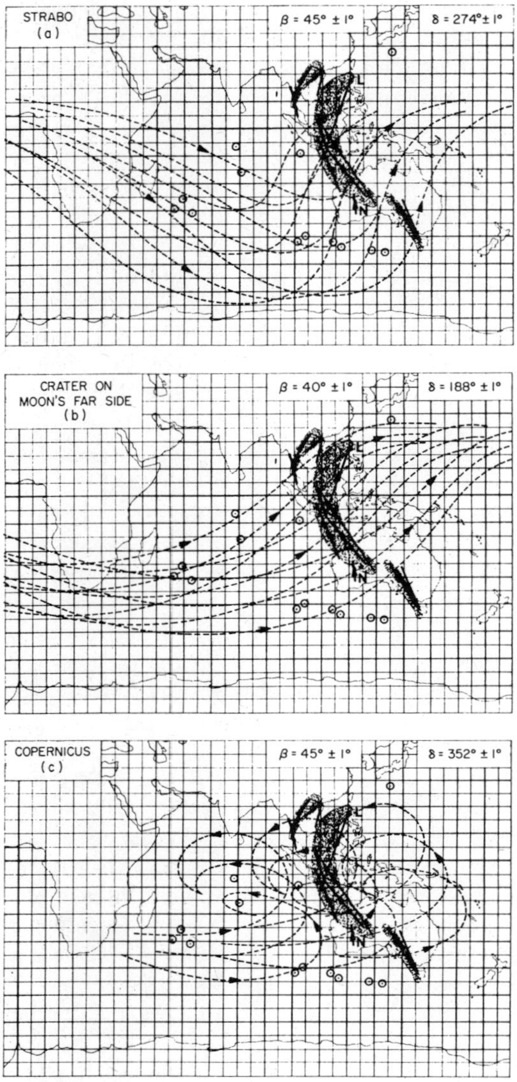

Fig. 8. Three representative examples of earth landing patterns for craters that failed to match the observed tektite distribution pattern.

Figure 9 illustrates an earth landing pattern for ejecta from Tycho. This pattern, for which the lunar longitude at the time of ejection was 267°, conforms with the tektite distribution pattern in all essential details. The over-all longitude spread of tektites from near Madagascar to south of Japan, as well as the corresponding spread in latitude, is compatible with the Tycho pattern; the northwesterly streak across Australia of the HCa australites is closely oriented with trail 4 of the Tycho pattern; the line of chemical match between Tasmania (Mount Darwin) and Central Australia (Alton Downs) virtually coincides with trail 4; the nearly northern streak in Western Australia, of normal australites with matching polygons of specific gravity, conforms with a trail interpolated between trails 8 and 9; the curved crescent zone of the HMg tektites is compatible with the curvature and positions of trails 1 and 5; the various streak lines of chemical matches within the HMg zone are also either in reasonable or in very close agreement with trails 1 and 5; the northeasterly streak lines across Southeast Asia are compatible with trails 2, 5, and 9; the nearly north streak from Malaya to West Thailand is parallel to, and just west of, trail 9 (interpolate between trails 9 and 6); the Nullarbor Plains match with normal philippinites in Luzon (*N* to *L* in Figure 9) falls along trail 1; and the various directional conditions for decreasing V_B, as designated by arrows, conform with the Tycho pattern at all areas of the strewnfield for which these conditions have been deduced from the tektite data. The chance that these many coincidences are merely accidental would represent an improbability of epic proportions.

This, however, is not the total evidence. As is noted in the upper right of Figure 9, the matching pattern represents material ejected from Tycho in one particular heading direction, namely, δ near E 19°N; and at one zenith angle, namely, β near 52°. A zenith angle of $\beta = 52°$, corresponding to ejecta leaving 38° from the lunar horizontal, is indeed a very reasonable angle for hypervelocity impact ejecta. A more crucial test, however, is whether or not a ray from Tycho was shot in the particular heading direction of $\delta = 19°$. In Figure 10 the great-circle direction represented by $\delta = 19°$ is designated by a short white line extending about three crater diameters from Tycho, and by the large arrow representing its continuation near the eastern limb of the moon. The heading E19°N from Tycho is seen to coincide, in fact, with one of the most prominent rays of Tycho, the 'Rosse ray,' which streams across the Sea of Mare Nectaris over the small crater Rosse in this sea. That the Australasian tektites came from the Rosse ray of Tycho is, in this writer's view, an inescapable conclusion.

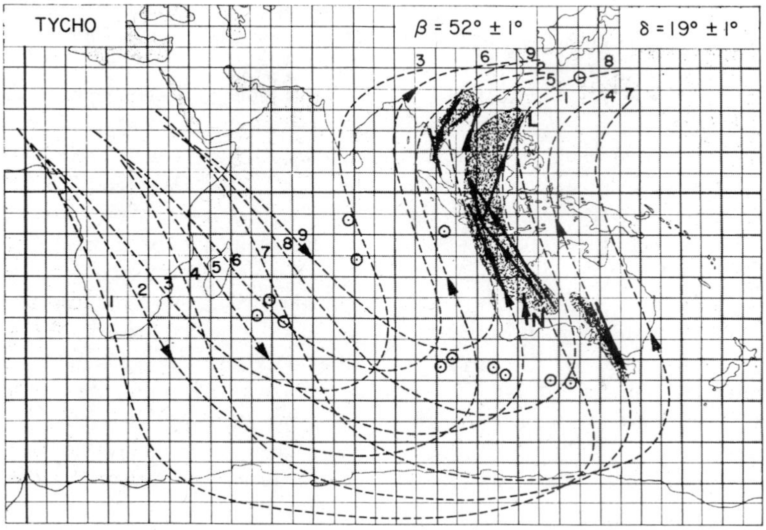

Fig. 9. Matching of earth landing pattern from Tycho with observed tektite distribution pattern.

CONNATE CRATER THEORY OF TEKTITE EVENTS

Certain evidence from tektite groups other than Australasian has led some to the conclusion that tektites originate from earth rather than moon craters. Current age data from both K-Ar and fission-track methods indicate that two craters on earth may have formed simultaneously with two tektite showers: the Ries crater in Germany simultaneously with the Czechoslovakian tektites about 15 m.y. ago, and the Bosumtwi crater in Ghana simultaneously with the Ivory Coast tektites about 1 m.y. ago [*Gentner et al.*, 1964, 1967, 1970; *Fleischer et al.*, 1965]. It is to be noted, however, that such an inference of a terrestrial origin for these tektites, which conflicts with the aerodynamic and trajectory evidence of a lunar origin for the Australasian tektites, is not the only possible inference which can be drawn from the age data.

It might be thought, for example, that the moldavites and Ivory Coast tektites are from the earth, while the Australasianites are from the moon. But, so extraordinary are the similarities in physical makeup of all tektites that this compromise interpretation seems untenable. Also, rocks having the same major and minor element composition as Ivory Coast tektites have not been found at Bosumtwi; likewise, rocks of moldavite composition have not been found at Ries. More devastating to the idea of a terrestrial origin for Ivory Coast tektites is the recent discovery of Ivory Coast microtektites in deep-sea sediments about 1300 km from Bosumtwi [*Glass*, 1968, 1969]. This distance is much too short to have produced any ablation droplets of tektite glass, yet is much too far for these objects (the largest are nearly 1 mm in size) to have traveled through the earth's atmosphere either by winds or by ballistic trajectories; and Bosumtwi crater is much too small to have temporarily removed the earth's atmosphere.

Still a different interpretation is that large, crater-forming objects impacted the earth simultaneously with the tektites. This broad possibility was noted, though without elaboration, by *Fleischer et al.* [1965]. A specific model advocated by *von Koenigswald* [1967] pictures both the earth and moon as having simultaneously encountered a swarm of large meteorites, some of which struck the moon and created tektites that later landed on earth, while others struck the earth and formed contemporaneous craters accidently near the tektites.

An alternate model compatible with both age and aerodynamic data is advanced herein. A tektite event is envisioned as originating from a burst of composite ejecta spewn from a meteoritic impact on the moon, ejecta comprising myriads of tektites together with one or more large fragments of the original meteoritic projectile. When this ejecta melange lands on earth, each large fragment would excavate a terrestrial crater 'connate' with, but not a parent of, the tektites. At first it would seem difficult to explain by this hypothesis how objects as large as those that excavated the Ries and the Bosumtwi craters could be thrown as fragments from a lunar crater. Around Arizona crater, iron meteorite fragments were found weighing of order 10^{-4} of the estimated mass of the original impacting meteorite. In order to form earth craters the size of Bosumtwi and Ries from a meteoritic fragment of mass 10^{-4} of that which struck the moon, the lunar-impacting meteorite would have to be of such size as would excavate a lunar crater roughly 50 to 100 km across. Consequently, tektite events forming connate earth craters the size of Bosumtwi or Ries can come only from large lunar craters, and therefore can occur only infrequently. Thus the 14 m.y. separating Ivory Coast and Czechoslovakian tektites, and the size of Bosumtwi and Ries craters, appear compatible with the connate crater idea.

In essence, the concept of connate craters inverts an apparently obvious association: it implies strangely that tektites are not on earth because certain large craters of identical age are, but vice versa. Past examples of inversion theories, which twist around seemingly evident causal associations, sometimes have succeeded eminently. The sun's apparent revolution daily around the earth, for example, is explained by the earth revolving, rather than the sun. The apparent creation of swarms of infusoria and bacteria seemingly out of nowhere from decaying matter is also explained by an inversion: by bacterial life causing decay, rather than decay creating life. Analogously, the present theory of tektite events associating a large entity causatively with a multitude of tiny objects attributes the existence of the large to the associated swarm of small forms, rather than the other way around.

A connate crater model for tektite events is constructed herein on the basis of a certain

Fig. 10. Source crater and ray of Australasian tektite origin.

interpretation of the visible structure of lunar rays. Telescope observations reveal that a lunar ray is not continuous, but is generally composed of a series of many elongate 'ray elements' [*Pickering,* 1892; *Kuiper,* 1965], each closely aligned with the main ray. These elements commonly resemble an exclamation mark (!), the linear streak appearing to spray radially outward from either a 'craterlet' or a cluster of craterlets. These craterlets are clearly secondary craters made by ejecta from the large primary crater. Near the primary crater, ray elements are superimposed so densely upon one another that a ray there appears virtually continuous. Farther out the ray elements become clearly separated. Figure 11 illustrates two such elements: one, comprising the craterlet Rosse and its associated 'spray,' is part of the prominent ray from Tycho that is associated herein with Australasian tektites; the other, comprising Messier A crater and its associated spray, is possibly part of a ray system from a crater situated on the moon's far side. *Whitaker* [1965] states that the Messier spray is precisely radial to an unnamed ray center photographed by Lunik 3 on the back side of the moon. *Billerbeck-Gentz* [1943] has emphasized that the Rosse ray element appears to be just one of four mutually associated elements aligned nearly end to end. An adaptation of his sketch is presented in Figure 12. The scale can be judged from the diameter of Rosse (~13 km) and the visible length of its associated spray (~220 km). These four ray elements are interpreted herein as the result of four large objects of crater-forming size being hurled from Tycho in nearly the same azimuthal direction, shock-accelerated and shock-heated crust being sprayed around and ahead of each object as it left Tycho.

The 'craterlet' Rosse is itself a large crater. Its size, 13 km in diameter, implies that ray elements can include very large objects. Situated approximately 1450 km from Tycho, it would have been excavated by a secondary object impacting at about 1.3 km/sec. If that same object were hurled to earth, however, it would impact at 11 km/sec, a much greater velocity, because of the earth's gravitational attraction. According to crater scaling laws, an earth impact of the Rosse object would excavate a terrestrial crater of diameter approximately 30 km. Thus, earth craters as large as Ries (24 km), or Bosumtwi (11 km), appear well within the possible size realm for craters originating as connates to a tektite strewnfield of lunar origin.

The large objects that trail a ray element upon ejection from a primary crater must be either blocks of lunar crust or fragments of the

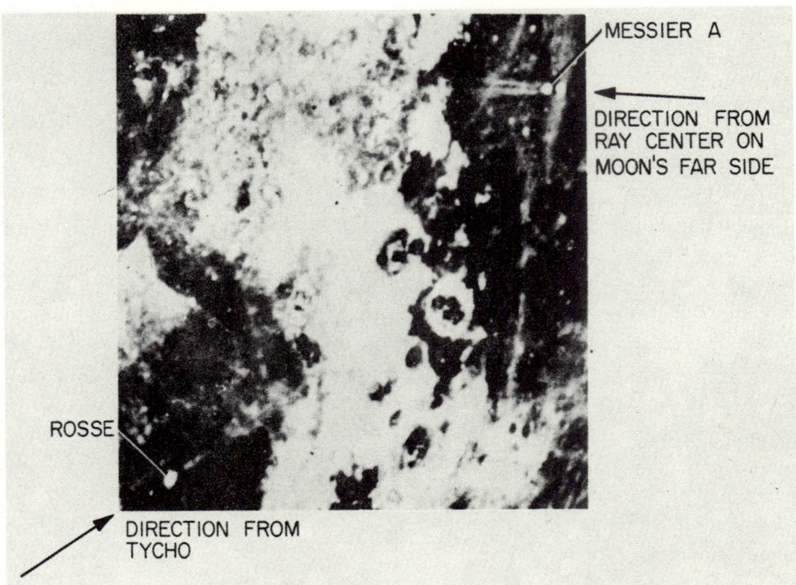

Fig. 11. Lunar ray elements Rosse and Messier.

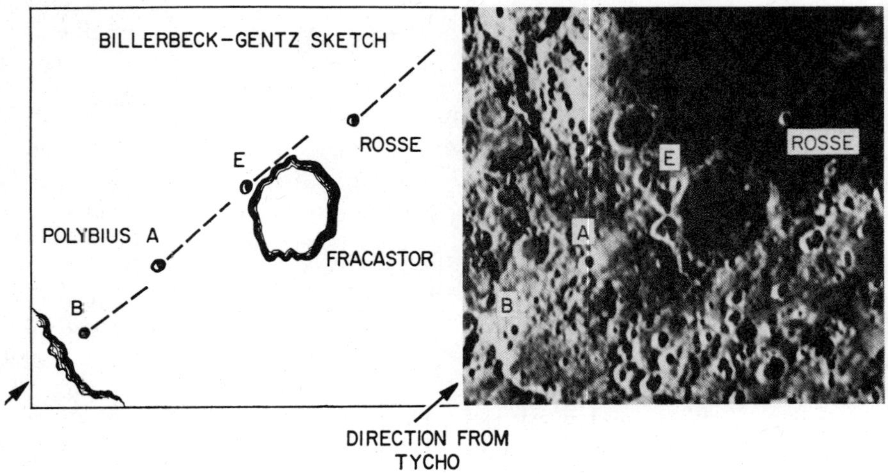

Fig. 12. Four nearly linearly aligned ray elements along Rosse ray of Tycho.

original cosmic projectile. Around the Arizona crater, large sandstone blocks are observed only relatively near the crater, not far out. The fragments of iron meteorite, in contrast, are observed out to much greater distances (~7 km) such as would imply, according to the equations of ballistic flight in the earth's atmosphere, a velocity of ejection of several kilometers per second (a velocity sufficient to escape the moon). A large projectile fragment can be ejected from a crater with greater velocity than a large crustal block, since a block of target crust must be accelerated from rest to its launch velocity, whereas a fragment of projectile need only be redirected, particularly in an oblique impact. Tycho appears to have been an oblique impact: its asymmetric splash pattern indicates an approach of the impacting body along virtually the same direction as that of the Rosse ray which contains the four large secondary craters. For these reasons, the large objects hurled as the trailing part of a ray element to great distances from their parent crater are not regarded as blocks of target material, but as fragments of the original meteoritic projectile.

The concept of a 'cometary' impact of underdense material appears to be hopeless in accounting for the survival and ejection to great distance of large projectile fragments. At cometary velocities, typically 20 to 40 km/sec, the impact pressure is so high (~30 mb) and the projectile temperature rise due to shock heating is so great (~10^4 °C), that the comet-head projectile would be vaporized, and therefore incapable of survival and ejection in the form of large solid fragments.

In contradistinction, sizable projectile fragments can be expected to survive a lunar impact of an iron mass coming from an asteroidal orbit. Statistically, objects from such orbits would impact the moon at relatively low velocity, a little over half in the range 3 to 7 km/sec, according to the calculations of *Arnold* [1965]. At 7 km/sec, the peak shock pressure would be about 1 mb [*Gault and Heitowit*, 1963; *Braslau*, 1970], at which pressure the residual temperature rise in iron would be about 650°C [*McQueen et al.*, 1962]. Since this is much less than the 1500°C melting temperature of iron, the fate of such an impacting iron projectile would be mainly fragmentation. Relatively little melting, and negligible vaporization, would occur.

That very large fragments can survive impact and be ejected far from their lunar crater is thus plausible for impacts of asteroidal iron. Meteoritic Fe-Ni inclusions have been found within Australasian tektites, within Ries glass, and within Bosumtwi glass. Consequently, the present model for tektite origin from lunar impact craters, such as Tycho, postulates an original cosmic projectile that was relatively strong and incompressible, probably iron or mainly iron, and that came from a relatively low-velocity orbit of asteroidal type.

It is envisioned that in forming ray ejecta the elevational angular spread ($\Delta\beta$), within the spray of a ray element, and the azimuthal angular spread ($\Delta\delta$), would be roughly the same. At the instant just after the impacting meteorite penetrates below the surface, when it has fragmented and begun to spread out radially, it is clear that the crustal movement cannot be perfectly symmetric. As large blocks of lunar crust begin to fracture, and to move outward, gaps between these crustal blocks will first appear at a discrete number of places around the embryo crater. It is hypothesized that a ray forms when the melange of fragmented meteorite and fused crust under high pressure squirts out through such gaps. During this squirting acceleration, fused crust of density 3 g/cm^3 is more mobile than iron fragments of density 8 g/cm^3, and thus will tend to move around each fragment and spray ahead. This process is suggested as a possible mechanism of forming ray elements. Since, on the average, such fragments would be about as high as wide, $\Delta\beta$ would average about the same as $\Delta\delta$.

Figure 13 illustrates the degree to which a departure from the approximation $\Delta\beta = \Delta\delta$ could affect the Tycho pattern. The earth landing domain for $\Delta\beta = 4°$ ($\beta = 50°$ to $54°$) is seen to be only slightly greater than for $\Delta\beta = 2°$, and not different enough to affect any conclusion. It is to be remembered that $\Delta\delta = 2°$ represents a dispersion broader than the observed width of most lunar rays.

An important feature of the present model is that tektite events can be of more than one type: a 'complete' event if both a large meteorite fragment and most of its associated spray lands on earth; an 'incomplete' event if part of the spray lands while the large fragment misses the earth; a 'truncated' event if the fragment together with only a small portion of spray lands on earth while the major portion of spray misses and travels off into space; and a 'compound' event if two or more fragments with their associated sprays land on earth. Thus a tektite event, if compound, may involve two or more connate craters on earth; if incomplete, no connate crater; if truncated, tektites may land on earth in only a very restricted geographical area near their connate crater.

A number of significant observations about tektites have long required explanation. From the present theory of tektite events, and from the moon-to-earth trajectory computations, explanations can now be given as to why tektites are not spread all over earth, why the moldavites and Ivory Coast tektites are not found all around their connate earth craters, and how it is possible for very restricted tektite distributions, such as the moldavites, to be compatible

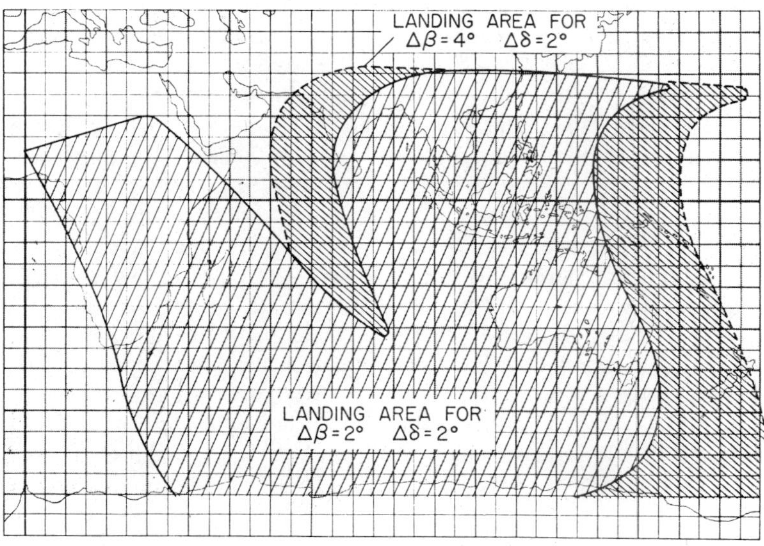

Fig. 13. Areal extension of Tycho landing pattern for $\Delta\beta = 4°$ compared to $\Delta\beta = 2°$.

with a lunar origin. From some recent experiments, an explanation also can be given as to why tektites are mainly acidic and not basic. These and some other explanations pertinent to tektite origin are presented in Appendix 3.

EVIDENCE FROM APOLLO ROCKS RELEVANT TO TEKTITES

Direct information about lunar rocks is limited as of this writing to the samples returned from Apollo 11 and 12 landings, both in maria areas. Since three-fourths of the lunar surface is covered by uplands, as yet unexplored, our present knowledge about lunar rocks must be far from complete. Although the Australasian tektites are concluded to originate from an upland area, whereas these Apollo landings were in maria, several observations pertinent to tektites have emerged from the Apollo data. First, glass in great abundance has been found in the mare soil. Much of this glass is of the size and form of microtektites. Since the mare glass yields unmistakable evidence of impact origin [*Chao et al.*, 1970; *Frederiksson et al.*, 1970; *von Engelhardt et al.*, 1970], we now have direct evidence that impacts have splashed large quantities of glass over the lunar surface. Undoubtedly some glass has splashed beyond the moon's weak gravitation, and hence there can be no reasonable doubt that lunar impact glass has long been here on earth.

Second, one of the samples from Apollo 12, rock 12013, has a composition remarkably similar in most respects to that of some tektites. This rock exhibits a light and a dark component. The light part contains up to 71% SiO_2 [*Morgan and Ehmann*, 1970] and is atypical of the main mass of baaltic rocks returned from the lunar maria. It may be an acidic differentiate of mare-like basalts [*Albee et al.*, 1970; *Laul et al.*, 1970; *Wakita and Schmitt*, 1970*a*]. Although termed 'granitic' [*Albee et al.*, 1970], the abundances of Co, Cr, and Ni in 12013 are strikingly different from earth granites (10 to 100 times higher); water and ferric iron are lacking; and MgO and CaO are comparatively high. Also, K_2O is considerably higher than Na_2O in 12013, whereas the sum $Na_2O + K_2O$ is significantly lower than in typical earth granites. These unusual characteristics are precisely those that long have typified tektites, particularly the HMg type. *O'Keefe* [1970] has pointed out the close similarity in major elements between the preliminary analysis of 12013 and the HMg Javanites of high SG (e.g., J86, J87). In Table 2 this comparison is amplified to include minor elements, some more recent analyses of 12013, and an additional tektite analysis (J2). The abundances of 20 out of 23 elements in 12013 fall in the tektite range (all but Ba, Y, Zr); and the abundances of the majority of elements are closely the same as the tektite values.

The over-all composition comparison between 12013 and the tektites is not sufficiently close or complete to constitute in itself strong evidence that tektites come from the moon. Rare earth elements in 12013, for example, exhibit a distinctive trend [*Schnetzler et al.*, 1970; *Hubbard et al.*, 1970] not found in tektites. In particular, Eu shows a pronounced deficiency relative to the other rare earth elements. Lunar anorthosites, in contrast, show a comparably pronounced Eu excess [*Wakita and Schmitt*, 1970*b*]. Thus it appears possible that still other lunar rocks may exhibit the intermediate case of a Eu abundance more like that found in tektites.

Further differences between the chemistry of tektites and that of the lunar rocks returned thus far are found in the oxygen isotopes ratio [*Taylor and Epstein*, 1970] and in the pattern of lead isotopes. Since isotope ratios can be affected by chemical and thermal fractionation processes, they may be different for material that comes from distant upland areas, such as the environs of Tycho, that are far from the mare areas explored thus far. In most lunar samples studied to date, lead is highly radiogenic, although the measurements by *Andersen et al.* [1970] on lunar ilmenite reported lead isotope ratios similar to common terrestrial lead. The Pb isotope pattern reported in several tektites is similar to terrestrial lead, but in at least one tektite is distinctly different [*Starik et al.*, 1962]. Thus far the amount of lead isotope data available on tektites, as well as the variety and number of sites sampled on the moon, are considered too meager to firmly exclude either a terrestrial or a lunar origin for tektites.

The over-all comparison between 12013 and HMg tektites appears sufficiently close to be of some significance. The HMg tektite group is a very distinct and well-defined chemical

TABLE 2. Composition Comparison between Apollo 12 Rock 12013 and Tektites (Oxide in weight per cent, elements in parts per million.)

	Apollo 12 Rock 12013				HMg Tektites		
	(1)	(2)	(3)	(4)	(5)	(6)	(7)
SiO_2	61	61	64	71	64.5	65	66
Al_2O_3	12	11	12	11	13	13	13
FeO	10	12	9.6	9.6	8.6	7.9	8.0
MgO	6.0				8.0	6.5	6.8
CaO*	6.3	5.2	6.4	3.9	2.9	3.7	3.3
Na_2O	0.69	1.16	1.36	1.25	0.72	0.77	0.75
K_2O*	2.0	2.0	2.1	3.6	1.34	1.56	1.44
TiO_2	1.2	1.3	0.8	0.8	0.7	0.8	0.8
MnO	0.12	0.16	0.14	0.12	0.18	0.13	0.14
B	15				11	21	21
Ba†	2150	2720	2750	3900	410	345	425
Co	13	35	20	28	56	48	52
Cr‡	1050	1770	1160	980	440	370	400
Ga		6	6	6	3		
La		50	60	59	60		
Ni	105				230	245	250
Pb*	30				<10		
Rb	33	50	49	99		66	
Sr	150				175	171	
Sc	21	28	22	20	16	16	
V	13	80	100	70	80	75	80
Y†	240				32	35	34
Zr†	2200				230	310	275

(1) *Lunar Sample Preliminary Examination Team* [1970].
(2) Fragment 37 + 24 from 12013; *Wakita and Schmitt* [1970], Si from *Morgan and Ehmann* [1970] Ga and Rb from *Laul et al.* [1970].
(3) Fragment 18 from 12013; ibid.
(4) Fragment 41 from 12013; ibid.
(5) Javanite J86, chemical data from USGS (M. Carron and C. Annell).
(6) Javanite J2, *Chapman and Scheiber* [1969].
(7) Javanite J87, unpublished data of Chapman and Scheiber.
* Apollo 12013 values conform closely with corresponding values for Australasian tektites other than J87, J86, J2.
† Apollo 12013 values not within factor of 2 of other tektites.
‡ Apollo 12013 values within factor of 2 of other tektites.

family exhibiting clearly the classical trends of an igneous differentiation sequence: analyses of this group [*Chapman and Scheiber*, 1969] show that as Si increases, Mg, Ca, Cr, Ni, Co, and Ni/Co decrease; Na, K, Sr/Ca, and Rb/Sr increase; Co-Mg and Co-Fe retain close coherence; Fe/(Fe + Mg) and normative pyroxene and Na/(Na + Ca) in normative plagioclase both increase; and the trends on an FMAlk ternary diagram conform to those of classical crystal fractionation. Yet the HMg tektite chemistry is clearly not that of earth igneous rock chemistry, and it is not that produced by vapor fractionation. For these reasons it has been concluded (ibid.) that this tektite chemistry must represent an extraterrestrial igneous rock chemistry. That the HMg tektite chemistry, which is so distinct from earth igneous rocks, happens to be so close in composition to the first acidic rock returned from the moon is surely an observation that should not pass unnoticed.

The Apollo missions also have revealed data that have been interpreted as constituting a major difficulty for the lunar impact origin of tektites. Thus far only old rocks with differentiation ages 3.3 b.y. or more have been returned from the maria landing sites of Apollo 11 and 12. In contrast, tektites are comparatively young, less than 2.5 b.y. The lunar uplands, it is reasoned, are even older than the maria, and therefore not a likely source of tektites.

Partly because of this apparent difficulty, *O'Keefe* [1970] has abandoned the widely accepted hypothesis of impact origin in favor of the older hypothesis of lunar volcanic origin. The latter, he suggests, circumvents the age difficulty, since the differentiation age of volcanic ejecta could be nearly the same as that of the final paroxysmic eruption which hurled the ejecta into space. However, an alternate explanation not incompatible with the lunar impact hypothesis also appears possible; the ancient uplands, even though comprised mainly of basic material, may contain acidic intrusions of various ages ranging from old to young. The combined physical processes of impact splash, atmosphere entry, and solution decomposition on earth (as explained in Appendix 3) could have resulted in dissolution of the basic glass and selective survival of the acidic glass. Thus the apparent differentiation ages of lunar impact glass which landed and survived on earth would be generally younger than the main mass of basic material in the moon's crust.

Inasmuch as Surveyor 9 landed near the north rim of Tycho and obtained a chemical analysis [*Turkevich et al.*, 1968; *Patterson et al.*, 1970], it is of interest to compare this with the tektite data. Crater ejecta are deposited in an inverted order to the original stratigraphic layering, hence, Surveyor 7 is presumed to have landed on low-velocity ejecta from the last stage of Tycho throw out. Such ejecta could have come from plutonic rock as deep as 10 to 15 km below the original surface [*Gault et al.*, 1968]. Tektites, on the other hand, are thought to originate as high-velocity fused ejecta from a relatively early and more highly shocked stage of the crater excavation process, and thus to represent rock originally from much shallower depths. The Surveyor 7 analysis resembles anorthosite, a rock with SiO_2 (46%) considerably lower, and with CaO (18%) and Al_2O_3 (22%) considerably higher, than in tektites. However, if an extrapolation to lower SiO_2 were made of the compositional trends for the family of HCa australites, which vary systematically over the range of SiO_2 from about 80% down to 66% [*Chapman and Scheiber,* 1969, Figure 9], a composition not greatly different from anorthosite would be obtained. Whether or not a single family of rocks will be found on the moon with such a wide variation in SiO_2, and with chemical characteristics which interlink HCa australites to anorthosite-like rocks, is something yet to be determined from future lunar explorations.

CONCLUDING REMARKS AND IMPLICATIONS FOR SELENOLOGY

Whereas a new theory in physical science can be verified repeatedly by direct evidence from laboratory experiments, any new theory in natural science about an event as ancient as the tektites must necessarily rest on circumstantial evidence. As in a courtroom trial, then, the present circumstantial evidence for the case of lunar tektite origin is summarized as follows: We have the aerodynamic ablation evidence that the Australasian tektites came through the earth's atmosphere as cosmic bullets from the moon; we have the evidence of a 'crater delecti,' Tycho, in the southern hemisphere of the moon within the vicinity where the computer trajectories fix the scene of the crime; we have further the evidence still visible that a shot was fired from Tycho in precisely the right direction; and we have the identifying fingerprint provided by matching the pattern of tektite distribution with the pattern of trajectory trails from Tycho. Moreover, from the Apollo data we have the evidence that glass of the size and form of microtektites has been splashed extensively over the moon; and that the composition of the first acidic rock returned from the moon closely resembles in most respects a tektite composition. In all, this is enough to prosecute for a conviction before the jury of the scientific community.

In view of such evidence, it seems warranted to outline some consequences to selenology that are implied by the conclusions about tektite origin. Inasmuch as tektites are high in silica, and the Australasianites are estimated to comprise as much as 10^8 tons of such glass, the concept of tektite origin by meteoritic impact on the moon clearly implies that:

1. Much glass has been splashed out of impact craters on the moon.
2. The moon has evolved rocks very high in silica, up to at least 85% SiO_2 (89% SiO_2 if Darwin glass is considered as tektite glass). Since it is generally accepted that high-silica rocks on earth are evolved in a late stage of

magnetic differentiation, we are led to the further implication that:

3. The moon's interior has been hot, at least in spots. The Rb-Sr isotope ages of the four known tektite events range from a few hundred million years for Australasian tektites to as much as 2.5×10^9 years for Ivory Coast tektites. It would be very improbable that a random sample of only four portions of the lunar crust would contain a specimen either of the youngest rock on the moon or of the oldest. Hence:

4. The formation of rocks on the moon (and thus lunar volcanism) must have spanned a very wide range in time, from geologically old to geologically young. The general chemistry of tektites in many respects is so similar to earth rock chemistry that geochemists have commonly preferred an earth origin for the tektites. If geochemists cannot clearly discern tektite chemistry from earth rock chemistry, then we cannot avoid the implication that:

5. The chemical origin of rocks in parts of the lunar crust is somehow closely related to the chemical origin of earth rocks. The principal conclusion of the present investigation—that the Australasian tektites came from the Rosse ray of Tycho—leads to two additional implications:

6. Tycho was formed at the same time as the Australasian tektites, about 700,000 years ago according to present age data.

7. The lunar upland areas near Tycho have evolved a number of different varieties of acidic igneous rock ranging widely in chemical composition.

Some, but not all, of these selenological implications have been confirmed by Apollo landings. As yet none have been demolished. Implication 1 has been decisively confirmed by the discovery in the lunar regolith of embarassingly large quantities of glass of microtektite size and shape. 'There are glass beads in every crater you come to and look in,' Apollo Astronaut Bean commented. Implication 2 has not yet been positively confirmed; but the discovery of glassy mesostasis containing as much as 78% SiO_2 within the interstices between crystals of the Apollo 11 basalts [*Roedder and Weiblen*, 1970; *Kushiro et al.*, 1970; *Keil et al.*, 1970; *Agrell et al.*, 1970; *Anderson et al.*, 1970] has led to the anticipation that granite-like rocks at least this high in silica will be found somewhere on the moon. Implication 3, long antithetical to the 'cold moon' idea, is consistent with Apollo 11 and Apollo 12 findings: the upwelling of basaltic magma of such vast extent as the maria, and the clear differentiation sequence of the Apollo 12 rocks [*Lunar Sample Preliminary Examination Team*, 1970] are compatible with this implication. Implication 4, though, has not yet been confirmed, inasmuch as only geologically old rocks have been identified thus far in the Apollo samples. Since basalt, anorthosite, and a 'granite,' all rock types common on earth, have already been returned by the astronauts, implication 5 is considered to be reasonably corroborated by the Apollo data. Implications 6 and 7, however, remain to be tested by future landings in the lunar uplands.

APPENDIX 1. PREVIOUS ATTEMPTS TO OUTLINE THE DISTRIBUTION PATTERN

It may be of interest to explain why previous attempts to define a geometric pattern within the strewnfield of Australasian tektites have led to patterns different from that constructed herein. From early measurements of the specific gravity of australites mainly from the three concentration centers of Victoria (2.41+ mode), Charlotte Waters (2.43+ mode), and Kalgoorlie (2.45+ mode), it was thought that there was a simple gradient pattern of silica decreasing from east to west across Australia [*Summers*, 1909; *Baker and Forster*, 1943]. But our more complete data on SG from more than thirty different australite localities now show that the population of highest SG is not in western Australia but is near the eastern coast (Uralla, 2.46+ mode), and the australite populations of lowest SG are not in eastern Australia but in central Australia (2.38+ mode). Thus these early SG data were simply inadequate in scope. A different pattern was suggested by *Beyer* [1933], who constructed from 25 chemical analyses then available a 'great-circle band' running northwest from Australia across Southeast Asia. His data also were inadequate in number. Still a different approach was taken by *Cohen* [1962]. Mainly on the basis of the Ni abundance at ten localities in the northern part of the strewnfield, he drew a two-lobe distribution pattern emanating from Thailand, one lobe low in Ni (less than 32 ppm) extending eastward to the Philippines, and one lobe high in Ni (over 125 ppm) extending southward to Java. But our subsequent measurements from about 20 times as many localities as were available to Cohen (including all ten localities used by Cohen) outline a different pattern (Figure 3). A number of localities with Ni over 125 ppm, for example, have now been found within his 'low Ni' lobe (e.g., Pasuquin, 325 ppm; South China Sea, 205 ppm; Plateau Van Hoa, 250 ppm; Buon M'Nar, 295 ppm; Kouang Tcheou Wan, 180 ppm). From a study of tektite internal structure, principally lechatelierite abundance and size, *Barnes* [1964] suggested that a concentric pattern exists in Southeast Asia, elongate in the northeast direction and centered around Thailand and Laos. He studied twelve localities in this area, but his limited data can also be interpreted as a northeast streak pattern—a pattern that would be compatible with that deduced from the more extensive chemical data of the present study. In essence, the raw data used in these four previous studies do not conflict with those of the present investigation: the strewnfield distribution pattern is simply too complex for even its principal features to have been discerned from the relatively meager data previously available.

Recently, from a study of the abundance pattern of australites, *McColl and Williams* [1970] observed that two principal lines of distribution could be discerned: one stretching from Port Campbell NW to Charlotte Waters, and one from the Nullarbor Plain NNW to Lake Wilson. Both of these lines are compatible with the distribution pattern deduced herein from the pattern of chemical variations.

Appendix 2. Examples of the Combined Use of Specific Gravity and Chemical Data

The data from SG polygons and chemical analyses were used in delineating fine structure among all but the uncommon or rare chemical types. For example, the SG polygons at Ban Tre Lai in North Vietnam, at Pahang in Malaya, and at Lumpoon in West Thailand are mutually quite similar, each exhibiting a prominent mode at 2.43+ of the type illustrated by Pahang in Figure 4c. The Ban Tre Lai specimens are all low Ni (four analyses) and include some HCu,B types, whereas the Pahang specimens are all medium-high Ni (three analyses). Hence, this particular pair of localities is not taken as a match. On the other hand, tektites in Malaya are of the same chemical type as those in West Thailand ('Chiang-Rai' type, CaO/MgO greater than or equal to unity, with medium-high Ni); their SG polygons are also of the same homogeneous type; and individual specimens from Batu Gajah in Malaya, and from Ban Mae Jong in West Thailand are found to chemically match (Table 1, match 8). This pair of localities in Malaya and West Thailand, therefore, is taken as a match. As another illustrative example, a specimen from Serpentine Lakes, South Australia, nearly matches the chemistry of a West Thailand specimen [*Chapman and Scheiber*, 1969, analyses 51 and 56]. The SG polygon at Serpentine Lakes, however, is of the heterogeneous type, without a prominent mode (Figure 4*b*), and with high SG specimens (above 2.46); whereas those in West Thailand are of the homogeneous 'spike' type with a very prominent mode, and without high SG specimens (Figure 4*c*). Hence Serpentine Lakes and West Thailand are not taken as matching localities. As a final example, chemical matches with Dalat tektites are found in North Cambodia and also in South China (Kouang Tcheou Wan). The SG polygons in North Cambodia, however, are relatively heterogeneous, containing high SG specimens (above 2.46), whereas the SG polygon for Dalat and South China cuts off at 2.45. The Dalat polygon has the same mode and is closely similar to the higher SG component polygon of the bi-modal polygon for Kouang Tcheou Wan; and chemical matches exist between these two localities (Table 1, match 3); hence Dalat and Kouang Tcheou Wan are taken as a match.

Appendix 3. Possible Explanations of Some Significant Observations about Tektites

Why tektites are glass. Crater ejecta shock-accelerated to a low velocity would comprise mainly warmed breccia; but ejecta shock-accelerated to lunar escape velocity would comprise thoroughly fused material. This follows from the Rankine-Hugoniot law for energy partition in a strong shock wave: $\Delta E = V_p^2/2$, where E is the internal energy per unit mass, and V_p is the particle velocity reached by shock acceleration. Shock acceleration to $V_p = 2.6$ km/sec, the most probable lunar ejecta velocity for hitting the earth, would impart $(2.6 \times 10^5)^2/2 = 3.4 \times 10^{10}$ erg/g of internal energy. After ejection into space, and adiabatic unloading of pressure, about half to two-thirds of this energy would remain as internal energy, corresponding roughly to 500 cal/g, or to temperatures of about 1800°C. The simultaneous combination of high temperature with very high shock pressure (hundreds of kilobars) is believed to be sufficient to thoroughly fuse silicates. We can thus explain why tektites are glass; why unfused mineral grains are rare in tektites; and why gobs of partially fused tektite material are not found on earth. An impact creates much more brecciated and partially fused material than glass; but such material, low in velocity, would remain near its source on the moon.

Why tektites are not spread all over the earth. The areal spread on earth of tektites from a single event is confined to only a portion of the globe by the pronounced effect of earth gravitational focusing. This can be illustrated by the moon-to-earth trajectories for the ejecta from the Rosse ray of Tycho. Shown to scale in Figure 14 are the positions of a cluster of 16 particles at three different times of 1, 2, and 2.9 days after leaving Tycho. Two views are depicted, one normal and one parallel to the ecliptic plane; the moon position is shown only at the instant of lunar impact. These 16 particles leave the moon with an azimuthal and an elevational velocity dispersion equal to 3% of the mean cluster ejection velocity (a dispersion equivalent to $\Delta\delta = 2°$ in azimuth, and $\Delta\beta = 2°$ in elevation). The longitudinal velocity dispersion, 2.55 to 2.73 km/sec, is 8% of the mean velocity, corresponding to several times the transverse dispersion. This is roughly the elongate proportions of typical 'ray elements' visible on the moon. Velocities greater than 2.73, or less than 2.55 km/sec, would miss the earth. It is seen that after one day the cluster has spread over dimensions several times the earth's diameter; and at 2.9 days, when the leading particle hits earth, the remaining particles are strewn in space over a third of the distance to the moon, stretching about 150,000 km along, and about 50,000 km transverse to, the mean trajectory path. Yet, at 3.5 days, all this highly strung-out cluster has landed on earth confined to a geographic spread of only 13,000 km in longitude by 9,000 km in latitude! Some scientists have disregarded the idea of lunar tektite origin on the unfounded belief that impact ejecta from the moon would spread all over earth. Such 'objecttions' overlook, among other things, this powerful effect of earth gravitational focusing. The earth's rotation, it should be noted, also happens to be in such a direction as to *compress* the geographic longitude spread of the ejecta relative to what it would be without rotation.

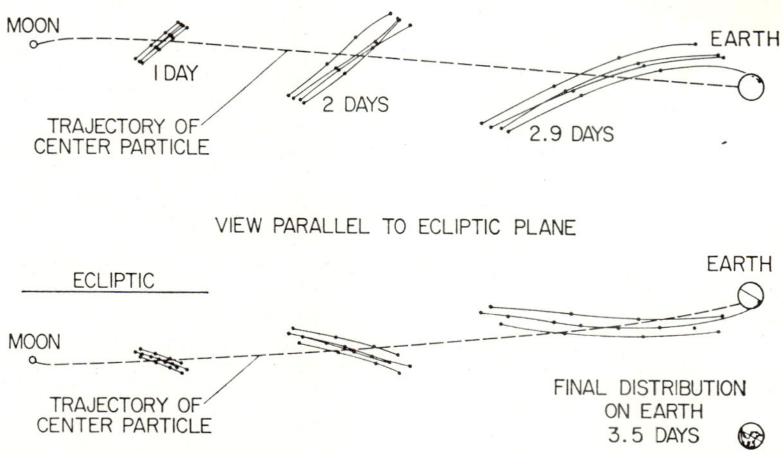

Fig. 14. Scale diagram of cluster position and spread in space during travel from Tycho to earth.

The major fraction of lunar crater ejecta that escapes the moon must miss the earth. It has been suggested that this fraction would later return and that glass in the recognizable form of tektites should be spread randomly over earth. Such returns, however, require on the average the order of 10^6 years or more before a probable encounter with earth; whereas the mean lifetime in space before destruction of tektite glass by meteoric bombardment, as *Gault and Wedekind* [1969] estimate, is several orders of magnitude less. The 'returns,' therefore, would not be tektites, but tektite dust comprising only a small fraction of the total meteoric dust that continually enters the earth's atmosphere. Some of the small particles may also be destroyed by the mechanism of 'rotational bursting' described by *Paddack* [1969].

Why moldavites and Ivory Coast tektites are not found all around their connate earth craters. An evident consequence of the present model for tektite events is that, when a ray element comprising spray and a trailing large fragment is hurled to earth, the tektite spray streams out mainly ahead of the fragment, and thus lands mainly on one side only of the connate crater—not uniformly all around it. This provides an explanation of why the moldavites are found on one side only of Ries crater, and why the Ivory Coast tektites and microtektites are found on one side only of Bosumtwi crater. Within the scope of this model, glass of tektite composition could be found right up to, and even in the immediate vicinity of, a connate crater.

How a very restricted tektite distribution, such as the moldavites, can be compatible with a lunar origin. As was noted earlier, the present model includes the possibility of a tektite event that is truncated. Few ray elements would travel dead-center toward earth. In a truncated event, most of the spray can miss the earth, in which case the small fraction that lands as tektites would land only in a restricted area near their connate crater. Also, a 2° spread of spray ($\Delta\beta = \Delta\delta = 2°$) like that used in most of the illustrations herein is considerably larger than for many ray elements observed on the moon. Some show only about 1/2° spread; these, if truncated, would land on earth in an area greatly confined both in length and breadth. Such circumstances, together with the processes of geological obscuration that have acted over the 15-million-year period since the moldavites fell, provide a possible explanation of the very restricted distribution known at present for this tektite group.

How it is possible for the australites to be restricted mainly to the southern part of the continent. It has long been known that few, if any, australites are found north of about 23°S latitude. Such a 'cut off' in latitude could result if the Australasian event were compound, and if many of the australites were associated with a connate crater that formed somewhere in the north of Australia. It is of interest, perhaps, to speculate on an earth crater that might be connate with the australites. Wolf Creek crater was formed by an iron meteorite, and is situated in northern Australia at a locality precisely aligned with the LSG–HCa streak. The HCa australites are not found elsewhere in the Australasian strewnfield. Unlike other meteoritic craters, the iron shale masses found around Wolf Creek commonly resemble teardrops, pears, disks, bananas [*McCall,* 1965], shapes typical of deformation and formation of iron in a plastic or highly viscous state. A meteoritic fragment of such size as that which formed Wolf Creek crater would cool little in a 3-day journey from moon to earth. Such a fragment, first shock heated throughout to the order of 650° during a lunar impact, would be shock heated a second time upon striking earth; and this time heated to temperatures at which iron softens and is readily deformed or extruded. Hence, if Wolf Creek crater is connate to the Australasian tektites, we would have an explanation for the curious shapes of iron shale masses found around this crater.

In Figure 15 the trajectory landing pattern is illustrated for a hypothetical compound event from Tycho comprising two ray elements: a major one, with $\Delta\beta = \Delta\delta = 1°$, and a minor one with $\Delta\beta = \Delta\delta = 0.8°$. The smaller projectile fragment associated with this minor ray element would leave Tycho at $V_B = 2.572$ km/sec, $\beta = 50.0°$, and $\delta = 17.5°$ in order to impact earth at Wolf Creek crater. These launch angles for β and δ are within 1° of those for the major fragment which could either have impacted somewhere in the Pacific Ocean or have missed the earth. Such a compound event

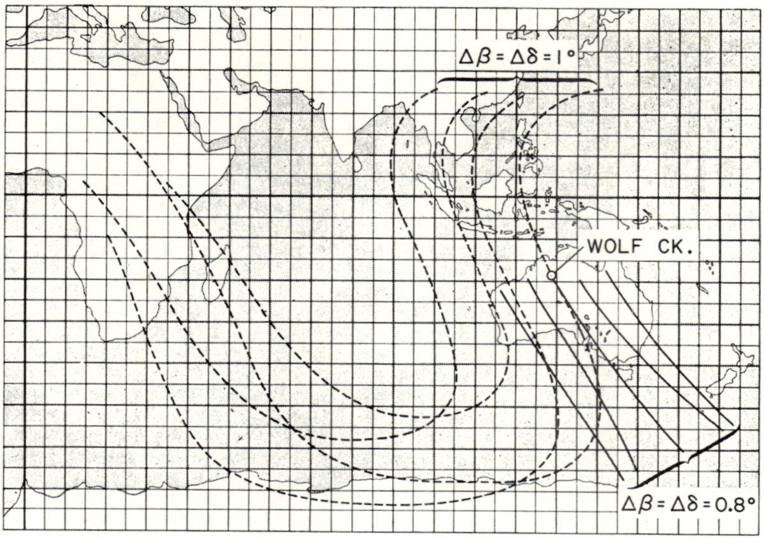

Fig. 15. Example of hypothetical compound event from Tycho comprising two ray elements: Projectile fragment for major element ($\Delta\beta = \Delta\delta = 1°$) lands either in Pacific or misses earth; projectile fragment for minor element ($\Delta\beta = \Delta\delta = 0.8°$) lands at Wolf Creek, Australia.

provides a possible, but as yet unsubstantiated, explanation for the observed latitude cut off of the australites. The age of Wolf Creek crater has not yet been determined; but its general appearance is such as to warrant investigation as a possible crater connate to the australites.

The trajectory trails in Figure 15 also illustrate that a 1° spray dispersion ($\Delta\beta = \Delta\delta = 1°$), which is representative of many lunar ray elements, is essentially adequate to explain the Australasian distribution pattern. The fact that a 2° spray dispersion was used for most illustrations herein is not critical to the principal conclusion that the Tycho ejecta pattern conforms to the tektite distribution pattern.

If the Australasian event were compound, it is also dynamically possible for one of the fragments to land in Africa near the Ivory Coast tektites. Present age data, however, apparently exclude such an association [*Gentner et al.*, 1970].

Why tektite shape variations are somewhat restricted. The law of energy equipartition for shock acceleration, combined with the relatively narrow range in lunar ejection velocity V_E that sends material to earth, provides a semiquantitative explanation of the most common variations in tektite shape. The Tycho ejecta that spreads from southern Australia to Southeast Asia, for example, represents V_E between 2.62 and 2.55 km/sec. The corresponding internal energies at ejection would vary by the factor $(2.62/2.55)^2 = 1.06$. This, in turn, corresponds to a temperature variation of about 130°C, which, for a given tektite composition, corresponds to a variation in viscosity by a factor of about 4. Since Australasianites commonly vary in SiO_2 from 70 to 80%, this introduces at the temperatures of formation additional variations in viscosity by another factor of 10. Variations in porosity of the pre-impact rock could introduce still further variations in viscosity for a given V_E. An over-all variation in viscosity by a factor of 40 is compatible with the observed variation in shapes of most tektites, such as from mainly small round forms in the Nullarbor Plains of Australia where SiO_2 averages about 70%, to mainly teardrops and more irregular shapes in Southeast Asia where SiO_2 averages about 75%. Experiments with glass ejection at various temperatures have shown that a variation in viscosity of formation by a factor of 44 [*Chapman*, 1964, Figure 12] produces shape and size variations from mainly round forms of small size to mainly teardrops and irregular forms of much larger size. Thus the Tycho trajectory calculations and corresponding energies provide reasonable explanation of the principal variations in Australasian tektite shape. Most of the ejecta that reaches earth from other lunar craters, such as those that produced the moldavites and the North American tektites, also would have left the moon in nearly the same velocity range (2.55 to 2.8 km/sec). Hence we also have an explanation of the striking conformity observed between the shapes of moldavites and of Southeast Asian tektites, as well as the general similarity in shapes between bediasites, Ivory Coast tektites, and certain Australasianites.

Why land tektites are acidic and not basic. Three factors are extremely selective for the survival of high-silica materials in a tektite event, and, hence, highly relevant to any deductions about the relative proportions of basic and acidic rock that may have existed at the crater impact site. The processes of impact splash, of atmosphere entry, and of terrestrial decomposition each act to selectively destroy basic glass exposed to the same severe physical environment that tektites have survived. First, because of its lower viscosity, basic glass splashed from an impact crater would solidify into smaller primary forms than acidic glass exposed to the same temperatures and forces of disruption. This has been demonstrated by experiments [*Chapman*, 1964] in which glasses of various viscosity were ejected from a pressurized tube, and recovered as splash forms—the lower the viscosity, the smaller the size recovered. On the basis of these experiments, it is to be expected that the same disruption process that produced tektite splash forms typically of centimeter size and rarely of 10-cm size, would produce basic glass forms typically of several mm size, and rarely of several cm size. Second, these smaller splash forms of basic glass, again because of their lower viscosity, would not survive the atmosphere entry that acidic tektites have survived. This has been proven by aerodynamic ablation experiments in which two spheres, one of tektite glass and one of basalt glass were placed in the same arc-jet stream. The amount of aerodynamic heating that turned a 2-cm tektite sphere into a typical australite button melted all but a small fraction of the sphere of basalt glass. This experiment is illustrated in Figure 16. Similar experiments also showed that the aerodynamic heating that turned a 1-cm tektite sphere into a button completely melted a 1-cm sphere of basalt glass. Hence, basic glass of several millimeters to 1 cm size would land on the ground only as a glass mist of ablation droplets a fraction of a millimeter in size. Third, small ablation droplets of basic glass

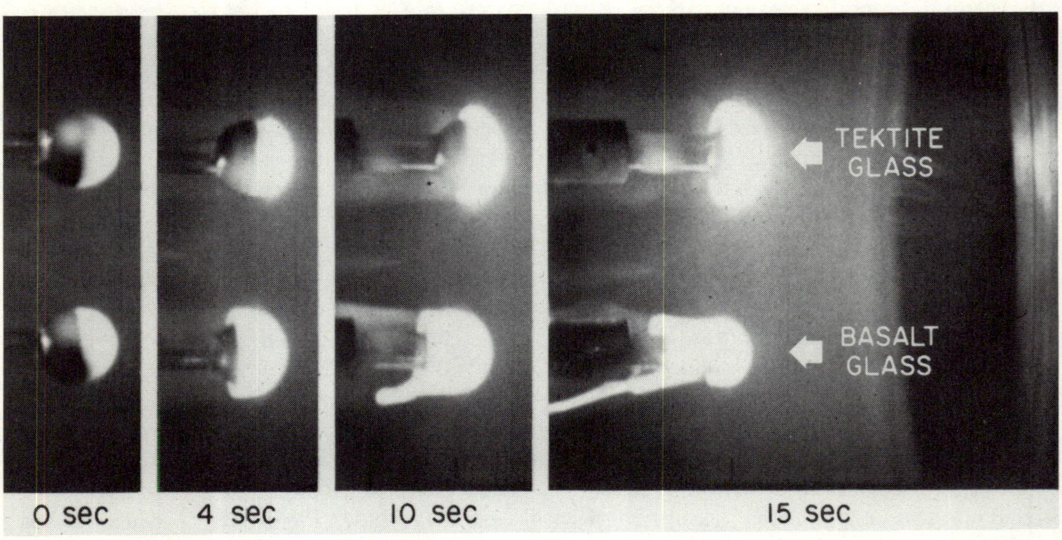

Fig. 16. Photographs taken during aerodynamic ablation experiment illustrate the difference between the ablation of tektite glass (*top*) and basalt glass (*bottom*). Arc-jet stream flows from right to left.

would not survive decomposition by ground waters. This has been established by some recent unpublished experiments of Mr. Frank Centolanzi who found that fresh water decomposes basaltic glass at ten times the rate of tektite glass. His measurements correspond to a decomposition rate at room temperature of 2 mm/m.y. for tektite glass, and approximately 2 cm/m.y. for basalt glass. Hence, basalt glass that is wet only a few days per year would decompose a few tenths of a millimeter in 700,000 years, enough to dissolve the ablation droplet remains of basalt glass. Because of these three powerful screening processes for selective survival of acidic glass, it follows that the Australasian tektite shower might have been accompanied by more basic than acidic glass, and we would have no way of knowing that today from observations on land.

In deep-sea cores, however, some basic microtektites have survived; and this is indeed consistent with Centolanzi's experiments. He observed that, whereas tektites and basalt glass lose weight by decomposing in fresh water, they do not in salt water. In fact, they gain a little weight in salt water. Some other glasses of different chemistry decompose in salt water. Depending upon the glass chemistry, therefore, and perhaps upon the chemistry of the sediment in which the glass is immersed, small droplets of basic glass (microtektites) in deep-sea sediments may or may not survive decomposition for 700,000 years. Thus, only a lower limit can be placed on the fraction of basic glass that originally accompanied the Australasian tektites; it was at least as large as the fraction recoverable today among microtektites, and possibly much larger.

Acknowledgments. I am singularly indebted to H. Julian Allen, a former colleague in aerodynamics and warm personal friend, who told me, two years before I could see how it possibly could be done, that if I were really any good as an aerodynamicist, I should be able to figure out from which crater and which ray the tektites came. I also thank Frank J. Centolanzi, Leroy C. Scheiber, and Norman B. Zimmerman, who have assisted in many phases of this research since its inception in 1960.

References

Agrell, S. O., J. H. Scoon, I. D. Muir, J. V. P. Long, J. D. C. McConnell, and A. Peckett, Mineralogy and petrology of some lunar samples, *Science, 167*, 583–586, 1970.

Albee, A. L., D. Burnett, A. Chodos, E. Haines, J. Huneke, D. Papanastassiou, F. Podosek, G. Russ, III, and G. Wasserburg, Mineralogic and isotopic investigations on lunar rock 12013, *Earth Planet. Sci. Lett., 9*, 137–163, 1970.

Andersen, C. A., J. R. Hinthorne, and K. Fredriksson, Ion microprobe analysis of lunar material from Apollo 11, *Proc. Apollo 11 Lunar Science Conférence, vol. 1*, 159–167, Pergamon, New York, 1970.

Anderson, A. T., Jr., A. V. Crewe, J. R. Goldsmith, P. B. Moore, J. C. Newton, E. J. Olsen, J. V. Smith, and P. J. Wyllie, Petrologic history of moon suggested by petrography, mineralogy and crystallography, *Science, 167*, 587–590, 1970.

Arnold, J. R., The origin of meteorites as small bodies, 2, The model, *Astrophys. J., 141*, 1536–1547, 1965.

Baker, G., and H. C. Forster, The specific gravity relationships of australites, *Amer. J. Sci., 241*, 377–406, 1943.

Barnes, V. E., Tektite strewn-fields, in *Tektites*, edited by J. A. O'Keefe, chap. 2, University of Chicago Press, Chicago, Ill., 1963*a*.

Barnes, V. E., Detrital mineral grains in tektites, *Science, 142*, 1651–1652, 1963*b*.

Barnes, V. E., Variation of petrographic and chemical characteristics of indochinite tektites within their strewnfield, *Geochim. Cosmochim. Acta, 28*, 893–913, 1964.

Beyer, H. O., The chemistry of tektites, and notes on their physical grouping, in *Phillippine Tektites*, vol. 1, part 1, pp. 59–86, 1933. (University of the Philippine Publications, 1961–1962).

Billerbeck-Gentz, F. Unterbrechungserscheinungen am mondstrahl Tycho-Polybius-Fracastor-Rosse, *Astron. Nachr. 274*, 140, 1943.

Braslau, D., Partitioning of energy in hypervelocity impact against loose sand targets, *J. Geophys. Res., 75*, 3987–3999, 1970.

Cassidy, W. A., B. Glass, and B. C. Heezen, Physical and chemical properties of Australasian microtektites, *J. Geophys. Res., 74*, 1008–1025, 1969.

Chao, E. C. T., E. J. Dwornik, and J. Littler, New data on the nickel-iron spherules from Southeast Asian tektites and their implications, *Geochim. Cosmochim. Acta, 28*, 971–980, 1964.

Chao, E. C. T., O. B. James, J. A. Minkin, J. A. Boreman, E. D. Jackson, and C. D. Raleigh, Petrology of unshocked crystalline rocks and shock effects in lunar rocks and minerals, *Science, 167*, 644–647, 1970.

Chapman, D. R., On the unity and origin of the Australasian tektites, *Geochim. Cosmochim. Acta, 28*, 841–880, 1964.

Chapman, D. R., A theory of tektite events and possible lunar crater of Australasian tektite origin, *Eos Trans. AGU, 49*(1), 244, 1968.

Chapman, D. R., and D. E. Gault, Critique of

'Cometary impact and the origin of tektites,' *J. Geophys. Res., 72,* 2695–2699, 1967.

Chapman, D. R., and H. K. Larson, On the lunar origin of tektites, *J. Geophys. Res., 68,* 4305–4358, 1963.

Chapman, D. R., and L. C. Scheiber, Chemical investigation of Australasian tektites, *J. Geophys. Res., 74,* 6737–6776, 1969.

Chapman, D. R., H. K. Larson, and L. C. Scheiber, Population polygons of tektite specific gravity for various localities in Australasia, *Geochim Cosmochim. Acta, 28,* 821–839, 1964.

Compston, W., and D. R. Chapman, Sr isotope patterns within the Southeast Australasian strewn-field, *Geochim. Cosmochim. Acta, 33,* 1023–1036, 1969.

Cohen, A. J., Asteroid-impact hypothesis of tektite origin, 3, The Southeast Asian strewn-fields, *Space Res., 3,* 950–973, 1962.

Darwin, C., *Geological Observations on the Volcanic Islands and Parts of South America Visited during the Voyage of H.M.S. Beagle,* pp. 44–45, 1844. (Republished by Appleton, New York, 1891.)

Fleischer, R. L., P. B. Price, and R. M. Walker, On the simultaneous origin of tektites and other natural glasses, *Geochim. Cosmochim. Acta, 29,* 161–166, 1965.

Fredriksson, K., J. Nelen, W. G. Melsen, E. P. Henderson, and C. A. Anderson, Lunar glasses and micro-breccias: Properties and origin, *Science, 167,* 664–666, 1970.

Gault, D. E., and E. D. Heitowit, The partition of energy for hypervelocity impact craters formed in rock, *Proceedings of Sixth Hypervelocity Impact Symposium, Cleveland, Ohio,* sponsored by the U.S. Army, vol. 2, 419–456, 1963.

Gault, D. E., E. M. Shoemaker, and H. J. Moore, Spray ejected from the lunar surface, *NASA Tech. Note D-1767,* 1962.

Gault, D. E., and J. A. Wedekind, The destruction of tektites by micrometeoroid impact, *J. Geophys. Res., 74,* 6780–6794, 1969.

Gault, D. E., J. B. Adams, R. J. Collins, G. P. Kuiper, H. Masursky, J. A. O'Keefe, R. A. Phinney, and E. M. Shoemaker, Lunar theory and processes, *Jet Propulsion Laboratory Tech. Rep. 32-1264,* chap. 9, 267–313, 1968.

Gentner, W., and J. Zähringer, Das Kalium-Argon Alter von Tektiten, *Z. Naturforsch., 152*(2), 93–99, 1960.

Gentner, W., H. J. Lippolt, and O. Muller, Das Kalium-argon-alter des Bosumtwi-Kraters in Ghana und die chemische Beschaffenheit seiner Glaser, *Z. Naturforsch., 192,* 150–153, 1964.

Gentner, W., B. Kleinmann, and G. A. Wagner, New K-Ar and fission track ages of impact glasses and tektites, *Earth Planet. Sci. Lett., 2,* 83–86, 1967.

Gentner, W., B. P. Glass, D. Storzer, and G. A. Wagner, Fission track ages and ages of deposition of deep-sea microtektites, *Science, 168,* 359–361, 1970.

Glass, B., Microtektites in deep-sea sediments, *Nature, 214,* 372–374, 1967.

Glass, B. P., Glassy objects (microtektites?) from deep-sea sediments near the Ivory Coast, *Science, 161,* 891–893, 1968.

Glass, B. P., Chemical composition of Ivory Coast microtektites, *Geochim. Cosmochim. Acta, 33,* 1135–1147, 1969.

Glass, B. P., Chromite and zircon crystals in a Muong Nong tektite, *Eos Trans. AGU, 51*(4), 341, 1970.

Hubbard, N. J., P. W. Gast, and H. W. Wiesmann, Rare earth, alkaline and alkaline metal and $^{87}Sr/^{86}Sr$ data for subsamples of lunar sample 12013, *Earth Planet. Sci. Lett., 9,* 181–184, 1970.

Keil, K., M. Prinz, and T. E. Bunch, Mineral chemistry of lunar samples, *Science, 167,* 597–599, 1970.

Kuiper, G. P., Interpretation of Ranger VII records, *Jet. Propulsion Laboratory Tech. Rep. 32-700,* 9–73, 1965.

Kushiro, I., Y. Nakamura, H. Haramura, and S. Skimoto, Crystallization of some lunar mafic magmas and generation of rhyolitic liquid, *Science, 167,* 610–612, 1970.

Laul, J. C., R. Keays, R. Ganapathy, and E. Anders, Abundance of 14 trace elements in lunar rock 12013, *Earth Planet. Sci. Lett., 9,* 211–215, 1970.

Lin, S. C., Cometary impact and the origin of tektites, *J. Geophys. Res., 71,* 2427–2437, 1966.

Lunar Sample Preliminary Examination Team, Preliminary examination of lunar samples from Apollo 12, *Science, 167,* 1325–1339, 1970.

McCall, G. J. H., Possible meteorite craters: Wolf Creek, Australia, and analogs, *Ann. N. Y. Acad. Sci., 123,* art. 2, 970–998, 1965.

McColl, D. H., and G. E. Williams, Australite distribution pattern in southern Central Australia, *Nature, 226,* 154–155, 1970.

McQueen, R. G., E. Zukas, and S. Marsh, Residual temperatures of shock-loaded iron, *ASTM Special Publ. 336,* 306–316, 1962.

Morgan, J. W., and W. D. Ehmann, Lunar rock 12013, 10; O, Si, and Fe abundances, *Earth Planet. Sci. Lett., 9,* 164–168, 1970.

O'Keefe, J. A., Tektite glass in Apollo 12 sample, *Science, 168,* 1209–1210, 1970.

Paddack, S. J., Rotational bursting of small celestial bodies: Effects of radiation pressure, *J. Geophys. Res., 74,* 4379–4381, 1969.

Patterson, J. H., A. L. Turkevich, E. J. Franzgrote, T. E. Economou, and K. P. Sowinski, Chemical composition of the lunar surface in a terra region near the crater Tycho, *Science, 168,* 825–828, 1970.

Pickering, W. H., An investigation of the systems of bright streaks visible upon the full moon, *Astron. Nachr., 130*(3111), 225–228, 1892.

Reynolds, J. H., Rare gases in tektites, *Geochim. Cosmochim. Acta, 20,* 101–114, 1960.

Roedder, E., and P. W. Weiblen, Silicate liquid immiscibility in lunar magmas, evidenced by melt inclusions in lunar rocks, *Science, 167,* 641-644, 1970.

Schieber, L. C., Comparison of tektite polygons of bulk specific gravity, true material specific gravity, and refractive index, *J. Geophys. Res., 75,* 7513-7515, 1970.

Schnetzler, C. C., J. A. Philpotts, and M. L. Bottino, Li, K, Rb, Sr, Ba and rare-earth concentrations and Rb-Sr age of lunar rock 12013, *Earth Planet. Sci. Lett., 9,* 185-192, 1970.

Starik, I. Y., E. V. Subotovich, M. M. Shats, and S. M. Grashchenko, On the question of the origin of tektites, NASA TTF-113, 1962. Translation of K voprosu o proiskhozhdenii tektitov, *Meteoritika, 22,* 97-102, 1962.

Summers, H. S., Obsidianities: Their origin from a chemical standpoint, *Proc. Roy. Soc .Victoria, 21*(2), 423-443, 1909.

Taylor, H. P., and S. Epstein, Correlations between O^{18}/O^{16} ratios and chemical compositions of tektites, *J. Geophys. Res., 74,* 6834-6844, 1969.

Taylor, H. P., and S. Epstein, Oxygen and silicon isotope ratios of lunar rock 12013, *Earth Planet. Sci. Lett., 9,* 208-210, 1970.

Turkevich, A. L., E. J. Franzgrote, and J. H. Patterson, Chemical analysis of the moon at the Surveyor 7 landing site: Preliminary results, *Science, 162,* 117-118, 1968.

Viste, E., and E. Anders, Cosmic-ray exposure history of tektites, *J. Geophys. Res., 67,* 2913-2919, 1962.

von Engelhardt, W., J. Arndt, W. F. Muller, and D. Stoffer, Shock metamorphism in lunar samples, *Science, 167,* 669-670, 1970.

von Koenigswald, G. H. R., Tektite studies, The origin of tektites, *Proc. Kon. Ned. Akad. Wetensch., Ser. B, 70,* 104-112, 1967.

Walter, L. S., Coesite discovered in tektites, *Science, 147,* 1029-1032, 1965.

Wakita, H., and R. A. Schmitt, Elemental abundances in seven fragments from lunar rock 12013, *Earth Planet. Sci. Lett., 9,* 169-176, 1970*a*.

Wakita, H., and R. A. Schmitt, Lunar anorthosites: Rare-earth and other elemental abundances, *Science, 170,* 969-974, 1970*b*.

Whitaker, E. A., Further observations on the Ranger VII records, *Jet Propulsion Laboratory Tech. Rep. 32-700,* 149-154, 1965.

(Received November 2, 1970.)

Erratum: Page 331, line 8 should read: "then northeast over the South China Sea, is"

Editors' Comments on Papers 40–45

The next six papers favor the terrestrial origin of tektites. These papers are arranged in chronological order, with three before the Apollo 11 mission and three following the mission. However, the paper by Taylor and Kaye* is in a transitional period of lunar knowledge, as they did have information from the Luna-10, and Surveyor V, VI, and VII lunar probes.

The first of the pre-mission papers is the one by Schwarcz* in which he systematically documents soil as the possible source material for tektites. This paper is followed by Faul's "Tektites Are Terrestrial,"* and then the chemical arguments of Taylor and Kaye* for the terrestrial origin of tektites. In addition to their chemical arguments, Taylor and Kaye* point out that information, provided by the similarity of australite compositions to that of Henbury glass and the absence of major compositional change during the impact melting at Henbury, conflicts with the data of Walter (1967) indicating that silica is volatile.

In the group of papers following Apollo 11, King, Martin, and Nance* strongly refute the paper by O'Keefe (1970) entitled "Tektite Glass in Apollo 12 Sample." Barnes*, in connection with the description of the largest tektite yet studied, suggested that the comae of comets might be the energy source for the formation of puddles of soil melt. However, Professor A. H. Delsemme of The University of Toledo, in a letter of June 5, 1972, in reference to the coma of a comet, states: "All effects of any shock wave from the coma interaction would indeed end at altitudes higher than 200 km." Nevertheless he also states: "Entirely different of course is the interaction of the nucleus itself with the atmosphere. A mass of 10^{15} to 10^{18} grams with a relative velocity between 12 and 72 km/sec must make a tremendous shock wave at its entry into the terrestrial atmosphere, but this is another story."

Appropriately the concluding paper of this Part is by Professor Harold C. Urey,* winner in 1934 of the Nobel Prize in Chemistry and member of the National Academy of Sciences. Dr. Urey, who originally proposed the cometary origin of tektites in 1957, also discussed the terrestrial and cometary origin of tektites in 1955, 1958a,b, 1959, 1962, 1963, and 1967.

Dr. Henry P. Schwarcz is now Professor, Department of Geology, McMaster University, Hamilton, Ontario, Canada.

Dr. Henry Faul is now Professor and Chairman, Department of Geology, University of Pennsylvania, Philadelphia, Pennsylvania 19104.

References

O'Keefe, J. A. (1970). Tektite glass in Apollo 12 sample. *Science,* **168,** 1209–1210.

Urey, H. C. (1955). On the origin of tektites. *Proc. (U.S.) Natl. Acad. Sci.,* **41,** 27–31.

Urey, H. C. (1957). Origin of tektites. *Nature,* **179,** 556–557.

Urey, H. C. (1958a). Origin of tektites. *Nature,* **181**, 1458.

Urey, H. C. (1958b). Origin of tekites. *Nature,* **182**, 1078.

Urey, H. C. (1959). Chemical composition of tektites. *Nature,* **183,** 1114.

Urey, H. C. (1962). Origin of tektites. *Science,* **137,** 746–748.

Urey, H. C. (1963). Cometary collisions and tektites. *Nature,* **197,** 228–230.

Urey, H. C. (1967). Moon, composition of. *In* "International Dictionary of Geophysics" (K. Runcorn, ed.) Vol. 2, pp. 997–1000. Pergamon Press, Oxford.

Walter, L. S. (1967). Tektite compositional trends and experimental vapor fractionation of silicates. *Geochim. Cosmochim. Acta,* **31,** 2043–2063.

40

(*Reprinted from Nature, Vol.* 194, *No.* 4823, *pp.* 8–10, *April* 7, 1962)

A POSSIBLE ORIGIN OF TEKTITES BY SOIL FUSION AT IMPACT SITES

By HENRY P. SCHWARCZ

Enrico Fermi Institute for Nuclear Studies, University of Chicago

TEKTITES are glassy objects found scattered on the Earth's surface in discrete arrays up to tens of thousands of kilometres across. They are composed of a highly siliceous and a luminous glass, devoid of crystalline phases except rare coesite[1] and nickel–iron[2] and containing particles of lechatelierite, pure silica glass.

A terrestrial versus extra-terrestrial origin for these objects has long been debated. As Viste and Anders[3] have shown, neither the presence of aluminium-26 nor any other distinctly extra-terrestrial property of tektites has yet been demonstrated. The chemical composition of tektites has been compared by various workers to various terrestrial rocks, especially sedimentary rocks, and contrasted with the compositions of meteorites and most igneous rocks. Spencer[4] has drawn analogies with impact-produced glass at known meteorite craters formed from the rocks of the impact site. Urey[5] and Barnes[6] have suggested that some impact process could produce the observed areal distribution of tektites.

I shall assume here, with Spencer[4], Urey[5] and Cohen[1], that tektites are formed on the Earth by the impact of some extra-terrestrial object. Tektites are presumably formed by the rapid fusion of rock during this impact. In this case the compositions of tektites should range widely over the possible range of terrestrial rocks, since the likelihood of (continental) impact on any given rock type should be proportional to the areal extent of the rock type. On the contrary, tektites are of restricted compositional range suggesting that the parental rock which has been fused to form them has a similar limited range of composition within which its components vary in some consistent fashion[7]. In a petrological sense they form a 'clan'.

Table 1

	A	*A′*	*B*	*C*	*D*
SiO_2	75·81	76·18	74·43	44·53	74·60
Al_2O_3	12·86	12·92	13·13	2·93	12·49
Fe_2O_3	4·18	—	—	—	—
FeO	—	3·77	4·27	14·57	4·11
MnO	0·13	0·13	0·12	0·29	0·13
TiO_2	0·93	0·94	0·83	0·13	0·90
CaO	1·33	1·33	1·76	2·28	2·04
MgO	0·88	0·89	2·01	33·76	2·01
K_2O	2·11	2·12	2·24	0·11	2·03
Na_2O	1·41	1·41	1·23	1·15	1·39
P_2O_5	0·30	0·30	nd	0·25	0·30

A, average of 11 soil analyses taken from Glinka (ref. 11) and Bear (ref. 12).

A′, same average recalculated with all Fe^{3+} reduced to Fe^{2+}.

B, average of 13 analyses of tektites taken from Barnes (ref. 25). Average includes 4 moldavites, 2 indochinites, 2 australites, 3 Ivory Coast tektites, 2 bediasites.

C, average chondrite excluding metal and sulphide (modified from average by Urey and Craig (ref. 14).

D, average soil plus 3·5 per cent chondrite. Proportion chosen to set MgO content equal to that in tektites.

A suitable mixture of shale and quartz sandstone[8] or an appropriate selection of igneous rocks[9] could account for the observed compositions; but it remains to be shown why just such a mixture or selection of rock types should be chosen as impact-site material to the exclusion of limestone, basalt, granite, etc.

A possible widely occurring rock type that may be the parental material of tektites is soil. It is worth while to see to what extent a soil parent rock will explain the observed properties of tektites. Soils formed within a given climatic zone on a diversity of bedrock types have remarkably similar composition[10]. The relations between the abundance of the chemical components of soil is largely a function of the weathering, dissolution and accumulation processes in the soil modified by the abundance ratios in the source bedrock. Therefore soils have a characteristic field of compositions and also form a clan, distinct from that of the parent rocks.

An average soil composition taken from analyses of temperate and humid zone soils is given in Table 1, column *A*. The data are from Glinka[11] and Bear[12]. For comparison I have constructed an average of 13 analyses from Barnes taken in roughly equal numbers from the 4 main areal (and age) groups, Ivory Coast, Central Europe, Australia–Asia and Texas. Each of these groups presumably represents a single impact site and parent material. This average is shown in column *B*. The similarity between these two averages is striking. Although one could construct a mixture of shale and sand or igneous rock and more closely approximate a tektitic composition this comparison with soil compositions is all the more striking because such a homogeneous rock-type does exist and is accessible to metamorphism into tektites by terrestrial impact processes. However certain discrepancies are noted between these averages:

(1) As in the sedimentary rock mixtures[8] iron is present in the starting material mostly as Fe^{3+} while it is mostly as Fe^{2+} in tektites. This point has been discussed by Friedman, Thorpe and Senftle[13], who conclude that the heating process alone could account for most of the reduction. The presence of lechatelierite particles indicates that temperatures as high as 1,700° C. have been reached during formation. For the reaction:

$$Fe_2O_3 + SiO_2 = Fe_2SiO_4 + \tfrac{1}{2}O_2 \qquad (1)$$

$\Delta F = -7{\cdot}8$ kcal. at 1,527° C. and decreases at higher temperatures so that even in the absence of any reducing agent the reduction of iron is favoured. No doubt the 2–8 per cent humic matter present in most soils would also contribute to the reduction of the iron.

(2) Magnesia and equivalent FeO (column A') are lower in soils than in tektites. The magnesium deficiencies are such that addition of a mixture of MgO and FeO to soil in the ratio of MgO : FeO = 1 : 0·45 would produce an average tektite composition. The ratio of these two oxides in stony meteorites

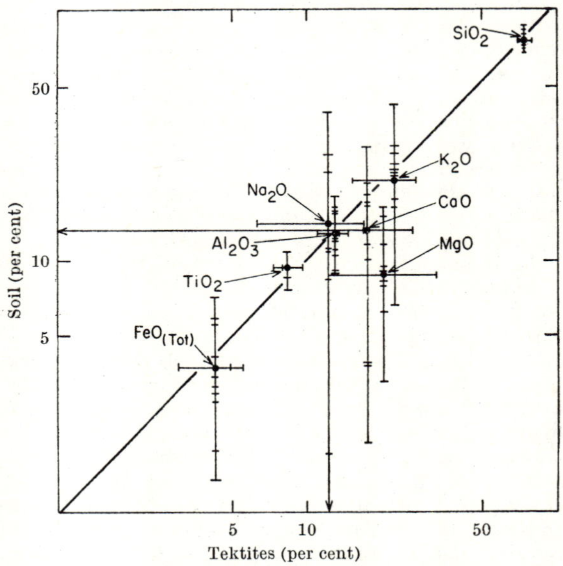

Fig. 1. Major element composition of eleven soils versus mean composition of each of four major tektite groups (moldavites, Asia–Australia, bediasites and Ivory Coast). K_2O, Na_2O, CaO, MgO and $TiO_2 \times 10$. Logarithmic plot

3

is 1:0·43 (ref. 14). In column *C* is shown a mixture of soil and 3·5 per cent chondrite. The proportions are chosen to give the appropriate MgO content (2·01 per cent). Such a mixture could occur in Nature if the tektites were produced by impact of a stony meteorite and there occurred some mixing of the silicate phases of the meteorite with the soil of the impact site. However, no appreciable mixing of the metal or sulphide components appears to have occurred. This is also indicated by the low nickel content in tektites observed by Ehmann[15].

Taylor, Sachs and Cherry[7] observed that of all major element abundances studied in tektites only that of calcium showed no significant correlation with any other element. Presumably the source of the parent tektite material contains varying amounts of a phase containing only CaO (or $CaCO_3$). This is probably material from the almost ubiquitous zone of calcium enrichment found at various depths in temperate-zone soil profiles. The varying CaO contents would correspond to varying depths of sampling by the impact process.

The trace-element abundances of tektites have been studied by Preuss[16], and more recently some individual elements have been measured by various workers (nickel, Ehmann[15]; rubidium, strontium, Schnetzler and Pinson[17]; lead, uranium, thorium, Tilton[18]; gallium, germanium, Cohen[19]; lithium, cæsium, Taylor and Sachs[7]). Using Preuss's data for the sake of its completeness I have compared the rank of the abundance-levels for tektites and two possible source rocks, soil and shale (Fig. 2). The abundance ranking for tektites agrees for most elements with that of soils, and less closely with that of shales, which are of course of related origin. Other possible source materials, such as igneous rocks and sandstones, agree more poorly.

Gases included in bubbles within tektites have been studied by F. E. Suess[20], who found that these cavities contained carbon dioxide, carbon monoxide and water. More recently H. Suess[21] observed that the gas pressure in these cavities was very low, but that on heating crushed tektites they did evolve carbon dioxide, carbon monoxide, hydrogen and water If the tektites were produced by fusion of soil these gases were presumably formed by the thermal decomposition of humic matter plus the adsorbed water of the soil. Note, however, that carbonaceous matter is also present in shales: an average of 44 shales compiled by Clarke[22] contained 0·8 per cent carbon.

Studies of lead, uranium and thorium in tektites

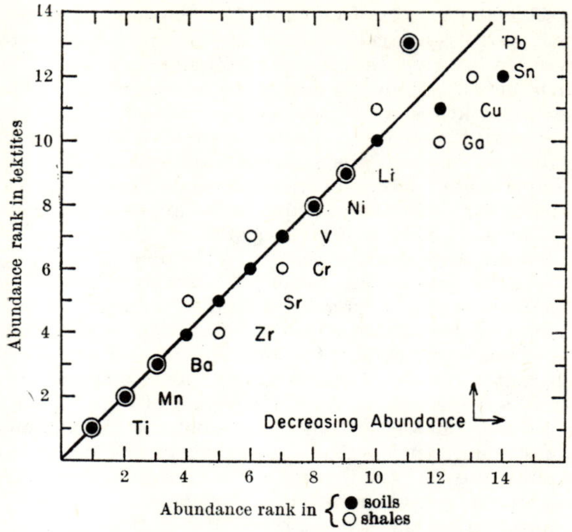

Fig. 2. Relative abundance rank of trace elements in tektites (Preuss) versus rank of same elements in shales (●) (Turekian, K. K., and Wedepohl, K. M., *G.S.A. Bull.*, **72**, 175, Table 2; 1961) and soils (○) (ref. 12)

by Tilton[18] showed that tektites contain modern lead such as would be derived from a typical crustal rock-system in which $^{238}U/^{204}Pb = 9{\cdot}8$ and $^{232}Th/^{204}Pb = 3{\cdot}9$. However, the observed values in tektites were 38·5 and 5·4 respectively. If the parent material of the tektites had had such high uranium and thorium contents for more than 50–100 million years a measurable deviation in the $^{206}Pb/^{207}Pb$ and $^{206}Pb/^{204}Pb$ ratios should have been observed. It is unlikely that thorium and uranium are enriched with respect to lead by the heating process alone. Therefore the tektites must have been derived from a rock of less than 100 m.y. age, that is, Cretaceous or Tertiary. In view of the limited surficial extent of Tertiary and Cretaceous rocks it seems an improbable coincidence that four (or more) random impacts, producing the four main families of tektites, should all have fallen on such young rocks. Soils, on the other hand, are known to be enriched in uranium and thorium, adsorbed into humic matter[23]. Such adsorbates would be short-lived since the humic matter is constantly being oxidized and carried out of the soil with a half-life of a few thousand years. The small amount of radiogenic lead produced in this interval by the decay of the adsorbed uranium is presumably carried away with the oxidized humic acid and would always be in low levels compared with the common and radiogenic lead of the mineral components of the soil.

With respect to the impact process itself, Bjork[24] has shown that on "hypervelocity" impact of a large iron meteorite the highest upward velocities are obtained by material originally located at and near the ground surface of the impact site. Material with such high velocities oriented at high enough angles to the horizon to follow intercontinental trajectories will therefore of necessity be near-surface material.

In conclusion, I have demonstrated that a possible parent rock for tektites is terrestrial soil. Soil forms a discrete, geochemically well-defined family of rock types which, with some *ad hoc* modifications, match the family of tektite compositions. Unlike the various mixtures of sedimentary or igneous rock types which have been proposed, soil is truly ubiquitous, covering all land surface except high mountains and deserts. Barnes[25] and Hawkins[26] have drawn comparisons between tektites and fulgurites. These latter would also be formed by fusion of soil, though by lightning rather than impact. I believe such a process is ruled out by the geographical distribution and the recently described occurrence of nickel–iron spherules in some tektites.

A test of the soil hypothesis would be some correlation between the local soil-chemistry at an impact site and the chemical composition of the resultant tektites. It must be assumed that climates have not changed appreciably since the time of impact since soil compositions would have changed accordingly. The Ivory Coast tektites are thought to have originated in an impact at Bosumtwi crater in Ghana[1]. If we can assume this area to have had a tropical or subtropical climate at the time of impact the soil of the impact site should have been enriched in Al_2O_3 and Fe_2O_3 and depleted in alkalis and CaO. Table 2 shows a comparison of these components with the average values. The alumina and iron oxide values are the highest for any tektites. The alkalis are not

Table. 2. PARTIAL ANALYSIS OF IVORY COAST TEKTITES AND COMPARISON WITH AVERAGE TEKTITE

	A	*B*
Al_2O_3	14·60	13·13
FeO	5·61	4·27
K_2O	1·53	2·12
Na_2O	1·71	1·41
CaO	1·67	1·76

A, partial analysis (average) of Ivory Coast tektites, data from Barnes.

B, average tektite. Average described in Table 1.

6

greatly depleted however. A second such comparison between source area and tektite is perhaps the Libyan desert glass for which no impact-site origin has been as yet proposed. The glass is close to pure silica in composition. This is the composition one would expect for a mature dune-sand such as that which forms the surficial cover over large parts of northern Africa. The rapid movement of these dunes may have buried and obliterated the impact site.

I thank Dr. R. N. Clayton and Dr. E. Anders for their advice.

This work was carried out under National Science Foundation contract *G*7292.

[1] Cohen, A. J., *Impact Crater Glasses: a Systematization*, Program, Forty-second Annual Meeting, Amer. Geophys. Union (1961).

[2] Chao, E. C. T., *Science*, **135**, 97 (1961).

[3] Viste, E., and Anders, E., *J. Geophys. Res.* (in the press).

[4] Spencer, L. J., *Mineralog. Mag.*, **23**, 387 (1933).

[5] Urey, H. C., *Nature*, **179**, 556 (1957).

[6] Barnes, Virgil F., *Sci. Amer.*, **205**, 58 (1961).

[7] Taylor, S. R., Sachs, M., and Cherry, R. D., *Geochem. Cosmochem. Acta*, **22**, 155 (1961).

[8] Cherry, R. D., and Taylor, S. R., *Geochem. Cosmochem. Acta*, **22**, 164 (1961).

[9] Mason, Brian, *Nature*, **183**, 254 (1959).

[10] Jenny, Hans, *Factors of Soil Formation* (McGraw-Hill Book Co. New York, 1941).

[11] Glinka, K., *Die Typen der Bodenbildung* (Geb. Borntraeger, Berlin, 1914).

[12] Bear, F. E., *Amer. Chem. Soc. Mon.* No. 126 (Reinhold Publishing Co., New York).

[13] Friedman, I., Thorpe, A., and Senftle, F., *Nature*, **187**, 1089 (1961).

[14] Urey, H. C., and Craig, H., *Geochem. Cosmochem. Acta*, **4**, 38 (1953).

[15] Ehmann, W. D., *Geochem. Cosmochem. Acta*, **19**, 149 (1960).

[16] Preuss, B., *Chemie Erde*, **9**, 365 (1935).

[17] Schnetzler, C. C., and Pinson, W. H., *Ninth Ann. Prog. Rep., U.S. AEC Contract AT* (31–1)–1381, 27–60, *Dep. Geol. and Geophys.*, Mass. Inst. Tech. (1961).

[18] Tilton, G., *Geochem. Cosmochem. Acta*, **14**, 323 (1958).

[19] Cohen, A. J., *Nature*, **188**, 653 (1960).

[20] Suess, F. E., *Verhandlung der K. K., Geologische Reichanstalt*, 387 (1898).

[21] Suess, H., *Geochem. Cosmochem. Acta*, **2**, 76 (1951).

[22] Clarke, F. W., *U.S. Geol. Surv. Bull.*, fifth ed., 770 (1924).

[23] Vine, J. D., Swanson, V. E., and Bell, K., *Surv. Raw Materials Resources*, **2**, 187, Intern. Conf. on Peaceful Uses of Atomic Energy, 2*d*, Geneva).

[24] Bjork, R. L., *J. Geophys. Res.*, **66**, 3379 (1961).

[25] Barnes, Virgil F., *North American Tektites*, 477 (Univ. Texas Publ. 3945 (1940).

[26] Hawkins. G. S., *Nature*, **181**, 1457 (1960)

Printed in Great Britain by Fisher, Knight & Co., Ltd., St. Albans.

Reprinted from *Science*, **152**, 1341–1345 (1966)

41

Tektites Are Terrestrial

Age determinations link the origin of some tektites to specific impact craters on the earth's surface.

Henry Faul

To anyone who has worked with them, tektites are probably the most frustrating stones ever found on earth. For almost 200 years these glassy objects have been scientifically collected, studied, and subjected to every conceivable kind of analysis. A vast amount of data has been accumulated, but one still can draw only a very crude outline of the physical and chemical processes that formed the tektites. It has been established that tektite-like glass is produced in hypervelocity impacts on rocks; that tektites formed and cooled enough to be rigid before they entered (or reentered) the earth's atmosphere; and that aerodynamic ablation partly melted them again on their passage through the atmosphere. The exact process of tektite formation remains largely a mystery, but age measurements are beginning to show *where* this formation occurred. The violent controversy over a terrestrial versus an extraterrestrial origin for tektites is nearing its end.

The age of a tektite can be measured in several different ways. One may determine the amount of Ar^{40} that has accumulated from the decay of K^{40}, or one may develop and count the tracks left in the tektite glass by the spontaneous fission of U^{238}, or one may determine the present-day Sr^{87}/Sr^{86} and Rb^{87}/Sr^{86} ratios in the glass and extrapolate to the initial Sr^{87}/Sr^{86} ratio, drawing the strontium isochron. These three ages have different meanings. The potassium-argon method dates the time when argon diffusion in the glass became negligible compared to the rate of its generation from potassium decay. The fission-track method establishes the time of last cooling to a temperature where the tracks in the glass remain stable with time. If a tektite has been through a grass fire, for example, the fission tracks will date the fire. That is one reason why ages measured by the fission-track technique are occasionally lower than ages determined by the potassium-argon method. The Sr^{87}/Sr^{86} and Rb^{87}/Sr^{86} ratios are not altered by melting, hence the strontium isochron age of tektites refers primarily to their parent material. If that was a crystalline rock, then the isochron dates the time when this rock was derived from the earth's mantle, directly or indirectly. If it was a sediment, then the isochron age is an average of the ages of the primary rocks from which the sediment was derived.

The author is professor of geophysics at the Graduate Research Center of the Southwest in Dallas, Texas.

Tektite Fields

There are basically four groups of tektites, undoubtedly formed in four separate events. The largest and youngest tektite field is in the southwest Pacific area, where tektites are found in the southern part of Australia and in Indochina, China, the Philippine Islands, and Indonesia (*1*). Australasian tektites differ slightly in physical and chemical properties from area to area; they are uniformly dark brown in transmitted light, and their ages, measured by the potassium-argon and fission-track methods, are about 700,000 years (*2*). The Australasian field is by far the largest known tektite field. It is so large, when its possible extension into the Indian and Pacific oceans is taken into account, that geographic arguments for a terrestrial origin of tektites hardly apply to it, as we shall see.

The next older tektites are found in the Ivory Coast Republic (Fig. 1). They occur in a relatively unexplored region and are rare. Both potassium-argon and fission-track dating methods give their age as about 1.5 million years (*2*). They are also dark brown.

Moldavites, the Czechoslovakian tektites named after the Moldau (the German name for the river Vltava), are next in age. They occur in a small area (about 50 kilometers long) in southern Bohemia and in an even smaller district in southern Moravia (Fig. 2). Again, there are slight differences in color and chemical composition (*3*). Bohemian moldavites are light green and Moravian ones are olive green or brownish, but both the potassium-argon and the fission-track dating methods yield ages close to 14.8 million years for all of them (*2*). The extent of the moldavite fields (*4*) is better known than the extent of any other tektite area, and a wealth of geologic detail (*5*) is available.

The oldest tektites are in America, in east-central Texas and south-central Georgia (Fig. 3). The Texas tektites (called bediasites after the Bedias or Bidai Indians) are generally brown, and the Georgia tektites are greenish and lighter in color, but ages of about 34 million years have been obtained for all of them. A single tektite was found on Martha's Vineyard in Massachusetts, but this solitary stone is indistinguishable from Georgia tektites and of exactly the same age. One cannot ignore the implication that it was brought to the island by man (*6*). The geology of Martha's Vineyard has been studied meticulously, and the area has been combed in search of another tektite, without success.

Geology

Tektites are always found on the surface or in sediments much younger than the measured age of the tektite glass (*7*). I am not aware of any valid report of a tektite's having been found in sediments of the same age. Bediasites are found in young gravels on the surface of the Jackson group of formations, usually considered to be of Late Eocene age. In Grimes County, Texas, they are found along the middle and lower parts of the Jack-

1

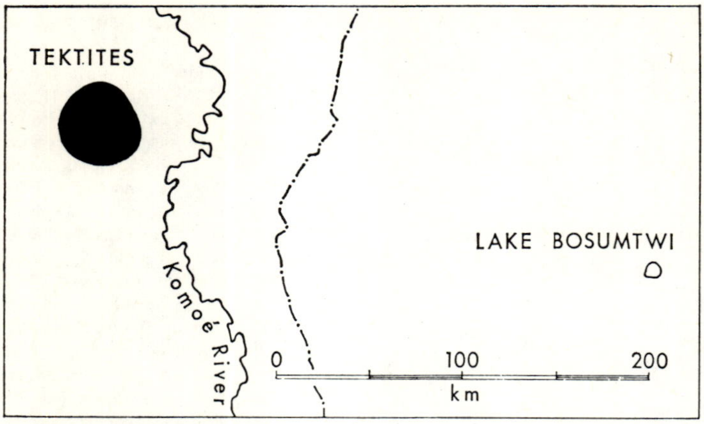

Fig. 1. The Bosumtwi crater in Ghana and (solid area) the approximate area where the Ivory Coast tektites are found.

son group, in Fayette County they occur along the upper part of the group, and elsewhere they lie in between. In view of the rarity of tektite-making events in general and the uniformity of bediasites in particular, one could hardly assume that more than one bediasite shower had occurred. The area now underlain by the Jackson group has undergone extensive uplift and erosion in post-Jackson time, as well as some warping, as shown by the relatively high dips of the Early Pleistocene Citronelle formation (*8*). Thus it appears probable that bediasites now found on the surface of the Jackson group have been moved by water.

Bediasites are now weathering out of the Jackson formations. If one accepts the view that the Jackson group belongs to the Late Eocene (about 40 to 45 million years ago), then one says, in effect, that Jackson time had ended more than 5 million years before the bediasites fell. That time interval is too large to be ascribed to errors in the age determination for the bediasites or to inadequacies of the Cenozoic time scale. Either Jackson rocks were exposed at the time of the bediasite fall, or else Jackson time extended well into the Oligocene.

Georgia tektites are exceedingly rare. They are found on the surface of young (Pliocene-to-Pleistocene) gravels that unconformably overlie Eocene, Oligocene, and Miocene rocks. They could have been reworked from any of them (*9*).

Moldavites had been thought to be Middle Miocene in age (*10*) long before their age was measured by the potassium-argon method. When this method gave an age of 14.8 million years, the result was hailed as confirming the stratigraphic correlation (*2*). It was even suggested that moldavites could be used as a tie point for the Helvetian stage in the Cenozoic time scale. Detailed geologic survey of the Moravian moldavite field, together with a study of the distribution of unusual and characteristic minerals of restricted provenance in the moldavite gravels (*5*), has shown that the earlier stratigraphic correlation is invalid. Moldavite gravels are all very young, Late Pliocene or younger. All known moldavites have been transported by water, even though the distances cannot have been very great (*4*).

Ivory Coast tektites are found in recent gravels on an ancient (2000-million-years-old) crystalline basement (*11*). Australasian tektites also occur on the surface or in strata generally younger than the tektites themselves, but occasionally there is room for doubt in correlating Pleistocene sediments. Hence the argument from stratigraphy carries less weight for these fields than for the older tektite fields.

The Age Paradox

It should not be surprising that tektites are not found where they landed. They are, first of all, geologically rare, even in places where they are most "abundant." In the tektite areas outside Asia it takes a trained man hours to find one.

Even in the Australasian field there are only a few places where tektites are sufficiently numerous to be picked up by the handful.

Being in the size range of gravel, tektites are readily concentrated by running water, especially by the intermittent water of streams in flood. Thus it is obvious why tektites occur in gravels. Furthermore, there is good evidence that tektites dissolve rapidly in pelitic sediments, especially where the *p*H is high. The Besednice brick pit occurrence in Bohemia is a good example. The tektites found in the white calcareous clay there are often corroded to mere slivers, yet the clay is quite young and the tektites were probably washed into it. There is no evidence that they fell there.

Hence the supposed paradox of stratigraphic versus radiometric age for tektites seems to be largely imaginary. A tektite is much more likely to have been transported than to lie where it fell. An analogy is the common occurrence of placer gold in stream gravels as opposed to the rarity of "mother lodes" upstream.

The Flight

It is an established fact that tektites fell from the sky. Aerodynamic ablation experiments with tektite glass (*12*) have gone a long way toward explaining how tektites entered the earth's atmosphere and how they interacted with it. Comparison of synthetically produced ablation forms with the morphology of some well-preserved australites has shown that australites entered the atmosphere at velocities around 10 kilometers per second, and at low angles. Strictly speaking, these results apply only to Australasian tektites, but it has been shown that at least one bediasite and one moldavite also have the flattening that is characteristic of aerodynamic ablation (*13*), and one may accept the conclusion that the entry of the less-well-preserved groups of tektites was not very different from the flight of the australites.

But what happened before the tektites entered the atmosphere? The lack of any evidence of cosmic-ray interaction with the tektite glass (*14*) shows that tektites were not long in space. They could not have come from far away, astronomically speaking. An object in the mass range of even the largest tektites will be slowed down to a halt after only a few kilometers' flight

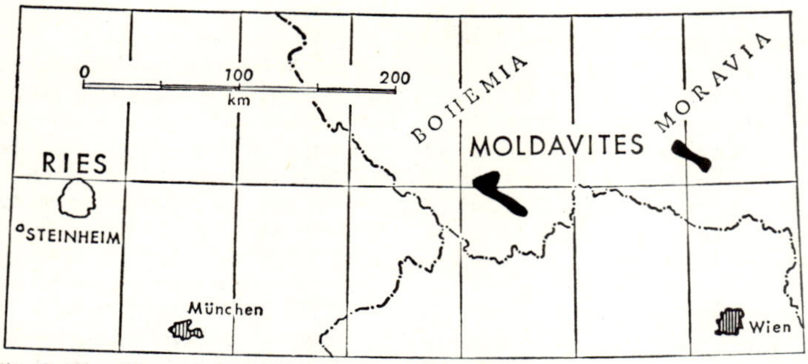

Fig. 2. The Ries and Steinheim craters in Germany and (solid areas) the tektite fields in Czechoslovakia.

in air at atmospheric pressure, no matter how great its initial velocity may have been. Thus we are left with two alternatives. If tektites originated on earth, then they had to fly up through an atmosphere that either moved along with them or was temporarily greatly rarefied, presumably by the event that produced the tektites (*15*). Alternatively, the tektites could have come from outside the earth's atmosphere, presumably from the moon.

Possibility of a Lunar Origin

There is much that motivates theorizing that tektites came from the moon. Both the composition and the morphology of tektites could be more easily understood if one could assume that the glass was made in a vacuum. Some tektites have bubbles in them, and many of these bubbles preserve a more or less high vacuum (*16*). Tektite glass contains an extremely small amount of water, 10 to 100 times less than man-made glass and 1000 times less than obsidian (*17*). The spherical and drop shapes of well-preserved tektites indicate some kind of splash of the molten glass and an absence of any strong forces acting on the still-molten particles. Solidification of molten tektite requires from a fraction of a minute to a few minutes, depending largely on its size. During this interval, surface tension seems to have been the dominant force in the shaping of the tektites. These processes can be most readily imagined as occurring in a vacuum. The lunar vacuum would seem to be ready-made for such purposes.

Furthermore, if it could be assumed that tektites are thermally altered samples of the lunar surface, then it would follow that the moon has a highly siliceous crust, with all the attendant implications of internal melting, magmatic differentiation, and an internal structure analogous to that of the earth. The tektite evidence then could be used to sweep aside the alternative hypothesis of a cold undifferentiated moon composed essentially of material similar to stony meteorites. A whole class of hypotheses concerning the origin and early history of the earth then could be dismissed.

One fatal objection has long stood in the way of the lunar origin hypothesis: the lack of any focusing mechanism that would keep the hypothetical tektite swarm together on its long journey from the moon. Only the Australasian tektite field is large enough to be attributable to a fall from an extraterrestrial source (*18*), and even there one would have to postulate very little mixing within the tektite swarm en route in order to account for the small but real differences observed in Australasian tektites from place to place.

No mechanism is known that would permit transportation of ejecta from some lunar crater into a target area as small as the tektite fields in Bohemia and Moravia or on the Ivory Coast. The various proposed mechanisms in-

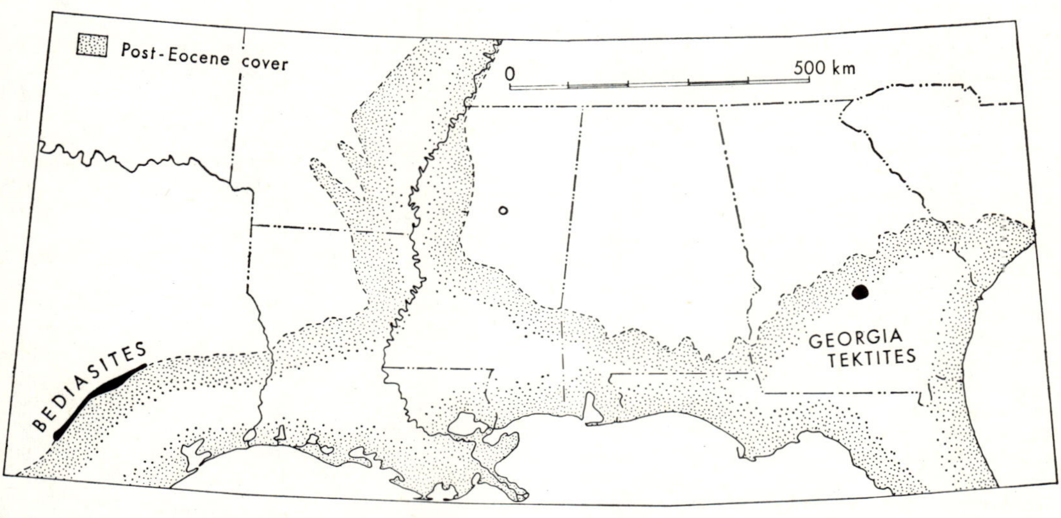

Fig. 3. (Solid areas) Regions where tektites are found in Texas and in Georgia. (Stippled areas) Regions covered by sediments deposited after the tektites fell. (Open circle) The Kilmichael crater in Mississippi.

volving intermediate parent bodies (*19*) also would produce strewn fields much too large to fit the facts. Theoretical analysis of the phenomena following the impact of a large body on the earth indicates that tektites could be formed in such an event (*15*). A body such as a comet with a geocentric velocity of several tens of kilometers per second, a density of the order of 0.1 gram per cubic centimeter, and a mass greater than about 5×10^{17} grams, colliding with the earth, would have sufficient energy to blow a momentary "bubble" in the earth's atmosphere and thus produce, for a moment, an environment suitable for the production and ejection of tektites from a terrestrial crater (see *15*).

Tektites and Craters

The proposal that moldavites are associated with the crater called Nördlinger Ries, in south-central Germany, was made even before it was known that impact glass from the Ries had the same age as the moldavites (*20*), which are found roughly 300 kilometers to the east. There were uncertainties about the validity of the first age measurements because of difficulties in completely extracting the radiogenic argon from the viscous tektite melt, but these technical problems have been resolved, and the age measured by the potassium-argon method has now been confirmed by the altogether independent method of fission-track counting. Within the limits of experimental error, the age of the moldavites and of the crater has been established, by these two independent methods, as 14.8 million years (*2*).

The glass in the debris around the Ries crater and the moldavites are of the same age, but their chemical compositions are different. The crater glass is chemically similar to granitic gneiss. Such gneisses are exposed in the Black Forest, a little more than 100 kilometers west of the Ries, and externally similar but severely shocked and altered rock fragments are common in the crater debris. The moldavites, like all tektites, have chemical and isotopic compositions reminiscent of weathered crystalline rocks or the sediments derived from them. In the Ries area, the crystalline basement rocks are overlain by sediments about 500 meters thick. The crater glass and the moldavites could not have come from the same melt but could have been produced by the melting of different rocks during separate phases of the same explosive event.

Another tektite field has a crater nearby. The Ivory Coast tektite area lies roughly 300 kilometers west of Lake Bosumtwi (in Ghana), an undoubted impact crater. Here again the two dating methods give the same result for glass from the impact area and for the tektite glass—about 1.3 million years (*2*). Again, there are some uncertainties about the results, but in this case the uncertainties stem mostly from the obvious fact that an age of 1.3 million years is more difficult to measure than one of 15 million years. As repeated measurements are made, the experimental error is reduced and it becomes clearer and clearer (*21*) that Lake Bosumtwi and the Ivory Coast tektites were formed at the same time.

The American tektites are not clearly associated with any known impact structure. The Gulf coastal plain region is geologically a poor place to look for an Oligocene crater. The geologic history of the coastal region makes for a vanishingly small probability that any trace of such a crater could be seen there today, had one existed. Further northward, the Kilmichael, Mississippi, structure is large enough and possibly old enough to be considered a possible source of bediasites and Georgia tektites, but it is a difficult one to study. For glass from the Clearwater Lake crater in northern Quebec, the fission-track dating method yields an age of 33.5 ± 4.5 million years (*2*), roughly in agreement with the age of the American tektites. This age has not been confirmed by potassium-argon dating. The rocks in the vicinity of Clearwater Lake are very ancient, and American tektites could not have been made from them. The strange glass found in great quantity in the Sand Sea of the Libyan Desert also has a similar age (*2*), but it is difficult to imagine what physical connection this glass could have with American tektites.

The size and geography of the Australasian tektite fields strain the imagination in considering a possible association of tektites and craters. The tektites might have come from more than one crater, and hypothetical locations in Antarctica and Indochina have been proposed (*22*). The craters would not have formed in the deep sea, but might have formed in the large areas of shallow water. Of the several known dry-land craters in Australasia, none seems to fit.

The Place

Ages yielded by the potassium-argon and fission-track methods link tektites with craters in terms of time but not necessarily in terms of place. The strontium isochron method, however, has the potential for showing *where* tektites were made.

Whatever the exact process of tektite formation may have been, it could not alter the average Sr^{87}/Sr^{86} ratio of the material, and it could only slightly reduce the Rb^{87}/Sr^{86} ratio by fractional volatilization of the silicates. Hence it is possible to make age determinations on tektites by the whole-rock strontium method. When plotted on an isochron diagram, the ages obtained in this way for American, Australasian, and Czechoslovakian tektites all fall roughly on the same line with a zero intercept (the initial Sr^{87}/Sr^{86} ratio) of about 0.705 (*11*). The slope of the line corresponds to an age of about 400 million years. If allowance were made for a rubidium loss of, say, 25 percent, the age would be correspondingly lower, or roughly 300 million years.

This curious agreement for the three groups of tektites could be interpreted in two entirely different ways. Either (i) these three kinds of tektites were formed from a common source material now 300 to 400 million years old, or (ii), in the case of the moldavites, this age would reflect the 300-million-year age of crystalline rocks of the German crystalline basement north of the Alps (*23*) but the apparent agreement of the American and Australasian tektite isochrons with the moldavite isochron would be, for the moment, unexplained.

For some time, the first interpretation seemed the more likely. It would favor the hypothesis of an extraterrestrial origin of tektites, in view of the difficulty of explaining how rocks at three regions on earth, "picked" effectively at random by the impacting bodies, could have the same strontium isochron age.

As it turns out, however, the second interpretation is probably correct. The crucial argument comes from recent age measurements, by the whole-rock strontium isochron technique, for the Ivory Coast tektites and for a suite of rocks from the vicinity of Lake Bosumtwi (*11*). The lake is located in a region of metasedimentary and granitic rocks with hardly any sedimentary cover. The rocks give a clearcut isochron with a slope equivalent to 2000 million years, and both the Ivory Coast tektite glass and the impact glass from Lake Bosumtwi fall convincingly on the same isochron, or a little to the left of it, as would be expected had there been a slight loss of rubidium. There can be little doubt that Ivory Coast tektites were made from rocks 2000 million years old, and that such rocks once lay on the spot now covered by Lake Bosumtwi.

Origin of Tektites

The accumulating geochronologic evidence indicates more and more convincingly that tektites were formed from terrestrial rocks in large meteoritic impacts on the earth. Ivory Coast tektites were formed about 1.3 million years ago, simultaneously with the Bosumtwi crater, and from rocks 2000 million years old. The rocks around the crater are of the same age. Moldavites were formed 14.8 million years ago, simultaneously with the Ries crater, and from rocks now roughly 300 million years old. Crystalline rocks throughout Germany north of the Alps are about 300 million years old. North American tektites were formed 35 million years ago, but no crater is known to be associated with them. Their age, as determined by the strontium isochron method, would be compatible with an origin from Appalachian granites, volcanic rock, or sediments derived from them.

Not much can be said about a parent rock for the Australasian tektites, except that it would be roughly 200 to 400 million years old. The principle of simplicity suggests that one might more profitably look for large concealed craters in that vast region than postulate an extraterrestrial origin for the tektites.

References and Notes

1. D. R. Chapman, *Geochim. Cosmochim. Acta* **28**, 841 (1964).
2. R. L. Fleischer and P. B. Price, *ibid.*, p. 755; ———, *ibid.* **29**, 161 (1965); J. Zähringer, in *Tektites*, J. A. O'Keefe, Ed. (Univ. of Chicago Press, Chicago, 1963), pp. 137–149.
3. J. A. Philpotts and W. H. Pinson, Jr., *Geochim. Cosmochim. Acta* **30**, 253 (1966).
4. C. W. Naeser, V. Bouška, H. Faul, *Trans. Amer. Geophys. Union* **46**, 546 (1965).
5. V. Bouška, *Geochim. Cosmochim. Acta* **28**, 921 (1964).
6. C. A. Kaye, C. C. Schnetzler, J. N. Chase, *Bull. Geol. Soc. Amer.* **72**, 339 (1961).
7. V. E. Barnes, in *Tektites*, J. A. O'Keefe, Ed. (Univ. of Chicago Press, Chicago, 1963), pp. 25–50.
8. J. A. Doering, *J. Geol.* **68**, 182 (1960).
9. E. A. King, Jr., *Georgia Mineral Newsletter* **15**, 84 (1962); *Geochim. Cosmochim. Acta* **28**, 915 (1964).
10. R. Janoschek, *Mitt. Geol. Ges. Wien* **29**, 329 (1937).
11. C. C. Schnetzler, W. H. Pinson, P. M. Hurley, *Science* **151**, 817 (1966).
12. D. R. Chapman and H. K. Larson, *J. Geophys. Res.* **68**, 4305 (1963); *ibid* **69**, 1937 (1964).
13. E. A. King, Jr., *ibid.* **69**, 4731 (1964); E. C. T. Chao, *Science* **146**, 790 (1964).
14. R. L. Fleischer, C. W. Naeser, P. B. Price, R. M. Walker, *J. Geophys. Res* **70**, 1491 (1965).
15. S. C. Lin, *ibid.*, in press.
16. R. Rost, *Geochim. Cosmochim. Acta* **28**, 931 (1964); J. A. O'Keefe, P. D. Lowman, Jr., K. L. Dunning, *Science* **137**, 228 (1962).
17. I. Friedman, *Geochim. Cosmochim. Acta* **14**, 316 (1958); ———, in *Tektites*, J. A. O'Keefe, Ed. (Univ. of Chicago Press, Chicago, 1963), pp. 130–136.
18. C. M. Varsavsky, *Geochim. Cosmochim. Acta* **14**, 291 (1958).
19. J. A. O'Keefe, in *Tektites*, J. A O'Keefe, Ed. (Univ. of Chicago Press, Chicago, 1963), pp. 167–188.
20. W. Gentner, H. J. Lippolt, O. A. Schaeffer, *Geochim. Cosmochim. Acta* **27**, 191 (1963).
21. W. Gentner, H. J. Lippolt, O. Müller, *Z. Naturforsch.* **19a**, 150 (1964).
22. A. J. Cohen, in *Space Research* (Proceedings of the Third International Space Science Symposium) (North-Holland, Amsterdam, 1963), vol. 3, pp. 950–973.
23. H. Faul, *Geol. Rundschau* **52**, 767 (1962).

24. This article is contribution No. 29 from the Geosciences Division of the Southwest Center for Advanced Studies of the Graduate Research Center of the Southwest.

42

Geochimica et Cosmochimica Acta, 1969, Vol. 33, pp. 1083 to 1100. Pergamon Press. Printed in Northern Ireland

Genetic significance of the chemical composition of tektites: A review

S. R. TAYLOR and MAUREEN KAYE
Department of Geophysics and Geochemistry, Australian National University
Canberra, A.C.T. 2600, Australia

(*Received* 1 *March* 1969; *accepted in revised form* 4 *April* 1969)

Abstract—Evidence from internal compositional variations and ratios, selective vaporisation and impact glass analogies, bearing on the relationship of the present composition to that of the parent material indicates that concentration changes during melting have been minor, thus enabling the parent material to be identified. Terrestrial igneous rocks are not suitable source material, but terrestrial sandstones (greywackes–subgreywackes–arkoses) show strong similarities in major and trace element abundances, and exhibit analogous inter-element variations to tektites.

A lunar origin of tektites is not compatible with the chemistry of the lunar surface as presently known. A terrestrial origin by impact on sandstones of appropriate composition is consistent with the chemical evidence.

INTRODUCTION

THE first systematic trace element determinations in tektites were made by PREUSS (1935), whose data are in general agreement with modern determinations. The next significant advance was by BARNES (1940) who noted the general resemblance between the major element chemistry of tektites and that of terrestrial sedimentary rocks. The distinction between the refractive index of tektites and igneous glasses had already been made by TILLEY (1922). CHERRY *et al.* (1960) and TAYLOR *et al.* (1961) pointed out the significance of the major element correlations with silica and concluded that they were distinct from those observed in igneous rock suites. A strong positive correlation among the five alkali elements in australites was noted by TAYLOR (1960), indicating that the alkali elements were behaving as constituents of a single phase. From this evidence CHERRY and TAYLOR (1961) postulated a parent material analogous to a quartz–shale mixture. Further study of trace element abundances and their correlations with SiO_2 indicated that the tektite composition resembled that of terrestrial sedimentary rocks such as subgreywacke or loess (TAYLOR, 1962; TAYLOR and SACHS, 1964) in which quartz and a fine grained clay matrix are major components, together with accessory phases such as calcite, and zircon. Most of the trace element abundances and ratios (e.g. K/Rb $\sim$ 200, Rb/Cs $\sim$ 30, Th/U $\sim$ 6, Ni/Co $\sim$ 3, Zr/Hf $\sim$ 50) are consistent with those observed in such terrestrial sediments. HASKIN and GEHL (1963) and TAYLOR (1966) found that the absolute and relative rare earth abundances in tektites resembled those of sedimentary rocks. TAYLOR and SACHS (1964) reviewed this evidence, concluding that the geochemical data were consistent with a terrestrial origin from sandstones and that "a lunar origin is not compatible with the chemical evidence" (p. 262).

The purpose of this paper is to review the chemical evidence since the 2nd Tektite Symposium, and to ask the following questions.

(1) What is the relationship of the present composition to that of the parent material?
(2) What is the nature and origin of the source material?
(3) Where is the location of the impact site?

Attention is directed towards the australites since the trace element data are most complete for this group. They also cover the widest range in composition, the greatest area, and are the youngest occurrence. Conclusions reached concerning these should be applicable to the other tektite groups, because of the broad similarities in chemistry of the several occurrences.

Relation of Present Compositon to Parent Material

The identification of the source material depends critically on the degree of change in composition during melting. Three lines of evidence provide information about such changes.

(a) *Internal compositional evidence*

Is there any evidence from the present composition of tektites which *demands* selective loss or enrichment of elements compared to possible parent materials? Taylor and Sachs (1964) showed that the alkali element ratios (K/Rb, Rb/Cs) are close to those observed in possible terrestrial parents. This is also the case for the ratios of involatile elements (e.g. Mg, Ti, Cr, Zr) to the alkali elements. These ratios are high for igneous, but similar to those in sedimentary rocks. The chalcophile elements as a group (Zn, Pb, Cu, Tl, In, Bi, Ag) are low in abundance in australites (Taylor, 1966) and appear to be low in other tektite groups where data are available (e.g. Zn, Greenland and Lovering, 1963). However, similar concentration levels are observed in terrestrial sandstones, although the data are very sparse (Pettijohn, 1963). More significantly the high ratios of alkali/chalcophile elements preclude much loss of volatiles unless the alkalies (e.g. Rb, Cs) are involatile and the chalcophile elements volatile during melting. The present chemical composition alone cannot be used to prove or disprove changes during melting but is consistent with hypotheses which call for only minor loss or gain.

(b) *Selective volatilization*

This process probably operates to a minor degree and may remove a few percent of elements volatile under the conditions of tektite formation. This will not greatly affect the concentrations of the other constituents. However the major problem is whether the observed variations and compositional trends are due to such a cause. Tektites are enriched in elements such as Mg, Ni, Co, Cr, V and Sc by large factors in comparison with igneous rocks containing high enough silica contents to be potential source materials (see Figs. 2–4).

Derivation from high silica ($>80\%$) sources as postulated by Walter (1967) is likewise difficult, unless these materials are of unusual composition. For example terrestrial material with this silica content is low in the ferromagnesian trace elements. The derivation of the high Ni–Cr tektites (Java, Lake Wilson australites) is particularly difficult to explain by enrichment from low Ni–Cr parents.

The data obtained by WALTER (1967) under oxidising conditions show positive SiO_2–K_2O relationships similar to those observed in bediasites, but in contrast to the negative SiO_2–K_2O correlations in australites. It could be expected that a process such as selective distillation would produce similar trends, in contrast to the differences observed among different tektite groups, not only for SiO_2–K_2O relationships, but for several other elements (see section on correlation coefficients). Deductions based on vaporisation experiments must be treated with caution because of our lack of understanding of conditions at the impact site, where near-vacuum conditions are commonly assumed. WALTER's (1967) experiments were conducted at one atmosphere under oxidising conditions.

(c) *Impact glass analogies*

Tektite and terrestrial impact glasses are broadly analogous in composition (high SiO_2–high MgO–low alkali glasses). This implies that only glasses of this composition survive the stresses of rapid propulsion during impact events, so that the apparent uniformity in composition, used by LOWMAN (1962) as an argument for lunar origin, may be due to this cause, as suggested by UREY (1963).

Both the parental material and the glass formed from it are available for study at terrestrial impact sites. The occurrences at the Ries crater (VON ENGELHARDT, 1967; CHAO, 1963). Bosumtwi Crater (SCHNETZLER *et al.*, 1966; KOLBE *et al.*, 1967) and Henbury (TAYLOR, 1966; TAYLOR and KOLBE, 1964, 1965; COMPSTON and TAYLOR, 1969) have been studied. A particular advantage of the Henbury glass is that it is close in composition to that of the australites (Fig. 1). Comparative studies of the impact glass and country rock have been extended to 50 elements. No evidence for loss of volatile elements such as Rb and Cs, the chalcophile elements, or boron is apparent. Increases in the concentration of Rb in the glass compared to the sediments may be due to facies variation in the parent material, or to preferential fusion of finer grained matrix of the subgreywacke sediments (COMPSTON and TAYLOR, 1969).

This interpretation conflicts with the data of WALTER (1967) indicating that SiO_2 is volatile. The conclusion from impact glass studies (TAYLOR, 1966) is that the present composition of the glass is similar to that of the parent material and that variations in composition are within a few percent of the concentration of most elements.

The principal limitation is how closely the conditions of glass formation resemble those during tektite formation. The scale of the events are vastly different. However, the pressure and temperature conditions of glass formation are possibly similar since the presence of lechatelierite indicates that tektite glasses have not been heated to extreme temperatures. The close similarity in composition between the australites and the Henbury impact glass, coupled with the absence of major compositional change at Henbury, suggests that the australite parent material was similar, to within a few percent for each constituent, to the present composition.

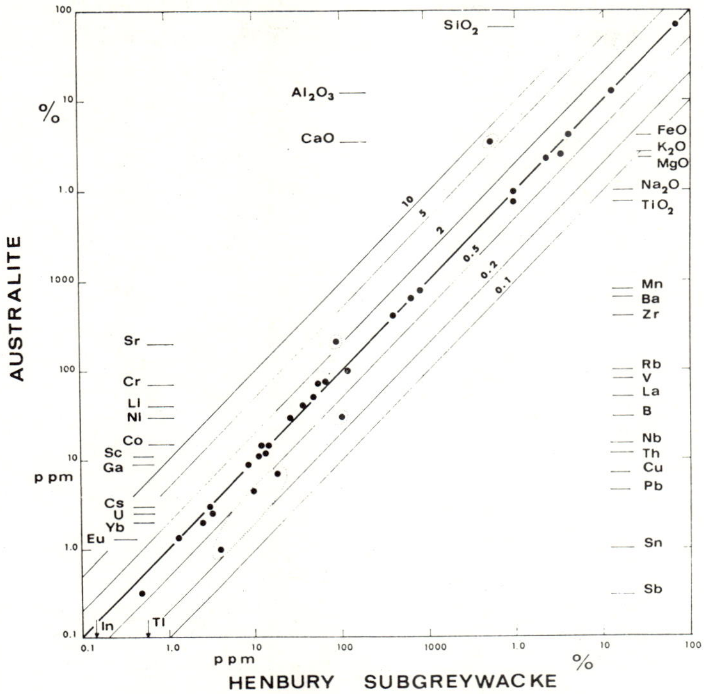

Fig. 1. Comparison of the average composition of australites (Taylor, 1968) with that of subgreywacke sandstone from Henbury, (Taylor, 1966). Note the close correspondence in composition, except for Ca, Sr, B and the chalcophile elements. The 45° diagonal lines give the australite/subgreywacke element abundance ratios.

Nature of the Parent Material

It is deduced from the discussion in the previous sections that the present composition of tektites is probably close to that of the parent material. Possible sources must provide not only an overall match in composition, but also must explain the observed variations both within and between tektite groups. A comparison of tektite compositions with those of available cosmic and terrestrial material immediately excludes all classes of meteorites as possible source materials. A possible lunar origin is considered in a later section.

(a) *Terrestrial igneous rocks*

Gross differences in composition exclude ultrabasic rocks, basalts and their differentiates and andesites as possible parental materials. A comparison of tektite composition with those of igneous rocks of comparable silica content reveals striking differences in composition. These are illustrated in Figs. 2–4 where average compositions of granites, leucogranites and calc-alkaline rhyolites are plotted against the average australite composition. The sources of the data are given in the figure captions.

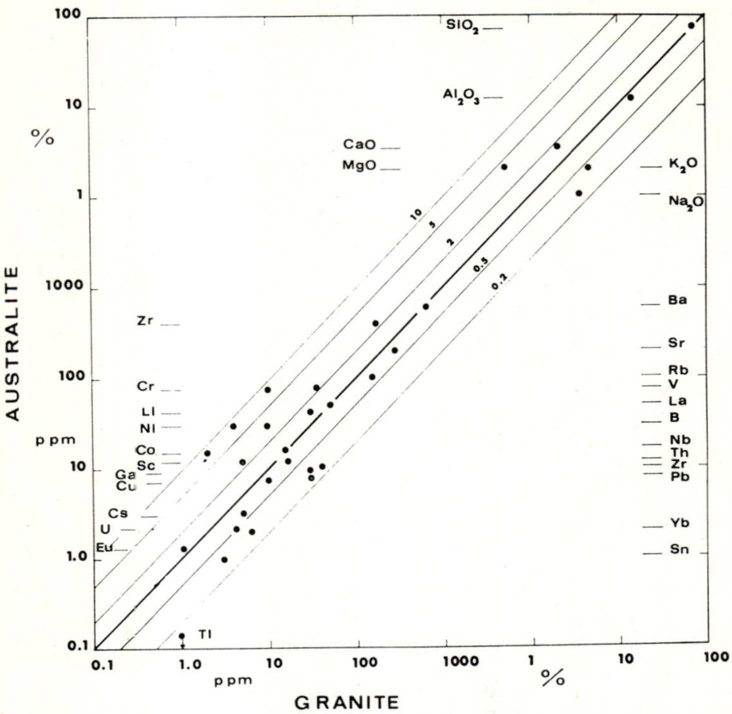

Fig. 2. Comparison of the average composition of australites (Taylor, 1968) with average granite (71% SiO_2) [Data from Turekian and Wedepohl (1961), modified by Taylor (1965), Kolbe and Taylor (1966). Haskin *et al.* (1966, Table 19) rare earths; Clark *et al.* (1966) Th, U; Norman and Haskin (1968), Sc]. Note the wide dispersion in element abundances. The 45° diagonal lines give the element abundance ratios of australites to granite. There is no simple correlation with properties such as volatility.

These figures clearly show the wide differences in concentration for most elements in acidic igneous rocks compared with australites. The most notable differences in composition are:

(1) MgO content is higher in australites by factors of 4 (granite) 10 (rhyolite) and 50 (leucogranites). Trace elements which show similar behaviour include Ni, Co, Cr, V, Sc, (Figs. 2–4).

(2) The K_2O content is lower by a factor of two in australites. The K/Rb average ratio of 190 (Taylor *et al.*, 1967) is typical of a few highly fractionated granites, but is lower than generally observed in granite rocks.

(3) The rare earth elements have relative and absolute abundances comparable with sedimentary rocks (Haskin and Gehl, 1963; Taylor, 1966). Highly fractionated igneous rocks such as leucogranites commonly show marked depletion in Eu but tektites do not exhibit this characteristic.

(4) Although there is some overlap at the low end, the boron content of tektites is in general higher than that observed in terrestrial granites.

(5) The upper limit of the SiO_2 range in tektites exceeds that commonly observed in igneous rocks.

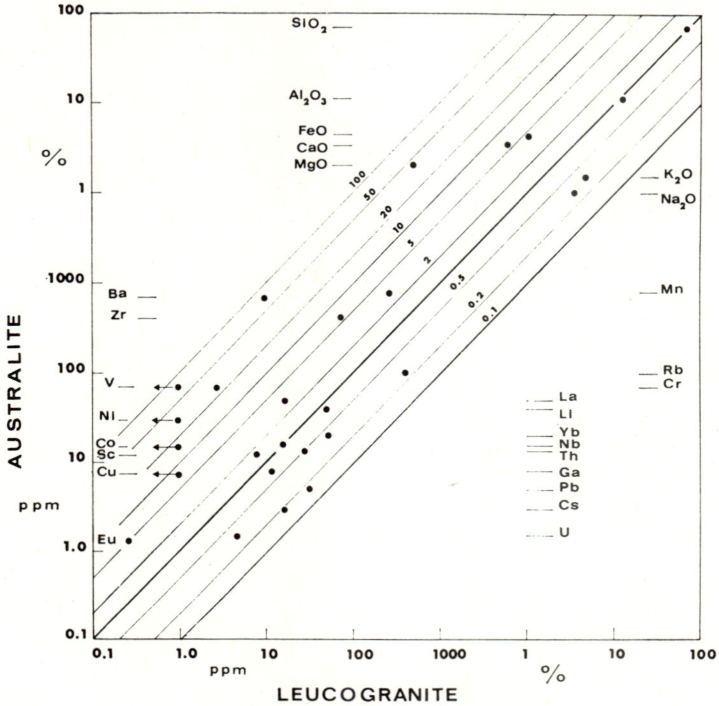

Fig. 3. Comparison of the average composition of australites (TAYLOR, 1968) with leucogranite (77% SiO_2). [Data from KOLBE and TAYLOR (1966)]. Note the extreme differences in concentration (up to factors of 100) for many elements.

(6) The absolute abundances of most elements in acid igneous rocks show no direct correlation with australite composition (Figs. 2–4) and there is no simple relation to volatility or other physical parameters.

(7) Inter-element correlations for acidic igneous rocks display a characteristic set of element correlations. A typical set is given in Table 1 for granodiorites and granites from the Snowy Mountains, Australia (KOLBE and TAYLOR, 1966) chosen as a well described suite ranging in SiO_2 content from 65 to 77% SiO_2. The most distinctive features are the strong negative SiO_2–ferromagnesian element correlations and the strong positive SiO_2–alkali (Na, K, Rb, Cs but not Li) correlations.

As CHAYES (1960) has pointed out, care must be exercised in interpreting correlations in a closed array and "in small samples, a correlation must be very strong to be significant". Many of the correlations are significant at $<0.01\%$. In the case of the alkali–SiO_2 correlations in acidic igneous rocks, the mathematical tendency toward negative correlation is overridden by fractionation processes producing strong positive SiO_2–alkali correlations. Of the differences between the australite and igneous rock correlations, the most significant are the negative alkali correlations, and the lack of significant SiO_2–CaO correlations. In contrast the North American tektites show positive SiO_2–K_2O and SiO_2–Rb. The significance of these differences

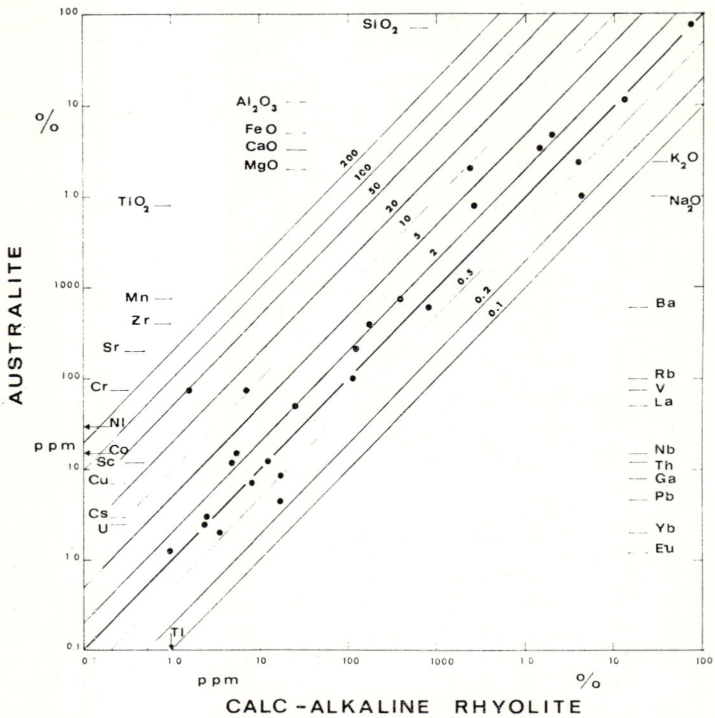

Fig. 4. Comparison of average composition of australites (TAYLOR, 1968) with that of calc-alkaline rhyolite (75% SiO_2) (Data from EWART *et al.*, 1968). Note the wide differences in composition for Cr, V, Ni, Co, Na and Mg, for Fe, Ca, Mg, Na, Ti, Cr, V, Ni, Co, Sc and Pb.

are discussed in the next section. In summary, there is little resemblance in detail between the chemistry of tektites and terrestrial igneous rocks.

(b) *Terrestrial sedimentary rocks*

Among the terrestrial sedimentary rocks, shales contain too low and quartzites too high a silica content to be considered *per se* as source material. The various classes of sandstones (greywacke, subgreywacke, arkose) show the closest approach to tektites among common terrestrial materials.

A major difficulty is the lack of trace element data for the various classes of sandstones (PETTIJOHN, 1963). Figures 1, 5 and 6 compare the average australite composition with subgreywacke, shale, and greywacke. Adequate trace element data for arkoses are lacking. In comparison with the igneous rock–australite plots, it is clear that sedimentary rocks and australites show close similarities, although the shale SiO_2 value is too low and the greywacke Na_2O value too high to make them suitable in detail. The scatter of the abundance data is usually with a factor of two.

Figure 1 shows the comparison between Henbury subgreywacke and australites. The very close fit (within 10%) for most elements is apparent. Those elements which show differences in composition are Ca and Sr, the chalcophile group (Cu, Pb etc.) and boron. TAYLOR (1966) has given a detailed discussion of these differences, concluding

Table 1. Correlation coefficients (r) vs. SiO_2 and levels of significance (1) for terrestrial rocks and tektites

n	Snowy Mountains granites (1) 28		North American tektites (8) (4) 29		Henbury subgreywackes (3) 9		Australites (2) 32		Indochinites (5) (6) 16		Philippinites (6) 13		Moldavites (7) 45	
	r	1	r	1	r	1	r	1	r	1	r	1	r	1
SiO_2 vs.														
Al_2O_3	−0·81	<0·01	−0·97	<0·01	−0·97	<0·01	−0·91	<0·01	−0·63	0·6	−0·89	<0·01	−0·69	<0·01
FeO	−0·94	<0·01	−0·98	<0·01	−0·83	0·15	−0·82	<0·01	−0·82	<0·01	−0·38	—	−0·29	—
MgO	−0·93	<0·01	−0·52	0·2	−0·99	<0·01	−0·96	<0·01	−0·81	0·01	+0·27	—	−0·42	0·3
CaO	−0·88	<0·01	−0·30	—	−0·04	—	−0·22	—	−0·29	—	−0·71	0·4	−0·50	0·03
Na_2O	+0·73	<0·01	−0·50	0·3	−0·08	—	−0·81	<0·01	−0·47	5	−0·30	—	−0·31	—
K_2O	+0·80	<0·01	+0·62	0·03	−0·87	0·04	−0·76	<0·01	−0·48	4	−0·58	3	−0·50	0·03
TiO_2	−0·87	<0·01	−0·94	<0·01	−0·65	—	−0·84	<0·01	−0·71	0·4	−0·75	0·2	+0·09	—
Cr	−0·65	0·01	−0·72	<0·01	−0·80	0·4	−0·64	0·01						
Li	+0·05	—	−0·65	<0·01	—	—	−0·60	0·03						
Ni	−0·66	0·01	−0·49	0·4	+0·17	—	−0·46	0·6						
Co	−0·85	<0·01	−0·68	<0·01	+0·38	—	−0·23	—						
Cu	−0·41	—	−0·53	0·1	−0·39	—	+0·47	0·5						
V	−0·93	<0·01	−0·76	<0·01	−0·82	0·2	−0·53	0·12						
Zr	−0·34	—	−0·41	1	−0·11	—	+0·55	0·09						
Sc	−0·88	<0·01	−0·48	0·3	−0·98	<0·01	−0·55	0·09						
Sr	−0·64	0·01	+0·59	0·02	−0·59	—	−0·02	—						
Ba	−0·51	0·5	−0·08	—	−0·40	—	−0·62	0·01						
Rb	+0·74	<0·01	+0·37	3	—	—	−0·79	<0·01						
Cs	+0·56	0·2			—	—	−0·85	<0·01						

Sources of data:

(1) Kolbe and Taylor (1966).
(2) Taylor (1962); Taylor and Sachs (1964).
(3) Taylor and Kolbe (1965); Taylor (1966, 1967).
(4) Cuttitta *et al.* (1967).
(5) Barnes (1964).
(6) Schnetzler and Pinson (1964).
(7) Rost (1968).
(8) Chao (1963).

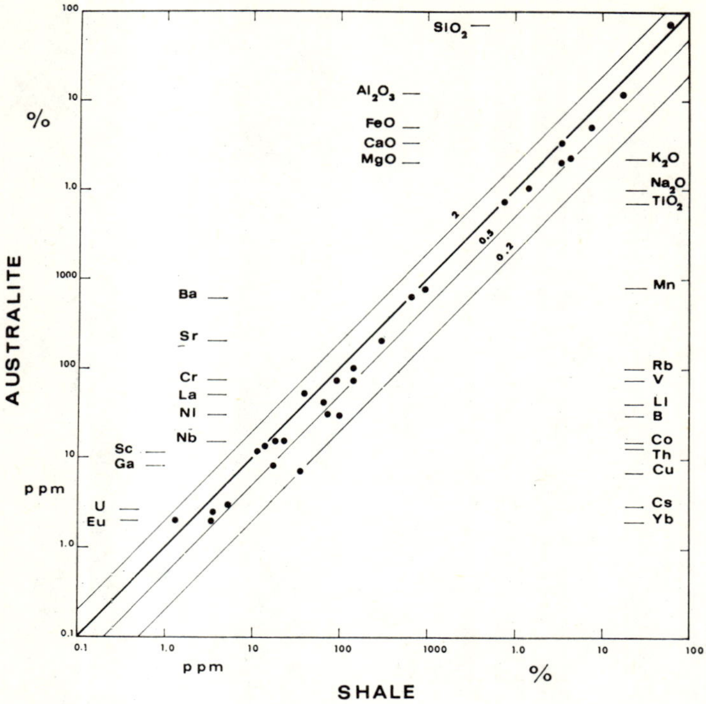

Fig. 5. Comparison of average composition of australites (TAYLOR, 1968) with average shale (63% SiO_2). [Data from PETTIJOHN (1957), for major elements, HASKIN *et al.* (1966, Table 28), rare earths; HARDER (1964) boron; PLILER and ADAMS (1964) Th, U; TUREKIAN and WEDEPOHL (1961)]. Note that the concentration of many elements shows much less scatter than for the igneous rock data. The scatter of data is larger than for the subgreywacke data.

that the Ca–Sr differences may be attributed to variations in a phase such as calcite, and those in the chalcophile group, to a low abundance of sulphide phase in the parent material. The boron content, as discussed later, is high for sandstones and attributable in part to detrital boron minerals. Thus the differences are attributable to variations in the minor phases.

Values for correlations with SiO_2 for the Henbury subgreywacke are given in Table 1. The samples selected cover a 10% spread in SiO_2 content (75–85%) slightly higher than the australite levels (68–80%).

The SiO_2 to Al_2O_3, FeO, MgO and K_2O correlations are all significantly negative at >0·1%. CaO and Na_2O show no correlation with SiO_2. Except for the Na_2O–SiO_2 relation, these correlations are similar to those observed in the australites, suggesting that the variations in tektite composition are analogous to those observed within sedimentary rocks of similar composition. The trace element data, again for a small sample number, shows analogies with the australite correlations.

In contrast, the North American tektites show different correlation patterns, more closely resembling those for igneous rocks, including positive SiO_2–K_2O and Rb correlations. The absolute abundances of SiO_2 and the alkali elements do not

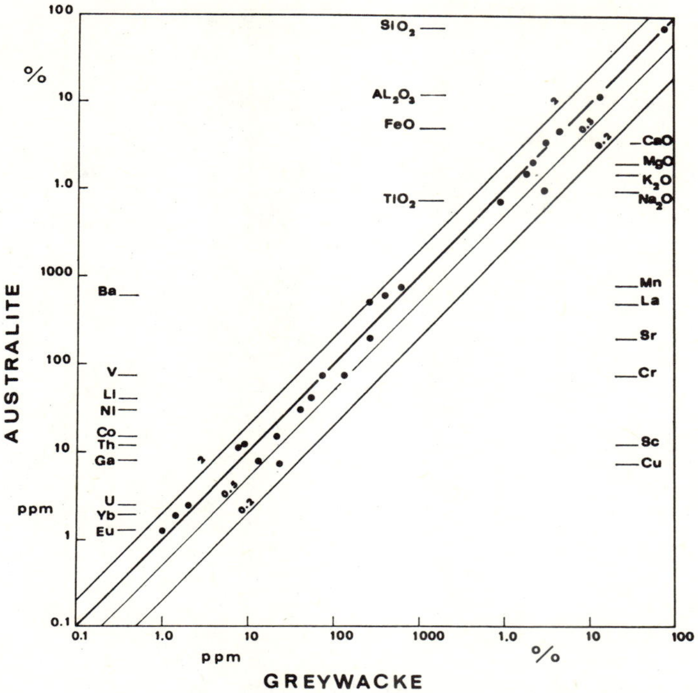

Fig. 6. Comparison of average composition of australites (Taylor, 1966) and greywacke. [(Data from Middleton, (1960) major elements; Haskin *et al.* (1966), Table 5) rare earths; Rogers and Richardson (1964) Th, U; Weber and Middleton (1961)]. Note that there is a small scatter in the data about the line of equal composition, except that Na is high in the greywacke.

resemble those of terrestrial rocks (SiO_2 is too high, K_2O too low) but are consistent with arkose compositions derived from weathering of acid igneous rocks.

The contrast between the bediasite and australite element abundances (Fig. 7) and element correlations is interpreted as due to differences in the parent material. Thus the bediasites are lower in the ferromagnesian trace elements, and also in Rb, consistent with differences expected between a micaceous sandstone (australites) and a feldspathic sandstone (bediasites). The variations within the group are consistent with those expected within sedimentary rocks covering the same range in composition.

The Ivory Coast tektites form a distinct group (Chao *et al.*, 1965). The major element composition resembles that of granodiorite but their high nickel ($\sim$100 ppm) and chromium ($\sim$300 ppm) are unknown in igneous rocks of such composition. The Rb–Sr isotopic evidence (Schnetzler *et al.*, 1966; Kolbe *et al.*, 1967) indicates a probable derivation from meta-sediments (greywacke) at the Bosumtwi crater site.

The chemistry of all classes of tektites is comparable to that of greywackes, subgreywackes or arkoses, and the internal variations in composition are consistent with this interpretation.

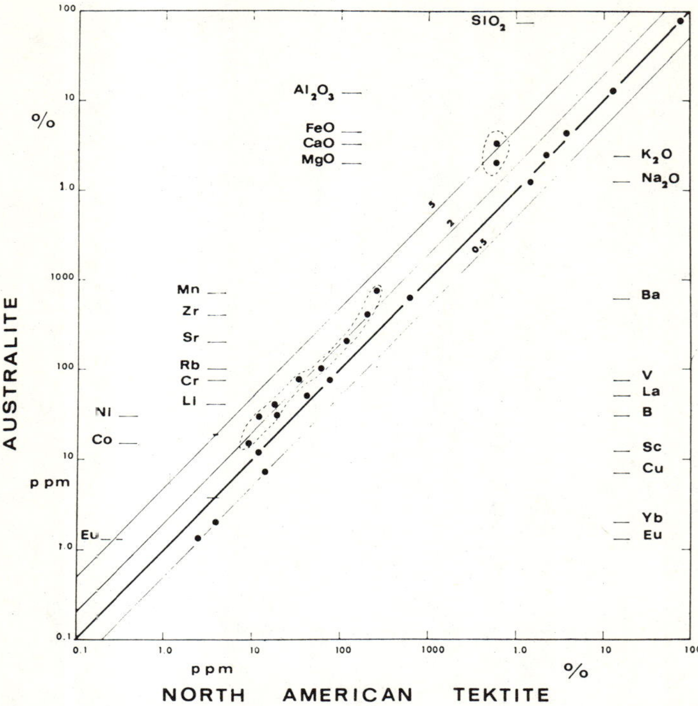

Fig. 7. Comparison of average composition of australites (TAYLOR, 1968) and North American Tektites. [Data from CHAO (1963), CUTTITTA *et al.* (1967)]. Note that the concentrations of Ca, Mg, Mn, Zr, Sr, Rb, Cr, Li, Ni and Co are higher, and Cu, Eu and Yb lower in australites, compared to bediasites.

SOURCE OF PARENT MATERIAL

The identification of the Bosumtwi crater rocks as the source of the Ivory Coast tektites is strongly supported by the isotopic evidence (SCHNETZLER *et al.*, 1966; KOLBE *et al.*, 1967). The chemistry of the moldavites and the Ries crater suevite is distinctly different, and country rock of moldavite composition does not appear to be present at Ries, although Early Tertiary sandstones which may have covered the crater site are a possible source (VON ENGELHARDT, 1967). The source area for the australites and South-east Asian tektites remains unknown.

(a) *Lunar source*

Information on the composition of the lunar surface obtained from Luna-10 (VINOGRADOV *et al.*, 1966) and Surveyors V (Mare Tranquillitas), VI (Sinus Medii), and VII (Tycho) (TURKEVICH *et al.*, 1967, 1968, a, b, c) provides no support for a lunar source for tektites. Upper limits for the abundances of K, U and Th are about the levels observed in terrestrial basalts (VINOGRADOV *et al.*, 1967). The maria compositions at the Surveyor V and VI sites resembles terrestrial tholeiitic basalt and that of the Tycho site resembles high-Al basalt (JACKSON and WILSHIRE, 1968). A comparison of these data with australite composition are given in Fig. 8 which clearly reveals the large compositional differences.

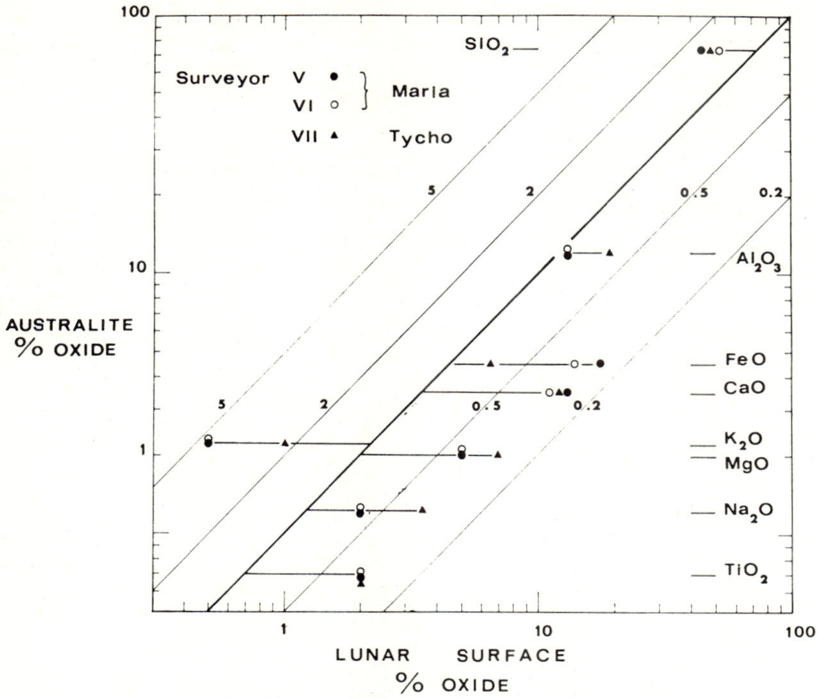

Fig. 8. Comparison of average major element composition of australites (TAYLOR, 1968) with that of the Surveyor analyses of the lunar surface (Data from JACKSON and WILSHIRE, 1968). Note the wide difference in composition for all elements except the Al_2O_3 maria values.

The iron content of the maria is not due to metallic iron (DE WYS, 1967) and is too high to support the suggestion of GILVARRY (1968) that the maria composition is similar to terrestrial mudstones. The heavy cratering in the vicinity of Tycho indicates that the Surveyor VII analysis may be typical of the uplands (TAYLOR, 1969). Derivation of tektites from the moon now calls for impact on a localised area of unusual chemistry.

(b) *Terrestrial origin*

The geochemical evidence has long pointed clearly to an origin on the earth. (BARNES, 1940; TAYLOR, 1962; TAYLOR and SACHS, 1964).

(1) *Land area.* 75% of the land area is underlain by sedimentary rocks, of which sandstones comprise 30% (PETTIJOHN, 1957). Subgreywackes comprise 20–35%, arkoses 15–33% and greywackes 10–26% of sandstones (PETTIJOHN, 1963). Thus about 20% of the land surface is underlain by suitable sediments to produce the range of composition of tektites. For the australites the figures is 6–10%.

(2) *Shallow shelf seas.* Element abundance criteria cannot readily distinguish the sediments in these areas from those exposed on land.

(3) *Deep sea.* Apart from the physical difficulties associated with impact in 5 km water, and the production of glasses of very low water content in such an event, the chemical composition of layer 1 of the oceanic crust is, not suitable parental

material, showing wide differences in Fe, Na, Ni, Co, Pb, Ba, Mn, B, Zn, Cu, Pb, Mo and other elements (Fig. 9). As with the lunar surface, small areas of appropriate composition may be present.

(c) *Boron as an environmental indicator*

The high concentration of boron in marine *argillaceous* sediments has been used to separate marine from non-marine *shales.* PREUSS (1935) and MILLS (1968) determined boron in tektites, and further determinations are reported here using an emission spectrographic analytical method. Precision is ±5% relative deviation.

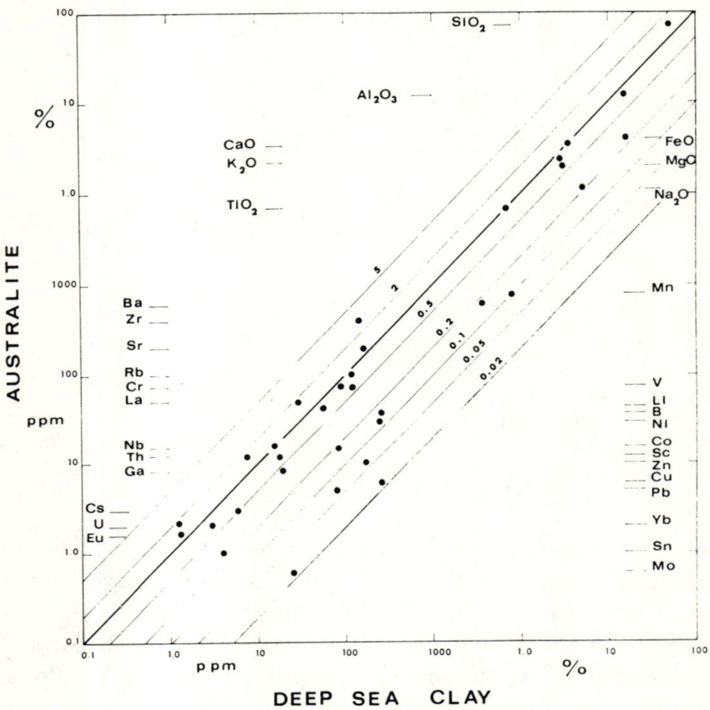

Fig. 9. Comparison of the average composition of australites (TAYLOR, 1968) with deep-sea clays [Data from TUREKIAN and WEDEPOHL (1961), WILDEMAN and HASKIN (1965, Table 3) rare earths]. Note the extreme differences in composition.

The analytical data and values used for standards are given in Table 2. Figure 10 shows the distribution of boron. The following conclusions may be drawn.

(1) There is no effective loss of boron during formation of the Henbury impact glass. Boron present in detrital tourmaline has been apparently incorporated in the glass.

(2) The boron content of tektites extends well beyond the range observed in terrestrial igneous rocks.

(3) The average concentration is similar to that of average sandstones or to loess.

(4) Shales of marine origin usually contain more than 100 ppm B and deep sea clays contain 130 ppm B. The lack of evidence of selective loss of boron excludes

Table 2. Boron determinations in tektites, impact glasses and sediments. (All data in ppm)

Australites*		Australites		Locality	Henbury impact glass†	
Sample no.	ppm	Sample no.	ppm		Sample no.	ppm
18	31	1200 A	27		11	115
11949 B	21	1200 B	38		1213	101
11947	15	1199	26		1415	119
415	13	1199 B	18	Lake Wilson S.A.	25	95
1176	30	LW33	55		28	109
541	22	LW34	40		29	111
602	24	LW36	41		30	104
414	3	MD37	32	Mt Davies airstrip	31	115
1026	19	MD38	42	Mt Davies airstrip	32	125
365	37	W40	33	Wingellina, W.A.	33	110
4	35				34	102
1004·9	26	Javenites			Ave.	110
230	27	16	45	Sangiran	Henbury subgreywacke†	
281	27	28	32	Sangiran	27	103
20	14	29	37	Sangiran	21	109
266	24	Moldavites			A 5	118
229	18	1	23	Locènice	B 3	111
11936	27	2	24	Locènice	17	111
1193	23	3	24	Nèchov	16	123
8	18	4	14	Lhenice	18	98
355	27	5	19	Vrábcè	20	103
1198	31	Darwin Glass			Quartzites†	
475	16	E	30		C 4	77
11945	34	I	23		19	54
15·1	47	M	33		D 6	57
15·2	36	W	46		F 9	49
15·3	37	W-2	24		G 8	69
11949 A	11	X-1	31		Soil†	
9	19	X-2	26	Mt. Darwin, Tas.	27 A	84
563	20	Y-1	36		Loess‡	
11943	33	Y-2	20		CY4-A	38
Standards		Z-1	24		CY4-B	45
G-1	1–2	Z-1	28		CY4-C	33
W-1	12	Z-3	34			
S-1	72					

* Localities and sample descriptions given by TAYLOR (1962) and TAYLOR and SACHS (1964).
† Localities and sample descriptions given by TAYLOR and KOLBE (1965) and TAYLOR (1967).
‡ Loveland loess unit, Kansas. Sample from Ada Swineford, Kansas Geological Survey.

these as parent materials. (The high boron content of the Henbury subgreywacke is due in part to a detrital tourmaline component.)

(5) The Lake Wilson and Java tektites, both high in nickel (PREUSS, 1935; TAYLOR, 1965) have higher average boron contents than the average for australites.

(6) In summary, the boron content of tektites is consistent with derivation from sandstones and excludes deep sea clays or marine shales as sources.

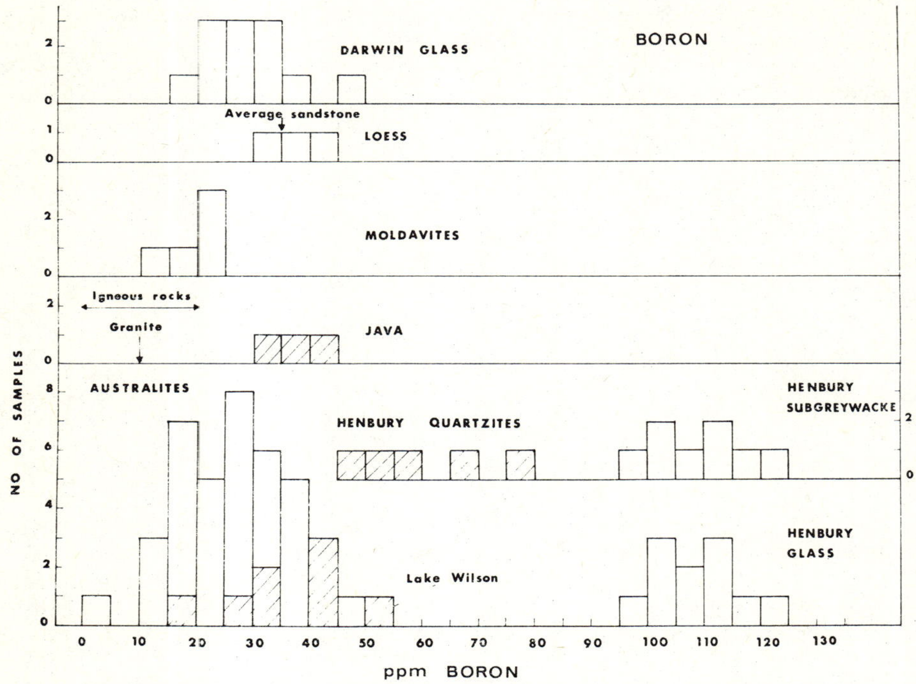

Fig. 10. Boron content of tektites, Henbury impact glass and sediments, Darwin Glass and loess. [Average sandstone value from PETTIJOHN (1963). Igneous rock data from TAYLOR (1965).] Note the similarities between Henbury impact glass and parent sediments, and between the Java and Lake Wilson tektites.

NATURE OF IMPACTING BODY

There is a general consensus that tektites originated during an impact event. CHAO *et al.* (1962, 1964) identified Fe/Ni spherules, (kamacite) schreibersite, and troilite. The low abundance contrasts with the common Fe/Ni spherules in terrestrial impact glasses, pointing to a highly energetic event, a low density of the impacting body or both. PRESTON (1967) records Fe, Ni, Cu, Cr, Mn, Ca, Na, and K emission spectra in the comet Ikeya-Seki and slight Fe/Ni contamination would not rule out cometary impact. UREY, (1957, 1963) and LIN (1966) have discussed the consequences of cometary impact.

CRITERIA FOR IMPACT SITE

These have been listed by TAYLOR (1969). The absence of a large Canyon Diablo type crater indicates that scaling from craters formed by impact of high density objects is probably invalid. The impact energy at Canyon Diablo crater (1·2 km diameter × 170 m deep) is given by SHOEMAKER (1963) as $6{\cdot}8 \times 10^{22}$ ergs (1·7 megaton TNT equivalent). In contrast the much more energetic Tunguska explosion, probably due to a comet, estimated at 8×10^{24} ergs (200 megaton TNT equivalent) (ROBEY, 1963) and occurring at a height of 5 km above the surface, produced no crater (KRINOV, 1966). A super-Tunguska event, advocated by UREY (1957, 1963) could produce a wide shallow crater, very difficult to detect, although it might be visible on satellite photographs. An origin for tektites by cometary impact on terrestrial sedimentary rocks of appropriate composition is consistent with the chemical evidence.

References

Barnes V. E. (1940) North American tektites. *Univ. Texas Publ.* 1940, (3945), 477–582.

Barnes V. E. (1964) Variation of petrographic and chemical characteristics of indochinite tektites within their strewn field. *Geochim. Cosmochim. Acta* **28,** 893–913.

Chao E. C. T. (1963) The petrographic and chemical characteristics of tektites. *Tektites,* (editor J. A. O'Keefe), Chap. 3, pp. 51–94. University of Chicago Press.

Chao E. C. T., Adler I., Dwornik E. J. and Littler J. (1962) Metallic spherules in tektites from Isabella, the Philippine Islands. *Science* **135,** 97–98.

Chao E. C. T., Cuttitta F., Carron M. K., Annell C. and Mount P. (1965) New data on some Ivory Coast tektites. (Abstract). *Trans. Amer. Geophys. Union* **46,** 427.

Chao E. C. T., Dwornik E. J. and Littler J. (1964) New data on the nickel–iron spherules from Southeast Asian tektites and their implications. *Geochim. Cosmochim. Acta* **28,** 971–980.

Chayes F. (1960). On correlation between variables of constant sum. *J. Geophys. Res.* **65,** 4185–4193.

Cherry R. D. and Taylor S. R. (1961) Studies of tektite composition II. Derivation from a quartz–shale mixture. *Geochim. Cosmochim Acta* **22,** 164–168.

Cherry R. D., Taylor S. R. and Sachs M. (1960) Major element relationships in tektites. *Nature* **187,** 680.

Clark S. P., Peterman Z. E. and Heier K. S. (1966) Abundances of uranium, thorium and potassium. *Geol. Soc. Amer. Mem.* **97,** 521–541.

Compston W. and Taylor S. R. (1969) Rb–Sr study of impact glass and country rock from the Henbury meteorite crater field. *Geochim. Cosmochim. Acta* **33,** 1037–1043.

Cuttitta F. Clarke R. S., Carron M. K. and Annell C. S. (1967) Martha's vineyard and selected Georgia tektites; New chemical data. *J. Geophy. Res.* **72,** 1343–1349.

De Wys Jane N. (1967) Surveyor V: Magnet experiment. *Science* **158,** 632–635.

Ewart A., Taylor S. R. and Capp A. C. (1968) Trace and minor element geochemistry of the rhyolitic volcanic rocks, Central North Island, New Zealand. *Contribs. Mineral. Petrol.* **18,** 76–104.

Gilvarry J. J. (1968) Observational evidence for sedimentary rocks on the moon. *Nature* **218,** 336–341.

Greenland L. and Lovering J. F. (1963) The evolution of tektites: elemental volatilization in tektites. *Geochim. Cosmochim. Acta* **27,** 249–259.

Harder H. (1964). To what extent is boron a marine index element. *Geochim. Int.* **1,** 105–112.

Haskin L. A., Frey F. A., Schmitt R. A. and Smith R. H. (1966) Meteoritic, solar and terrestrial rare earth distributions. *Phys. Chem. Earth* **7,** Chap. 3, 169–321.

Haskin L. and Gehl M. (1963) The rare-earth distribution in tektites. *Science* **139,** 1056–1058.

Jackson E. D. and Wilshire H. G. (1968) Chemical composition of the lunar surface at the Surveyor landing sites. *J. Geophys. Res.* **73,** 7621–7629.

Kolbe P., Pinson W. H., Saul J. M. and Miller E. W. (1967) Rb–Sr study on country rocks of the Bosumtwi crater, Ghana. *Geochim. Cosmochim. Acta* **31** 869–875.

Kolbe P. and Taylor S. R. (1966) Major and trace element relationships in granodiorites and granites from Australia and South Africa. *Beitr. Mineral. Petrol.* **12,** 202–222.

Krinov E. L. (1966) *Giant Meteorites.* Pergamon.

Lowman P. D. (1962) Tektites vs terrestrial rocks: A comparison of variance in compositions. *Geochim. Cosmochim. Acta* **26,** 561–579.

Lin C. S. (1966) Cometary impact and the origin of tektites. *J. Geophys. Res.* **71,** 2427–2437.

Middleton G. V. (1960) Chemical composition of sandstones. *Bull. Geol. Soc. Amer.* **71,** 1011–1026.

Mills A. A. (1968) Boron in tektites. In *Origin and Distribution of the Elements,* (editor L. H. Ahrens), pp. 521–531. Pergamon.

NORMAN J. C. and HASKIN L. A. (1968) The geochemistry of Sc: A comparison to the rare earths and Fe. *Geochim. Cosmochim. Acta* **32,** 93–108.

PETTIJOHN F. J. (1957) *Sedimentary Rocks*, 2nd edition. Harper.

PETTIJOHN F. J. (1963) Chemical composition of sandstones. In *Data of Geochemistry*, 6th edition (editor M. Fleischer), Chap. 5. *U.S. Geol. Surv. Prof. Paper* 440-S.

PLILER R. and ADAMS J. A. S. (1962) The distribution of thorium and uranium in a Pennsylvanian weathering profile. *Geochim. Cosmochim. Acta* **26,** 1137–1146.

PRESTON G. W. (1967) The spectrum of comet Ikeya-Seki (1965f). *Astrophys. J.* **147,** 718–742.

PREUSS E. (1935) Spektralanalytische Untersuchung der Tektite. *Chem. Erde* **9,** 365–416.

ROBEY D. H. (1963) General Dynamics/Astronautics Report No. GD/A 63–0064.

ROGERS J. J. W. and RICHARDSON K. A. (1964) Thorium and uranium contents of some sandstones. *Geochim. Cosmochim. Acta* **28,** 2005–2011.

ROST R. (1968) Moldavites. Int. Geol. Congress, 23rd Session, Prague, 1968.

SCHNETZLER C. C. and PINSON W. H. (1964) A report on some recent major element analyses of tektites. *Geochim. Cosmochim. Acta* **28,** 793–806.

SCHNETZLER C. C., PINSON W. H. and HURLEY P. M. (1966) Rubidium–strontium age of the Bosumtwi Crater area, Ghana, compared with the age of the Ivory Coast Tektites. *Science* **151,** 817–819.

SHOEMAKER E. M. (1963) Impact mechanics at Meteor Crater, Arizona. *The Moon Meteorites and Comets, Vol. IV. The Solar System*, (editors B. M. Middlehurst and G. P. Kuiper), Chap. 11, pp. 301–336. University of Chicago Press.

TAYLOR S. R. (1960) Abundance and distribution of alkali elements in australites. *Geochim. Cosmochim. Acta* **20,** 85–100.

TAYLOR S. R. (1962) The chemical composition of australites. *Geochim. Cosmochim. Acta* **26,** 685–722.

TAYLOR S. R. (1965) Nickel rich tektites from Australia. *Nature* **201,** 281–282.

TAYLOR S. R. (1966) Australites, Henbury impact glass and subgreywacke: A comparison of the abundances of 51 elements. *Geochim. Cosmochim. Acta* **30,** 1121–1136.

TAYLOR S. R. (1967) Composition of meteorite impact glass across the Henbury strewn field. *Geochim. Cosmochim. Acta* **31,** 961–968.

TAYLOR S. R. (1968) Geochemistry of Australian impact glasses and tektites (Australites). In *Origin and Distribution of the Elements*, (editor L. H. Ahrens), pp. 533–541. Pergamon.

TAYLOR S. R. (1969) Criteria for terrestrial origin of australites. *Chem. Geol.* in press.

TAYLOR S. R., ERLANK A. and GURNEY J. J. (1967) K/Rb ratios in australites. *Geochim. Cosmochim. Acta* **31,** 953–960.

TAYLOR S. R. and KOLBE P. (1964) Henbury impact glass: Parent material and behaviour of volatile elements during melting. *Nature* **203,** 390–391.

TAYLOR S. R. and KOLBE P. (1965) The geochemistry of Henbury impact glass. *Geochim. Cosmochim. Acta* **29,** 741–754.

TAYLOR S. R. and SACHS M. (1964) Geochemical evidence for the origin of australites. *Geochim. Cosmochim. Acta* **28,** 235–264.

TAYLOR S. R., SACHS MAUREEN and CHERRY R. D. (1961) Studies of tektite composition. I Inverse relation between SiO_2 and the other major constituents. *Geochim. Cosmochim. Acta* **22,** 155–163.

TILLEY C. E. (1922) Density, refractivity and composition relations of some natural glasses. *Mineral. Mag.* **19,** 275–294.

TUREKIAN K. K. and WEDEPOHL K. H. (1961) Distribution of the elements in some major units of the Earth's crust. *Bull. Geol. Soc. Amer.* **72,** 175–192.

TURKEVICH A. L., FRANZGROTE E. J. and PATTERSON J. H. (1967) Chemical analysis of the Moon at the Surveyor V landing site. *Science* **158,** 635–637.

TURKEVICH A. L., FRANZGROTE E. J. and PATTERSON J. H. (1968a) Chemical analysis of the lunar surface at the Surveyor landing sites. *Science* **160,** 443–444.

TURKEVICH A. L., PATTERSON J. H. and FRANZGROTE E. J. (1968b) Chemical analysis of the moon at the Surveyor VI landing site: preliminary results. *Science* **160,** 1108–1110.

Turkevich A., Franzgrote E. J. and Patterson J. H. (1968c) Chemical analysis of the moon at the Surveyor VII landing site: preliminary results. *Science* **162**, 117–118.

Urey H. C. (1957) The origin of tektites. *Nature* **179**, 556–557.

Urey H. C. (1963) Cometary collisions and tektites. *Nature* **197**, 228–230.

Vinogradov A. P., Surkov Yu. A., Chernov G. M., Kirnozov F. F. and Nazarkina G. B. (1966) Measurements of gamma-radiation from the surface of the moon by the Cosmic station "Luna-10". *Geochemistry* **3**, 707–715. (English Trans.)

Von Engelhardt W. (1967) Chemical composition of Ries glass bombs. *Geochim. Cosmochim. Acta* **31**, 1677–1689.

Walter L. S. (1967) Tektite compositional trends and experimental vapor fractionation of silicates. *Geochim. Cosmochim. Acta* **31**, 2043–2063.

Weber J. N. and Middelton G. V. (1961) Geochemistry of the turbidites of the Normanskill and Charny formations. I, II. *Geochim. Cosmochim. Acta* **22**, 200–288.

43

Reprinted from
9 October 1970, volume 170, pages 199-200

SCIENCE

Tektite Glass Not in Apollo 12 Sample

O'Keefe (*1*) has reported the supposed discovery of tektite glass in lunar sample 12013 on the basis of information from the Apollo 12 preliminary examination (*2*) and other communications. This report is based on faulty data and assumptions, and the major conclusion is incorrect. Although portions of the sample might appear to be vitreous by cursory examination, neither fragment of sample 12013 analyzed during the preliminary examination was glass. X-ray diffraction analyses of both fragments (Fig. 1) were performed prior to chemical analysis and gave the following mineral identi-

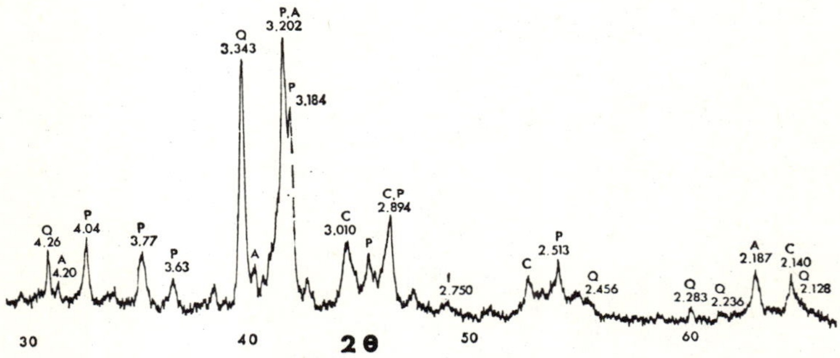

Fig. 1. A portion of the x-ray diffractogram of one of the analyzed fragments of lunar sample 12013. The fragment is obviously not glass. Peaks are identified as follows: *A*, alkali feldspar; *C*, clinopyroxene; *I*, ilmenite; *P*, plagioclase; *Q*, quartz. This diffractogram is approximately the same as the one obtained for the second fragment. (CrKα radiation).

Fig. 2. Sawed slab of lunar sample 12013, in which the heterogeneity and complex petrography of the sample are illustrated. The scale is 0.5 cm [NASA, Manned Spacecraft Center photograph No. S-70-40833, sample No. 12013,9].

Table 1. Composition of two fragments (total weight of both fragments combined is 55 mg) from lunar sample 12013. These values should not be taken as representative of the total sample or of any lithologically distinct portion of the sample.

Oxide or element	Percent by weight	
	A*	B†
SiO_2	61.2	62
Al_2O_3	12.0	10
Total Fe as FeO	12.6	10
MgO	8.0	6.0
CaO	6.3	6.2
TiO_2	1.17	1.2
Na_2O	1.20	0.9
K_2O	1.96	2.0
MnO	0.19	0.16
Cr		0.12
Ni		0.012
Zr		0.15
Co		0.0018
Y		0.006
Sc		0.005
Ba		0.45
V		0.022
Cu		0.0040
Nb		0.018
Total	104.62	99.24

* (A) John Allen, analyst; silica by light absorption spectroscopy; other values by atomic absorption spectroscopy. † (B) R. Martin, analyst; optical emission spectrograph.

fications and estimated modal mineral percentages: calcic plagioclase, 35 percent; clinopyroxene, 20 percent; alkali feldspar, 15 percent; low (alpha) quartz, 12 percent; tridymite, < 1 percent; ilmenite, 3 percent. O'Keefe further implies that the analysis of sample 12013 (*2*) is somehow representative of the sample or is some distinct lithologic portion of sample 12013. This sample is an extremely complex and heterogeneous breccia (Fig. 2). Individual mineral grains and included fragments range in size from more than 3 mm to less than 10 μm. The analyzed samples weighed only 30 and 25 mg, respectively. Examinations of two thin sections (12013,5 and 12013,7), as well as low magnification examination of the main mass of this sample (*3*), indicate that it is not probable that samples of this size were representative of any major or distinct lithologic portion of sample 12013. In fact, it seems likely that similarly sized fragments from various parts of the sample will have a wide range of major element compositions, because the modal mineralogy shows a wide range at the scale of a few millimeters and less. Glass is only a minor phase in the thin sections examined, and minor glass is indicated by the higher than usual background of the diffractogram. Moreover, the relative refractive indices of the glass observed and comparison of the estimated mineral modal percentages with the analysis (Table 1) indicate a pyroxene-like composition.

Refinements of the preliminary analytical data have resulted in minor revisions of the data originally reported and in the determination of additional elements (Table 1). Chemical inhomogeneities between the two fragments may be responsible in part for the discrepancies in values between the different analytical techniques. These analyses are no more similar to tektites than they are to possible mixtures of common terrestrial minerals or rock types. They do not resemble either tektites or common terrestrial rocks very closely. Furthermore, the comparison of compositions of fragments of lunar sample 12013 with rare chemical varieties of tektites (*1*) does not adequately treat the abundant isotopic, minor and trace element data (*4*) that seem to be characteristic of all lunar samples now analyzed. These data are overwhelmingly striking for their *contrasts* with all published tektite analyses. However, such analyses have not yet been performed on lunar sample 12013 or these rare tektites. With so little data now available on this petrographically and mineralogically complex sample, the significance of sample 12013 is not yet known.

There is *no* existing chemical or mineralogical observation or data that uniquely support the idea that tektites originate from the moon. There *are* abundant chemical data that support a close genetic relationship between tektites and terrestrial rock materials. These data have been presented or summarized by numerous authors (*5*). Tektite glass has not been found in samples from Apollo 12 (or Apollo 11).

ELBERT A. KING, JR.
Department of Geology, University of Houston, Houston, Texas 77004

R. MARTIN, WELDON B. NANCE
Brown and Root-Northrop, Chemical Analysis Group, Lunar Receiving Laboratory, NASA Manned Spacecraft Center, Houston 77058

References and Notes

1. J. A. O'Keefe, *Science* **168**, 1209 (1970).
2. Lunar Sample Preliminary Examination Team, *ibid.* **167**, 1325 (1970).
3. The help of the NASA Lunar Receiving Laboratory curator's office in furnishing documentary photography, thin sections, and the main mass of sample 12013 for examination is gratefully acknowledged.
4. S. Epstein and H. P. Taylor, Jr., in *Proceedings of the Apollo 11 Lunar Science Conference* (Pergamon, Oxford, 1970), vol. 2, p. 1085; J. A. Philpotts and C. C. Schnetzler, *ibid.*, p. 1471; A. A. Smales *et al.*, *ibid.*, p. 1575; M. Tatsumoto, *ibid.*, p. 1595; H. P. Taylor, Jr., and S. Epstein, *ibid.*, p. 1613; and others.
5. H. Faul, *Science* **152**, 1341 (1966); H. C. Urey, *ibid.* **137**, 746 (1962); G. S. Hawkins, *J. Geophys. Res.* **68**, 895 (1963); V. E. Barnes, *Sci. Amer.* **205**, 58 (1961); E. A. King, Jr., *Guidebook 32nd Annual Meeting* (Meteoritical Society, Houston, 1 November 1969).

13 July 1970

Reprinted from *Chemie der Erde,* **30**, 13–19 (1971)

44

Description and Origin of Large Tektite from Thailand[1)]

By **Virgil E. Barnes**
The University of Texas at Austin
Austin/Texas 78712 (U.S.A.)

With 2 figures

Abstract

The refractive index of a 12.8-kg layered tektite collected in Thailand is in the range 1.510 to 1.514 with an indicated average silica content of about 73 percent. This specimen is the largest known tektite and is a segment from a much larger mass postulated to have solidified from a puddle of melted soil formed by comet impact. Melt from such puddles, whipped into the air by atmospheric turbulence, probably was the source of shaped indochinites.

Introduction

Layered indochinites were first recognized to be different from shaped tektites by Lacroix (1935) who described "tectites sans formes figurées" from a locality near Muong Nong, Laos, in which 67.5 kg of tektites were excavated from an area 100 m². The largest specimen from this locality weighed 3.2 kg. Prior to the present find the largest tektite of which I am aware and have examined weighs 7.4 kg, was collected in the Libyan Desert, and is in the Geological Museum, Cairo. This Libyan Desert glass is a silica-glass; it has been described by Clayton and Spencer (1934). It, too, is a layered tektite showing "brown streaks and parallel bands in the paler coloured glass" (Spencer 1939). The significance of layering in the interpretation of the origin of tektites has been discussed by Barnes (1964).

Layered tektites already have provided much data of importance in deciphering the origin of tektites, and probably much remains to be learned by examination of additional layered tektites. For this reason I have taken every opportunity to obtain additional layered tektites and to examine their occurrence in the field.

---

1) Dedicated to Prof. Dr. F. Heide on the occasion of his 80th birthday.

While I was in Thailand, Mr. Louis S. Gardner, U. S. Geological Survey, wrote me (letter dated January 2, 1965) that he had a piece of tektite a couple of inches across from the vicinity of Savannakhet which apparently was broken from a projection on a large tektite. He had been unable to obtain the entire specimen before leaving Thailand but gave me as much information as he had concerning its location. He also told of two large tektites having been reported in the vicinity of Chumpae west of Kohn Kaen. I was able to investigate that reported occurrence but the material turned out to be slag, not tektite.

After I left Thailand, Mr. Akanit Suwanasing, Economic Geology Division, Geological Survey Department, Royal Bureau of Mines, who accompanied me to Chumpae, continued investigating the other occurrence reported by Mr. Gardner. In a letter of February 28, 1967, Mr. Suwanasing stated: "Now I have in hand a big specimen from Amphoe Khemarat. It is 12.8 kg in weight, specific gravity 2.37, refractive index 1.495, and hardness 6." After Mr. Suwanasing left Thailand to do graduate work at McGill University, Mr. Sangob Kaewbaidhoon, Chief of the Economic Geology Division, forwarded the specimen to me through the office of Mr. Dallas D. Fowler, U.S. Agency for International Development. This specimen is not the one referred to by Mr. Gardner as no recent break is present.

Petrologic data

The tektite described herein is 5.4 kg heavier than the largest previously described tektite; it measures 20.0 $\times$ 29,7 cm and is 19.0 cm in height perpendicular to the layering. When sitting on its flat side, which is probably the correct orientation, (Fig. 1) approximately the lower one-third of the specimen is a layer of relatively dense glass, followed by lenses of bubbly glass, then folded relatively dense glass to the top of the specimen. Within the thicker layers, less distinct thinner layers can be seen which are more evident in plates cut from the specimen.

The upper part of the specimen is synclinal, emphasized by a linear depression in the surface of the specimen. One limb dips gently and the other steeply with individual layers thickening greatly in the steeper limb. One interpretation is that melt representing the upper portion of the specimen was flowing from left to right (Fig. 1) over a substrate that shortly before may have arrived from another direction. As the upper melt cooled the forward part of the flow became viscous and slowed and the portion following met resistance, thickened, and became folded.

The size of the original mass from which this specimen came is conjectural, but if it covered an area 2 m in diameter and was uniformly 20 cm thick it would weigh about 1300 kg. It is safe to predict that layered tektites exist that are several times larger than the one being described.

In addition to the furrows outlining the layering, much of the surface is rough, with pits up to thumb size. Typical finely etched tektite sculpture lines the pits and covers most of the specimen.

For microscopic examination, plates were cut from the upper, middle, and lower portions of the specimen. The cuts were made across the layering and normal to the surface. The glass is dark brown by transmitted light making it difficult, because of weak signals, to measure its refractive index on the Abbe refractometer. Little confidence can be placed in the results; the values obtained for the three plates using the refractometer appear to be about the same, around 1.509 ± 0.003. In view of the unsatisfactory refractometer measurements and the earlier reported refractive index of 1.495, crushed samples were examined in refractive index oils using sodium light.

Fig. 1. Layered tektite, weighing 12.8 kg, from Amphoe Khemarat, Thailand. Scale is 20 cm long

In an oil with a refractive index of 1.5118, some particles of the lower sample were higher, some were lower, and a few were the same. All particles were higher in refractive index than 1.5109 and lower than 1.5145. For the middle sample, most particles were higher than 1.5116, a few were lower, and some with strong flow structure exhibited schlieren in part higher and in part lower than the index oil. For the top sample some particles were higher, some lower, and a few equal to 1.5114. A crushed chip removed from massive glass between the lower and middle cuts has a refractive index around 1.5105, one from the main frothy layer around 1.5116, and one from a large recess at the large end of the specimen around 1.5142. From these results it is obvious that the values obtained with the Abbe refractometer are too low.

The massive glass is somewhat lower in refractive index than the glass in the recessed area that has an earthy brownish-gray appearance as if the glass is weathered. From the refractive index of the glass in the recessed area the silica content is judged to be about 71 percent. The silica content of the frothy layer is about 74 percent, and the specimen as a whole probably averages about 73 percent silica.

Microscopically the glass is similar to other layered tektite glass previously illustrated (Barnes 1964) in which the numerous wavy elongate glassy threads are probably derived from thinly drawn out melted mineral grains. Bubbles vary widely in abundance from layers that are opaque because of bubble content to others in which bubbles are fairly abundant to abundant. In the lower sample the two largest bubbles measure 1.0 and 1.3 mm, an elliptical bubble measures 0.5 $\times$ 0.7 mm, a few bubbles measure 0.04 mm, and most are 0.02 mm ranging down to 0.002 mm and smaller. A few layers contain bubbles that are elongate and branching to spinose. In the middle sample bubbles are up to 0.6 mm in size, and in the upper sample a segment of an elliptical bubble is 2.75 mm across and another complete bubble is 1.3 $\times$ 2.1 mm.

Lechatelierite is scarce, mostly frothy to bubbly, and very thinly drawn out. The bubbles in the lechatelierite also are mostly thinly drawn out and where abundant cause the lechatelierite to appear opaque. Clear lechatelierite is very scarce; the largest piece seen, 0.02 $\times$ 0.14 mm, is in the upper sample.

Tiny spherules occur sparingly in the middle sample in schlieren-like sheets, occasionally as roe-like clusters, and as individuals. The largest spherules are 0.015 mm in size; and many are 0.001 mm and smaller. These spherules superficially resemble baddeleyite from the Martha's Vineyard tektite and from impact glasses, illustrated by Clarke and Wosinski (1967) and by El Goresy (1965), respectively. However, baddeleyite is anisotropic and these spherules, so far as can be seen with the petrographic microscope, are entirely isotropic as shown by absence of birefringence, including strain birefringence. The absence of strain birefringence indicates that the spherules have about the same coefficient of expansion as the enclosing glass.

Experiments with solar-furnace melts to learn how various minerals behave when mixed with synthetic tektite glass included three runs using zircon. In solar-furnace melting the highest temperature is reached at the focal point and drops rapidly outward where unmelted mineral is incorporated by remelted synthetic tektite glass. A 30-second run with synthetic tektite glass to zircon in the ratio of 3:1 gave an opaque glass up to the edge of the zone of unmelted zircon, as did a second run using a ratio of 5:1. A third sample having a ratio of synthetic tektite glass to zircon of 5 :1 was melted for 60 seconds on a hazy day when the temperature was lower than normal at the focal point. The volume of melt was small, and the zircon within it remained as discrete areas of opaque brown with zones next to some unmelted zircon grains and in some of the brown areas that are very finely granular. The granules appear to be birefringent when the condenser lens is in, but this is not certain as the glass is too opaque to transmit light with the condenser removed. These granules do not resemble the spherules found in the middle sample of the large tektite but do closely resemble the baddeleyite illustrated by Clarke and Wosinski.

Other minerals melted with synthetic tektite glass in the solar furnace include garnet, muscovite, apatite, microcline, plagioclase, hornblende, biotite, vermiculite, and quartz. Rayed bubbles formed with muscovite, biotite, and possibly with hornblende, and quartz became lechatelierite, but nothing was found resembling the spherules in the middle sample of the large tektite.

Discussion

Some workers believe that layered tektites are meteorites, others believe that they are terrestrial in origin. For example, Centolanzi (1969), in discussing maximum tektite size, mentioned the deposit of layered tektites near Muong Nong, Laos, in which 67 kg of tektite glass was found "within such a small area as to indicate that a 67-kg tektite impacted the ground at this particular spot". Centolanzi disregards the abundant evidence for the local origin of such tektite glass. That layered tektites formed as puddles of local melt and were the source of shaped tektites (Barnes 1968) is supported by the following lines of evidence:

1. The Sr^{87}/Sr^{86} ratios for layered tektites in two localities and for the soil in which they rest are almost identical (Barnes 1969).
2. Quartz grains in a few layered tektites are identical in size and shape to quartz grains in the associated soil (Barnes 1963). Some of these grains have been changed to coesite (Walter 1965).
3. Glass (1970) found quartz and shocked zircon and chromite in a layered tektite and concluded: "The presence of chromite and zircon together suggests that the Muong Nong-type tektite was produced from sedimentary material."
4. Chemically, layered tektites and shaped tektites are identical (Barnes and Pitakpaivan 1962).
5. Some shaped tektites in my collection have sagged and otherwise have been deformed, showing that they were still plastic when they landed. Nininger and Huss (1967, Fig. 1) illustrated similar tektites.
6. Some layered tektites were faulted, and quartz grains were introduced (Barnes 1964). That the melt was still very hot at the time of faulting is shown by incipient devitrification along fault planes and perfect welding of the melt around sand grains.
7. Refractive index histograms of layered tektites from two localities in northeastern Thailand show two peaks, Fig. 2, indicating the presence of two kinds of glass having a difference in silica content of about 8 percent. A likely interpretation is that puddles of low-silica melt from argillaceous bedrock exposed in shallow drainages accumulated first, followed by more slowly moving, viscous high-silica melt from surrounding siliceous soil.

Atmospheric turbulence would have whipped the fluid, low-silica melt into the air to produce shaped tektites before the viscous high-silica melt would have reached the puddles.

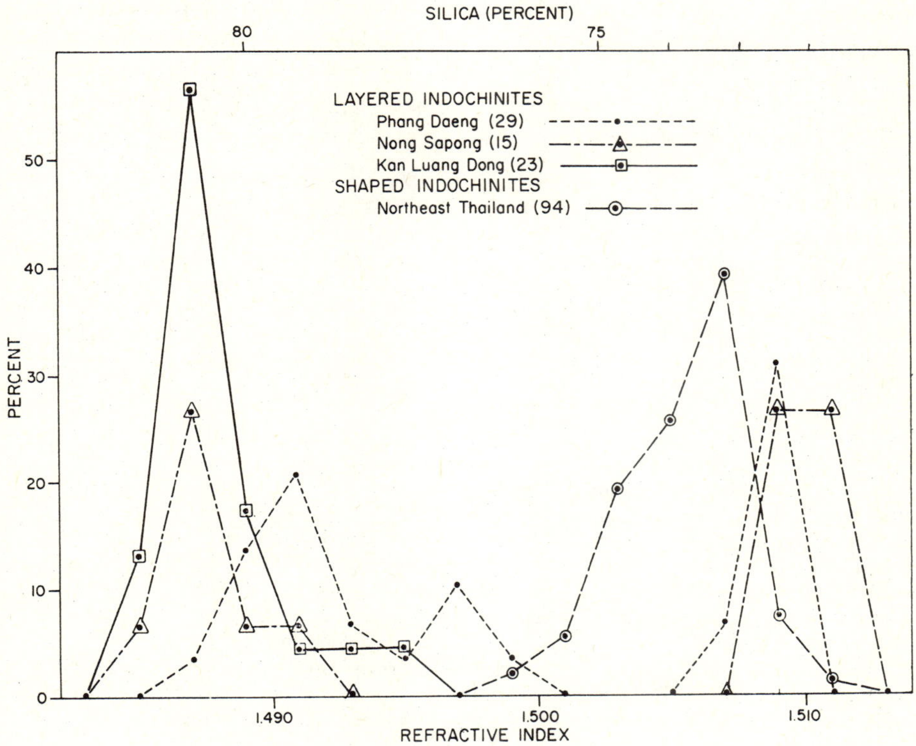

Fig. 2. Refractive index histograms of northeast Thailand tektites

From this evidence it seems likely that indochinites were formed by impact, as shown by the presence of coesite and shocked zircon and chromite, and that they were probably formed locally, as shown by shaped forms that sagged upon landing and by the correspondence of layered tektites and local soil in Sr^{87}/Sr^{86} ratios. Specimens of layered tektite glass show faulting and trapping of sand grains along the faults. This is strongly indicative that sharp earth movements continued for several minutes after the initial shock.

A specific interpretation of the comet impact hypothesis (Urey 1957) successfully explains the origin of indochinites and the distribution of microtektites. The comae of the comet would have produced heat and shock sufficient to melt the soil and to change quartz to coesite and shock other minerals over an area hundreds of kilometers across. The nucleus, if compact, might have produced a crater with local throwout; otherwise, an impact crater might not be present.

Whatever the source of heat to form the melt and of pressure to form the coesite, the melted rock and soil would have flowed to low areas, forming puddles. The more fluid melt, which accumulated first, would have been whipped into the air by tornadic winds to produce shaped tektites and a spray of microtektites. The violent turbulence would have carried microtektites for thousands of miles to be deposited in the ocean where they are now found.

In conclusion, it is believed that the 12.8-kg tektite specimen described herein is a segment from a puddle several square meters in size formed of local soil melt and that in the future fragments larger than this one will be found.

Acknowledgments

I am greatly indebted to Dr. Kaset Pitakpaivan, Mr. Sangob Kaewbaidhoon, and Mr. Akanit Suwanasing, Geological Survey Department, Royal Bureau of Mines, Thailand, for their aid in my tektite research; to Mr. Louis S. Gardner, U.S. Geological Survey, Washington, D.C., for locality information concerning large tektites; and to the National Science Foundation for support of my tektite research through Grants G-10236, GP-408, and GA-914.

Literature

Barnes, V. E., Detrital mineral grains in tektites. Science **142** (1963), 1651—1652.

Barnes, V. E., Terrestrial implication of layering, bubble shape and minerals along faults in tektite origin. Geochim. Cosmochim. Acta **28** (1964), 1267—1271.

Barnes, V. E., Petrography of new tektite localities in Thailand and the Philippines. Proc. 23rd Inter. Geol. Congress, Prague, **13** (1968), 27—36.

Barnes, V. E., Petrology of moldavites. Geochim. Cosmochim. Acta **33** (1969), 1121—1134.

Barnes, V. E., and Pitakpaivan, K., Origin of indochinite tektites. Proc. U. S. Acad. Sciences **48** (1962), 947—955.

Centolanzi, F. J., Maximum tektite size as limited by thermal stress and aerodynamic loads. Jour. Geophys. Research **74** (1969), 6723—6736.

Clarke, R. S., Jr., and Wosinski, J. F., Baddeleyite inclusion in the Martha's Vineyard tektite. Geochim. Cosmochim. Acta **31** (1967), 397—406.

Clayton, P. A., and Spencer, L. J., Silica-glass from the Libyan Desert. Mineral Mag. **23** (1934), 501—508.

El Goresy, A., Baddeleyite and its significance in impact glasses. Jour. Geophys. Research **70** (1965), 3453—3456.

Glass, B. P., Zircon and chromite crystals in a Muong Nong-type tektite. Science **169** (1970), 766—769.

Lacroix, A., Les tectites sans formes figurées de l'Indochine. C. R. Acad. Sciences, France, **200** (1935), 2129—2132.

Nininger, H. H., and Huss, G. I., Tektites that were partially plastic after completion of surface sculpturing. Science **157** (1967), 61—62.

Spencer, L. J., Tektites and silica-glass. Mineral Mag. **25** (1939), 425—440.

Urey, H. C., Origin of tektites. Nature **179** (1957), 556—557.

Walter, L. S., Coesite discovered in tektites. Science **147** (1965), 1029—1032.

Reprinted from *Science*, **171**, 312–314 (1971)

45

Tektites from the Earth

Recently, O'Keefe (*1*) has published another report in which he tries to maintain that tektites come from the moon, and, as usual, he has attempted to answer my argument for the low probability of objects coming from the moon and arriving on the earth in a localized area (*2*). He compares an unusual rock of the moon—in fact, parts of the unusual rock of the moon —with some unusual tektites. He finds rough agreement for the more abundant elements and no evidence for agreement for the less abundant ones. He mentions Taylor's (*3*) work in a reference but does not discuss his results. Taylor found a rock in Australia, a subgraywacke (which is a muddy sandstone), which for some 45 or 50 elements agrees in composition remarkably well with the abundant class of tektites found in that area. Taylor did not maintain that this rock was the particular one that produced the tektites, nor did he maintain that the tektites came from material in Australia. Subgraywacke is rather a common form of sandstone. O'Keefe's handling of the data, as has been usual for 10 years both by him and by others, is of a very partisan character, and he has not considered the high improbability of the lunar origin. The rocks of the moon, at the present time, would seem to indicate that tektites have not come from the moon.

O'Keefe proposes that tektites are propelled from the moon by volcanoes (*1*). It would seem likely that this process, requiring a velocity of at least 2.38 km/sec for the objects expelled, would probably be produced only by rather large, vigorous volcanoes and only in a vertical direction from the lunar surface. At a velocity of 2.38 km/sec, the objects expelled would travel in orbits near that of the moon, since they would have the angular momentum of the moon.

If the object were propelled in the forward direction from the moon at a velocity greater than 2.60 km/sec, it would leave the earth-moon system. On the other hand, if it were propelled in the backward direction at a velocity of 2.60 km/sec, it would fall directly to the earth. If we assume that it were expelled by a volcano in other areas of the moon and in various directions at velocities between 2.38 and 2.60, it would surely remain in the earth-moon system with orbits quite different from that of the moon. If the objects were to hit the earth, very special directional velocity considerations are required to make a direct hit on the first pass. My remarks are, of course, based on the assumption that the moon moves in a circular orbit, which is approximately true. If the object should be propelled with high velocity, it is quite easy to calculate the probability that it would be captured by the earth, since it is only a matter of the angle subtended by the earth at the moon. In this case, with an equal probability of the objects' being propelled in all directions, it would mean that about one in ten thousand would arrive at the earth. As O'Keefe states, there is a focusing effect of the earth's gravitational field for low velocities; hence, the probability would be somewhat greater than this. What appears to be true is that one must expect that a great variety of objects of various kinds would be propelled from the moon on the basis of a reasonable probability and would move in a great variety of orbits in the earth-moon system. They would cross the orbit of the moon, of course, having originated at the moon, and pass near it; they would then be captured by the moon, thrown out of the earth-moon system by interaction with its gravitational field, or thrown into orbits such that they would hit the earth. On the basis of this probability, we would expect tektites to be found in terrestrial deposits of all ages on all parts of the earth. Also, one must expect to find many well-preserved large volcanoes on the moon, which is certainly not true.

For some years, collisions of objects with the moon have been proposed as a method of getting tektites from the moon. We see from the ray craters on the moon that materials thrown out in these great collisions travel with rather low velocities. It may be that there is evidence for circumlunar objects propelled from Tycho as a means of explaining displaced rays relative to the center of the crater. These must have moved with velocities between 1.68 and 2.38 km/sec. However, the rays are mostly rather short, and, hence, velocities were mostly less than 1.68 km/sec. Thus, collisions have produced some rather slow-moving objects from the moon. It is, of course, probable that some of them escaped, and, if they did, my remarks in regard to the directional effects for the proposed volcanic origin apply here. They could escape from other parts of the moon and, if they had the correct direction and correct velocity, could collide with the earth. Again, this possibility is exceedingly improbable, and one would surely expect to find many Tychos or many recent large craters on the moon of this type if the four sets of tektites that are generally recognized on the earth's surface originated in this way, and many collisions of similar magnitude should have occurred on the earth. As stated above, it seems most improbable that tektites are coming from the moon. This is true on the basis of the mechanics but, also, on the basis of chemical composition as we know from the Apollo 11 data.

I should like to mention again briefly the possibility that cometary collisions with the earth are responsible for tektites (*2*). They must occur occasionally. One small comet apparently collided with the earth in 1908 in Siberia. Comets would produce an enormous blast of gases and, if they were large enough to penetrate the atmosphere of the earth to its surface, would produce substantial collisional effects on the earth; a blast of gases that should carry small objects to great distances would subject them to high temperatures, probably producing melting, reduction of iron (even perhaps to metallic iron), degassing of water, and so forth. Tilton (*4*) showed that the isotopic composition of the lead of tektites approximates the composition of modern terrestrial lead and thus is consistent with a terrestrial origin of tektites, as he concluded. Recent work on lunar samples shows that tektite lead is not similar to lunar lead (*5*). Barnes (*6*) has discussed the mineralogy and general geology of tektites in great detail, and he concluded that they are most probably of terrestrial origin. Moreover, Taylor's studies show that the chemical composition of materials found on the earth will agree approximately with those of tektites. The more glassy or acid ones with high silica content and high aluminum content would be very viscous liquids and would, in the melted state, hold together under the blast of gases. Also, the more basaltic type of rocks would probably be melted and disrupted into exceedingly fine particles, just as are found in the microtektites that have recently been discussed. It seems there is no reasonable objection to the cometary origin of the larger tektites from terrestrial materials of the graywacke type and related types of rocks, which occur in abundance on the surface of the earth, or of the microtektites from terrestrial basaltic rocks. The mechanics of this process were discussed some years ago by Lin (*7*), who is an exceedingly capable person in this particular field of high velocity gaseous physical phenomena. The failure to find a crater somewhere in or near Vietnam, where massive tektites have been found, is an unsolved puzzle. Faul (*8*) reviewed much evidence in regard to this problem and showed that the dates of the moldavites and the Ries and Steinheim craters and the dates of the Ivory Coast tektites and Lake Bosumtwi were closely the same. Zähringer (*9*) shows that the gases in tektites have terrestrial compositions and not those characteristic of lunar samples.

To me, it has been most depressing that so much effort has been put into the very improbable "tektites from the moon," and that this has discouraged efforts in regard to the undoubtedly correct origin from the earth and the possible mechanisms of distribution. However, the evidence at last shows that the correct origin was outlined in the 1950's and possibly now the outstanding work of Barnes, Tilton, Faul, Lin, and others, as well as the suggestions of the writer, will be recognized.

It seems to me that the problem of the lunar origin of tektites might be closed out, since we now have definite information in regard to the composition of the moon, due to the work of Turkevich *et al.* (*10*) and to the many people who have worked on the lunar Apollo 11 and 12 samples (*11*). It has seemed unnecessary to me to discuss this subject, and I am surprised that it is again being brought up by anyone.

HAROLD C. UREY

University of California
at San Diego, La Jolla 92037

References and Notes

1. J. O'Keefe, *Science* **168**, 1209 (1970).
2. H. C. Urey, *Proc. Nat. Acad. Sci. U.S.* **41**, 27 (1955); *Nature* **179**, 556 (1957); *ibid.* **182**, 1078 (1958); *Science* **137**, 746 (1962).
3. S. R. Taylor, *Geochim. Cosmochim. Acta* **30**, 1121 (1966).
4. G. R. Tilton, *ibid.* **14**, 323 (1958).
5. See, for example, M. Tatsumoto, *Proceedings of the Apollo 11 Lunar Science Conference* (Pergamon, Oxford, 1970), vol. 2, p. 1610.
6. V. E. Barnes, *Geochim. Cosmochim. Acta* **39**, 1121 (1969), and references given therein.
7. C. C. Lin, *J. Geophys. Res.* **71**, 2427 (1966).
8. H. Faul, *Science* **152**, 1341 (1966). Faul outlined many geological facts indicating a terrestrial origin for tektites.
9. J. Zähringer, in *Radioactive Dating* (International Atomic Energy Agency, Vienna, 1963), p. 289.
10. A. L. Turkevich, E. J. Franzgrote, J. H. Patterson, *Science* **165**, 277 (1969); J. H. Patterson, A. L. Turkevich, E. J. Franzgrote, T. E. Economou, K. P. Sowinski, *ibid.* **168**, 825 (1970).
11. *Science* **167**, No. 3918 (1970).

8 September 1970

Bibliography Prior to 1959

VII

Editors' Comments on Paper 46

Although most of the papers published since 1959 are listed in one or more of the preceding papers and their commentaries, many published prior to 1959 are not listed. Therefore, Baker's comprehensive bibliography is reproduced because of its world-wide scope and completeness of coverage to tektite literature. This bibliography is also used as the reference list for papers prior to 1959 cited in the introductory chapter and for the chapter on tektite sculpture.

In a letter of February 10, 1941, Dr. L. J. Spencer noted some corrections to my 1940 bibliography which have been perpetuated in Baker's bibliography. Dr. Spencer states:

"In your very complete bibliography of tektites the date 1787 in reference 46a (copied from C. Fenner!) belongs to reference 129. A. Dufrénoy was not born until 1792 and he gave the name moldavite in 1859 (Traité de Mineralogie, Paris, Vol. 4, p. 50).

"Page 567, reference 139a—Nováček.

"Page 567, reference 140a—Opik."

It should also be noted that the page citation in Baker's reference for Lacroix, 1932b, should read 139-236.

Krinov (1958) lists the following papers not in Baker's bibliography:

Malyouga, D. P. (1949). The cognition of the nature of tektites. *Meteoritika* (collected articles), **6,** 92–100.

Reprinted from *Memoirs of the National Museum of Victoria, Melbourne*, **23**, 231–246 (1959)

46

BIBLIOGRAPHY.

The following references provide as comprehensive a list of the literature on natural glass objects from various parts of the world, as the writer has been able to accumulate over the past twenty years. The articles on tektites include papers and monographs available to 1957. The list is partly compiled from bibliographies published by Walcott (1898), F. E. Suess (1900, 1914), Lacroix (1932), Fenner (1938), Spencer (1939) and Barnes (1940), with the addition of recently published articles and a few earlier, hitherto overlooked references. The list includes a few papers on fulgurites, meteorite craters and meteorite crater glasses (" impactites ").

Abel, O., 1901. Uber sternförmige Erosionskulpturen auf Wüstengesteinen K. K. geol. Reichsanstalt, Wien Jahrb., 51, 25-40.

Adams, J. A. S., 1956. Uranium Contents and Alpha-Particle Activities of Tektites. Section 11b, 20th Internat. Geol. Cong. Publ.

Adams, J. A. S., Saunders, D. F. and Zeller, E. J., 1953. Uranium Content, Alpha Particle Activity, and K_2O, Na_2O, CaO Analysis of Obsidians, Pitchstones and Tektites. Geol. Soc. America, Abstracts for November Meeting, Toronto, Canada.

Ahrens, L. H., Pinson, W. H. and Kearns, Margaret K., 1952. Association of Rubidium and Potassium and their abundance in common igneous rocks and meteorites. Geochimica et Cosmochimica, Acta, II., 229-242.

Alderman, A. R., 1932. The meteorite craters at Henbury, Central Australia (addendum by L. J. Spencer). Min. Mag., 23, 19-32.

————, 1933. The meteorite craters at Henbury, Central Australia (addendum by L. J. Spencer). Smithsonian Inst. Ann. Rept. 1932, 223-234.

Aminoff, G., 1929. Om Meteoriter. Bonniers sma handboker i vetenskapliga ämnen, Stockholm.

Anderson, A. E., 1925. Sand Fulgurites from Nebraska. Nebraska State Mus., Bull. 7, vol. I., 49-86.

Anonymous, 1934. News and Views:—Origin of Tektites. Nature, 133, 605.

————, 1937. Queensland Tektites. The Gemmologist 7 (77), 310. (Goldsmith's Journal and Gemmologist, London). (Reviewed by Tromnau in Neues Jahrbuch, Referate I., 498, 1938).

————, 1941, News and Views:—Origin of Tektites, Nature 147, 242. (A review of F. A. Paneth's Halley Lecture on "The Origin of Meteorites" delivered May, 1940.)

Armitage, R. W., 1906. Natural History Notes—Obsidian Bombs. Vic. Naturalist, 23, 100.

Baertschi, P., 1950. Isotopic Composition of the Oxygen in Silicate Rocks, Nature, 166, 112-113. (Review in Bull. Geol. Soc. America, 64, part 3, March, 1953).

Baillaud, J., 1935. Sur l'origine des tectites. Rev. Scientifique, 73, 602.

Baker, G., 1937. Tektites from the Sherbrook River District, east of Port Campbell. Roy. Soc. Vic., Proc., 49 (2), 165-177.

————, 1938. Article on "Port Campbell" in "Walkabout" (Australia). 4, no. 9, 36.

————, 1940a. Some australite structures and their origin. Min. Mag., 25, 487-494.

————, 1940b. An Unusual Australite Form. Roy. Soc. Vic., Proc., 52, (2), 312-314.

————, 1944. The Flanges of Australites. Mem. nat. Mus. Vict., 14 (1), 7-22.

————, 1946. Some Unusual Shapes and Features of Australites (Tektites). Mem. Nat. Mus. Vict., 14 (2), 47-51.

————, 1950. Geology and Physiography of the Moonlight Head District, Victoria. (Section on Australites, p. 35). Roy. Soc. Vic., Proc., 60.

————, 1953. Natural Sinters from Mt. Remarkable and Tempe Downs. Roy. Soc. S. Aust., Trans., 76, 27-33.

————, 1955a. Curvature—Size Relationships of Port Campbell Australites. Roy. Soc. Vic., Proc., 67, 165-219.

————, 1955b. Australites from Harrow, Victoria. Min. Mag. 30, 596-603.

————, 1956. Nirranda Strewnfield Australites, South-east of Warrnambool, Western Victoria. Mem. nat. Mus. Vict., 20, 59-172.

Baker, G., 1956a. Natural Black Glass Resembling Australite Fragments. Mem. nat. Mus. Vict., 20, 173-189.

———, 1957. Alleged Meteorite from Horsham, Victoria. Mem. nat. Mus. Vict., 21, 72-78.

———, 1957. The Rôle of Australites in Aboriginal Customs. Mem. nat. Mus. Vict., 22 (8), 1-26.

———, 1958. The Rôle of Aerodynamical Phenomena in Shaping and Sculpturing Australian Tektites. Am. Journ. Sci. 256, 369-383.

Baker, G., and Forster, H. C., 1943. The Specific Gravity Relationships of Australites. Amer. Journ. Sci., 241, 377-406.

Baker, G., and Gaskin, A. J., 1946. Natural Glass from Macedon, Victoria, and its Relationships to Other Natural Glasses. Journ. Geol. liv, 88-104.

Baker, R. T., 1900. Note on an obsidian "bomb" from New South Wales. Roy. Soc. N.S.W., Journ. and Proc., 34, 118-120.

Bares, J., 1899. Hornini archaického utvaon a vltavín. Casopis pro průmysl chemicky. Jahrg. 9, 118-123. (Further discussion between Bares and Slavik on vltavín, Casopis pro průmysl chemicky, Jahrg. 9, 223-225 and 264-266.).

Barnes, V. E., 1940a. North American Tektites. Contributions to Geology; Part 2, Univ. of Texas Publ. no. 3945, 477-582.

———, 1940b. Distribution and Origin of Tektites. Bull. Geol. Soc. America, 51, part 2, 1919-1920 (Abstracts), and Amer. Min., 26, 194.

———, 1951. New Tektite Areas in Texas. Bull. Geol. Soc. America, 62, 1422 (Abstract), (Reprinted in "Popular Astronomy", 59, 538, and Contrib. Meteoritical Soc. (1952), 5 (1), 97-98).

———, 1957. Tektites. Geo. Times Mag. (Amer. Geol. Inst.), 1 (12), 6-7 and 16-17.

———, 1958. Origin of Tektites, Nature, 181, 1457. (Criticism of Tektite Origin from the Moon.)

Barton, R. F., 1946. The Religion of the Ifugaos. Mem. Am. Anthrop. Assoc., 65, pp. 39, 77, 125. (Reference to tektites in Northern Luzon as "buga" (= magic-stones)).

Basedow, H., 1905. Trans. Roy. Soc. South Australia, 29, 89. (Reference to use of australites by aborigines).

Bayer, J., 1919. Zur Frage der Herkunft der Tektite. Mitt. Geol. Gesell. Wien, 11, 248-251.

Beck, R., 1910. Über die in Tektiten eingeschlossenen Gase. Deutsche geol. geseil. Monatsber., 62, 240-245.

Belben-Flux, J., 1950. Tektites. Monthly Notes of Astron. Soc. S. Africa, 9, 6-8.

Belot, E., 1933. Le mystère des tectites; larmes Bataviques tombées du ciel. Rev. Scientifique, 71, 577-581.

Berwerth, F., 1903. Verzeichnis der Meteoriten in K. K. Hofmuseum. Annalen der Naturhist, Hofmuseum, Wien.

———, 1910. Oberflächenstudien an Meteoriten. Tschermak's min. petr. Mitt., neue Folge, 29, 153-168.

———, 1912. Fortschritte in der Meteoritenkunde seit 1900. Fortschr. Mineralogie, Jena, 1, 283.

———, 1914. Meteoriten. Handwörterbuch Naturwiss., Jena, 6, 859.

———, 1917. Können die Tektite als Kunstprodukte gedeutet werden? Centralbl. Mineralogie, 240-254. (Eine Bejahung).

Beudant, F. A., 1822. Voyage Minéralogique et Géologique en Hongrie, pendant l'année 1818, 2, 214, Verdiére, Paris.

Beyer, H. O., 1928. Tektites in Luzon. Schreibmaschinenschriftl. Mitt., 20 pp. (November). (Reviewed by H. Himmel in Neues Jahrbuch, Referate 1, 615 (1934).

———, 1934. A Brief Statement of Some Essential Facts regarding Philippine and other Indo-Malaysian tektites; with notes on recent theories of tektite origin. Schreibmaschinenschriftl. Mitt. (Reviewed by H. Himmel in Neues Jahrbuch, Referate 1, 615-616).

———, 1935. Philippine Tektites. Philippine Mag. 32, 534, 542-543 and 581-582. (Also Book: Manila, 1934).

———, 1936. General notes on the Santa Mesa tektite site. Schreibmaschinenschriftl. Mitteilung (December). (Reviewed by H. Himmel, Neues Jahrbuch, Referate 1, 313.)

Beyer, H. O., 1939. A bibliography of tektites; Manila.

————, 1940. Philippine tektites and the tektite problem in general. Popular Astron., 48, 43-48. (Reprinted in Smithsonian Inst. Ann. Rept., 253-259, 1942.)

————, 1947. Outline Review of Philippine Archaeology by Islands and Provinces. Phil. Journ. Sci., 77 (3-4), 205-374. (Various references to tektites and their uses.)

————, 1948. Philippine and East Asian Archaeology. Phil. Nat. Research Council, Bull. 29. (References to flaked tektites on pp. 13-14, &c.).

————, 1955a. The relation of Tektites to Archaeology. Phil. Nat. Research Council, Univ., Phil. Publ., Dillman, Quezon City.

————, 1955b. Additional Notes on Tektite Lore. (Multigraph Publication), Manila, 1st Nov., 1955.

Blatchford, T., 1899. The Geology of the Coolgardie Goldfield. W. Aust. Geol. Surv., Bull. 3, 36.

Boon, J. D., and Albritton, C. C., Jr., 1938. Established and Supposed Examples of Meteorite Craters and Structures. Field and Laboratory, Chap. 6, 44-56.

Bowley, H., 1945. Australite Observed to Fall at Cottesloe—a Correction. Roy, Soc. W. Aust., Journ., 29, 163.

Brandes, G., 1905. Zwei Hallische Meteoritenfälle. Zeitschr. Naturw., 76, 459.

Breithaupt, A., 1823. Vollständige Charakteristik des Mineral-Systems, 223-224, Arnold'schen Buchhandlung, Dresden.

Brezina, A., 1904. Über Tektite von beobachtetem Fall. K. Akad. Wiss. Wien, Anzeiger, 112, Abt., 2a, 41.

Brown, H. Y. L., 1893. Catalogue of South Australian Minerals, &c., 25.

Buch, L. von, 1809. Geognostiche Beodachlugen auf Reisen durch Deutchland und Italien. Band Italien, 2, 51.

Buddhue, J. D., 1939. Age of tectites. Mineralogist, Portland, Oregon, 7, 405.

————, 1941. Tektites. Puzzle of Science. Scientific American 164, 354-356.

————, 1946. The Abundance of the Chemical Elements in Meteorites and Tektites. Contrib. Soc. for Research on Meteorites, 3 (5), 262-265.

Busick, R., 1937. Rizalites—Philippine Tektites—with a description of the Pugad-Babuy Site. Mich, Acad. Sci., Papers, 23, 21-27. (Also special reprint, 1938.)

Campbell, D. A., 1954. Tektites. Journ. Brit. Astron. Soc., 64 (8), 410-411.

Campbell, W. D., 1906. The Geology and Mineral Resources of the Norseman District, Dundas Goldfield. W. Aust. Geol. Surv., Bull., 21, 22-23.

Campbell Smith, W., 1931. Review of "Obsidian buttons" in "Open Air Studies in Australia", by F. Chapman. Min. Abstr., IV, 264.

Campbell, Smith, W., and Hey, M. H., 1952a. Le Verre de Silice d'Aouelloul, (Adrar-Sahara Occidental). Direct. des Mines, Bull., 15, Gouv. Gen. de l'Afrique Occ. Franç., 443-446.

————, 1952b. The Silica Glass from the crater of Aouelloul (Adrar, Western Sahara). Bull. Inst. Franc. Afrique Noire, XIV, (3) 762-776.

Canham, —., 1880. Roy. Soc. S. Aust., Proc., 4, 4, 148.

Card, G. W., 1902. Handbook to the Mining and Geological Museum, New South Wales. N.S.W. Dept. of Mines and Agriculture, 172.

————, 1903. Mineralogical Notes, No. 8. N.S.W. Geol. Surv., Records, 7, (3), 218.

————, 1907. Annual Report of the Curator and Mineralogist. N.S.W. Dept. of Mines and Agriculture, Ann. Rept. for 1907, 190 and 197.

————, 1919. Mineralogical and Petrographical Notes, No. 13, N.S.W. Geol. Surv., Records, 9 (4), 183.

Cassidy, W. A., 1956. Australite Investigations and their Bearing on the Tektite Problem. Meteoritics, 1 (4), 426-437.

————, 1957. Australia's Fiery Rain. Aust. Mus. Mag., XII, (6), 182-186.

Cassidy, W. A., and Segnit, E. R., 1955. Liquid Immiscibility in a Silicate Melt. Nature. 176, 305.

Chandler, J., 1880. Roy. Soc. S. Aust., Trans., 4, 149.

Chapman, F., 1929. Obsidian buttons; an Australian riddle. In "Open Air Studies in Australia", 144-149, J. M. Dent and Sons, London.

————, 1933. Origin of Tektites. Nature, 131. 876.

Chloupek, J., 1929. Die Herkunft der Moldavite. Die Naturwissenschaften, 17, 598-600. (A German abstract of the Czech paper by F. Hanůs entitled "Über Moldavie von Böhmen und Mähren", 1928.)

Clarke, F. W., 1904. Analyses of Rocks from the laboratory of the United States Geological Survey, 1880-1903. U.S. Geol. Surv, Bull., 228, 276.

Clarke, W. B., 1855. On the Occurrence of obsidian bombs in the auriferous alluvia of New South Wales. Geol. Soc. London, Quart. Journ., 11, 403.

————, 1857. Additional Notes on the occurrence of Volcanic Bombs in Australasia. (Abstract.) Geol. Soc. London, Quart. Journ., 13, 188.

Clayton, P. A., and Spencer, L. J., 1934. Silica Glass from the Libyan Desert. Min. Mag., 23, 501-508.

Codazzi, R. L., 1915. Contribucion el estudio de los minerales de Colombia; Bogota.

————, 1925. Nota sobra las tektitas de Colombia. Bibliotheca del. Mus. Nac., República de Colombia, 35-38 and 89; Bogota.

————, 1929. Notas adicionales sobre las minerales y las rocas de Colombia. Bibliotheca del Mus. Nac., República de Colombia, 1-51; Bogota.

Conder, H., 1934. Darwin Glass. Industrial and Mining Standard of Australia, 89, 329-330.

Cross, F. C., 1948. A New Glass of Possibly Extraterrestrial Origin. Contrib. Meteoritical Soc., 4 (2), 154-157.

Curr, E. M., 1886-1887. The Australian Race, Vol. III., p. 547. Government Printer, Melbourne, Australia.

Czjzek, —, 1859. Zepharovich für das Kaiserthum Oesterrich, 1, 290, Wien.

Damour, A., 1844. Sur une Obsidienne de L'Inde qui a éclaté avec dêtonation, au moment où on la sciait. Acad. Sci. Paris, Comptes rendus, 13, 4.

Darwin, Charles, 1844. Geological Observations on coral reefs, volcanic islands, p. 38. London (p. 44, second edition, 1876.)

————, 1851. Geological Observations on coral reefs, volcanic islands, and in South America, 190-191 (reprint 1890).

David, Sir T. W. E., Summers, H. S., and Ampt, G. A., 1927. The Tasmanian tektite—Darwin Glass. Roy. Soc. Vic., Proc., 39 (2), 167-190.

Dawson, J., 1881. Australian Aborigines. Geo. Robertson, Melbourne, Sydney and Adelaide.

de Boer, K., 1929. Uber die Fundortverteilung der Glasmeteoriten. Astronom. Nachr., 234, 135-6.

————, 1932. (Great Circle Distribution of Tektites.) Astr. Nach., 246, 201-204.

de Groot, C., 1880. Referat über obigen Aufsatz. Indische Gids., 495-496.

Dennaeyer, M. E., 1944. Les tectites de L'Indochine. Bull, Soc. Belge Geol., Bruxelles, 53, 45-49.

Dittler, E., 1933. Beitrag zur chemischen Systematik der Tektite. Centralblat. Mineralogie, Abt. A, 214-219.

Dodwell, G. F., and Fenner, C., 1943. The Kybunga Daylight Meteor. Roy. Geogr. Soc. of Australasia (South Australian Branch), Proc. xliv, 6-19.

Döring, T., and Stutzer, O., 1928. Kolumbianische Glasmeteorite. Centralblat Mineralogie, Abt. A, 35-41.

Dubey, V. S., 1933. The Origin of Tektites. Nature, 132, 678.

Dufrenoy, A., 1844. Analysis of moldavite glass, in Treatise on Mineralogy, vol. 4.

Dunn, E. J., 1908a. Rock and Mineral Analyses. Victoria, Ann. Rept. Sec. Mines for 1907, 63.

————, 1908b. Obsidian Buttons. Victoria, Geol. Surv. Records, 2 (4), 202-207.

————, 1911. Pebbles. pp. 34 and 64. Robertson and Co., Melbourne.

————, 1912a. The Mt. William Goldfield, Grampians. Section on "Obsidianites" (= Australites). Victoria, Geol. Surv. Records III, (2), 119.

————, 1912b. Australites. Victoria, Geol. Survey. Bull., 27, 3-23.

————, 1914. Further notes on australites. Victoria, Geol. Surv. Records, 3 (3), 322-326.

————, 1916. Additional notes on australites; Darwin Glass. Roy. Soc. Vic., Proc., 28 (2), 223-227.

Dunn, E. J., 1935. Australites. Geol. Mag., 72, 139-140.

Dunstan, B., 1913. Queensland Mineral Index and Guide. Queensland Dept. of Mines, Publication No. 241, 795.

Dvorak, R., 1929. O vltavínech, Zvlastni otisk. Casopisu Od Horacka k Podyji, Year 5.

Dvorsky, F., 1880. Über einige in der Umgebung von Trebitsch vorkommende Felsarten und Mineralien, 3, Programm des Staatsuntergymnasiums zu Trebitsch.

————, 1883. Die am Iglavaflusse abgesetzten Moldavitquartzgerölle. Ein Beitrag zur Bouteillensteinfrage. Programm des Gymnasiums in Trebitsch, 2-17. (Abstract—K.K. geol. Reichsanstalt Verh., 219, 1883.)

————, 1898. O vltavínech moravskych. Mus. Francisceum Annales, Brünn, 55.

————, 1914. Zpráva o dvou novych nalezitich vltavinů. Casopis Morav. Museum, Brünn, 1.

Easton, N. Wing, 1921. The billitonites, an attempt to unravel the tektite puzzle. K. Akad. Wetensch. Amsterdam Verh., Tweede Sectie, 22 (2), 1-32.

————, 1923. Zur tektitfrage. Centralblat Mineralogie, 33-42.

Eggen, O. J., 1953. Tektites: Glass from the Moon. Science Digest, 33 (1), 53-55.

Ehmann, W. D., 1957. Cosmic-Ray-Induced Radioactivities in Meteorites and Tektites. U.S. Atomic Energy Comm. Rept. No. NYO-6634.

Eichstädt, F., 1908. En egendomlig av rent glas bestaende meteorit funnen i Skane. Geol. Fören. Stockholm Förh., 30 (5), 323-330.

Eitel, W., 1927. Kolumbianische Glasmeteorite von O. Stutzer.

Eppler, A., 1914. Glasmeteoriten. Himmel u. Erde, Leipzig, 231-234.

Erben, B., 1892. Moldavit. Naturwiss. Zeitschr. "Vesmír", Jahrg. 21, p. 123, Prague. (Referate—J. J. Jahn, K.K. geol. Reichsanstalt, Verh., 85, (1893).)

Erdmann, O. L., 1832. Chemische Untersuchung einiger Obsidiane des Spharoliths und eines ähnlichen Minerals, des Pechsteines oder Perlsteines. Tech.-ökon. Chemie, Liepzig, Jour., 15, 35.

Escher, B. G., 1925. Geëtste Botsfiguren op Billitonieten. Geol. mijnb. genootsch. Nederland en Kolonien Verh., Geol. ser., 8, 155-160.

Fenner, C., 1933a. Origin of Tektites. Nature, 132, 571.

————, 1933b. Bunyips and billabongs, an Australian out of doors, 39-46, Angus and Robertson, Sydney.

————, 1934. Australites, Part I. Classification of the W.H.C. Shaw Collection. Roy. Soc. S. Aust., Trans. and Proc., 58, 62-79.

————, 1935a. Australites, Part II. Numbers, forms, distribution and origin. Roy. Soc. S. Aust., Trans. and Proc., 59, 125-140.

————, 1935b. The forms and distribution of australites (abstract) Australian Assoc. Adv. Sci., Report, 22, 143.

————, 1937a. Australites; are they glass meteorites? Popular Astronomy, 45, 504-507.

————, 1937b. Australites; a unique shower of glass meteorites. (Abstract.) British Assoc. Adv. Sci., Rept., p. 356.

————, 1938a. Australites; a unique shower of glass meteorites. Min. Mag., 25, 82-85.

————, 1938b. Australites, Part III. A contribution to the problem of the origin of tektites. Roy. Soc. S. Aust., Trans., 62, 192-216.

————, 1939. Blackfellows' Buttons, the remarkable glass meteorites of Australia. The Sky—Magazine of Cosmic News, 3 (VIII), 16-17 and 27.

————, 1940. Australites, Part IV. The John Kennett Collection, with notes on Darwin Glass and Bediasites. Roy. Soc. S. Aust., Trans., 64, 305-324.

————, 1949a. Sandtube Fulgurites and their bearing on the Tektite Problem. S. Aust. Mus. Records, ix, 127-142.

————, 1949b. Australites, Part V. Tektites in the South Australian Museum, with some notes on theories of origin. Roy. Soc. S Aust., Trans., 73, 7-21.

————, 1950. A note on the origin of Tektites: a correction. Popular Astronomy, 58, 518.

————, 1953a. Australites and Other Tektites. South Australian Naturalist, 27, (4), 1-8.

————, 1953b. Glass Meteorites. Walkabout (Australian Geogr. Mag.), 19 (12), 29-30.

Fenner, C., 1955. Australites, Part VI. Some Notes on Unusually Large Australites. Roy. Soc. S. Aust., Trans., 78, 88-91.

Friedlaender, I., 1927. "Tektite" von Columbien. Centralblat Mineralogie, Abt. A., 67-69.

Friedman, I., 1955. Bull. Geol. Soc. Amer., 66, 1562.

Friedman, I., Kohman, T. P., and Cassidy, W. A., 1958. Tektites. Science, vol. 127, 91-93.

Fromaget, J., 1932. La date probable de la chute des Tectites. Prehistorica Asiae Orientalis, Vol. I, 47-61. (Hanoi, 1932.)

Geological Survey Progress Report, Victoria, iii, 286.

Gerling, E. K., and Yashchenko, M. L., 1952. (On the Age of Tektites). Doklady Acad. Sci., U.S.S.R., 83, 901-902.

Glocker, E. R., 1848. Ueber die ursprüngliche Lagerstätte des chrysolithartigen Obsidians. Annalen der Physik (Pogendorff), 75, 458.

Goldschmidt, V., 1918. Ueber erosion und lösung. Goldschmidt's Beitr. Krist. Mineralog., 1, 183-198.

———, 1921. Himmelgläser. Zeitschr. Kristall., 56, 420.

———, 1924. Über Meteorgläser, ihre Bildung und Gestalt. Goldschmidt's Beitr. Krist. Mineralog., 2, 148-155.

Grant, K., 1909. Obsidianites—their origin from a physical standpoint. Roy. Soc. Vic., Proc., 21, (2), 444-448.

———, 1937. Problems of the Tektites. Origin and Composition of Mysterious Little Black Stones. Science for All. Advertiser Newspapers Ltd., 56-59.

Gregory, J. W., 1912. The Making of the Earth. Home University Series, 36, Williams and Norgate, London.

Grewingk, O., and Schmidt, C., 1864. Über die Meteoritenfälle von Pillistfer, Buschoff und Igast. Arch. Naturh., Liv. Esth. und Kurl., 3, 421-553.

Gutenberg, G., 1951. Internal Constitution of the Earth. 2nd edit., 439 pp., 43 figs. Dover Publications Inc., New York.

Habermann, J., 1881a. Chemische Analyse des "Bouteillensteines". Naturwiss. Ver. Brünn Verh., Sitzungsber., 20, 21.

———, 1881b. Weiters Bermerkungen uber den "Bouteillenstein". Naturwiss. Ver. Brunn Verh., Sitzungsber., 20, 26.

Hammond, C. R., 1950. The Chemical Composition and Some Physical Characteristics of Tektites. · Contrib. Meteoritical Soc., 4 (4), whole number 16, 271-275

Hanamann, J., 1893. O povaze ceského vltavinu. Bömische Zeitschr. für chemische Industrie. Jahrg. 3, 365, Prague. (Referate—J. J. Jahn, K.K. geol. Reichsanstalt, Verh., 194, 1894).

Hanus, F., 1909. Neue Moldavitfundstätten bei Budweis. K.K. geol. Reichsanstalt Wien Verh., 364-366.

———, 1928a. Les moldavies (tektites) de la Bohême et de la Moravie. Rozpravy ces. Akad. II, (37), 1-83. (Internat. Acad. Sci. Bohême, Bull., Prague.)

———, 1928b. Über Moldavie von Böhmen und Mähren. Ibid., 37, (24), 1-83 (Cechisch. französisches Resumé im Internat. Akad. Sci. Bohême, Bull., Prague.)

Hardcastle, H., 1926. The origin of australites. Plastic sweepings of a meteorite. New Zealand Journ. Sci. Techn., 8 (2), 65-75.

Hassell, E., 1936. Notes on the Ethnology of the Wheelman Tribe of South-western Australia. Anthropos, 31, 706-707.

Hauer, F. von, 1880. Bouteillenstein (Obsidian) von Trebitsche. K.K. geol. Reichsanstalt Wien Verh., 282-284.

Hauer, K. von, 1854. Bouteillenstein (Obsidian) von Moldawa in Böhmen. K.K. geol. Reichsanstalt, Wien, Jahrb., 868.

Heide, F., 1936a. Seltene Elemente in den Tektiten. Forsch. u. Fortschr. 12, 232.

———, 1936b. Neue Kristallführende Gläser von Macusani in Peru. Die Naturwissenschaften, 24, 281-282.

———, 1936c. Das Tektitproblem. Frankfurter Ztg., 18, No. 421.

———, 1938a. Über tektite von den Philippinen. Centralbl. Mineralogie, Abt. A, 289-293.

———, 1938b. Über tektite von Siam. Centralbl. Mineralogie, Abt. A, 359-360.

Heide, F., 1938c. Nickel im Bimsstein von Köfels. Die Naturwissenschaften, 26, 495. (Reviewed in Neues Jahrb., 494, 1939.)

———, 1939. Über tektite von Java. Centralbl. Mineralogie, Abt. A, 199-206.

Helmhacker, R., 1873. Mineralogische Beobachtungen aus dem Böhmerwalde. Tschermak's min. petr. Mitt., K.K. geol. Reichsanstalt, Wien Beilage z. Jahrb., 281.

Henshall, B. D., 1956. Research at Hypersonic speeds. Discovery, Oct. 1956.

Hill, J., 1947. Something new in fulgurites. Rocks and Minerals, Peekskill, New York, vol. 22, 923.

Hillebrand, W. F., 1910. Analyses of Tasmanian Tektites. U.S. Geol. Surv., Bull., 419, 181.

Hills, L., 1915. Darwin Glass, a new variety of tektites. Geol. Surv. Tasmania, Records, No. 3, Dept. Mines, 1-16.

Himmel, H., 1938. Review of "Tektites from the Sherbrook River District, east of Port Campbell, Victoria", by G. Baker (1937), Neues Jahrb., 609.

Himpel, K., 1938. Zur Entstehung der Tektite. Gerlands Beitr. Geophysik, 54, 21-28.

Hodge Smith, T., 1932. Obsidianites in the Philippines Islands. Philippine Journ. Sci., 48, 115-143 and 581-585.

———, 1934. Tektites. Australian Mus. Mag., 5, 225-227.

———, 1939. Australian Meteorites. Australian Mus. Memoir VII, 65-70.

Hoffet, J. H., 1933. Notes sur la géologie du territoire de Kouang-Tchéou-Wan, Indo-China. Serv. Géol. B, 20, 1-11.

Hoffleit, D., 1955. Origin of Tektites. Sky and Telescope, 14 (7), 281.

Högbohm, A. G., 1900. Eine meteorstatistische Studie. Upsala Univ., Geol. Inst. Bull. 9, vol. 5 (1), 132-144.

———, 1922. Om Tektiterna sällsamma stenar fran en främmande värld. Populär Astronomisk tidskrift. Jahrg. 3, Stockholm.

Horne, G., and Aiston, G., 1924. Savage Life in Central Australia, pp. 60, 135-136. Macmillan and Co., London.

Houziaux, L., 1956. Spectres d'absorption infra-rouge de quelques verres naturels entre 2 et 24 microns. Geochim. et Cosmochim. Acta, 9, 298-300.

Hövig, P., 1923. Over billitonieten, ertslaag en woestijnklimaat. Geol. Mijnb. Genootsch. Nederland en Kolonien Verh. Geol. ser., 7, 1-13.

Howchin, W., 1909. Report of Annual Meeting, 5th October, 1909. Trans. Roy. Soc. South Australia, 33, 349. (Reference to use of australites by aborigines.)

Hubbard, D., Krumrine, E. M., and Stair, R., 1956. Australite (Meteoric) Glass. Trans. Amer. Geophys. Union, 37 (6), 767-778.

Humboldt, A. de, 1823. A Geognostical Essay on the Supposition of Rocks in Both Hemispheres, 433. Longman, Hurst, Rees, Orme, Brown and Green, London.

Jahn, J. J., 1899a. Ueber das Vorkommen der Moldavite in den Nordböhmischen Pyropensanden. K.K. geol. Reichsanstalt Wien Verh., 81-85.

———, 1899b. O vltavinu. Casopis pro průmysl chemicky, Jahrg. 9, Prague.

Janoschek, R., 1934. Das Alter der Moldavitschotter in Mähren. Anzeiger der Akad. d. Wiss. in Wien, Maths.-naturwiss. Klasse, 71 (17), 195-197.

———, 1937. Die Moldavitschotter in Mähren. Geol. Gesell. Wien, Mitt., 29, 329-356.

Jeans, J. H., 1919. Problems of Cosmogony and Stellar Dynamics. Cambridge University Press.

———, 1928. The configurations of rotating liquid masses. In Astronomy and Cosmogony, Chapter 8, Cambridge University Press.

Jensen, H. I., 1915. Bulletin of the Northern Territory, 14, 16 (Dec.).

Jezek, B., 1910. Der heutige Stand der Moldavitfrage. "Prirodovědecky Klub", Prague, Jahresber., 23-33.

———, 1911. O vltavínech (= on moldavites). Priroda, Brno, 10, 31-33.

———, 1911a. O povrchu vltavínovém. Predneseno na Schůzi Priro. sboru Musea král. ceského, 27, dubna 1911, 9, 295-304. (= on the surface of moldavites.)

———, 1911b. Dnesní stav otázky vltavínové, 41, vyrocni zpráva klubu prir. v Praze, 1911 se 16 obr.

———, 1912. Referate über die Tektite-Literature in tschechischer Sprache Casopis Morav. mus. zem Praha, 1, 116-123.

Jezek, B., and Woldrich, J. N., 1910a. Prispěvek k resěni otázky tektitkové (= contribution toward the solution of the tektite question). Rozpravy ces Akad. Series II, vol. 19 (30), 1-12, and p. 265.

———, 1910b. Beitrag zur lösung der Tektitfrage. Internat. Acad. Sci. Bohême, Bull., 15, 232-245.

John, C. von, 1889a. Ueber den Moldavit oder Bouteillenstein von Radomilic in Böhmen. K.K. geol. Reichsanstalt Wien. Jahrb., 39, 473.

———, 1899b. Ueber die chemische Zusammensetzung der Moldavite. K.K. geol. Reichsanstalt Wien. Verh., 179-182.

Johnson, A., 1910. Beiträge zur Kenntnis natürlischer und Künstlicher Gläser. Phys.-ökon Gesell. Konigsberg Schr., 47, 105-110.

Julien, A. A., 1901. A Study of the Structure of Fulgurites. Journ. Geol. ix (2), 673-693.

Kaspar, J., 1938. Czechoslovakian tektites and the problem of their origin; an up-to-date resumé of the question. Popular Astronomy, 46, 47-51.

Khan, M. A. R., 1947. Atomic Bombs, the Tektite Problem and "Contraterrene" Meteorites. Contrib. Meteoritical Soc., 4 (1), 35-36.

Klaproth, M. H., 1816. Chemische Untersuchung des Pseudo-Chrysoliths von Thein an der Moldau. Gesell. Naturf. Freunde zu Berlin Magazin, Jahrg. 7, 86-88.

Kluge, K. E., 1860. Handbuch der Edelsteinkunde für Mineralogen, Steinschneider und Juweliere, 425.

Koenigswald, G. H. R. von, 1935. Vorlaüfige Mitteilung über das Vorkommen von Tektiten auf Java. K. Akad. Wetenschappen, Amsterdam Verh., Tweede Sectie, 38 (3), 287-289. (Reviewed by F. Musper—Neues Jahrbuch, Referate III, 783, 1935.)

———, 1936. Das glasmeteorieten van Nederlandsch-Indie. Natuurkundig Tijdschrift Ned.-Indie, Batavia, Deel 96, 283-296.

———, 1956. Meeting Prehistoric Man, pp. 104-105. Thames and Hudson, London.

———, 1957. Tektites from Java, Proc. Ned. Akad. Sci. Ser. B, vol. 60 (5), 371-382.

———, 1958. A Tektite from the Island of Flores (Indonesia). Koninkl. Nederl. Akad. Wetensch., Amsterdam, Proc., Series B, 61 (1), 44-46.

Kohman, T. P., 1958. "Are Tektites Extra-Solar-System Meteorites". Nature, 182, 252-253.

Koomans, K., 1938. On tektites and pseudo-tektites from Dutch East Indies and Philippines. Leidsche Geol. Meded., 10 (1), 63-81.

Kopal, Z., 1958. Origin of Tektites, Nature, 181, 1457. (Criticism of Tektite Origin from the Moon.)

Kraus, E. H., and Slawson, C. B., 1939. Gems and Gem Materials, 221 and 266. McGraw-Hill Book Co., New York and London.

Krause, F. M., 1874. Report, Cape Otway District. Geol. Surv. Vic., Rept., Progress, No. 1, 99-109.

———, 1896. Introduction to Mineralogy, p. 214.

Krause, P. G., 1898a. Obsidianbomben aus Niederländisch-Indien. Jaarb. mijnwezen Ned.-Indie, Amsterdam, Wetensch. Gedeeltke, 17-31.

———, 1898b. Obsidianbomben aus Niederländisch-Indien. Geol. Reichsmus., Leiden Samml. ser. 1, 5, 237-252.

Krinov, E. L., 1946. Tektity. Priroda, Moskva, 35 (12), 15-27.

Kuiper, G. P., 1953. Satellites, Comets and Interplanetary Material. Proc. Natl. Acad. Sci., Washington, 40, 1153-1158.

———, 1954. On the Origin of the Lunar Surface Features. Proc. Natl. Acad. Sci., Washington, 40, 1096-1112.

Lacroix, A., 1929a. Sur l'existence de tectites au Cambodge; leur morphologie. Acad. sci. Paris Comptes rendus, 188, 117-121.

———, 1929b. Sur la composition chemique des tectites, et en particulier de celles du Cambodge. Acad. sci. Paris Comptes rendus, 188, 284-289.

———, 1930. Nouvelles observations sur les tectites de l'Indochine. Acad. sci. Paris Comptes rendus, 191, 893-899.

———, 1931a. Les tectites des Philippines. Acad. sci. Paris Comptes rendus, 193, 265-267.

———, 1931b. Nouvelles observations sur les tectites de l'Indo-chine; discussion sur leur origine. Acad. sci. Paris Comptes rendus, 192, 1685-1689.

Lacroix, A., 1932a. Tectites du Cambodge. Paris Comptes rendus, 12, 377-378.

———, 1932b. Les Tectites de l'Indochine. Mus. nat. histoire nat., Archives, 6th ser., 8, 193-236.

———, 1934a. Nouvelles observations sur la distribution des tectites en Indochine et dans les pays voision. Acad. sci. Paris Comptes rendus, 199, 6-9.

———, 1934b. Sur la découverte de tectites à la Côte d'Ivoire. Acad. sci. Paris Comptes rendus, 199, 1539-1542.

———, 1934c. The tektites of Indo-China and the East Indian Archipelago. 5th Pacific sci. Congress, Canada, Proc., 3, 2543-2545, University of Toronto Press.

———, 1935a. Les tectites sans formes figurées de l'Indochine. Acad. sci. Paris Comptes rendus, 200, 2129-2132.

———, 1935b. Les tectites de l'Indochine et de ses abords et celles de la Côte d'Ivoire. Mus. nat. histoire nat., Archives, 6th ser., 12, 151-170.

La Paz, L., 1938. The great circle distribution of the tektites. Popular Astronomy, 46, 224-230.

———, 1944. On the Origin of Tektites. Popular Astronomy, 52, 194-200.

———, 1948. The Valverdites: A Weathered Obsidian Form Superficially Resembling Certain Tektites. Contrib. Meteoritical Soc., 4 (2), 157-163.

Leonard, F. C., 1955. A Large Tektite from the Philippines. Meteoritics, 1 (3), 357-358.

Leonhard, C. C. von, 1826. Handbuch der Oryktognosie, 416. Heidelberg.

Leontieva, A. A., 1941. Vyazkost meteoritov i tektitov (= viscosity of meteorites and tektites). Akad. nauk, SSSR., 1, 313-315.

Lewis, A. D., 1936. Fulgurites from Witsands, Kalahari. South African Geog. Journ., 19, 50. (Reviewed in Research Items, Nature, 140, 368, 1937.)

Linck, G., 1924. Aufbau des Erdballs. G. Fischer, Jena.

———, 1926a. Ein neuer kristallführender Tektit von Paucartambo in Peru. Chemie der Erde, 2, 157-174.

———, 1926b. Über das Achserverhaltnis des Sillimanit. Centralbl. Mineralogie, Abt. A, 385-387.

———, 1928. Oberfläche und Herkunft der meteorische Gläser. Neues Jahrb., 57, 223-236.

———, 1934a. Tektite. Handwörterbuch Naturwiss., 2d. Aufl., 9, 901-906.

———, 1934b. Ueber den Tektit von Paucartambo. (A reply to E. Dittler, 1933.) Centralbl. Mineralogie, Abt. A, 13-15.

Lindaker, J. T., 1792. Einige Nachträge und Zusätze zu den böhmischen Topasen und Chrysolithen. Sammlung physikalischer Aufsatze besonders die böhmische Naturgeschichte betreffend, von einer Gesellschaft böhmischer Naturforscher; herausgegeben von Dr. Johann Mayer, 2nd. Band, 272, Dresden.

Lindemann, F. A., 1926. Meteors and the Constitution of the Upper Air. Nature, 118, 195-198.

Loewinson-Lessing, F. J., 1935. De la composition chimique des tectites. Acad. sci. U.R.S.S., Akad. Nauk., Comptes rendus, Dokl., 3, 181-185.

Mahony, D. J., 1910. Some bodies resembling obsidianites. (Read but not published.) Notice in Annual Report of the Council for the year 1908. Roy. Soc. Vic., Proc., 22 (2), 336.

Makowsky, A., 1881a. Ueber den "Bouteillenstein" von Trebitsch. Naturwiss. Ver. Brünn Verh., Sitzungber., 20, 21.

———, 1881b. Wietere Bermerkungen über den "Bouteillenstein". Ibid., 26.

———, 1881c. Ueber die "Bouteillensteine" von Mähren und Böhmen. Tschermak's min. petr. Mitt., Neue Folge, 4, 43 (1882).

Martin, K., 1881. Referat über van Dijk. Neues Jahrbuch, 2, 380.

Martin, R. W., 1934a. Tektieten hun aard en oorsprung. Natur. en Mensch, 54, 261-265 and 295-298.

———, 1934b. Are the "amerikanites" tektites? Leidsche Geol. Meded., 6 (2), 123-132.

Mason, H. C., 1930. The Origin of Meteorites. African Journ. Sci., 27.

Mathew, J., 1910. Two Representative Tribes of Queensland, pp. 174-175. T. Fisher Unwin, London and Leipsic.

Maurain, C, 1931. Sur la vitesse de chute des météorites. France, Bull. Soc. Min., 54, 80-95.

Mayer, J., 1788. Ueber die böhmischen Gallmeyarten, die grüne Erde der Mineralogen, die Chrysolithen von Thein und die Steinart von Kuchel. Böhmischen Gesell. Wiss. Abh. Jahr. 1787, 265-268. Prague und Dresden.

McIvor, E. A., 1937. McCoy Society Reports (Lady Julia Percy Island). Roy. Soc. Vic., Proc., 49 (2), 348.

Merrill, G. P., 1911. On the supposed origin of the Moldavites and like sporadic glasses from various sources. U.S. Nat. Mus., Proc., 40, 481-486. (Abstract—Bull. Geol. Soc. America, 22, 736, 1911.)

Michel, H., 1913. Zur Tektitfrage. K. Naturh. Hofmus. Wien Annalen, 27, 1-12.

———, 1922. Fortschritte der Meteoritkunde seit 1900. Fortschr. Mineralogie, 7, 316, Jena.

———, 1925. Die Entstehung der Tektite und ihre Oberfläche. Naturh. Mus. Wien, Annalen, 38, 153-161.

———, 1939. Tektite. Fortschr. Min. Krist. Petr., 23, cxliii-cxlv.

Michel, H., and Rledl, G., 1925. Die praktische Auswertung der Absorptionsverhältnisse der Edelsteinzuiher Erkennung und Bestimmung. Staatlich tech. Versuchsamt Wien Mitt., 14, 46-51.

Mingaye, J. C. H., 1916. Analyses of obsidianites from the Uralla District and Charlotte Waters. New South Wales Geol. Surv. Records, 9, 170-171.

Mitchell, S. R., 1949. Stone Age Craftsmen. Tait Book Co., Pty. Ltd., Melbourne, Australia.

Monod, Th. et Pourquié, A., 1951. Le Cratère d'Aouelloul (Adrar, Sahara Occidental). Bull. de l'Inst. Franc. d'Afrique Noire, 13, 293-304.

Moore, E. S., 1916. "Pelée's Tears" and their bearing on the origin of australites. Bull. Geol. Soc. America, 27, 51-55.

Morey, G. W., 1938. Properties of Glass. Chem. Soc. American Mon. Ser., No. 77, Reinhold Publishing Corporation, New York.

Moulden, J. C., 1896. Petrographical observations upon some South Australian rocks. Roy. Soc. S. Aust., Trans., 19, 77.

Mueller, F. P., 1915. Tektite from British Borneo. Geol. Mag., dec. 6, 2, 206-211.

Nemec, F., 1933. Druhé sklo s povrchem vltavínovym z Trebic (= a second glass with moldavite-like surface from Trebitsche.) Priroda, Brno, 26 (9), 259-261.

Nininger, H. H., 1940. The Moon as a source of tektites. Bull. Geol. Soc., America, 51, 12 (2), 1936. (Abstract.)

———, 1941. The Moon as a source of tektites. Am. Min., 26, 199.

———, 1943. The Moon as a source of tektites. (Part I. and II.) Sky and Telescope, 2 (4), 12-15 and (5), 8-9.

———, 1947. Chips from the Moon. Part I, The Battle-Scarred Moon. Part II, Tektites. El Centro, 36 pp. Desert Press Inc.

Novácek, R., 1932a. Analysy ctyr vltavínû ceskych a Moravskych (= Analyses of four Bohemian and Moravian moldavites.) Casopis Národního Musea Praha, 106, 68.

———, 1932b. Ergebnisse der chemischen und physikalischen Untersuchung einiger Moldavite aus Bohmen und Mahren. Rozpravy ces. Akad. Trida II, 42 (31), 1-12. (res. Internat. Acad. sci. Bohême, Bull., Prague).

Oakley, K. P., 1952. Dating the Libyan Desert Silica-Glass. Nature, 170, No. 4324, 447-449.

Ohashi, R., 1936. A meteorite or a pseudo-meteorite. Geol. Soc. Tokyo, Journ., 43, 407-410.

O'Keefe, J., Varsovsky, C. M., and Gold, T., 1958. Origin of Tektites. Nature, Vol. 181, pp. 172-174.

Opik, E., 1937. Researches on the physical theory of meteor phenomena, III. Observatoire Astronomique Université Tartu Pub., 29, (5), 69 pp.

Oswald, J., 1935. O vzniku povrchu meteorickyck skel (= on the origin of the surface of meteoric glasses). věda Prirodni, Praha, 16, 177-184. (Discussion by V. Rosicky and F. Slavik, Idem., 288-289).

———, 1936a. Stará a nová nalezistě vltavinu Moravskych a ceskych. Zvlastniotisk z Casopisu Národního Musea, Prague, 1-20.

———, 1936b. Meteoritic glasses—tectites. The Mineralogist, 4 (6), 5-6, June. (Formerly Oregon Mineralogist).

Oswald, J., 1942. Meteorické sklo (Meteoritic glass). Nakladem ceska akademi věd a uměni, Prague, 95 pp., 7 pl.

Palliardi, J., 1897. Mitt. d. praehistor. Kommiss. de Kon. Akad. Wiss. en Wien, I (4), 249. (Reference to flaked moldavites among Neolithic finds in Moravia and Lower Austria).

Paneth, F. A., 1940. The Origin of Meteorites (being the Halley Lecture, May, 1940). Clarenden Press, Oxford.

Paneth, F. A., Orry, W., and Koch, W., 1930. Zur frage des ursprunges der Meteoriten. Zeitschr. angew. physikal chemie, band 36.

Paneth, F. A., Peterson, K. W., and Chloupek, J., 1929. Helium Untersuchungen, VI; Über den Heliumgehalt von moldaviten und Künstlichen Gläsern. Deutsche chem. Gesell. Ber., 62, 801-809.

Park, J., 1914. A Text-Book of Geology (Section on Fulgurites). Chas. Griffin & Co., Ltd., London.

Patte, E., 1934. Les tectites d'Hai-nan. Géologie et folklore. Soc. géol. France, Comptes rendus, Seances f. 10-12, 159-161.

Petterd, W. F., 1903. The Minerals of Tasmania, p. 6.

———, 1910. Obsidianites or Australites: acid meteorites. The Minerals of Tasmania, 125-128, Mines Dept., Hobart.

Philby, H. St. J., 1932. Wabar Craters. Nature, 129, 932.

———, 1933. The empty quarter, 157-180. Constable & Co., London. (With appendix by L. J. Spencer, 365-370.)

Pinson, W. H., Ahrens, L. H. and Franck, Mona L., 1953. The Abundance of Li, Sc, Sr, Ba and Zr in Chondrites and Some Metamorphic Rocks. Geochimica et Cosmochimica, Acta, 4, 251-260.

Pinson, W. H. (Jr.), Herzog, L. F., and Cormier, R. F., 1956. Age Study of a Tektite. Bull. Geol. Soc. Amer., 67, 1725-1726.

Preuss, E., 1934. Chrom und Nickel in Tektiten. Die Naturwissenschaften, 22, 480.

———, 1935. Spektralanalytische Untersuchung der Tektite. Chemie der Erde, 9, 365-418.

Prior, G. T., 1927. Tektites. Nat. Hist. Mag. (Brit. Mus.), 1, 8-13.

Pruett, J. H., 1939. The "Tree Meteorite" of La Pine, Oregon. Popular Astronomy, 47, 150-151.

Rankama, K., and Sahama, Th. G., 1950. Appendix to Chapter I, Composition and Structure of Meteorites, 30-31, in "Geochemistry".

Reiss, W., and Stübel, A., 1899. Reisen in Südamerika. Geologische Studien in der Republik Colombia, II Teil. Petrographie, II Teil. A, Asher, Berlin.

Richards, H. C., 1934. Commentary in "Extraterrestrial Gems". Notes and News, Queensland Government Mining Journal, 35, 222.

Richly, H., 1901. Über zwei neuentdeckte Fundstätten von Moldaviten bei Neuhaus-Wittingau. K.K. geol. Reichsanstalt Wien, Verh., 40-43.

Rinne, F., 1914. Beitrag zur optische Kenntnis der Kolloidalen Kiesselsaure. Neues Jahrb., 39, 388-414.

———, 1924. Röntgenographische Untersuchungen an einigen feinzerteilten Mineralien; Künstprodukten und dichten Gesteinen. Zeitschr. Kristallographie, 60, 55-69.

Rosicky, V., 1934. Jak vznikl povrch vltavínů (= how had the surface of moldavites been formed?). Príroda, Brno, 27, 41-49.

———, 1935. Ueber den Ursprung der Tektitoberfläche. Centralbl. Mineralogie, Abt. A, 270-277.

Rufus, C. W., 1940. An astronomical theory of Tektites. Popular Astronomy, 48, 49-51, and Contrib. Soc. Research on Meteorites, 2 (3), 163-165. Supplement, p. 166.

Rutley, F., 1885. On Fulgurite from Mt. Blanc, with a note on Bouteillenstein or Pseudochrysolite of Moldauthein in Bohemia. Geol. Soc., London, Quart. Journ., 41, 152-156.

Rutten, L.M.R., 1927. Voordrachten over der Geologie van Nederlandsch Oost-Indie, (Die Billitonieten, 341).

Rzehak, A., 1897. Zur Geschichte des Glases in Mähren. Mährischen Gewerbe-Mus., Brünn Mitt., 9, 69.

2392/58.—**16**

Rzehak, A., 1898. Ueber die Herkunft der Moldavite. K.K. geol. Reichsanstalt Wien, Verh., 415-419.

————, 1899. Eine Neue Art von Meteoriten? "Prometheus," illustrirte und Wochenschr. Fortschr. in Gewerbe, Industrie und Wiss., Jahrg., 10, 369-371, Berlin.

————, 1901. Glasmeteoriten. "Prometheus," 2, 691-696.

————, 1909. Die angeblichen Glasmeteoriten von Kuttenberg. Centralbl. Mineralogie, 452-462.

————, 1912a. Über die von Prof. Weinschenk als Tektite gedeuteten Glaskugeln. Märischen Landmuseums Zeitschr., 12 (1), 40-75.

————, 1912b. Chemische Analyse eines Glases mit Rindenbildung. Centralbl. Mineralogie, 23-26.

Sacco, F., 1935. Notiziario di astronomia (Meteoriti e tectiti). Urania, Barcelona, 2, 56.

Saurin, E., 1935. Sur quelques gisements de tectites de l'Indochine du Sud. Acad. sci. Paris Comptes rendus, 200, 246-248.

Schoof, D., 1935. When the heavens rained glass. Junior Astron. News, 4 (Dec.).

Schrauf, A., 1882. Beiträge zur Kenntnis des Associationskreises der Magnesiasilicate. Zeitschr. Kristallographie, 4, 345. Anmerkung.

Schreiter, R., 1911. Die Meteoriten des Kgl. Mineralogischen Museums in Dresden. Naturwiss. Gesell. Isis Dresden Sitzungsber. Abh., 2, 58-75.

Schwantke, A., 1909. Die Brechungskoeffizienten des Moldavit. Centralbl. Mineralogie, 26-27.

Scoular, G., 1879. The Geology of the Hundred of Munno Para. Philos. Soc. S. Aust., Trans., 2, 68, Adelaide.

Scrivenor, J. B., 1909. Obsidianites in the Malay Peninsula. Geol. Mag. dec. 5, 6, 411-413.

————, 1916. Two large obsidianites from the Raffles Museum, Singapore, and now in the Geological Department. Geol. Mag. dec. 6, 3, 145-146.

————, 1931. The Geology of Malaya. 181-183, London.

————, 1933. Tektites. Nature, 132, 678.

Selga, M., 1930. Meteorites in the Philippines. Manila Observatory Pub., 1 (9), 3-52. (Tektites on p. 50).

Selwyn, A. R. C. et alia, 1868. Descriptive catalogue of the rock specimens and minerals in the National Museum, collected by the Geological Survey of Victoria, 79-80.

Sigamony, A., 1944. The Magnetic Behaviour of a Tektite. Proc. Indian Acad. sci., Bangalore City, 20, sec. A, 15-17.

Sigmund, A., 1911. Neue Mineralvorkommen in Steiermark und Niederösterreich. Naturwiss. Ver. Steiermark, Graz. Mitt., Jahrg., Band 48, 239-247.

Silverman, S. R., 1951. The Isotope Geology of Oxygen. Geochimica et Cosmochimica, Acta, II, 26-42.

Simon, R., 1955. Príspěvek kotácze původu vitavínů. Rise hvězd, 36 (6), 121-124.

Simpson, E. S., 1902. Obsidianites. Notes from the Departmental Laboratory, Western Australian Geol. Surv., Bull. 6, 79-85.

————, 1916. Analyses of Western Australian Rocks, Meteorites and Natural Waters. Western Australian Geol. Surv., Bull. 67, 16 and 135.

————, 1935. Note on the Australite observed to fall in Western Australia. Roy. Soc. Western Australia, Journ., 21, 37-38.

————, 1939. A Second Australite observed to fall in Western Australia. Roy. Soc. Western Australia, Journ., 25, 99-101.

Singleton, F. A., 1939. Über drei Australite von ungewöhnlicher Form. Centr. Mineral. Geol. Abt. A, 1, 32.

Skeats, E. W., 1915a. Notes on the so-called obsidian from Geelong and from Taradale, and on Australites. Roy. Soc. Vic., Proc., 27, 333-341.

————, 1915b. Description of three unusual forms of australites from Western Victoria. Roy. Soc. Vic., Proc., 27, 362-366.

Smith, G. A. 1926. Contribution to the Mineralogy of New South Wales. New South Wales Dept. of Mines; Mineral Resources, no. 34, 62.

Smyth, R. Brough, 1878. The Aborigines of Victoria, Vols. I and II. Government Printer, Melbourne, Australia.

Spencer, L. J., 1932. Meteorite Craters. Nature, 129, 781-784.

———, 1933a. Origin of Tektites. Nature, 131, 117-118 and 876.

———, 1933b. L'Origine des tectites. Acad. sci. Paris Comptes rendus, 196, 710-712.

———, 1933c. El origen de las tectitas. Ibérica Barcelona, an 20, 40, 383-384.

———, 1933d. Two New Gem Stones. The Gemmologist, 3, 110-113.

———, 1933e. Answer to Fenner's "Origin of Tektites". Nature, 132, 571.

———, 1936a. The Tektite Problem. Popular Astronomy, 44, 381-383.

———, 1936b. A Key to Precious Stones. Blackie, London and Glasgow. Section on Silica Glass, 205-208.

———, 1937a. Meteorites and the craters on the moon. Nature, 139, 655-657.

———, 1937b. The tektite problem. Min. Mag., 24, 503-506.

———, 1939. Tektites and silica glass. Min. Mag., 25, 425-440.

———, 1940. Tektites and silica glass. Am. Min., 25, 154.

Spencer, L. J. and Hey, M. H., 1933. Meteoric iron and silica glass from the meteorite craters of Henbury (Central Australia) and Wabar (Arabia). Min. Mag., 23, 387-404.

Spencer, Sir W. B. and Gillen, F. J., 1912. Across Australia, 1, 92-93, Macmillan and Co., London.

Stair, R., 1954. Tektites and the Lost Planet. Smithsonian Inst., Annual Rept., Publication No. 4190, Washington, pp. 217-230, 4 plates. (Also Scientific Monthly, 83, 3-12, 1956.)

———, 1955. The Spectral-Transmission Properties of Some of the Tektites. Geochim. Cosmochim., Acta 7, 43-50.

———, 1956. Tektites, Meteoric Glass. Discovery, Oct., 1956, 408-413.

Stark, M., 1904. Über den Zusammenhang der Brechungsexponenten Natürlicher Gläser mit Ihrem Chemismus. Tschermak's Min. Petr. Mitt., 23, 536.

Stelzner, A. W., 1893a. Ueber eigenthümliche Obsidianbomben aus Australien. Deutsche Gesell. Zeitschr., 45, 299.

———, 1893b. Supplementary notes on rock specimens. Roy. Soc. S. Aust., Trans., 16, 112.

Stephens, T., 1897. Notes on a specimen of basaltic glass (tachylyte) from near Macquarie Plains, Tasmania, with remarks on obsidian "buttons". Roy. Soc. Tasmania, Papers and Proc., 54-58 (1898).

———, 1902. A further note on obsidian buttons. Roy. Soc. Tasmania, Proc., 42-44.

Sternberg, C. G., 1826. Rede des Präsidenten in der öffentlichen Sitzung des böhmischen Museums am 15 März, 1826. Gesell. vaterländischen Mus. Prague Verh., 42.

Streich, V., 1893. Elder Expedition: Geology. Roy. Soc. S. Aust., Trans., 16, 74-115, (Part II, pp. 84 and 106).

Stutzer, O., 1926. Kolumbische Glas-Meteorite (Tektite). Centralbl. Mineralogie, Abt. A, 137-145.

Suess, F. E., 1898a. Ueber die Herkunft der Moldavite aus dem Weltraume. K. Akad. Wiss. Wien, Anzeiger, nr. 24, 2.

———, 1898b. Ueber den Kosmischen Ursprung der Moldavite. K.K. geol. Reichsanstalt Wien Verh., 387-403.

———, 1900. Die Herkunft der Moldavite und verwandter Gläser. K.K. geol. Reichsanstalt Wien Jahrb., 50 (2), 193-382.

———, 1901. Die Moldavite, eine neue Gattung von Meteoriten. Monats. wiss. Klub, Wien, 22, 85-88.

———, 1909a. Notizen über Tektite. Centralbl. Mineralogie, 462-467.

———, 1909b. Über Gläser Kosmischer Herkunft. Gesell. Deutsch. Naturfor. und Äartze, Verh., 3-16. (Abgedrucht in "Naturwiss. Rundschau", Braunschweig, 4, 573-585, 1909).

———, 1914. Rückschau und Neueres über die Tektitfrage. Geol. Gesell. Wien Mitt., 7 (I, II), 51-121.

———, 1916. Können die tektite als Kunstprodukte gedeutet werden? Centralbl. Mineralogie, 569-578.

Suess, F. E., 1922. Zu Wing Easton's Versuch einer lösung des Tektiträtsels. Centralbl. Mineralogie, 227-232.

———, 1932. Zur Beleuchtung des Meteoritenproblems. Geol. Gesell. Wien Mitt., 25, 115-143.

———, 1933. Wie gestaltet sich das Gesamtproblem der Meteoriten durch die Einreihung der Tektite under die meteorischen Körper? Die Naturwissenschaften, 21 (49), 857-861.

———, 1935. Australites. Geol. Mag., 72, 288.

———, 1936. Der Meteor-Krater von Köfels bei Umhausen im Ötzale, Tirol. Neues Jahrb., Min. Abt., 72, 98-155.

Suess, H. E., 1938. Bemerkungen zum Meteoritenproblem; die Radioaktivität das kaliums als Mittel zur Bestimmung des relativen Alters der Elemente. Die Naturwissenschaften, 26, 411-412.

———, 1951. Gas content and Age of Tektites. Geochimica et Cosmochimica, Acta, II, 76-79.

Suess, H. E., Hayden, R. J., and Inghram, M. G., 1951. Age of Tektites. Nature, 168, 432.

Summers, H. S., 1909. Obsidianites—their origin from a chemical standpoint. Roy. Soc. Vic., Proc., 21 (2), 423-443.

———, 1913. On the composition and origin of australites. Australian Assoc. Adv. Sci., Report, 14, 189-199.

Talbot, H. W. B., 1910. Geological observations in the country between Wiluna, Hall's Creek and Tanami. Western Australian Geol. Surv., Bull. 39, 29.

Tate, R., 1879. Anniversary address of the president. Philos. Soc. S. Aust., Trans., 2, 70, Adelaide.

Tate, R., and Watt, J. A., 1896. Report on the work of the Horn Scientific Expedition to Central Australia, Part III, Geology and Botany, 70-71.

Tesch, P., 1903. Over den brekingsindex van gesteinteglazen. K. Akad. Wetensch., Amsterdam Verh.

Thorp, C. G., 1913. A theory of the method of the formation of Australites. Nat. Hist. Sci. Soc., Journ., Western Australia.

———, 1914. A contribution to the study of australites. Nat. Hist. Sci. Soc., Journ., Western Australia, 5, 20-43.

Tilley, C. E., 1922. Density, refractivity and composition relations of some natural glasses. Min. Mag., 19, 275-294.

Trechmann, C. T., 1938. Relics of the Mount Pelée eruption of May 8, 1902. Nature, 141, 435-436.

Twelvetrees, W. H., 1906. Record of obsidianites, or obsidian buttons, in Tasmania. Ann. Report Sec. Mines for 1905, Tasmania, 60 (1906).

Twelvetrees, W. H., and Petterd, W. F., 1897. On the occurrence of obsidian "buttons" in Tasmania. Roy. Soc. Tasmania, Papers and Proc., 39-46, (1898).

———, 1898. The Igneous Rocks of Tasmania. Australasian Inst. Min. Eng., Trans., 5, 107-108.

Ulrich, G. H. F., 1866. Mineral Species of Victoria. Essay—Notes on the physical geography, geology and mineralogy of Victoria. Melbourne Intercolonial Exhibition Catalogue, 65.

———, 1875. A descriptive catalogue of the specimens in the Industrial Museum (Melbourne), illustrating the rock system of Victoria, 35.

Urey, H. C., 1955. On the Origin of Tektites. Proc. Nat. Acad. Sci., Washington, 41, 27-31.

———, 1957. Origin of Tektites. Nature, March 16th., 556-7.

———, 1958. Origin of Tektites, Nature, 181, 1458. (Criticism of Tektite Origin from the Moon.)

Van der Veen, R. W., 1919. Het onstaan der secundaire tinerts afzetlingen op Banka en Billiton. De Ingenieur, no. 10.

———, 1923. Origin of the tectite sculpture and some consequences. Geol. Mijnb. Genootsch. Nederland en Kolonien Verh., Geol. ser., 7, 15-41.

———, 1925. Nogiets over Billitonieten. Geol. Mijn. Genootsch. Nederland en Kolonien Verh., Geol. ser., 8, 551-552.

Van der Veen, R. W., 1927. Wieteres uber Billitonite. Gleiches Gedenkbuch, 551-552.

Van Dijk, P., 1879. Obsidaan van Billiton. Jaarb. mijnwezen Ned. Oost-Indie, Amsterdam, 8 (2), 225-230.

Van Eek, D., 1939. The tektites of Coco Grove. (Marsman Trading Co.), Marsman Mag., Manila, 4 (2), 10-12. (Abstract in Neues Jahrb., Min. Referate I, 51-52, 1940.)

Van Lier, R. J., 1933. The problem of the tektites. K. Akad. Wetensch. Amsterdam, Sec. Sci., Pr., 36 (4), 454-463.

Varigny, H. de, 1933. L'énigme des tectites. Rév. gen. sci. pures et appl., 44 (4), 115-117.

Varsavsky, C. M., 1957. Smithsonian Institution Astrophysical Observatory, Tech. Rep. No. 4.

Verbeek, R. D. M., 1897a. Over Glaskögels van Billiton. Verslagen van de vergadering der Wissen Natuurkundige Afdeeling. K. Akad. Wetensch. Amsterdam, Verh., 5, 421.

————, 1897b. Glaskögels van Billiton. Jaarb. mijnwezen Ned. Oost-Indie., Jahrb. 36, 235-272.

Vogt, T., 1935. Notes on the origin of the tektites. I. Tektites as aerial fulgurites. K. norske vidensk. selsk. Forh. Trondhjem, 8 (3), 9-12.

Volarovich, M. P. and Leontieva, A. A., 1939. An investigation into the viscosity of meteorites. Acad. sci., U.R.S.S., Comptes rendus (Doklady), n.s., 22 (9), 589-591.

————, 1941. Issledovaniye vyazkosti meteoritov i tektitov (=a study of the viscosity of meteorites and tektites). Meteoritika, 1, 33-42.

Wahl, W. A., 1909. Beiträge zur Kenntnis des Tektiten von Källna in Skane. Geol. fören. Stockholm, Förh, 31 (6), 364 and 471-478. (Ref. Neues Jahrb., 1911).

————, 1910. Beiträge zur Chemie der Meteoriten. Zeitschr. für Anorganische Chemie, 69, 52.

Walcott, R. H., 1898. The Occurrence of so-called Obsidian Bombs in Australia. Roy. Soc. Vic., Proc., 11 (1), 23-53.

————, 1915. Description of the Victorian meteorites with notes on Obsidianites. Nat. Mus. Melbourne, Mem., 6, 3-66.

Washington, H. S., 1917. Chemical analyses of igneous rocks. U.S. Geol. Surv., Prof. Paper, 99. (Contains some tektite analyses).

————, 1939. The crust of the earth and its relation to the interior—in "Physics of the Earth", VII, 111-113. McCraw-Hill Book Co., New York. Also reprinted as Paper no. 1008 of the Geophysical Laboratory, Carnegie Institution of Washington.

Watson, F. (Jr.), 1935. Origin of tectites. Nature, 136, 105-106.

Weil, R., and Siat, A., 1947. Catalogue de la collection de météorites de l'Institut de Minéralogie et Pétrographie de l'Université de Strasbourg, Strasbourg (mimeographed).

Weinschenk, E., 1908. Die kosmische Natur der Moldawite und verwandter Gläser. Centralbl. Mineralogie, 24, 737-742.

————, 1909. Zum Streit uber die "Echtheit" der Moldawite. Centralbl. Mineralogie, 545-550.

Weinschenk, E. und Steinmetz, H., 1911. Weitere Mitteilungen über den Neuen Typus der Moldavite. Centralbl. Mineralogie, 231-240.

Wenzliczke, A. 1880. Chemische Analyse des Bouteillensteins von Trebitsche in Mähren. Naturwiss. Ver. Brunn Verh. Abh., 19, 9.

Wichmann, A., 1882. Beiträge zur Geologie Ostasiens und Australiens. Gesteine von Timor. Geol. Reichsmus. Leiden, Samml., 2, 22-23. Anmerkung.

————, 1893. Protokoll der Sitzung vom August 14, 1893. Deutsche geol. Gesell. Zeitschr., 45, 518-519.

————, 1913. On the pseudometeorite of Igast in Livonia. K. Akad. Wetensch. Amsterdam Verh., 16, 292-296.

Wilford, G. E., 1957.—Geology of Brunei and the Adjoining Areas of Sarawak. Ann. Rept. Geol. Surv. Dept. for 1957, pp. 121 to 124, British Territories in Borneo.

Wiman, C., 1941. Über den falschen Tektit aus Källna in Schönen. Geol. Inst. Univ. Upsala, Bull., 28, 3-16.

Winderlich, R., 1940. Boten aus dem Weltenraum; Tektite von den Philippinen. Freude am Leben, Berlin, Jg. 17 (3), 42-45.

———, 1948. Glas-Meteorite. Natur und Volk, 78 (7/9), 110-116.

Woldrich, I., 1908. O otázce vltavínové. Věstník 5, sjezdu prirodopytu a lékaru césych v Praze, 9, 430.

Woldrich, J. N., 1886. Ueber das Vorkommen einiger Mineralien in Südböhmen. K. K. geol. Reichsanstalt Wien, Verh., 455.

———, 1888. Ueber Moldavite von Radomilic. K. K. geol. Reichsanstalt Wien, Verh., 164.

———, 1893. Príspěvek k seznání budějovické pánve permskě a třetíhorní. Böhmischen Gesell. Wiss. Sitzungsber., 4, 11, Prague.

———, 1898. Príspěvek k otázce o vltavínech. Věstník Ceske Akademie cisare Frantiska Josefa. Jahrg. 7, 643.

———, 1898a. Beiträg zur Moldavitfrage. Intern. Acad. Sci. Bohême, sci., math. et nat., Bull. 5, 84-87.

Woodward, H. P., 1894. Mining Handbook of the Colony of Western Australia, 34.

Wright, F. E., 1915. Obsidian from Hrafntinnuhryggur, Iceland: its lithophysae and surface markings. Geol. Soc. America, Bull. 26, 255-286.

Yashchenko, M. L., and Gerling, E. K., 1953. Akad. Nauk. S.S.S.R. Lab. Geol. Dokembriya Trudy, 2, 232.

Zahálka, S. B., 1904. Otázka moldavitu neboli vltavínu. Vesmir, Praha, 33, 196-198 and 206-207.

Zenzen, A., 1940. Bermerkungen über den sogenannten Schönite, den falschen Tektite aus Källna in Schönen. Geol. För. Förh., Stockholm, 62, 161-172.

Zippe, F. M., 1831. Uebersicht der Gebirgformationen in Böhmen. Böhmischen Gesell. Wiss. Abh., Prague, 72. (Auch Böhmens Edelsteine Aus den Vortragen bei der 1 Jubelfeier am 14 September, 1836, ebenda. Neue Folge 4 (4), 26 and 49).

———, 1840. Die Mineralien Böhmens nach ihren geognostischen Verhältnissen und ihrer Aufstellung des vaterländischen Museums geordnet und beschrieben. Gesell vaterländischen Mus., Prague, Verh. Beilage-Band.

Author Citation Index

Subject Index

Numbers in *italics* indicate pages with illustrations.